Mobile Wireless Communications

Wireless communication has become a ubiquitous part of modern life, from global cellular telephone systems to local and even personal-area networks. This book provides a tutorial introduction to digital mobile wireless networks, illustrating theoretical underpinnings with a wide range of real-world examples. The book begins with a review of propagation phenomena, and goes on to examine channel allocation, modulation techniques, multiple access schemes, and coding techniques. GSM and IS-95 systems are reviewed and 2.5G and 3G packet-switched systems are discussed in detail. Performance analysis and accessing and scheduling techniques are covered, and the book closes with a chapter on wireless LANs and personal-area networks. Many worked examples and homework exercises are provided and a solutions manual is available for instructors. The book is an ideal text for electrical engineering and computer science students taking courses in wireless communications. It will also be an invaluable reference for practicing engineers.

Mischa Schwartz joined the faculty of Electrical Engineering at Columbia University in 1974 and is now Charles Batchelor Professor Emeritus. He is the author and co-author of ten books, including best-selling books on communication systems and computer networks. His current research focuses on wireless networks. He is a Fellow and former Director of the IEEE, past President of the IEEE Communications Society, and past Chairman of the IEEE Group on Information Theory. He was the 1983 recipient of the IEEE Education Medal and was listed among the top ten all-time EE educators, IEEE survey, 1984. He also received the 2003 Japanese Okawa Prize for contributions to Telecommunications and Engineering Education and the New York City Mayor's Award in 1994 for contributions to computer communications.

Mobile Wireless Communications

Mischa Schwartz

Department of Electrical Engineering
Columbia University

PUBLISHED BY THE PRESS SYNDICATE OF THE UNIVERSITY OF CAMBRIDGE
The Pitt Building, Trumpington Street, Cambridge, United Kingdom

CAMBRIDGE UNIVERSITY PRESS
The Edinburgh Building, Cambridge, CB2 2RU, UK
40 West 20th Street, New York, NY 10011–4211, USA
477 Williamstown Road, Port Melbourne, VIC 3207, Australia
Ruiz de Alarcón 13, 28014 Madrid, Spain
Dock House, The Waterfront, Cape Town 8001, South Africa

http://www.cambridge.org

First published 2005

Printed in the United Kingdom at the University Press, Cambridge

Typefaces Times 10/13 pt. and Formata *System* $\LaTeX\ 2_{\varepsilon}$ [TB]

A catalog record for this book is available from the British Library

Library of Congress Cataloging in Publication data

ISBN 0 521 84347 2 hardback

To my wife Charlotte

Contents

Preface

It is apparent, even to the most casual observer, that a veritable revolution in telecommunications has taken place within recent years. The use of wireless communications has expanded dramatically worldwide. Cell phones are ubiquitous. Although most such mobile terminals still carry voice principally, more and more users are sending and receiving data and image applications. Wi-Fi, an example of a wireless local area network (LAN), has caught on spectacularly, joining the major cellular networks deployed throughout the world.

This book, designed as an introductory textbook in wireless communication for courses at the Senior and first-year graduate level, as well as a self-study text for engineers, computer scientists, and other technical personnel, provides a basic introduction to this booming field. A student or reader of the book should come away with a thorough grounding in the fundamental aspects of mobile wireless communication, as well as an understanding of the principles of operation of second- and third-generation cellular systems and wireless LANs. To enhance an understanding of the various concepts introduced, numerical and quantitative examples are provided throughout the book. Problems associated with each chapter provide a further means of enhancing knowledge of the field. There are many references to the current technical literature appearing throughout the book as well. The author considers these references an integral part of the discussion, providing the reader with an opportunity to delve more deeply into technical aspects of the field where desired.

After the introductory Chapter 1, which provides a history of mobile communications, followed by a detailed description of the book, chapter by chapter, the book breaks down roughly into two parts, the first part containing Chapters 2 to 8, the second part Chapters 9 to 12. Chapters 2 to 7 provide an introduction to the fundamental elements of wireless mobile communications, with Chapter 8 then providing a detailed discussion of the second-generation systems, GSM, IS-95, and IS-136 or D-AMPS, in which these basic concepts are applied. Specifically, Chapter 2 treats the propagation phenomena encountered in communicating over the wireless medium, while Chapter 3 introduces the cellular concept. Power control, channel assignment, modulation, coding, and access techniques are then discussed in the six chapters following. This material in the first part of the book is covered in the first semester of a full-year course on wireless communication at Columbia

University. The second-semester course following this first course then covers material from the second part of the book, as well as additional material based on current reading and research.

The second part of the book provides more advanced material. It begins in Chapter 9 with a thorough discussion of the performance analysis of wireless systems, building on some of the items touched on only briefly in earlier chapters. Chapter 10 then describes the third-generation systems W-CDMA, cdma2000, and GPRS in depth, again building on the earlier discussion of second-generation systems in Chapter 8. The focus of this chapter is on data and multimedia wireless communication using packet-switched technology. A brief review of the concept of layered architectures is included, showing, in particular, how GPRS interfaces with Internet-based packet networks. Chapter 11 then discusses the access and scheduling techniques used or proposed for use in cellular systems. The book concludes with a discussion of wireless LANs (WLANs) and Personal-Area Networks (PANs). The WLAN discussion focuses on Wi-Fi and extensions thereof to much higher bit-rate wireless LANs; the PAN discussion deals with the Bluetooth system.

The book can, therefore, be used for either a single-semester course on wireless systems or for a full-year course. A single-semester course might cover the first eight chapters, as has been the case at Columbia, or might use as examples of current wireless systems material drawn from sections of Chapters 8, 10, and 12. A full-year course would cover the entire book. The material in the book could also be used in conjunction with a course on communication systems, providing the application of communication technology to the wireless area.

The only prerequisites for the book, aside from a certain technical maturity commonly available at the Senior or first-year graduate level, are a knowledge of basic probability and linear algebra. There is no prior knowledge of communication theory and communication systems assumed on the part of the reader, and the material in the first part of the book dealing with modulation and coding, for example, is meant to be self-contained. Those readers with some prior knowledge of communication systems should find the tutorial discussion in the coding and modulation chapters a useful review, with the specific application of the material to wireless systems further solidifying their knowledge of the area.

The author would like to acknowledge the help of a number of individuals with whom he worked during the writing of this book. In particular, the author is indebted to the students who took the wireless course at Columbia, using the preliminary notes covering the material, and to the instructors with whom he shared the teaching of the course. He would like to particularly acknowledge the help and co-teaching support of Professor Andrew Campbell of the Electrical Engineering department at Columbia; Professor Tom LaPorta, formerly of Bell Labs and currently with the Computer Science department of Pennsylvania State University; and Dr. Mahmoud Naghshineh of IBM. The author has, at various times, carried out research in the wireless area with each of these colleagues as well. The questions raised and answered while teaching the course, as well as conducting research in the area, were invaluable in writing this book.

Further thanks are due to Sanghyo Kim for help with preparing the book figures, to Edward Tiedemann of Qualcomm for help with the material on cdma2000, Dr. Chatschik Bisdikian of IBM for help with the section on Bluetooth, and Dr. Krishan Sabnani for providing such hospitable support during a sabbatical with his wireless group at Bell Laboratories. Thanks are also due to Dr. Li-Hsiang Sun, one of the author's former doctoral students, and now with Florida State University, who carried out some of the performance calculations in Chapter 9. The author is thankful as well to the anonymous reviewers of the book manuscript who pointed out a number of errors and made suggestions for improving the book. The author can only hope the reader will enjoy reading and studying this book, and learn from it as much as the author has enjoyed and gained from the task of writing it.

Chapter 1

Introduction and overview

This book provides a tutorial introduction to digital mobile wireless networks. The field is so vast, and changing so rapidly, that no one book could cover the field in all its aspects. This book should, however, provide a solid foundation from which an interested reader can move on to topics not covered, or to more detailed discussions of subjects described here. Much more information is available in the references appended throughout the book, and the reader is urged to consult these when necessary. There is a host of journals covering current work in the field, many of them referenced in this book, which will provide the reader with up-to-date research results or tutorial overviews of the latest developments.

Note the use of the word *digital* in the first line above. The earliest wireless networks used analog communication, as we shall see in the historical section following. We shall provide a brief description of one of these analog networks, AMPS, that is currently still deployed, later in that section. But the stress in this book is on modern digital wireless networks. There are, basically, two types of digital wireless networks currently in operation worldwide. One type is the class of cellular networks, carrying voice calls principally, but increasingly carrying data and multimedia traffic as well, as more cell phones or other cell-based mobile terminals become available for these applications. These digital wireless networks are now ubiquitous, being available worldwide to users with cell phones, although, as we shall see, different cellular systems are not compatible with one another. The second type of digital wireless networks is the class of local- and personal-area networks. Much of this book is devoted to the first type, digital cellular networks. A number of the topics covered, however, are equally applicable to both types of network, as will become apparent to readers of this book. In addition, Chapter 12 provides a comprehensive introduction to local- and personal-area networks, focusing on the increasingly widespread "Wi-Fi" networks of the local-area type, and "Bluetooth" as an example of the personal-area type.

Consider cellular networks, to be described in detail in the chapters following. In these networks, user cell phones connect to so-called base stations, each covering a geographic "cell." The base stations are, in turn, connected to the wired telephone network, allowing user calls, in principle, to be transmitted to any desired location, worldwide. The first cellular networks to be developed were analog, as we shall see in the historical section following. They were succeeded by digital cellular networks, the ones now most commonly used throughout the world. The analog networks are still used, however, to provide a backup when a digital cellular connection cannot be made. This is particularly the case in North America where a number of competing and incompatible digital cellular networks may preclude a connection to be made in a given geographic area or region. The analog networks are often referred to as "first-generation" cellular networks; the currently deployed digital networks are referred to as "second-generation" networks. As noted above, the stress in this book is on digital cellular networks, although some of the tutorial material in the chapters following applies to any type of cellular network, as well as to local-area and personal-area networks, as stated above. There are three principal types of second-generation networks in operation throughout the world: GSM, D-AMPS or IS-136, and IS-95. We provide a detailed description of these three networks in Chapters 6 and 8.

Much work has gone on in recent years to upgrade the second-generation networks to so-called "third-generation" networks. These networks are designed to transmit data in a *packet-switched* mode, as contrasted to second-generation networks, which use circuit-switching technology, in exact conformity with digital wired telephone networks. We discuss the third-generation networks in detail in Chapter 10. The distinction between *circuit switching* and *packet switching* is made clear in that chapter.

1.1 Historical introduction

It is of interest to precede our detailed study of mobile wireless communications in this book with a brief historical survey of the field. Much of the emphasis in the material following is on work at Bell Laboratories, covering developments in the United States. Further details appear in Frenkiel (2002) and Chapter 14 of O'Neill (1985). A brief discussion of European activities in mobile wireless communications from 1981 on concludes this section. Details appear in Paetsch (1993).

Ship-to-shore communication was among the first applications of mobile telephony. Experimental service began on coastal steamers between Boston and Baltimore in the United States in 1919; commercial service using AM technology at 4.2 and 8.7 MHz began in 1929. This was roughly the same period at which AM radio broadcasting began to capture the public's attention. Note the wavelengths at these frequencies are about 70 and 35 meters, respectively, making ships suitable vehicles for carrying antennas of these lengths. Ships were also suitable for the size and weight of the radio equipment required to be carried, as well as being able to provide the power required. Police communication was begun at about the same time. In 1928, the Detroit police department introduced land mobile communications using small, rugged radios. By 1934, 5000 police cars from 194 municipal and 58 police systems in the US had been equipped for, and were using, mobile

Frenkiel, R. 2002. "A brief history of mobile communications," *IEEE Vehicular Technology Society News*, May, 4–7.

O'Neill, E. F. 1985. *A History of Engineering and Science in the Bell System: Transmission Technology (1925–1975)*, ed. E. F. O'Neill, AT&T Bell Laboratories.

Paetsch, M. 1993. *Mobile Communications in the US and Europe: Regulation, Technology, and Markets*, Boston, MA, Artech House.

radio communications. These early mobile communications systems used the frequency band at 35 MHz. It soon became apparent, however, that communication to and from automobiles in urban areas was often unsatisfactory because of deleterious propagation effects and high noise levels. Propagation effects in urban environments were an unknown quantity and studies began at Bell Laboratories and elsewhere. Propagation tests were first carried out in 1926 at 40 MHz. By 1932, tests were being conducted at a number of other frequencies over a variety of transmission paths with varying distances, and with effects due to such phenomena as signal reflection, refraction, and diffraction noted. (Such tests are still being conducted today by various investigators for different signal propagation environments, both indoor and outdoor, and at different frequency bands. Chapter 2 provides an introduction to propagation effects as well as to models used in representing these effects.) In 1935, further propagation tests were carried out in Boston at frequencies of 35 MHz and 150 MHz. Multipath effects were particularly noted at this time. The tests also demonstrated that reliable transmission was possible using FM rather than the earlier AM technology. These various tests, as well as many other tests carried out over the years following, led to the understanding that propagation effects could be understood in their simplest form as being the combination of three factors: an inverse distance-dependent average received power variation of the form $1/d^n$, n an integer greater than the usual free-space factor of 2; a long-term statistical variation about the average received power, which is now referred to as shadow or log-normal fading; a short-term, rapidly varying, fading effect due to vehicle motion. These three propagation effects are discussed and modeled in detail in Chapter 2 following.

The advent of World War II interrupted commercial activity on mobile wireless systems, but the post-war period saw a rapid increase in this activity, especially at higher frequencies of operation. These higher frequencies of operation allowed more user channels to be made available. In 1946, the Federal Communications Commission, FCC, in the USA, granted a license for the operation of the first commercial land-mobile telephone system in St Louis. By the end of the year, 25 US cities had such systems in operation. The basic system used FM transmission in the 150 MHz band of frequencies, with carrier frequencies or channels spaced 120 kHz apart. In the 1950s the channel spacing was reduced to 60 kHz, but, because of the inability of receivers to discriminate sufficiently well between adjacent channels, neighboring cities could only use alternate channels spaced 120 kHz apart. A 50-mile separation between systems was required. High towers covering a range of 20–30 miles were erected to provide the radio connections to and from mobile users. Forty channels or simultaneous calls were made available using this system. The FCC divided these radio channels equally between the local telephone companies (Telcos) and newly established mobile carriers, called Radio Common Carriers (RCCs). These early mobile systems were manual in operation, with calls placed through an operator. They provided half-duplex transmission, with one side of a connection only being able to communicate at any one time: both parties to a connection used the same frequency channel for the air or radio portion of the connection, and a "push-to-talk" procedure was required for the non-talking party to "take over" the channel. With the number of channels available set at 40, the system could accommodate 800–1000 customers in a given area, depending on the length of calls. (Clearly, as user conversations increase in length, on the average, the number of customers that can be accommodated must be reduced. We shall quantify this obvious statement later in this book, in discussing the concept of "blocking probability.") As the systems grew in popularity, long waiting lists arose to obtain a mobile telephone. These systems thus tended to become somewhat "elitist," with priorities being established for people who desired to become customers; priority might be given to doctors, for example.

The introduction of new semi-conductor devices in the 1960s and the consequent reduction in system cost and mobile phone-power requirement, as well as the possibility of deploying more complex circuitry in the phones, led to the development of a considerably improved mobile telephone service called, logically enough, IMTS, for Improved Mobile Telephone Service. Bell Laboratories' development of this new service was carried on from 1962 to 1964, with a field trial conducted in 1965 in Harrisburg, PA. This service had a reduced FM channel spacing of 30 kHz, it incorporated automatic dialing, and operated in a full-duplex mode, i.e., simultaneous operation in each direction of transmission, with each side of a conversation having its own frequency channel. A mobile phone could also scan automatically for an "idle" channel, one not currently assigned to a user. This new system could still only accommodate 800–1000 customers in a given area however, and long waiting lists of as many as 25 000 potential customers were quite common! In addition, the limited spectrum made available for this service resulted, quite often, in users experiencing a "system busy" signal, i.e. being blocked from getting a channel. The "blocking probability" noted above was thus quite high.

Two solutions were proposed as early as 1947 by Bell Laboratories' engineers to alleviate the sparse mobile system capacity expected even at that time: one proposal was to move the mobile systems to a higher frequency band, allowing more system bandwidth, and, hence, more user channels to be made available; the second proposal was to introduce a cellular geographic structure. The concept of a cellular system is quite simple, although profound in its consequences. In a cellular system, a given region is divided into contiguous geographic areas called *cells*, with the total set of frequency channels divided among the cells. Channels are then reused in cells "far enough apart" so that interference between cells assigned the same frequencies is "manageable." (We shall quantify these ideas in Chapter 3.) For example, in one scheme we shall investigate in Chapter 3, cells may each be assigned one-seventh of the total number of channels available. This appears, at first glance, to be moving in the wrong direction – reducing the number of channels per cell appears to *reduce* the number of users that might be simultaneously accommodated! But, if the area covered by each cell is small enough, as is particularly possible in urban areas, frequencies are reused over short-enough distances to more than compensate for the reduction in channels per cell. A problem arising we shall consider in later chapters, however, is that, as mobile users "roam" from cell to cell, their on-going calls must be assigned a new frequency channel in each cell entered. This process of channel reassignment is called "handoff" (an alternative term used, particularly in Europe, is "handover"), and it must appear seamless to a user carrying on a conversation. Such a channel reassignment also competes with new calls attempting to get a channel in the same cell, and could result in a user disconnection from an ongoing call, unless special measures are taken to either avoid this or reduce the chance of such an occurrence. We discuss the concept of "handoff dropping probability," and suggested means for keeping it manageable, quantitatively, in Chapter 9, and compare the tradeoffs with respect to "new-call blocking probability."

We now return to our brief historical discussion. In 1949 the Bell System had requested permission from the FCC to move mobile telephone operations to the 470–890 MHz band to attain more channels and, hence, greater mobile capacity. This band was, at that time, intended for TV use however, and permission was denied. In 1958 the Bell System requested the use of the 764–840 MHz band for mobile communications, but the FCC declined to take action. By this time, the introduction of the cellular concept into mobile systems was under full discussion at Bell Laboratories. By 1968, the FCC had decided to allocate spectral capacity in the vicinity of 840 MHz for mobile telephony, and opened the now-famous "Notice of Inquiry and Notice of Proposed Rule-Making," Docket 18262, for this purpose. The Bell System responded in 1971, submitting a proposal for a "High-Capacity Mobile Telephone System," which included the introduction of cellular technology. (This proposed system was to evolve later into AMPS, the Advanced Mobile Phone Service, the first-generation analog cellular system mentioned earlier.) A ten-year conflict then began among various parties that felt threatened by the introduction by the Bell System of a mobile service in this higher frequency range. The broadcasters, for example, wanted to keep the frequency assignment for broadcast use; communication manufacturers felt threatened by the prospect of newer systems and the competition that might ensue; the RCCs feared domination by the Bell System; fleet operators wanted the spectrum for their private mobile communications use. It was not until 1981 that these issues were resolved, with the FCC finally agreeing to allocate 50 MHz in the 800–900 MHz band for mobile telephony. (Actually, a bandwidth of 40 MHz was allocated initially; 10 MHz additional bandwidth was allocated a few years later.) Half of this band of frequencies was to be assigned to local telephone companies, half to the RCCs. By this time as well, the widespread introduction of solid-state devices, microprocessors, and electronic telephone switching systems had made the processes of vehicle location and cell handoff readily carried out with relatively small cells.

During this period of "skirmishing and politicking" (O'Neill, 1985) work continued at Bell Laboratories on studies of urban propagation effects, as well as tests of a cellular-based mobile system in an urban environment. A technical trial for the resulting AMPS system was begun in Chicago in 1978, with the first actual commercial deployment of AMPS taking place in that city in 1983, once the FCC had ruled in favor of moving ahead in 1981. (Note, however, that the AT&T divestiture took place soon after, in 1984. Under the Modification of Final Judgment agreed to by AT&T, the US Justice Department, and the District Court involved with the divestiture, mobile operations of the Bell System were turned over to the RBOCs, the seven Regional Bell Operating Companies established at that time.)

The AMPS system is generally called a first-generation cellular mobile system, as noted above, and is still used, as noted earlier, for cell-system backup. It currently covers two 25 MHz-wide bands: the range of frequencies 824–849 MHz in the uplink or reverse-channel direction of radio transmission, from mobile unit to the base station; 869–894 MHz in the downlink or forward-channel direction, from base station to the mobile units. The system uses analog-FM transmission, as already noted, with 30 kHz allocated to each channel, i.e. user connection, in each direction. The maximum frequency deviation per channel is 12 kHz. Such a communication system, with the full 25 MHz of bandwidth in each direction of transmission split into 30 kHz-wide channels, is called a frequency-division multiple access (FDMA) system. (This frequency-division strategy, together with TDMA, time-division multiple access, and CDMA, code-division multiple access, will be discussed in detail in Chapter 6.) This means 832 30 kHz frequency channels are made available in each direction. In the first deployments of AMPS in the US, half of the channels were assigned to the RBOCs, the Regional Bell Operating Companies, and half to the RCCs, in accordance with the earlier FCC decision noted above. One-seventh of the channels are assigned to a given cell, also as already noted. Consider, for example, a system with 10-mile radius cells, covering an area of about 300 sq. miles. (Cells in the original Chicago trials were about 8 miles in radius.) Twenty-five contiguous cells would therefore cover an area of 7500 sq. miles. Consider a comparable non-cellular system covering about the same area, with a radius of about 50 miles. A little thought will indicate that the introduction of cells in this example increases the system capacity, the number of simultaneous user connections or calls made possible, by a factor of 25/7, or about 3.6. The use of smaller cells would improve the system capacity even more.

Despite the mobile capacity improvements made possible by the introduction of the cellular-based AMPS in 1983, problems with capacity began to be experienced from 1985 on in major US cities such as New York and Los Angeles. The Cellular Telecommunications Industry Association began to evaluate various alternatives, studying the problem from 1985 to 1988. A decision was made to move to a digital time-division multiple access, TDMA, system. This system would use the same 30 kHz channels over the same bands in the 800–900 MHz range as AMPS, to allow some backward compatibility. Each frequency channel would, however, be made available to three users, increasing the corresponding capacity by a factor of three. (TDMA is discussed in Chapter 6.) Standards for such a system were developed and the resultant system labeled IS-54. More recently, with revised standards, the system has been renamed IS-136. It is often called D-AMPS, for digital-AMPS, as well. This constitutes one of the three second-generation cellular systems to be discussed in this book. These second-generation digital systems went into operation in major US cities in late 1991. A detailed discussion of IS-135 is provided in Chapters 6 and 8.

In 1986, QUALCOMM, a San Diego communications company that had been developing a code-division multiple access, CDMA, mobile system, got a number of the RBOCs, such as NYNEX and Pacific Bell, to test out its system, which, by the FCC rules, had to cover the same range of frequencies in the 800–900 MHz band as did AMPS and D-AMPS. This CDMA system, labeled IS-95, was introduced commercially in the US and in other countries in 1993. This system is the second of the second-generation systems to be discussed in this book. It has been adopted for the 2 GHz PCS band as well. (We shall have little to say about the 2 GHz band, focusing for simplicity in this book on the 800–900 MHz band.) In IS-95, the 25 MHz-wide bands allocated for mobile service are split into CDMA channels 1.25 MHz wide. Code-division multiple access is discussed in detail in Chapter 6, and, again, in connection with third-generation cellular systems, in Chapter 10. Details of the IS-95 CDMA system are provided in Chapters 6 and 8.

Note that, by the mid and late 1990s, there existed two competing and incompatible second-generation digital wireless systems in the United States. If one includes the later adoption of the 2 GHz PCS band for such systems as well, there is clearly the potential for a great deal of difficulty for mobile users in maintaining connectivity while roaming far from home in an area not covered well by the designated mobile carrier. Dual-mode cell phones are used to help alleviate this problem: they have the ability to fall back on the analog AMPS system when experiencing difficulty in communicating in a given area. The situation in Europe was initially even worse. The first mobile cellular systems were introduced in the Scandinavian countries in 1981 and early 1982. Spain, Austria, the United Kingdom, Netherlands, Germany, Italy, and France followed with their own systems in the period 1982–1985. These systems were all analog, but the problem was that there were eight of them, all different and incompatible. This meant that communication was generally restricted to one country only. The problem was, of course, recognized early on. In 1981 France and Germany instituted a study to develop a second-generation digital system. In 1982 the Telecommunications Commission of the European Conference of Postal and Telecommunication Administrations (CEPT) established a study group called the *Groupe Speciale Mobile* (GSM) to develop the specifications for a European-wide second-generation digital cellular system in the 900 MHz band. By 1986 the decision had been reached to use TDMA technology. In 1987 the European Economic Community adopted the initial recommendations and the frequency allocations proposed, covering the 25 MHz bands of 890–915 MHz for uplink, mobile to base station, communications, and 935–960 MHz for downlink, base station to mobile, communications. By 1990 the first-phase specifications of the resultant system called GSM (for either Global System for Mobile Communications or the GSM System for Mobile Communications) were frozen. In addition, that same year, at the request of the United Kingdom, work on an adaptation of GSM for the 1.8 GHz band, DCS1800, was begun. Specifications for DCS1800 were frozen a year later, in 1992. The first GSM systems were running by 1991 and commercial operations began in 1992. GSM has since been deployed throughout Europe, allowing smooth roaming from country to country. A version of GSM designed for the North American frequency band has become available for North America as well. This, of course, compounds the incompatibility problem in the United States even more, with three different second-generation systems now available and being marketed in the 800–900 MHz range, as well as systems available in the 2 GHz PCS band. Canada, which initially adopted IS-136, has also seen the introduction of both IS-95 and GSM.

Japan's experience has paralleled that of North America. NTT, Japan's government-owned telecommunications carrier at the time, introduced an analog cellular system as early as 1979. That system carried 600 25 kHz FM duplex channels in two 25 MHz bands in the 800 MHz spectral range. Its Pacific Digital Cellular (PDC) system, with characteristics similar to those of D-AMPS, was introduced in 1993. It covers the same frequency bands as the first analog system, as well as a PCS version at 1.5 GHz. The IS-95 CDMA system, marketed as CDMAOne, has been introduced in Japan as well.

As of May 2003, GSM was the most widely adopted second-generation system in the world, serving about 864 million subscribers worldwide, or 72% of the total digital mobile users throughout the world (GSM World). Of these, Europe had 400 million subscribers,

Asia Pacific had 334 million subscribers, Africa and the Arab World had 28 million subscribers each, and North America and Russia had 22 million subscribers each. IS-95 CDMA usage was number two in the world, with 157 million subscribers worldwide using that type of system. D-AMPS had 111 million subscribers, while PDC, in Japan, had 62 million subscribers.

1.2 Overview of book

With this brief historical survey completed, we are ready to provide a summary of the chapters to come. As noted above, propagation conditions over the radio medium or air space used to carry out the requisite communications play an extremely significant role in the operation and performance of wireless mobile systems. Multiple studies have been carried out over many years, and are still continuing, of propagation conditions in many environments, ranging from rural to suburban to urban areas out-of-doors, as well as a variety of indoor environments. These studies have led to a variety of models emulating these different environments, to be used for system design and implementation, as well as simulation and analysis. Chapter 2 introduces the simplest of these models. The discussion should, however, provide an understanding of more complex models, as well as an entrée to the most recent literature on propagation effects. In particular, we focus on a model for the statistically varying received signal power at a mobile receiver, given signals from the base station transmitted with a specified power. This model incorporates, in product form, three factors summarizing the most significant effects experienced by radio waves traversing the air medium. These three factors were mentioned briefly earlier. The first term in the model covers the variation of the average signal power with distance away from the transmitter. Unlike the case of free-space propagation, with power varying inversely as the square of the distance, the transmitted signal in a typical propagation environment is found to vary inversely as a greater power of the distance. This is due to the effect of obstacles encountered along the signal propagation path, from base station to mobile receiver, which serve to reflect, diffract, and scatter the signal, resulting in a multiplicity of rays arriving at the receiver. A common example, which we demonstrate in Chapter 2 and use in later chapters, is variation as $1/(\text{distance})^4$, due to the summed effect of two rays, one direct from the base station, the other reflected from the ground.

The second term in the model we discuss in Chapter 2 incorporates a relatively large-scale statistical variation in the signal power about its average value, sometimes exceeding the average value, sometimes dropping below it. This effect, covering distances of many wavelengths, has been found to be closely modeled by a log-normal distribution, and is often referred to as shadow or log-normal fading. The third term in the model is designed to incorporate the effect of small-scale fading, with the received signal power varying statistically as the mobile receiver moves distances the order of a wavelength. This type of fading is due to multipath scattering of the transmitted signal and is referred to as Rayleigh/Rician fading. A section on random channel characterization then follows. We also discuss in Chapter 2 the rate of fading and its connection with mobile velocity, as well as the impact of fading on the information-bearing signal, including the condition for frequency-selective fading. This condition occurs when the delay spread, the differential delays incurred by received multipath rays, exceeds a data symbol interval. It results in signal distortion, in particular in inter-symbol interference. Chapter 2 concludes with a discussion of three methods of mitigating the effects of multipath fading: equalization techniques designed to overcome inter-symbol interference, diversity procedures, and the RAKE receiver technique used effectively to improve the performance of CDMA systems.

Chapter 3 focuses on the cellular concept and the improvement in system capacity made possible through channel reuse. It introduces as a performance parameter the signal-to-interference ratio or SIR, used commonly in wireless systems as a measure of the impact of interfering transmitters on the reception of a desired signal. In a cellular system, for example, the choice of an acceptable lower threshold for the SIR will determine the reuse distance, i.e., how far apart cells must be that use the same frequency. In discussing two-dimensional systems in this chapter we focus on hexagonal cellular structures. Hexagons are commonly used to represent cells in cellular systems, since they tessellate the two-dimensional space and approximate equi-power circles obtained when using omni-directional antennas. With the reuse distance determined, the number of channels per cell is immediately found for any given system. From this calculation, the system performance in terms of blocking probability may be found. This calculation depends on the number of mobile users per cell, as well as the statistical characteristics of the calls they make. Given a desired blocking probability, one can readily determine the number of allowable users per cell, or the cell size required. To do these calculations, we introduce the statistical form for blocking probability in telephone systems, the Erlang distribution. (We leave the actual derivation of the Erlang distribution to Chapter 9, which covers performance issues in depth.) This introductory discussion of performance in Chapter 3, using SIR concepts, focuses on average signal powers, and ignores the impact of fading. We therefore conclude the chapter with a brief introduction to probabilistic signal calculations in which we determine the probability the received signal power will exceed a specified threshold. These calculations incorporate the shadow-fading model described earlier in Chapter 2.

Chapter 4 discusses other methods of improving system performance. These include dynamic channel allocation (DCA) strategies for reducing the call blocking probability as well as power control for reducing interference. Power control is widely used in cellular systems to ensure the appropriate SIR is maintained. As we shall see in our discussions of CDMA systems later in the book, in Chapters 6, 8, and 10, power control is critical to their appropriate performance. In describing DCA, we focus on one specific algorithm which lends itself readily to an approximate analysis, yet is characteristic of many DCA strategies, so that one can readily show how the use of DCA can improve cellular system performance. We, in fact, compare its use, for simple systems, with that of fixed-channel allocation, implicitly assumed in the discussion of cellular channel allocations in Chapter 3. Our discussion of power control algorithms focuses on two simple iterative algorithms, showing how the choice of algorithm can dramatically affect the convergence rate. We then show how these algorithms can be written in a unified form, and can, in fact, be compared with a simple single-bit control algorithm used in CDMA systems.

Chapter 5 continues the discussion of basic system concepts encountered in the study of digital mobile wireless systems, focusing on modulation techniques used in these systems. It begins with a brief introductory section on the simplest forms of digital modulation, namely phase-shift keying (PSK), frequency-shift keying (FSK), and amplitude-shift keying or on–off keyed transmission (ASK or OOK). These simple digital modulation techniques are then extended to quadrature-amplitude modulation (QAM) techniques, with QPSK and 8-PSK, used in third-generation cellular systems, as special cases. A brief introduction to signal shaping in digital communications is included as well. This material should be familiar to anyone with some background in digital communication systems. The further extension of these techniques to digital modulation techniques such as DPSK and GMSK used in wireless systems then follows quite readily and is easily described. The chapter concludes with an introduction to orthogonal frequency-division multiplexing (OFDM), including its implementation using Fast Fourier Transform techniques. OFDM has been incorporated in high bit rate wireless LANs, described in Chapter 12.

In Chapter 6 we describe the two major multiple access techniques, time-division multiple access (TDMA) and code-division multiple access (CDMA) used in digital wireless systems. TDMA systems incorporate a slotted repetitive frame structure, with individual users assigned one or more slots per frame. This technique is quite similar to that used in modern digital (wired) telephone networks. The GSM and D-AMPS (IS-136) cellular systems are both examples of TDMA systems. CDMA systems, exemplified by the second-generation IS-95 system and the third-generation cdma2000 and WCDMA systems, use pseudo-random coded transmission to provide user access to the cellular system. All systems utilize FDMA access as well, with specified frequency assignments within an allocated spectral band further divided into time slots for TDMA transmission or used to carry multiple codes in the case of CDMA transmission. We provide first an introductory discussion of TDMA. We then follow with an introduction to the basic elements of CDMA. (This discussion of CDMA is deepened later in the book, specifically in describing IS-95 in Chapter 8 and the third-generation CDMA systems in Chapter 10.) We follow this introduction to CDMA with some simple calculations of the system capacity potentially provided by CDMA. These calculations rely, in turn, on knowledge of the calculation of bit error probability. For those readers not familiar with communication theory, we summarize classical results obtained for the detection of binary signals in noise, as well as the effect of fading on signal detectability. Error-probability improvement due to the use of diversity techniques discussed in Chapter 2 is described briefly as well.

These calculations of CDMA capacity using simple, analytical models enable us to compare, in concluding the chapter, the system-capacity performance of TDMA and CDMA systems. We do stress that the capacity results obtained assume idealized models of cellular systems, and may differ considerably in the real-world environment. The use of system models in this chapter and others following do serve, however, to focus attention on the most important parameters and design choices in the deployment of wireless systems.

Chapter 7, on coding for error detection and correction, completes the discussion of introductory material designed to provide the reader with the background necessary to understand both the operation and performance of digital wireless systems. Much of this chapter involves material often studied in introductory courses on communication systems and theory. A user with prior knowledge of coding theory could therefore use the discussion in this chapter for review, while focusing on the examples provided of the application of these coding techniques to the third-generation cellular systems discussed later in Chapter 10. We begin the discussion with an introduction to block coding for error correction and detection. We focus specifically on so-called cyclic codes used commonly in wireless systems. The section on block coding is followed by a discussion of convolutional coding, with emphasis on the Viterbi algorithm and performance improvements possibly due to the use of convolutional coding. We conclude Chapter 7 with a brief discussion of turbo coding adopted for use with third-generation CDMA cellular systems.

Chapter 8 begins the detailed discussion of specific digital wireless systems described in this book. This chapter focuses on GSM, D-AMPS or IS-136, and IS-95, the three second-generation cellular systems already mentioned a number of times in our discussion above. For each of these systems, we describe the various control signals transmitted in both directions across the air interface between mobile and base station required to register a mobile and set up a call. Control signals and traffic signals carrying the desired information are sent over "channels" defined for each of these categories. The control signals include, among others, synchronization signals, paging signals asking a particular mobile to respond to an incoming call, and access signals, used by a mobile to request a traffic channel for the transmission of information, and by the base station to respond to the request.

For the TDMA-based GSM and IS-136 systems, the various channels defined correspond to specified bit sequences within the time slots occurring in each frame. We describe for each of these systems the repetitive frame structure, as well as the bit allocations within each time slot comprising a frame. We describe as well the various messages transmitted for each of the control channel categories, focusing on the procedure required to set up a call. For the CDMA-based IS-95, these various channels correspond to specified codes. We begin the discussion of IS-95 by providing block-diagram descriptions of the traffic channel portion of the system used to transmit the actual call information. We then move on to the control channels of the system, used by the mobile to obtain necessary timing information from the base station, to respond to pages, and to request access to the system. This discussion of the various code-based channels and block diagrams of the system sub-structure used to implement them, serves to deepen the discussion of CDMA systems begun in Chapter 6, as noted earlier. One thus gains more familiarity with the concepts of CDMA by studying the specific system IS-95. This knowledge of the basic aspects of CDMA will be further strengthened in Chapter 10, in discussing third-generation systems, again as noted above. Once the various IS-95 channel block diagrams are discussed, we show how the various channels are used by a mobile to acquire necessary phase and timing information, to register with the base station, and then to set up a call, paralleling the earlier discussion of setting up a call in the TDMA case. The formats of the various messages carried over the different channels are described as well.

The discussion thus far of the material in Chapter 8 has focused on signals carried across the air or radio interface between mobile station and base station. Much of the material in the earlier chapters implicitly focuses on the radio portion of wireless systems as well, but the expected mobility of users in these systems leads to necessary signaling through the *wired* networks to which the wireless systems are generally connected. We have already alluded implicitly to this possibility by mentioning earlier the need to control handoffs occurring when mobile stations cross cell boundaries. Mobile management plays a critical role in the operation of wireless cellular systems, not only for the proper control of handoffs, but for the location of mobiles when they roam, and for the appropriate paging of mobiles within a given cell, once they are located. These three aspects of mobile management, requiring signaling through the wired network as well as across the mobile base station air interface, are discussed in detail in Chapter 8. The discussion covers, first, handoff control, both across cells (inter-cellular handoff) and between systems (inter-system handoff). It then moves on to a description of location management and paging, and concludes by showing, through simple considerations, that the implementation of these two latter two functions require necessary tradeoffs in performance.

Second-generation cellular systems have been used principally for the purpose of transmitting voice calls, although there has been a growing use for the transmission of data. We therefore provide, in the concluding section of Chapter 8, a detailed discussion of the processing of voice signals required to transmit these signals effectively and efficiently over the harsh transmission environments encountered. We note that the voice signals are transmitted at considerably reduced rates as compared with the rates used in wired telephone transmission. This requires significant signal compression followed by coding before transmission. All second-generation systems have adopted variations of a basic voice compression technique called linear predictive coding. We describe this technique briefly in this chapter and then go on to discuss the variations of this technique adopted specifically for the GSM, IS-136, and IS-95 systems.

In Chapter 9 we provide a detailed discussion of performance issues introduced in earlier chapters. We do this in the context of describing, in a quantitative manner, the processes of admission and handoff control in wireless networks, and the tradeoffs involved in adjudicating between the two. We start from first principles, beginning with an initial discussion of channel holding time in a cell, using simple probabilistic models for call duration and mobile dwell time in a cell. Examples are provided for a variety of cell sizes, ranging from macrocells down to microcells. Average handoff traffic in a cell is related to new-traffic arrivals using flow–continuity arguments. We then obtain equations for new-call blocking probability and handoff-dropping probability based on an assumption of Poisson statistics and the exponential probability distribution. A by-product of the analysis is the derivation of the Erlang distribution, introduced in earlier chapters. This analysis, relying on flow–continuity arguments, involves no specific cell-model geometry. We follow this analysis by a detailed discussion of system performance for one- and two-dimensional cellular geometries, using various models of user–terminal mobility. We show how to obtain important statistical distributions, such as the cell dwell-time distributions for both new and handoff calls, and, from these, the channel holding time distribution. Results are compared with the exponential distribution. A comparison is made as well of a number of admission/handoff control strategies proposed in the literature.

Chapter 10 returns to the discussion of specific cellular systems, focusing on third-generation (3G) wireless systems. These systems have been designed to handle higher bit-rate packet-switched data as well as circuit-switched voice, providing a flexible switching capability between the two types of signal traffic, as well as flexibility in switching between different signal transmission rates. Our discussion in this chapter is geared primarily to the transmission of packet data. The concept of quality of service or QoS plays a critical role in establishing user performance objectives in these systems, and the specific QoS performance objectives required are noted for each of the systems discussed. Three 3G systems are described in this chapter: wideband CDMA or W-CDMA (the UMTS/IMT-2000 standard); the cdma2000 family of CDMA systems based on, and backward-compatible with, IS-95; and GPRS, the enhanced version of GSM, designed for packet transmission. We begin our discussion of the two third-generation CDMA standards by describing and comparing a number of techniques used in CDMA to obtain the much higher bit rates required to handle Internet-type multimedia packet data. We then describe how these techniques are specifically used in each of the standards to obtain the higher bit rates desired. We discuss details of operation of these standards, describing the various logically defined channels used to transport the packet-switched data traffic, as well as the various signaling and control channels used to ensure appropriate performance. This discussion parallels the earlier discussion in Chapter 8 of the various control and traffic channels defined for the second-generation systems. In the case of W-CDMA, the various QoS-based traffic classes defined for that system are described as well. The discussion of the cdma2000 family includes a brief description of the 1xEV-DV standard, designed specifically to handle high-rate packet data.

The discussion of third-generation systems in Chapter 10 concludes with a description of enhancements to GSM, designed to provide it with packet-switched capability. Two basic sets of standards have been defined and developed for this purpose. One is GPRS (General Packet Radio System), mentioned earlier; the other is EDGE. GPRS consists of two parts, a core network portion and an air interface standard. EDGE is designed to provide substantially increased bit rates over the air interface. These TDMA-based 3G systems are often referred to as 2.5G systems as well, since they were designed to provide GSM with a packet-switching capability as simply and rapidly as possible. The core network portion of GPRS uses a layered architecture incorporating the Internet-based TCP/IP protocols at the higher layers. To make this book as self-contained as possible, the discussion of the GPRS layered architecture is preceded by a brief introduction to the concept of layered architectures, using the Internet architecture as a specific example. Quality of service (QoS) classes defined for GPRS are described as well. The air–interface discussion of GPRS includes a description of various logical channels, defined to carry both traffic and the necessary control signals across the air interface. It includes as well a brief description of four coding schemes, defined to provide varying amounts of coding protection against the vagaries of propagation conditions during the radio transmission. This discussion of GPRS is followed by a discussion of the enhanced bit rate capacity offered by the introduction of EDGE. Included are simulation results, taken from the literature, showing the throughput improvement EDGE makes possible as compared with GSM.

Obtaining the QoS performance objectives described and discussed in Chapter 10 requires the appropriate control of packet transmission. This control is manifested, first, by controlling user access to the system, and then by appropriately scheduling packet transmission once access is approved. Chapter 11 focuses on access and scheduling techniques proposed or adopted for packet-based cellular systems. The discussion in this chapter begins by describing, and determining the performance of, the commonly adopted slotted-Aloha access strategy. Although the stress here is on packet access control, it is to be noted that the Aloha strategy is commonly used as well in the circuit-switched second-generation systems discussed in Chapter 8. The analysis provided in Chapter 11 incorporates the model of Poisson statistics introduced earlier in Chapter 9. The effect of fading on the access procedure, leading to the well-known capture effect, is described as well. Following this discussion of the slotted-Aloha access procedure, improved access procedures are described, some of which have been adopted for 3G systems. The discussion in the chapter then moves to multi-access control, the access control of multiple types of traffic, each with differing traffic characteristics and different QoS requirements. A prime example is that of jointly controlling the access of individual data packets and voice calls, which require access to the transmission facility for the time of a complete call. Multi-access control techniques proposed in the literature all build on the basic slotted-Aloha technique. The first such strategy proposed in the literature, and described in this chapter, was PRMA, designed to handle voice and data. Improved versions of PRMA such as PRMA++ have been proposed and a number of these access schemes are discussed in this chapter. The PRMA and related access strategies have all been designed for TDMA-based systems. As is apparent from the summary of Chapter 10 presented above, the focus of cellular systems appears to be moving to CDMA-based systems. We therefore conclude our discussion of multi-access control with a description of a number of access strategies proposed specifically for CDMA systems. Most of these are derived from the basic PRMA scheme, as adapted to the CDMA environment.

Chapter 11 concludes with a detailed discussion of the process of scheduling of streams of data packets in wireless systems, both uplink, from each mobile in a cell to the cell base station, and downlink, from base station to the appropriate mobile terminal. The purpose of scheduling is two-fold: to provide QoS performance guarantees to users, where possible, propagation conditions permitting, and to ensure full use of the system resources, most commonly link bandwidth or capacity. The concept of scheduling packets for transmission from multiple streams has long been studied and implemented in wired packet networks. Some of the scheduling algorithms adopted or proposed for use in packet-based wireless systems are, in fact, variations of those originally studied for use in wired networks. A number of these algorithms are described and compared in this chapter, with reference made to simulation performance results appearing in the literature.

The final chapter in the book, Chapter 12, is devoted to a detailed discussion of wireless local-area networks (WLANs) carrying high bit rate data traffic, as well as wireless personal-area networks (WPANs), exemplified here by the Bluetooth standard. WLANs provide coverage over distances of 100 meters or so. Personal-area networks are designed to provide wireless communication between various devices at most 10 meters apart. The focus of the discussion of WLANs in this chapter is the IEEE 802.11standard, with emphasis on the highly successful and widely implemented 802.11b, commonly dubbed "Wi-Fi." The 802.11b standard is designed to operate at a nominal data rate of 11 Mbps. The higher bit rate WLAN standards, 802.11g and 802.11a, allowing data transmission rates as high as 54 Mbps, are discussed as well. These higher bit rate WLAN standards utilize the OFDM technique described earlier in Chapter 5. The Bluetooth standard, also adopted by the IEEE as the IEEE 802.15.1 WPAN standard, is designed to operate at a rate of 1 Mbps. Both standards use the 2.4 GHz unlicensed radio band for communication.

The first section in Chapter 12 is devoted to the IEEE 802.11 wireless LAN standard. It begins with a brief introduction to Ethernet, the popular access technique adopted for wired LANs. The access strategy adopted for 802.11, *carrier sense multiple access collision avoidance* or CSMA/CA, is then described in detail. The collision-avoidance mechanism is necessary because of contention by multiple users accessing the same radio channel. The Ethernet and 802.11 access strategies both operate at the medium access control, MAC, sub-layer of the packet-based layered architecture introduced in Chapter 10. Actual data transmission takes place at the physical layer of that architecture and the physical layer specifications of the 802.11b, 802.11g, and 802.11a standard, the latter two using OFDM techniques, are described in the paragraphs following the one on the MAC-layer access control. This description of the 802.11 standard is followed by an analysis of its data throughput performance, with a number of examples provided to clarify the discussion.

The second and concluding section of Chapter 12 describes the Bluetooth/IEEE 802.15.1 wireless PAN standard in detail. Under the specifications of this standard, Bluetooth devices within the coverage range of 10 meters organize themselves into *piconets* consisting of a master and at most seven slaves, operating under the control of, and communicating with, the master. There is thus no contention within the system. The discussion in this section includes the method of establishment of the piconet, a description of the control packets used for this process, and the formats of the various types of data traffic packets used to carry out communication once the piconet has been organized. A discussion of Bluetooth performance based on simulation studies described in the literature concludes this section. A number of these simulations include the expected impact of shadow- and Rician-type fast fading, the types of fading described earlier in Chapter 2. Scheduling by the master of a piconet of data packet transmission plays a role in determining Bluetooth performance, and comparisons appearing in the literature of a number of scheduling techniques are described as well.

Chapter 2

Characteristics of the mobile radio environment–propagation phenomena

In Chapter 1, in which we provided an overview of the topics to be discussed in this book, we noted that radio propagation conditions play a critical role in the operation of mobile wireless systems. They determine the performance of these systems, whether used to transmit real-time voice messages, data, or other types of communication traffic. It thus behooves us to describe the impact of the wireless medium in some detail, before moving on to other aspects of the wireless communication process. This we do in this chapter. Recall also, from Chapter 1, that the radio or wireless path normally described in wireless systems corresponds to the radio link between a mobile user station and the base station with which it communicates. It is the base station that is, in turn, connected to the wired network over which communication signals will travel. Modern wireless systems are usually divided into geographically distinct areas called *cells*, each controlled by a base station. We shall have more to say about cells and cellular structures in later chapters. (An exception is made in Chapter 12, the last chapter of this book, in which we discuss small-sized wireless networks for which the concepts of base stations and cells generally play no role.) The focus here is on one cell and the propagation conditions encountered by signals traversing the wireless link between base station and mobile terminal.

Consider, therefore, the wireless link. This link is, of course, made up of a two-way path: a forward path, *downlink*, from base station to mobile terminal; a backward path, *uplink*, from mobile terminal to base station. The propagation or transmission conditions over this link, in either direction, are in general difficult to characterize, since electromagnetic (em) signals generated at either end will often encounter obstacles during the transmission, causing reflection, diffraction, and scattering of the em waves. The resultant em energy reaching an intended receiver will thus vary randomly. As a terminal moves, changing the conditions of reception at either end, signal amplitudes will fluctuate randomly, resulting in so-called *fading* of the signal. The rate of fading is related to the relative speed of the mobile with respect to the base station, as well as the frequency (or wavelength) of the signal being transmitted. Much work has gone into developing propagation models appropriate to different physical environments, that can be used for design purposes to help determine the number of base stations and their locations required over a given geographic region to best serve mobile customers. Software packages based on these models are available for such purposes, as well as for assessing the performance of wireless systems once installed.

In this book we focus on the simplest of models to help gain basic insight and understanding of the mobile communications process. We start, in the first section of this chapter, with a review of free-space propagation. We then describe the propagation model used throughout this book which captures three effects common in wireless communication: average power varying as $1/(\text{distance})^n$, $n > 2$, $n = 2$ being the free-space case; long-term variations or fading about this average power, so-called shadow or log-normal fading; short-term multipath fading leading to a Rayleigh/Rician statistical model which captures power variations on a wavelength scale. The discussion of Rayleigh fading leads us to introduce a linear impulse response model representing the multipath wireless medium, or channel, as it is often called. The chapter continues with a discussion of fading rate, its relation to mobile velocity, and the impact of fading on the information-bearing signal being transmitted. In particular, we discuss the condition for frequency-selective fading, relating it to the time dispersion or delay spread of the signal as received. Frequency-selective fading, in the case of digital signals, leads to inter-symbol interference. We conclude this chapter with a brief discussion of methods used to mitigate against signal fading. These include such techniques as channel equalization, diversity reception, and the RAKE time-diversity scheme.

2.1 Review of free-space propagation

It is well-known that ray optics may be used to find the power incident on a receiving system located in the far field region of an antenna. Specifically, consider, first, an omnidirectional, isotropically radiating antenna element transmitting at a power level of P_T watts. The received power density at a distance d m is then simply $P_T/4\pi d^2$ w/m^2. This em energy falls on a receiving system with effective area A_R. The received power P_R is then given by

$$P_R = \frac{P_T}{4\pi d^2} A_R \eta_R \tag{2.1}$$

with $\eta_R < 1$ an efficiency parameter. Figure 2.1(a) portrays this case.

Most antennas have gain over the isotropic radiator. This is visualized as providing a focusing effect or gain over the isotropic case. In particular, given a transmitting antenna gain $G_T > 1$, the power captured by the receiver of area A_R now becomes

$$P_R = \frac{P_T G_T}{4\pi d^2} A_R \eta_R \tag{2.2}$$

This gain parameter G_T is proportional to the effective radiating area A_T of the transmitting antenna. (The solid angle subtended by the transmitted em beam is inversely proportional to the area.) But it is the antenna size in wavelengths that determines the focusing capability: the larger the antenna size in wavelengths, the narrower the beam transmitted and the greater the energy concentration. Specifically, the gain–area relation is given by the expression $G_T = 4\pi \eta_T A_T/\lambda^2$, $\eta_T < 1$ the transmitting antenna efficiency factor. The effective receiver area A_R obeys a similar relation, so that one may write, in place of A_R in (2.2), the receiver gain expression $G_R = 4\pi \eta_R A_R/\lambda^2$. One then gets the

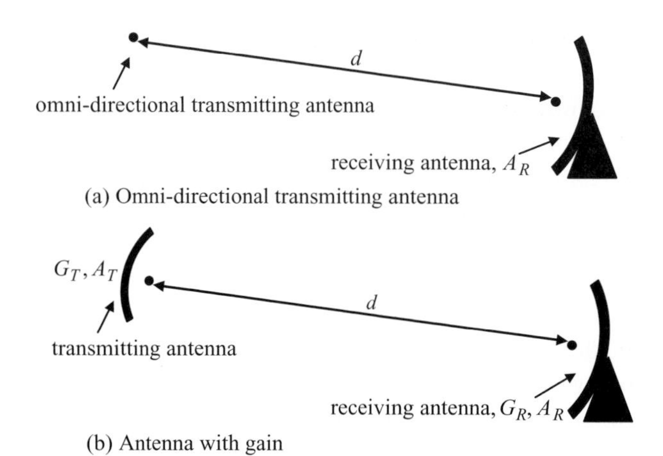

Figure 2.1 Free-space communication

well-known free-space received power equation

$$P_R = P_T G_T G_R \left(\frac{\lambda}{4\pi d} \right)^2 \tag{2.3}$$

This expression is used in determining power requirements for satellite communications, as an example. Figure 2.1(b) portrays the case of beamed communications.

2.2 Wireless case

We now turn to electromagnetic propagation in the wireless environment. We focus on the downlink, base station (BS) to mobile case, for simplicity. Similar considerations apply in the uplink, mobile to BS, case. As noted earlier, em energy radiated from the BS antenna in a given cell will often encounter a multiplicity of obstacles, such as trees, buildings, vehicles, etc., before reaching a given, possibly moving, wireless receiver. The radiating em field is reflected, diffracted, and scattered by these various obstacles, resulting commonly in a multiplicity of rays impinging on the receiver. The resultant em field at the receiver due to the combined interference among the multiple em waves varies in space, and hence provides a spatially varying energy pattern as seen by the receiver. In particular, as the receiver moves, the power of the signal it picks up varies as well, resulting in signal fades as it moves from a region of relatively high power ("good" reception) to low power received ("poor" reception). As a drastic example, consider a mobile receiver, whether carried by an individual or mounted inside a vehicle, turning a corner in a crowded urban environment. The em waves striking the antenna will probably change drastically, resulting in a sharp change in the characteristics of the signal being processed by the receiver.

The effect of this "harsh" radio environment through which a mobile wireless receiver has to move is to change the free-space power equation considerably. One finds several effects appearing:

1 Typically, the far-field *average* power, measured over distances of many wavelengths, decreases inversely with distance at a rate greater than d^2. A common example, to be demonstrated shortly, is a $1/d^4$ variation.
2 The actual power received, again measured over relatively long distances of many wavelengths, is found to vary randomly about this average power. A good approximation is to assume that the power measured in decibels (dB) follows a gaussian or normal distribution centered about its average value, with some standard deviation, ranging, typically, from 6 to 10 dB. The power probability distribution is thus commonly called a *log-normal* distribution. This phenomenon is also commonly referred to as *shadow fading*. These first two effects are often referred to as *large-scale fading*, varying as they do at relatively long distances.
3 At much smaller distances, measured in wavelengths, there is a large variation of the signal. As a receiver moves the order of $\lambda/2$ in distance, for example, the signal may vary many dB. This is particularly manifested with a moving receiver, these large variations in signal level with small distances being converted to rapid signal variations, the rate of change of the signal being proportional to receiver velocity. This *small-scale* variation of the received signal is attributed to the destructive/constructive phase interference of many received signal paths. The phenomenon is thus referred to as *multipath fading*. For systems with relatively large cells, called macrocellular systems, the measured amplitude of the received signal due to multipath is often modeled as varying randomly according to a *Rayleigh* distribution. In the much smaller cells of microcellular systems, the *Ricean* distribution is often found to approximate the random small-scale signal variations quite well. We shall have much more to say about these distributions later.

Putting these three phenomena together, the statistically varying received signal power P_R may be modeled, for cellular wireless systems, by the following equation

$$P_R = \alpha^2 10^{x/10} g(d) P_T G_T G_R \tag{2.4}$$

The terms $10^{x/10}$ and α^2, to be discussed further below, represent, repectively, the shadow-fading and multipath-fading effects. Both x and α are random variables. The term $g(d)$, also to be discussed in detail below, represents the inverse variation of power with distance. The expression $g(d)P_TG_TG_R$ represents the average power, as measured at the receiver at a distance d from the transmitter. Specifically, then, the average received power is given by

$$\overline{P}_R = g(d) P_T G_T G_R \tag{2.5}$$

In free space the term $g(d)$ would just be the $1/d^2$ relation noted earlier. We shall show shortly that, for a common two-ray model, $g(d) = kd^{-4}$, k a constant. More generally, we might have $g(d) = kd^{-n}$, n an integer. This relation is often used in evaluating macrocellular system performance. For microcells, one finds some investigators using the expression for $g(d)$ given by (suppressing the constant k)

$$g(d) = d^{-n_1}\left(1 + \frac{d}{d_b}\right)^{-n_2} \tag{2.6}$$

Table 2.1 *Empirical power drop-off values*

City	n_1	n_2	$d_b(m.)$
London	1.7–2.1	2–7	200–300
Melbourne	1.5–2.5	3–5	150
Orlando	1.3	3.5	90

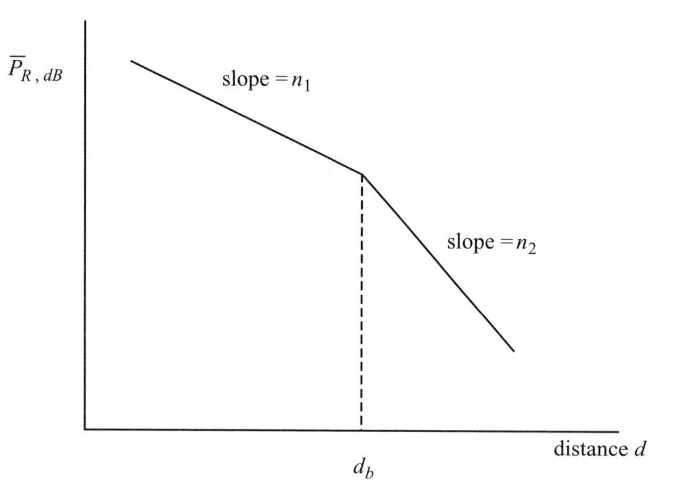

Figure 2.2 Two-slope received signal model, wireless communication

with n_1, n_2 two separate integers and d_b a measured breakpoint. Table 2.1 shows different values for n_1, n_2, and d_b fitted to measurements in three different cities (Linnartz, 1993).

Another simple model for the expression $g(d)$ is given by

$$\begin{aligned} g(d) &= d^{-n_1} & 0 \leq d \leq d_b \\ &= d_b^{-n_1}(d/d_b)^{-n_2} & d_b \leq d \end{aligned} \tag{2.6a}$$

For the case of (2.6a), the average received signal power measured in decibels (dB)

$$\overline{P}_{R,dB} \equiv 10 \log_{10} \overline{P}_R$$

would have the two linear-slope form of Fig. 2.2, with a break in the slope occurring at distance d_b.

As noted above, the expression $\overline{P}_R = P_T g(d) G_T G_R$ in (2.4) provides a measure, to within a constant, of the *average* power (or, equivalently, power density or energy density) measured at the receiving system. This power term is sometimes called the *area-mean* power. As also indicated earlier, the actual instantaneous received power P_R as given by (2.4) is found to vary statistically about this value. The large-scale statistical variation, shadow or log-normal fading, is accounted for by the $10^{x/10}$ term. Small-scale variation or multipath fading is incorporated in the α^2 term. We discuss each of these terms in the

Linnartz, J. P. 1993. *Narrowband Land-Mobile Radio Networks*, Boston, MA, Artech House.

paragraphs below, beginning with shadow fading, followed by long-term path loss and small-scale fading, in that order.

Shadow fading

Consider the shadow-fading term $10^{x/10}$ of (2.4) first. As noted earlier, this type of fading is relatively slowly varying, being manifested over relatively long distances (many wavelengths). The somewhat peculiar form in which this term is written in (2.4) is explained by noting again that it is found that the received signal power measured in dB can be fairly well-approximated by a gaussian random variable. Specifically, consider P_R in dB

$$\begin{aligned} P_{R,dB} &= 10\log_{10} P_R \\ &= 10\log_{10}\alpha^2 + x + 10\log_{10} g(d) + P_{T,dB} + 10\log_{10} G_T G_R \\ &\equiv 10\log_{10}\alpha^2 + p_{dB} \end{aligned} \tag{2.7}$$

The shadow-fading random variable x, measured in dB, is taken to be a zero-mean gaussian random variable with variance σ^2

$$f(x) = \frac{e^{-\frac{x^2}{2\sigma^2}}}{\sqrt{2\pi\sigma^2}} \tag{2.8}$$

This term provides the statistical variation measured about the *average* dB received power or area-mean power

$$\overline{P}_{R,dB} = P_{T,dB} + 10\log_{10} g(d) + 10\log_{10} G_T G_R$$

ignoring for the time being the second random term α^2, modeling the effect of multipath fading. The term p_{dB} in (2.7), the random variable (rv) representing the statistically varying, long-term received power in dB due to shadow fading is then modeled as a gaussian rv with average value $\overline{P}_{R,dB}$. (This thus neglects the multipath fading which varies much more rapidly with mobile motion, as will be noted later. Multipath fading changes on the order of wavelengths. Shadow fading varies on the order of meters, as was noted earlier, and as will be seen from the discussion below.) The term p_{dB} is sometimes referred to as the *local-mean* power, to distinguish it from the area-mean power, and to indicate that it differs from the instantaneous received power P_R which incorporates the effect of the more rapidly varying multipath fading. The probability density function for p_{dB} is thus written as follows

$$f(p_{dB}) = e^{-(p_{dB}-\overline{P}_{R,dB})^2/2\sigma^2}/\sqrt{2\pi\sigma^2} \tag{2.9}$$

Figure 2.3 shows a typical received dB power plot as a function of distance d between BS and mobile receiver. It shows the power points scattering statistically, the shadow-fading phenomenon, about the average power, and ignores the small-scale multipath fading. As noted earlier, typical values of σ range from 6 to 10 dB. Figure 2.3 is drawn for the case $g(d) = kd^{-n}$, n an integer. The average power-distance curve is thus obviously an inverse straight line, as shown in the figure.

Long-term shadow fading is, in general, caused by variations in radio signal power due to signal encounters with terrain obstructions, such as hills, or man-made obstructions, such as buildings. Measured signal power may thus differ substantially at different

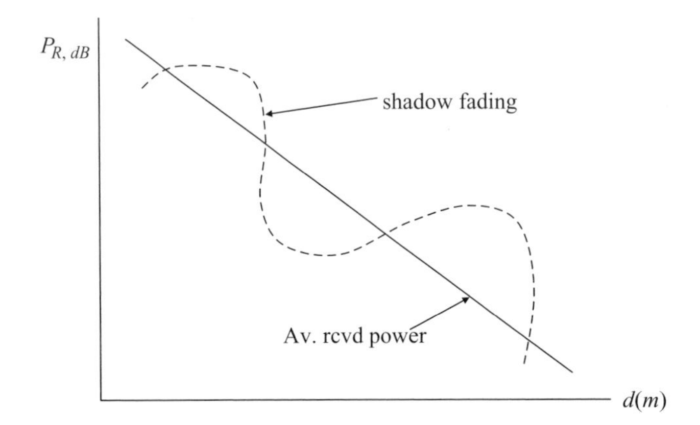

Figure 2.3 Large-scale propagation effects in wireless transmission

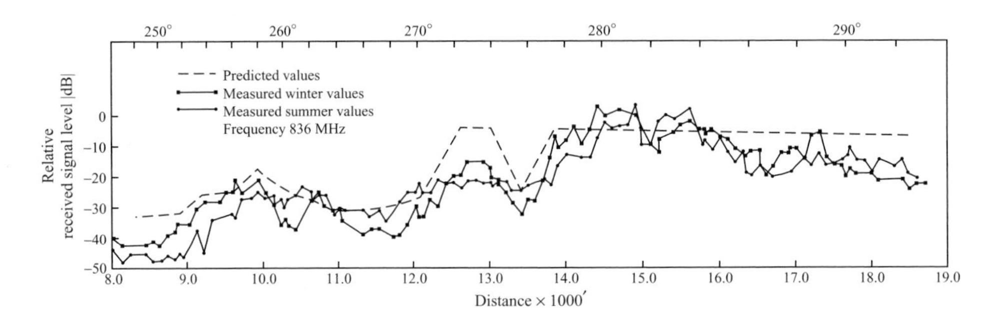

Figure 2.4 Measured signal level, fixed distance from transmitter, 836 MHz (from Jakes, 1974: Fig. 2.2–24)

locations, even though at the same radial distance from a transmitter. Long-term fading results in fluctuations of the field strength of the radio signal around the average power, the area-mean power, which is given by the expression $P_T g(d) G_T G_R$, as noted earlier. We can say that shadowing represents medium-scale fluctuations of the radio field strength occurring over distances from tens to hundreds of meters. Many empirical studies, such as those performed by Reudink and co-workers (Black and Reudink, 1972) and Egli (1957) many years ago, demonstrate that the received mean power of the radio signal fluctuates about the area-mean power with a log-normal distribution, thus validating the model of (2.7)–(2.9). Such studies and empirical results are based on measurements of the received signal strength at a fixed distance from a transmitter where the measurement is repeated every few hundreds of meters while the vehicle travels along a circular or semi-circular arc. Figures 2.4 and 2.5 show the results of such experiments performed at an approximately two-mile distance from a transmitter (Jakes, 1974). One interesting remark to make from these figures is that there is a 10 dB difference in received radio signal strength

Black, D. M. and D. O. Reudink. 1972. "Some characteristics of radio propagation at 800 MHz in the Philadelphia area," *IEEE Transactions on Vehicular Technology*, 21 (May), 45–51.
Egli, J. J. 1957. "Radio propagation above 40 Mc/s over irregular terrain," *Proc. IRE*, October, 1383–1391.
Jakes, W. C. ed. 1974. *Microwave Mobile Communications*, AT&T, 1995 edition, New York, IEEE Press.

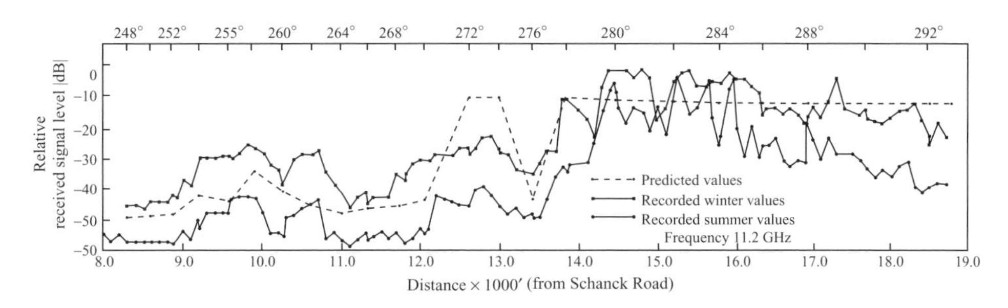

Figure 2.5 Measured signal level, fixed distance from transmitter, 11.2 GHz (from Jakes, 1974: Fig. 2.2–25)

between measurements performed in summer and winter, due to the presence or absence of tree leaves!

As mentioned above, shadowing is mainly caused by terrain and buildings. As a result, since each geographical area has its own unique terrain, buildings, building shapes and density, as well as different building materials, the standard deviation σ in the gaussian model of (2.8) and (2.9) depends on the environment. Due to this phenomenon, it is highly important to perform field measurements and carefully characterize an environment prior to deployment of cellular systems. In this book we assume a known value of σ, in the range of 6–10 dB, for simplicity in modeling. The reader is thus cautioned that results of analysis we will obtain in chapters following are to be used with care, since they are generally based on simplified models.

Path-loss: two-ray model

We now discuss the average power propagation term $g(d)$ in more detail, focusing on a common two-ray propagation model for which one finds, in a straightforward way, $g(d) = kd^{-4}$. This model is the one most often used in the literature in carrying out cellular system performance calculations. We shall assume this simple model as well in carrying out performance calculations throughout this book.

The two-ray propagation model is the simplest one that one could adopt in demonstrating the effect on the average received power of multiple rays due to reflection, diffraction, and scattering, impinging on a receiver. It treats the case of a single reflected ray. Despite the relative simplicity of this model, it has been found to provide reasonably accurate results in macrocellular systems with relatively high BS antennas and/or line of sight conditions between BS and receiver. The two-ray model assumes that the transmitted em wave reaches the (non-moving) receiver directly through a line-of-sight path, and indirectly by perfect reflection from a flat ground surface. Figure 2.6 portrays the geometry involved: the transmitting antenna, located at the base station, is shown radiating from a height h_t above a perfectly reflecting, flat ground surface. The receiving antenna, a free-space distance d m away, is shown situated at a height h_r above the ground. For simplicity's sake both antennas may be assumed, initially, to be isotropic radiators. Appropriate gain factors to accommodate directional antennas and to compare with the case of free-space propagation are incorporated later. The indirect ray is shown traveling a distance d_1 before being reflected at the ground surface. The reflected ray then travels a distance d_2 further

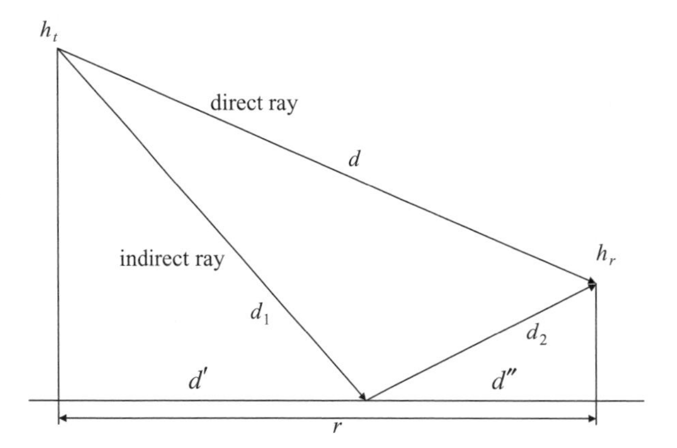

Figure 2.6 Two-ray propagation

before reaching the receiving antenna. The angle θ is the angle of incidence and reflection. The parameter r represents the ground distance between transmitting and receiving antennas, as shown.

Say the electric field appearing in the far-field of the transmitter is a sinewave at frequency f_c with amplitude E_T. In complex notation it is written in the usual form

$$\tilde{E}_T = E_T e^{j\omega_c t}$$

with $\omega_c = 2\pi f_c$. The transmitted power P_T is proportional to the magnitude squared of the electric field and is thus just KE_T^2, K a proportionality constant. Now consider the direct wave impinging on the receiving antenna. In complex phasor form this is given by

$$\tilde{E}_{R,D} = \frac{E_T}{d} e^{j[\omega_c(t-\frac{d}{c})]} \tag{2.10}$$

with c the velocity of light. The received indirect wave, assuming perfect reflection at the ground, appears in a similar form, except that its total distance traveled is $d_1 + d_2$, while, with perfect reflection, it undergoes an added π radians phase change. It is thus written in complex phasor form as

$$\tilde{E}_{R,I} = -\frac{E_T}{d_1 + d_2} e^{j\left[\omega_c\left(t - \frac{d_1+d_2}{c}\right)\right]} \tag{2.11}$$

The total received em field is the sum of these two phasors. It is thus given by

$$\tilde{E}_R = \frac{E_T e^{j\omega_c(t-\frac{d}{c})}}{d}\left[1 - \frac{d}{d_1 + d_2} e^{-j\omega_c\left(\frac{d_1+d_2-d}{c}\right)}\right] \tag{2.12}$$

Noting that the average received power P_R is proportional to the magnitude squared of the electric field, we find from (2.12), extending this result to the case of directional antennas with gain, that the average received power is given by

$$\overline{P}_R = K\,|E_R|^2 = P_T G_T G_R \left(\frac{\lambda}{4\pi d}\right)^2 \left|1 - \left(\frac{d}{d_1 + d_2}\right) e^{-j\Delta\phi}\right|^2 \tag{2.13}$$

Here $\Delta\phi \equiv (d_1 + d_2 - d)\omega_c/c = 2\pi\,\Delta d/\lambda$, with $\Delta d \equiv (d_1 + d_2 - d)$ and $\lambda = c/f_c$, the wavelength of the em waves. Details are left to the reader.

We now use Fig. 2.6 and some simple algebra to show that $d_1 + d_2$ may be rewritten in terms of the transmitter–receiver spacing d and the antenna heights h_t and h_r. Under a reasonable assumption that d is large compared with the antenna heights, (2.13) simplifies to the desired $1/d^4$ dependence. Specifically, note first that $d_1^2 = d'^2 + h_t^2$ while $d_2^2 = d''^2 + h_r^2$, with d' and d'' the distances indicated in Fig. 2.6. Since the angle of incidence θ of the indirect ray equals the angle of reflection, as shown, we also have $d' = d''h_t/h_r$. Note also that $d' + d''$ is the ground distance r between transmitter and receiver. It is then left for the reader to show, replacing d' and d'' by r, that

$$d_1 + d_2 = \sqrt{r^2 + (h_t + h_r)^2} \tag{2.14}$$

From the geometry of Fig. 2.6, however, it is clear that we also have $d^2 = (h_t - h_r)^2 + r^2$. We can thus eliminate r in (2.14) to obtain

$$d_1 + d_2 = d\sqrt{1 + \frac{4h_t h_r}{d^2}} \tag{2.14a}$$

We now make the assumption that $d^2 \gg 4h_t h_r$. As a check, note that if the transmitting antenna height h_t is 50 meters and the receiving antenna height h_r is 2 meters, this assumption implies that $d \gg 20$ m. If the transmitting antenna height is increased to 100 meters, we must have $d \gg 30$ m. The assumption thus appears quite reasonable. It implies that $(d_1 + d_2)/d$ in (2.13) is approximately 1. Now consider the effect of this assumption on $\Delta\phi$ in (2.13). Note first that $\Delta d = (d_1 + d_2 - d) \cong 2h_t h_r/d$. Then $\Delta\phi \cong 4\pi h_t h_r/\lambda d$. We now assume this phase difference term is small. Specifically, take $\Delta\phi \le 0.6$ radian. This implies that $d \ge 21 h_t h_r/\lambda$, generally a tighter restriction on d than the one assumed above. Specifically, take the frequency of operation to be $f_c = 800$ Mhz ($\lambda = 3/8$ m), the frequency around which most first- and second-generation cellular systems cluster. Then, for $h_t = 50$ m and $h_r = 2$ m, as above, we get $d \ge 5.6$ km. If a higher frequency of 2 Ghz is used, with a corresponding wavelength of $\lambda = 0.15$ m, we get $d \ge 14$ km. Under this assumption of small $\Delta\phi$ and letting $(d_1 + d_2)/d = 1$, we find

$$\left|1 - \frac{d}{d_1 + d_2} e^{-j\Delta\phi}\right|^2 \cong (\Delta\phi)^2$$

Equation (2.13) then simplifies to

$$\overline{P}_R = P_T G_T G_R \left(\frac{\lambda}{4\pi d}\Delta\phi\right)^2 = P_T G_T G_R \frac{(h_t h_r)^2}{d^4} \tag{2.13a}$$

Note that this final equation for the average received power has the desired $1/d^4$ dependence on distance. The term $(h_t h_r)^2/d^4$ is just the term $g(d)$ appearing in (2.4) and (2.7). As noted earlier, the $1/d^4$ average power dependence is often assumed in performance studies of cellular systems. We shall provide many examples of its use in subsequent material in this book. Note further that under the assumptions made here

$$r \cong d \cong d_1 + d_2$$

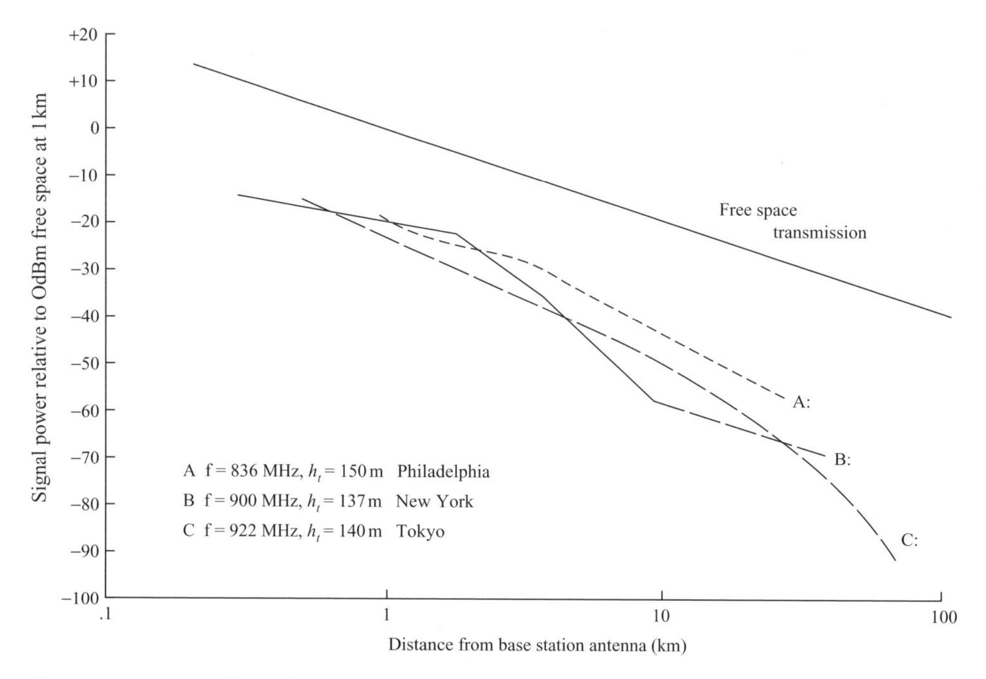

Figure 2.7 Received signal power as function of distance (from Jakes, 1974: Fig. 2.2–10)

Hence the ground distance r between the transmitting (base station) and receiving (mobile terminal) systems may be used in place of the distance d in determining average received power. We shall, in fact, always refer to the *ground* distances between systems.

Now consider some measurements roughly validating these two-ray results. Figure 2.7 shows received signal power as a function of distance from the transmitters in New York, Philadelphia, and Tokyo at 900 MHz using relatively high antennas (Jakes, 1974). These measurements are compared with the free-space transmission model of (2.1) in which the power drops off as $1/d^2$. Figure 2.8 shows the effect of base station antenna height (indicated by the symbol h_b in the figure) on received signal strength measured in Tokyo at a frequency of 922 MHz, again taken as a function of distance from the base station. Note that these curves agree roughly with the simple expression (2.13a), obtained using the two-ray model. They indicate that the propagation loss does fall off more rapidly than that in the free-space case; they also show that higher base station antennas increase the average received signal power, as predicted by (2.13a).

Multipath fading: Rayleigh/Ricean models

Recall that we indicated earlier that we would be discussing in this section three significant propagation effects due to the scattering, diffraction, and reflection of radio waves that appear in wireless systems. All three effects appear in the simple received power equation (2.4). One effect was the $1/d^n$ dependence of average power on distance. We have just provided the commonly used $n = 4$ case, developed from a two-ray propagation model.

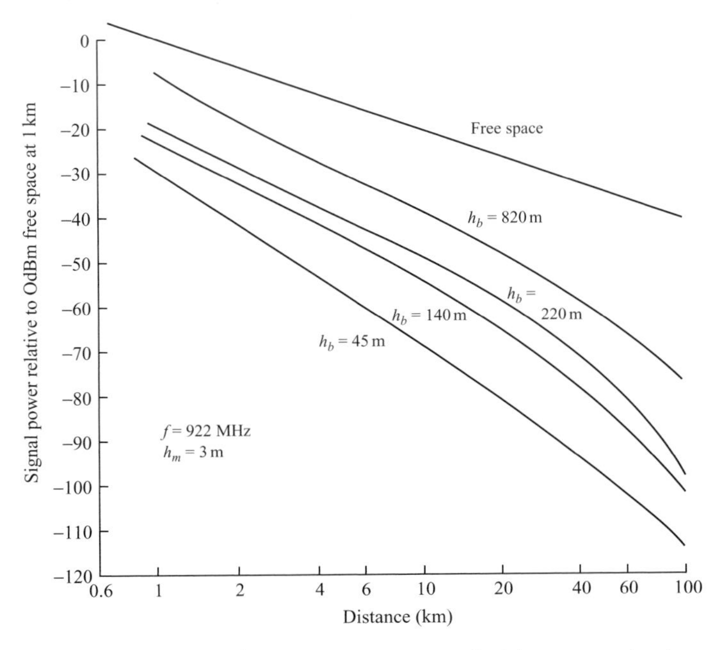

Figure 2.8 Effect of base station antenna height on received power (from Jakes, 1974: Fig. 2.2–11)

We discussed prior to this log-normal or shadow fading, used to represent large-scale random variations about the average power. We now move on to the third phenomenon, that of small-scale multipath fading, which we shall show can be modeled by using Rayleigh/Ricean statistics. As noted earlier, amplitude and power variations occur over distances of wavelengths, hence the reference to small-scale fading. Specifically, with Rayleigh statistics, the probability density function $f_\alpha(\alpha)$, $\alpha \geq 0$, of the random variable α in (2.4) is given by

$$f_\alpha(\alpha) = \frac{\alpha}{\sigma_r^2} e^{-\alpha^2/2\sigma_r^2} \qquad \alpha \geq 0 \tag{2.15}$$

with σ_r^2 an adjustable Rayleigh parameter. It is readily shown that the second moment of the Rayleigh distribution is $E(\alpha^2) = 2\sigma_r^2$. The Rayleigh function is shown sketched in Fig. 2.9. Ricean statistics will be discussed later. In this case of modeling the effect of received power by (2.4) it is readily shown that we must set $\sigma_r^2 = 1/2$. To demonstrate this value of $\sigma_r^2 = 1/2$ in this case, it suffices to return to the basic propagation equation of (2.4). We have indicated earlier, a number of times, that the average received power is just $P_T g(d) G_T G_R$. This implies taking the expectation of the instantaneous *received* signal power P_R with respect to the two random variables x and α. Doing so, we find that $E(\alpha^2) = 1$. Hence we must set $\sigma_r^2 = 1/2$, as indicated. An alternative approach is to take expectations with respect to x and α in the dB received power expression of (2.7). It is clear then that $E(x) = 0$ and $E(\alpha^2) = 1$. We shall also show that the instantaneous local-mean received power P_R is, in this case, an exponentially distributed random variable.

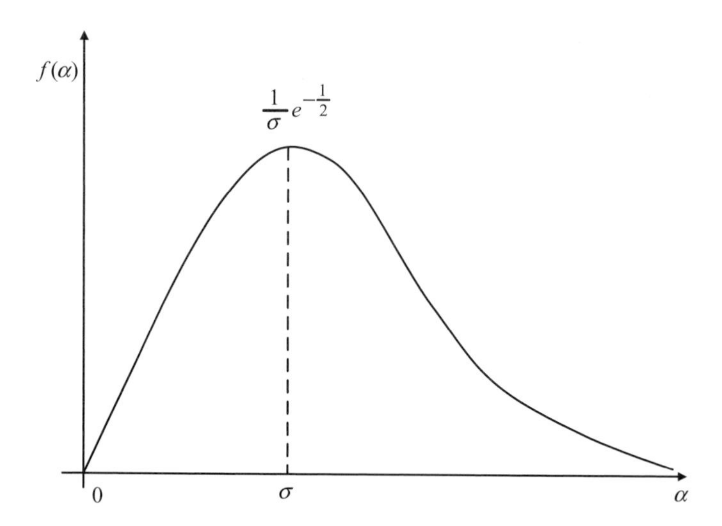

Figure 2.9 Rayleigh distribution

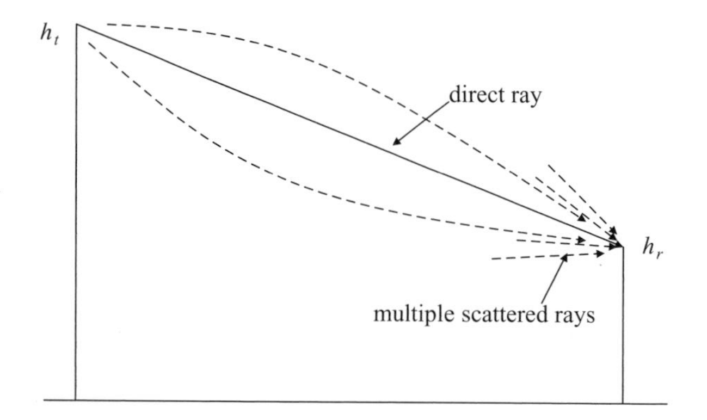

Figure 2.10 Multiple received rays due to scattering

The Rayleigh distribution arises naturally out of random multipath considerations. Measurements made in the early days of mobile telephony were found to agree with the Rayleigh model (Jakes, 1974). In the discussion above, we focused on a particular two-ray model. Each ray is, in reality, made up of the superposition of multiple rays, due to scattering by buildings and other obstacles along a given path. Consider the direct ray in Fig. 2.6, for example. It will actually be made up of many rays, due to scattering multiple times by buildings and other objects encountered along its path, all traveling, on average, about the same distance d. This is shown schematically in Fig. 2.10. Each of these rays appearing at the receiver will differ randomly in amplitude and phase because of the scattering. Alternatively, consider the shadow-fading example of Fig. 2.3. Each power point along the randomly varying shadow-fading curve is, in turn, actually varying randomly at distances the order of wavelengths because of multiple-ray scattering at each of these points. It is this multiple-ray scenario on which we focus. Say there are L such rays. If the transmitted signal is again taken to be an unmodulated sinewave at frequency f_c Hz, we can write the superposition of the L rays in a complex phasor form extending that of (2.10). The resultant phasor appearing at the receiver is then given by

$$\tilde{S}_R(t) = \sum_{k=1}^{L} a_k e^{j[\omega_c(t-t_0-\tau_k)+\theta_k]} \tag{2.16}$$

The term $t_0 = d/c$ represents the average delay encountered previously, appearing in (2.10). Each of the L rays has a random variation about this delay, denoted by the random variable τ_k. We assume the random phase $\omega_c\tau_k$ is uniformly distributed between 0 and 2π, as is the random additional phase θ_k. The sum of the two random phases, $\theta_k - \omega_c\tau_k \equiv \phi_k$, is then uniformly distributed between 0 and 2π as well. The amplitude term a_k of the kth ray is taken to be a real number and random as well. (A mobile's movement will introduce additional Doppler shift phase terms, to be discussed shortly.)

Now consider the actual received normalized signal $S_R\,(t)$ given by the real part of (2.16). This is just

$$S_R(t) = \sum_{k=1}^{L} a_k \cos[\omega_c(t-t_0)+\phi_k] \tag{2.17}$$

The instantaneous local-mean received power P_R is proportional to the time average of $S_R^2\,(t)$. From (2.17), this is just

$$P_R = \frac{c}{2}\sum_{k=1}^{L} a_k^2 \tag{2.18}$$

with c a proportionality constant. This is precisely the power given by (2.4), which includes the three effects of propagation distance, shadow fading, and multipath fading, the last the effect on which we are focusing. Expanding (2.17) by trigonometry, we get

$$\begin{aligned} S_R(t) &= \sum_{k=1}^{L} a_k \cos\phi_k \cos\omega_c(t-t_0) - \sum_{k=1}^{L} a_k \sin\phi_k \sin\omega_c(t-t_0) \\ &= x\cos\omega_c(t-t_0) - y\sin\omega_c(t-t_0) \end{aligned} \tag{2.17a}$$

where

$$x \equiv \sum_{k=1}^{L} a_k \cos\phi_k \quad \text{and} \quad y \equiv \sum_{k=1}^{L} a_k \sin\phi_k \tag{2.19}$$

It is important to note now that for large L the random variables x and y, each defined as the sum of L random variables, become gaussian distributed, by the Central Limit Theorem of probability. In particular, since a_k and ϕ_k are independent random variables, and the ϕ_ks were assumed to be uniformly distributed, with zero average value, it is left to the reader to show that the expectations of the various rvs are given by

$$E(x) = E(y) = 0, \quad \text{while } E(x^2) = E(y^2) = \frac{1}{2}\sum_{k=1}^{L} E\left(a_k^2\right) \equiv \sigma_R^2$$

The gaussian variables x and y are thus zero-mean and have the same variance σ_R^2. In addition, it is readily shown from the orthogonality of $\cos\phi_k$ and $\sin\phi_k$ that $E(xy) = 0$. Here we have taken expectations with respect to the short-distance (order of wavelengths) random variations among the various path components only. We are essentially conditioning on the much longer-distance variations due to shadow fading.

Now refer back to our basic instantaneous power equation (2.4). We can write this expression in the form $P_R = \alpha^2 p$, with p the local-mean power term due to shadow fading defined earlier in its dB form by (2.7). Taking the expectation with respect to the multipath, shorter-distance, random variation only (i.e., conditioning on the longer-distance shadow-fading variations), we have, from (2.18) and the definition of σ_R^2 above

$$c\sigma_R^2 = E(P_R) = E(\alpha^2)p = p \tag{2.20}$$

We have used the expectation symbol $E(\ )$ here to indicate, implicitly, averaging with respect to the multipath term only, to obtain the local-area mean power p. Further averaging with respect to this power would then give us the average received power $\overline{P}_R$. This verifies the statement made earlier that $E(\alpha^2) = 1$, as well as the shadow-fading probability density function expression (2.9) in which we have the average of p_{dB} given by the dB average power.

We continue our analysis. Since gaussian uncorrelated variables are independent, x and y, as defined by (2.19), are independent as well. Again using simple trigonometry, we rewrite (2.17a) in the form

$$S_R(t) = a\cos[\omega_c(t - t_0) + \theta] \tag{2.17b}$$

where $a^2 = x^2 + y^2$, and $\theta = \tan^{-1} y/x$. Hence $E(a^2) = E(x^2) + E(y^2) = 2\sigma_R^2$. Note that a is the (random) envelope or amplitude of the received signal $S_R(t)$. Given x and y zero-mean gaussian, it can be shown that the signal envelope a is Rayleigh distributed (Schwartz, 1990), as given by

$$f_a(a) = \frac{a}{\sigma_R{}^2} e^{-a^2/2\sigma_R^2} \tag{2.21}$$

Note that this is precisely of the form of (2.15), but with σ_r^2 replacing σ_r^2. In particular, we must have, comparing (2.17), (2.17a), (2.17b), and (2.18), and again averaging $S_R^2(t)$ over time

$$P_R = ca^2/2 = \alpha^2 p \tag{2.22}$$

c again a proportionality constant. Since $\alpha = \sqrt{\frac{c}{2p}}a$, and a is Rayleigh distributed, it is clear that α must be Rayleigh distributed as well, as given by (2.15). Note that averaging (2.22) with respect to α one obtains (2.20), as expected. Equation (2.20) relates the variance of the amplitude rv a to the local-mean received power p.

The relation (2.22), together with (2.15), enables us to find the probability distribution of the instantaneous power P_R. In particular, from simple probability theory, we have

$$f_{P_R}(P_R) = f_\alpha\left(\sqrt{\frac{P_R}{p}}\right)\left|\frac{d\alpha}{dP_R}\right| = \frac{1}{p}e^{-P_R/p} \tag{2.23}$$

The instantaneous received power is thus exponentally distributed with its average value given by the local-mean power p, as stated earlier. The exponential distribution is sketched in Fig. 2.11.

Schwartz, M. 1990. *Information Transmission, Modulation, and Noise*, 4th edn, New York, McGraw-Hill.

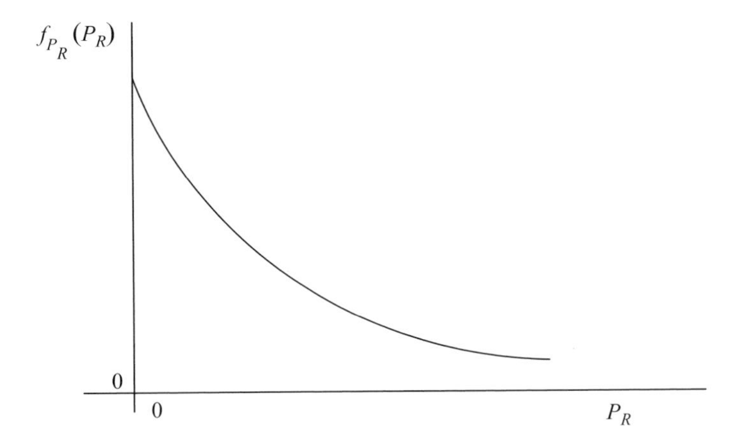

Figure 2.11 Exponential distribution

It was noted above that the gaussian statistics for the random variables x and y arise from the Central Limit Theorem as the number of scattered rays L becomes large. How large must L be? Work done many years ago showed that the sum of as few as six sine-waves with independent random phases gives rise to a resultant function whose amplitude closely obeys Rayleigh statistics (Schwartz, Bennett, and Stein, 1966). It is for this reason that it is found that in macrocellular wireless systems the Rayleigh multipath-fading model discussed here is often a fairly accurate one. In microcellular systems, Ricean fading, to be discussed next, appears to be a more valid model. In that type of system, with much shorter distances between transmitter and receiver, it is more likely that one of the multiple rays will strike the receiver directly, dominating the reception. As we see next, this is exactly the model used in obtaining Ricean statistics.

Consider Fig. 2.10 again in which the set of multiple direct rays between transmitter and receiver is shown, due, as noted above, to scattering from obstacles on the path between the two antennas. Here, however, assuming the distance between the two antennas is not too large, i.e., the microcellular case, the two antennas are taken to be in line of sight of one another, so that a portion of the em energy is directly received, represented by one of the rays labeled *direct ray*. Let the amplitude of the normalized signal representing this direct ray have a value A. (But note again that we are implicitly assuming the shadow-fading case: the amplitude of this "direct" ray A includes the effect of shadow fading.) Proceeding as above, in the Rayleigh case, in which the rays were represented as having randomly distributed amplitudes and randomly varying phases due to the scattering processes, the actual received signal $S_R(t)$ may now be written

$$\begin{aligned} S_R(t) &= A\cos\omega_c(t-t_0) + \sum_{k=1}^{L} a_k \cos[\omega_c(t-t_0-\tau_k)+\theta_k] \\ &= (A+x)\cos\omega_c(t-t_0) - y\sin\omega_c(t-t_0) \\ &= a\cos[\omega_c(t-t_0)+\theta] \end{aligned} \tag{2.24}$$

Schwartz, M., W. R. Bennett, and S. Stein. 1966. *Communication Systems and Techniques*, New York, McGraw-Hill; reprinted, IEEE Press, 1996.

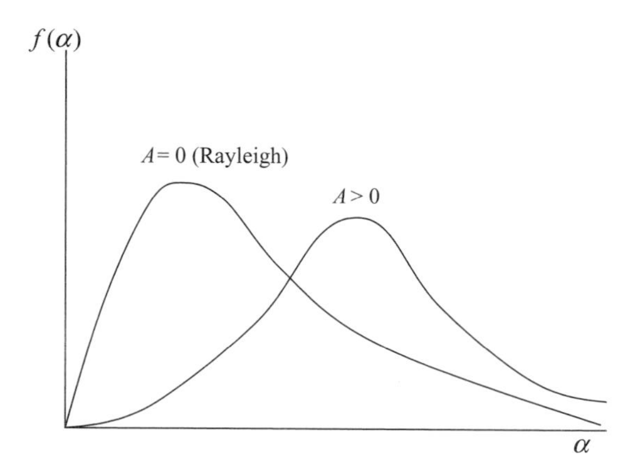

Figure 2.12 Ricean distribution

just as was done in (2.16) and (2.17), but adding the extra carrier term representing the line-of-sight direct ray. The random variables x and y are precisely those defined in (2.19): they are gaussian and independent, with variance σ_R^2. Since phase is relative, all phase terms have been referenced to the phase of the received direct ray, taken to be zero for simplicity.

Note that the random amplitude a now differs from its earlier definition in the Rayleigh case above. It is now given by $a^2 = (A+x)^2 + y^2$. An analysis similar to that carried out in the Rayleigh case shows that, with x and y independent gaussian random variables, the probability distribution of the random amplitude a is given by the following *Ricean* distribution (Schwartz, 1990):

$$f_a(a) = \frac{a}{\sigma_R^2} e^{-\left(\frac{a^2+A^2}{2\sigma_R^2}\right)} I_0\left(\frac{aA}{\sigma_R^2}\right) \tag{2.25}$$

$I_0(z)$ is the modified Bessel function of the first kind and zero order, defined as

$$I_0(z) \equiv \frac{1}{2\pi} \int_0^{2\pi} e^{z\cos\theta} d\theta \tag{2.26}$$

The Ricean distribution is sketched in Fig. 2.12 and compared there with the Rayleigh distribution, which is just the case with $A = 0$. The Rayleigh distribution is thus a special case of the Ricean distribution. It may readily be shown that, for $A^2/2\sigma_R^2 \gg 1$, the Ricean distribution approaches the gaussian distribution, centered about the value A. This is as to be expected from (2.24), since, for A large enough, the contribution of the quadrature term involving y may be neglected. We then simply have $S_R \cong ((A+x) \cos \omega_c(t - t_0))$. The amplitude a is thus just $A + x$, a gaussian random variable with average value A.

These remarks are readily quantified from the properties of the modified Bessel function $I_0(z)$. It is clear from (2.26) that, for z $\ll 1$, $I_0(z) \to 1$. This corresponds, of course, to the case $A^2/2\sigma_R^2 \ll 1$ in (2.25), resulting in the Rayleigh distribution. Now consider the other extreme, with $z \gg 1$. Then it may be shown that $I_0(z) \to e^z/\sqrt{2\pi z}$. Using this form for the

Bessel function in (2.25) and noting that $A \gg \sigma_r$, it is readily seen that $f_a(a)$ peaks at about A, and, in the vicinity of that value of the amplitude, is closely gaussian, verifying the comments above. We shall return to the Rayleigh statistics, due to small-scale multipath variations, during the discussion of the effects of receiver mobility in the next section.

The probability distribution of the instantaneous power P_R, varying again about the local-mean power p, may be found from (2.25), using the relation (2.22) connecting P_R and the signal envelope or amplitude a. It is left for the reader to show that this distribution, in the Ricean model case, is given by

$$f_{P_R}(P_R) = \frac{(1+K)e^{-K}}{p} e^{-\frac{1+K}{p}P_R} I_0\left(\sqrt{\frac{4K(1+K)}{p}P_R}\right) \tag{2.27}$$

The parameter $K \equiv A^2/2\sigma_R^2$ is called the Ricean K-factor. Use has been made, in deriving this distribution from (2.25) and (2.22), of the fact that, since $a^2 = (A+x)^2 + y^2$, $E(a^2) = A^2 + 2\sigma_R^2 = 2\sigma_R^2\,(1+K)$. The Ricean K factor is related very closely to the ratio of the average received line-of-sight or direct-ray signal power to the average received signal power through scattered rays. Measurements made in microcellular environments show K falling in the range of 6 dB–30 dB.

2.3 Random channel characterization

Before proceeding further with our discussion of signal fading, we pause briefly to further characterize the wireless medium over which signals are transmitted. This characterization will be found useful in understanding the transmission of digital signals over the channel, as well as studying ways, later in the chapter, of mitigating some of the effects of the fading process on signal transmission. Note that, in writing an expression such as (2.16), we have indicated that the transmission of a carrier signal over the wireless medium, often referred to as the *wireless channel*, results in a received random multipath signal due to scattering of the transmitted signal by multiple scatterers along the transmission path. The scattering introduces random amplitude, phase, and time delay terms as indicated in (2.16). More generally, a modulated carrier signal would be transmitted, the modulation carrying the information to be transmitted. This more general transmitted signal can be written in complex phasor form as

$$\tilde{S}_T(t) = s(t)e^{j\omega_c t} \tag{2.28}$$

with $s(t)$ the complex phasor representation of the modulating (information-bearing) signal. The modulating signal is often referred to as well as the baseband signal. The complex phasor form $s(t)$ of the signal is often called the complex envelope of the signal. $s(t)$ is assumed to have a baseband bandwidth $B/2$. The bandwidth of the transmitted carrier signal $\tilde{S}(t)$ is then B, i.e., double the baseband bandwidth. The received signal, generalizing (2.16), can then be written as

$$\begin{aligned}\tilde{S}_R(t) &= \sum_{k=1}^{L} a_k s(t-t_k)e^{j[\omega_c(t-t_o-\tau_k)+\theta_k]} \\ &= a(t)e^{j\omega_c(t-t_0)}\end{aligned} \tag{2.29}$$

Here $a(t)$ represents the complex amplitude of the received baseband signal, and is given by

$$a(t) = \sum_{k=1}^{L} a_k s(t - t_k) e^{j(\theta_k - \omega_c \tau_k)} \tag{2.30}$$

Equation (2.29) generalizes the earlier form (2.17b) of the received signal for an unmodulated signal, shown there in its real-signal form. Implicit in the representations (2.16) and (2.29) of multipath scattering is that the amplitude and phase terms introduced by the scatterers do not vary with time. The wireless channel can then be modeled as a linear time-invariant one. More generally, if the amplitude and phase terms vary with time, the channel turns out to be a time-varying linear channel. As an example, introducing receiver motion, neglected in the representation thus far, introduces time variation into the channel model. Both the time-invariant case and the time-variant one are considered further in the material following.

The time-invariant linear channel model follows directly from a comparison of (2.29) and (2.30). In particular, we note that the received and transmitted baseband signals, $a(t)$ and $s(t)$, respectively, may be readily related by the common linear time-invariant channel model in which the output signal $a(t)$ appears as the convolution of the input signal $s(t)$ with the channel impulse function $h(t)$

$$a(t) = \int_{-\infty}^{+\infty} h(\tau) s(t - \tau) d\tau \tag{2.31}$$

The impulse response $h(t)$ in this case represents the (random) channel response at baseband. For the scattering channel defined here, we have, comparing (2.31) and (2.30), $h(\tau)$ given by the discrete signal representation

$$h(\tau) = \sum_{k=1}^{L} a_k e^{j(\theta_k - \omega_c \tau_k)} \delta(t - \tau_k) \tag{2.32}$$

The function $\delta(t - \tau_k)$ appearing in (2.32) is the usual unit impulse or delta function defined to be zero everywhere except at $t = \tau_k$, where it equals ∞, with $\int \delta(t) dt = 1$.

Now let the amplitude, delay, and phase terms introduced by the L scatterers change with time. As noted above, these variations in time might be due to terminal motion to be discussed in the next section. They could also be introduced by movement of the scatterers themselves. An example of the latter might be one in which the scatterers are large trucks, buses, or trains. In this case of the amplitude, delay, and phase changing with time, we have $a_k(t)$, $\tau_k(t)$, and $\theta_k(t)$ appearing in (2.29) and (2.30). The linear representation of the wireless channel takes on the more general time-variant form of the convolution integral given by

$$a(t) = \int h(t, \tau) s(t - \tau) d\tau \tag{2.33}$$

The impulse response $h(t, \tau)$ is now time varying and, in particular, becomes a random process whose instantiation at any time is a random variable (Papoulis, 1991). For the

Papoulis, A. 1991. *Probability, Random Variables, and Stochastic Processes*, 3rd edn, New York, McGraw-Hill.

case under consideration here, we have, as the impulse response of the linear time-varying channel,

$$h(t, \tau) = \sum_{k=1}^{L} a_k(t) e^{j[\theta_k(t) - \omega_c \tau_k(t)]} \delta(t - \tau_k(t)) \tag{2.34}$$

The impulse response $h(t, \tau)$, written as a function of the two time variables t and τ, is defined to be the response at time t to an impulse applied τ units of time before, i.e., at time $t - \tau$. Studies carried out in the planning of the GSM cellular system to be discussed at length in later chapters used a variety of channel impulse response models to estimate system performance in a variety of environments, including rural areas, hilly environments, and urban areas (Steele, 1992). Six- and 12-component models were used for this purpose. The effect of receiver motion was incorporated in these models. As we shall see in the next section, this motion is manifested by Doppler shifts in the received frequency.

The random process $h(t, \tau)$ is a function of two variables, arising as it does as a representation of a time-varying linear channel. Such processes in the context of random wireless channels were extensively studied by Bello in the early 1960s (Bello, 1963). To simplify his studies, he assumed these channels were of the *wide-sense stationary* (WSS) type. Random processes of this type are characterized by the assumption that their first and second moments are independent of time. This assumption simplifies the study of these processes immensely. (Random processes *all* of whose moments are independent of time are called *strict-sense stationary* (Papoulis, 1991). The conditions for a WSS process are thus weaker.) To understand the application of the WSS assumption to the wireless channel, it is useful first to discuss the meaning of wide-sense stationary for a random process involving one time variable only. Consider, therefore, a random process $x(t)$ varying with time. At any selected time t the instantiation of this process is a random variable, with some known probability distribution function $F_x(x)$. A measure of the variation of the random process $x(t)$ with time is given by its autocorrelation function $R_x(t_1, t_2)$ defined as the ensemble average of the product of $x(t)$ with itself at two times t_1 and t_2. Thus we have

$$R_x(t_1, t_2) = E[x(t_1)x(t_2)] \tag{2.35}$$

The E or expectation operation just means taking the average of the product within the brackets, as indicated. This correlation function measures, in a rough way, the "connectedness" of the two instantiations of $x(t)$, one measured at time t_1, the other at time t_2. In particular, if the two quantities are uncorrelated, we get $R_x(t_1, t_2) = E[x(t_1)]E[x(t_2)]$, i.e., the product of the two average values. If these two values are both zero, the (uncorrelated) autocorrelation is zero. (Often a correlation function is defined with the average value subtracted from each random variable. The uncorrelated function then automatically takes on the value 0.) Now consider a WSS process. In this case the autocorrelation function is defined to be invariant with time, depending only on the *difference* in the times at which

---

Steele, R. ed. 1992. *Mobile Radio Communications*, London, Pentech Press; New York, IEEE Press.

Bello, P. A. 1963. "Characterization of randomly time-variant linear channels," *IEEE Transactions on Communication Systems*, 12, CS-11 (December), 360–393.

it is evaluated, not the actual times themselves. The average or mean value is also defined to be constant with time, taking on the value $E(x)$. The autocorrelation function for this WSS process is thus given by

$$R_x(\tau) = E[x(t+\tau)x(t)] = E[x(t)x(t-\tau)] = R_x(-\tau) \tag{2.36}$$

Here $\tau = t_1 - t_2$, the difference in the two times. As the time difference τ increases, the correlation approaches the product of the mean values, hence the mean value squared, and the two random variables $x(t+\tau)$ and $x(t)$ are said to be uncorrelated. For zero-mean value, the autocorrelation function thus becomes zero for a large enough spacing in time. It is to be noted that two uncorrelated random variables (rvs) are not necessarily independent. Independence requires much stronger conditions to be satisfied. Gaussian rvs, however, are independent if uncorrelated. Conversely, any independent rvs are uncorrelated. The WSS process is particularly significant, since the power spectral density, providing a measure of the variation of the power of the random process with frequency, turns out to be the Fourier transform of the autocorrelation function in this case (Papoulis, 1991).

The definition of the autocorrelation function may be extended to that of a time-varying random process such as that of the complex phasor form of the impulse response $h(t, \tau)$ of the time-varying wireless channel, as given by (2.34). In particular, assuming a wide-sense stationary (WSS) channel, one has

$$R_h(\tau, \tau_1, \tau_2) = E[h(t+\tau, \tau_1)h^*(t, \tau_2)] \tag{2.37}$$

The * operation indicates the complex conjugate of h. The autocorrelation function in the case of a time-varying channel is a function of three parameters in this case: it represents the average of the product of the two impulse responses, one measured at time $t+\tau$, the other at time t, τ units of time earlier, and, just as in the case of the WSS process (2.36), depends only on the time τ separating the two measurements, not on the specific time at which the evaluation takes place. The impulse responses at times $t+\tau$ and t are, in turn, functions of the time intervals τ_1 and τ_2 prior to these times, respectively, at which the impulses were applied. All three times, τ, τ_1, and τ_2 must be considered in evaluating the autocorrelation function in this case. The form of $h(t, \tau)$ represented by (2.34), arising from the multipath model of L scatterers, is a special case of a WSS process and was defined by Bello as an example of a *wide-sense stationary uncorrelated scattering* (WSSUS) process (Bello, 1963).

In the next section we specifically introduce the effect of terminal motion into the multipath received signal model of (2.16) and (2.17). This leads to a discussion of the rate of fading experienced by moving terminals. In Section 2.5 following we introduce a simple two-frequency model for the transmitted signal of (2.28) that enables us to show frequency-selective fading arising in a multipath environment with consequent dispersion of the signal in time.

2.4 Terminal mobility and rate of fading

Note that we have ignored terminal mobility to this point. We consider the motion of terminals in this section, and, most particularly, the effect of the motion on the signal

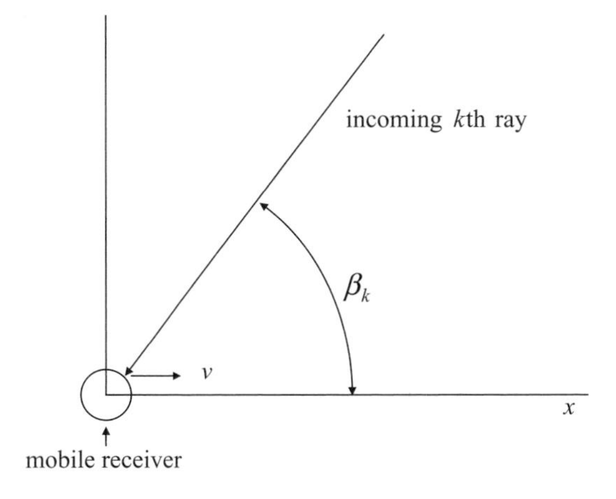

Figure 2.13 Geometry doppler effect

fading experienced. Specifically, say a vehicle or a person carrying a mobile terminal is moving at a velocity of v km/hr in the x direction. Assume the radio waves impinging on the terminal are vertically polarized, the kth ray hitting the mobile at an angle β_k from the horizontal, as shown in Fig. 2.13. Then the vehicle motion with respect to the incoming ray introduces a Doppler frequency shift $f_k = v \cos \beta_k / \lambda$ Hz. If $\beta_k = 0$, i.e., the ray is directed opposite to that of the vehicle's motion, $f_k = v/\lambda$, and the frequency of the received signal is increased to $f_c + v/\lambda$, f_c the unmodulated carrier frequency, as noted in the previous section. If $\beta_k = \pi$ rad., i.e., the wave is in the direction of motion of the vehicle, the frequency is decreased to $f_c - v/\lambda$. These values are as expected from Doppler velocity considerations. Note, however, that these shifts in frequency are *very* small. For example, even for a 100 km/hr (60 miles/hr) vehicle and a 1 GHz radio system, with $\lambda = 0.3$ m, $v/\lambda = 90$ Hz. Then $(v/\lambda)/f_c \approx 10^{-7}$! A mobile velocity of 5 km/hr corresponding to a person carrying a cell phone and walking at that speed gives rise to a v/λ ratio of 4.5 Hz. Although these Doppler shifts are quite small, they can play a significant role in cellular system performance, as noted below, in discussing fade rate due to terminal motion.

This Doppler shift due to the movement of the terminal with respect to the multiple received rays manifests itself as an added phase shift $2\pi f_k t = \omega_k t = (2\pi v \cos\beta_k)t/\lambda$ for the kth ray. Focusing on the Rayleigh multipath-scattering case of (2.17), the received signal $S_R\,(t)$, in the case of a moving terminal, is now given by

$$S_R(t) = \sum_{k=1}^{L} a_k \cos[\omega_c(t - t_0) + \phi_k + \omega_k t] \tag{2.38}$$

We can use the discussion of Rayleigh statistics and the Doppler shift to make some quantitative comments about the fade rate and the effect of multipath and the Doppler shifts on signal transmission. Consider fading first. We have already stated qualitatively that the rate of fading of signals and the depth of the fades must be related: first, to the small-scale multipath phenomena leading to the Rayleigh model, and, second, to the velocity of the mobile terminal. In particular, since fades, due to successive constructive and destructive interference of the multiple rays, vary over distances the order of a wavelength, the rate of change of a fade must be related to v/λ. For example, if a vehicle is moving at a speed of $v = 100$ km/hr $= 28$ m/sec, and the frequency of signal transmission f_c is 1 GHz or $\lambda = 0.3$ m, the example cited earlier, the fade rate should be about $v/\lambda = 90$ fades/sec. The fade duration is the order of λ/v, or about 11 msec. For the case of a person carrying a cell phone and walking at a speed of 5 km/hr, the fade is, of course, slower by a factor of 20, the fade rate being the order of 4.5 fades/sec and the fade duration being of the order of 0.2 sec. One can be more precise, however, using Rayleigh statistics. In particular, a detailed analysis indicates that, for the case of a vertically polarized electric field, as measured at the mobile, the average fade duration τ_f is given by the following expression (Jakes, 1974)

$$\tau_f = \frac{e^{\rho^2} - 1}{\rho f_m \sqrt{2\pi}} \tag{2.39}$$

Here $f_m \equiv v/\lambda$, just the maximum Doppler shift, and $\rho \equiv a/\sqrt{E(a^2)}$, a the Rayleigh-distributed amplitude of the received signal. So the average fade duration depends on the amplitude level at which it is measured. The parameter ρ is the amplitude a normalized to its rms value $\sqrt{E(a^2)}$. Equation (2.39) has been plotted in both Jakes (1974) and Yacoub (1993). As an example, say that the amplitude level is the rms level. Then $\rho = 1$ and, from (2.39), the average fade duration $\tau_f \cong 0.7/f_m = 0.7\,\lambda/v$. The average fade rate is $1/\tau = 1.4$ v/λ. These are close to the values we found intuitively. For the example noted above of a 100 km/hr vehicle and operation at 1 Ghz, the average fade duration at an amplitude with $\rho = 1$ is about $\tau_f = 8$ msec. If $\rho = 0.3$, i.e., if the received signal amplitude a is chosen to be 0.3 of its rms value, the average duration of a fade drops to $0.1/f_m$ or 1 msec for this same example. For the 5 km/hr speed of motion, these numbers increase, respectively, to 0.12 sec and 20 msec.

How does one derive (2.39)? The analysis is rather lengthy, so we shall just summarize the approach used, leaving details to the reference Yacoub (1993). Given a specific amplitude $a = R$ of the signal at the mobile receiver, the average rate R_c of crossing of that amplitude, also referred to as the level crossing rate, is found to be given by the following expression (Yacoub, 1993):

$$R_c \equiv E[\dot{a}/a = R] = \sqrt{2\pi}\, f_m \frac{R}{\sqrt{2}\sigma_R} e^{-\frac{R^2}{2\sigma_R^2}} \tag{2.40}$$

f_m is again the maximum Doppler shift v/λ and σ_R^2 is the Rayleigh parameter introduced earlier. (Recall that $E(a^2) = 2\sigma_R^2$.) Figure 2.14 portrays a typical fading pattern as a function of time, with a level R and the rate of crossing (slope) at that value indicated. Note, from (2.40), that the average rate of crossing R_c is itself in the form of a Rayleigh function. For $R \ll \sigma_R$, the level crossing rate is small; for $R > \sigma_R$, it is again small. It is readily shown that R_c reaches a maximum value of $1.08\, f_m$ at an amplitude of $R = \sigma_R$.

The expression for the average fade duration τ_f is now obtained from (2.40) as follows. Consider a long time interval T. Say the signal amplitude fades, or drops, below a level R,

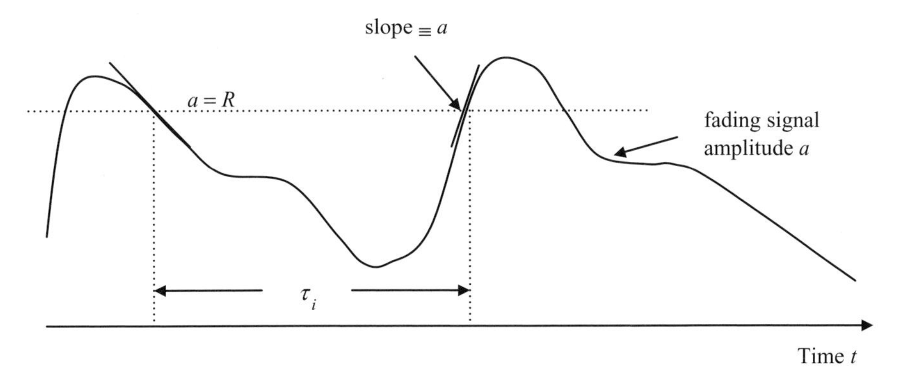

Figure 2.14 Fading signal and level crossing

n times in this interval, the ith fade lasting τ_i sec. An example of a fade duration lasting τ_i sec is shown in Fig. 2.14. Summing these fade times and normalizing to T, the quantity $\Sigma_i \tau_i / T$ represents the fraction of time the amplitude is below the value R. For large n this quantity provides a good approximation to the probability the level of a given fade is below R. This is just the probability the Rayleigh-distributed signal amplitude a is less than the value R. From (2.21), this is calculated to be

$$P(a \leq R) = \int_0^R f_a(a)da = 1 - e^{-\frac{R^2}{2\sigma_R^2}} \tag{2.41}$$

Now note that the quantity $\Sigma_i \tau_i / T$ may also be written as $(\Sigma_i \tau_i / n) n/T$. The term n/T is just the number of crossings of the level R and approaches the average number of level crossings R_c for large T, while $\Sigma_i \tau_i / n$ approaches the average fade duration τ_f. We thus find

$$\tau_f R_c = 1 - e^{-\frac{R^2}{2\sigma_R^2}} \tag{2.42}$$

using (2.41). The desired expression (2.39) for the average fade duration τ_f is now obtained by writing (2.40) for R_c in (2.42) and solving for τ_f.

2.5 Multipath and frequency-selective fading

The discussion in the previous section has focused on signal fading, the rapidity with which fades occur, and, the converse, the duration of a fade, related to the mobile's velocity as embedded in the Doppler shift parameter $f_m = v/\lambda$. The discussion there also focused on the simplest case of a single, unmodulated carrier being transmitted. In actuality, of course, real communication requires transmission and reception of information-carrying modulated carrier signals. Recall that in Section 2.3, on channel characterization, we did modify the initial discussion of multipath reception in Section 2.2 to include the transmission of a baseband signal $s(t)$ with corresponding baseband bandwidth $B/2$. The transmitted signal in this case was then given by (2.28) in complex phasor form. We

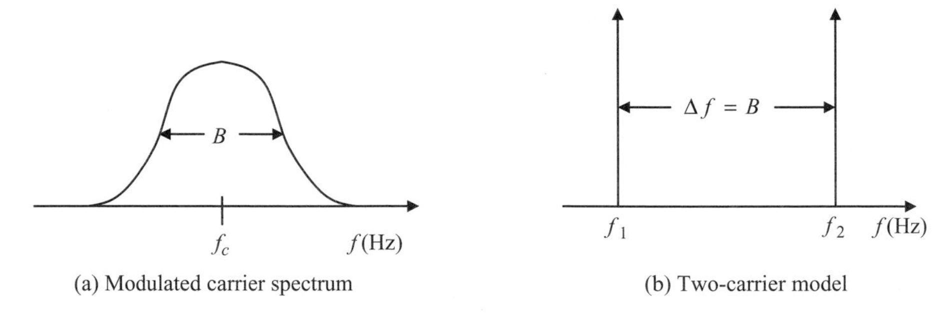

(a) Modulated carrier spectrum

(b) Two-carrier model

Figure 2.15 Signal bandwidths

also noted there that the transmission bandwidth was double the baseband bandwidth or B Hz. (Implicit in the discussion, represented by the form of (2.28), is the assumption we are dealing with amplitude modulation. Other types of signals, particularly those used in mobile wireless systems, are discussed in Chapter 5 on modulation. In the case of FM, the transmission bandwidth is a more complicated function of the baseband bandwidth (Schwartz, 1990).) We now return to this more realistic model of an information-bearing signal being transmitted. An example of the frequency spectrum of a typical transmitted signal, showing the variation of amplitude with frequency, appears in Fig. 2.15(a). Note that the transmission bandwidth about the carrier f_c is shown as having the value B.

The question we discuss in this section is the effect of multipath fading on the reception of information-carrying signals. It turns out that this effect depends on the signal bandwidth. For relatively large bandwidth one encounters frequency-selective fading, with different frequency components of the signal being handled differently over the propagation path, leading to signal distortion. For the case of digital signals, the case we deal with in this book, this distortion introduced by frequency-selective fading manifests itself in *inter-symbol interference* (ISI), with successive digital symbols overlapping into adjacent symbol intervals. For narrower signal bandwidths, one gets non-selective fading, or *flat fading*, as this type of fading is often called. In the opposite extreme case, that of very wide bandwidths, multipath replicas of the same signal may arrive at distinctly different times, allowing the separate signals arriving to be individually resolved, thus enhancing the detectability of the original signal. This phenomenon is, in fact, used to enhance the performance of CDMA signals, to be discussed later in this book. We discuss this phenomenon briefly at the end of this chapter.

To approach this question of the impact of the propagation path on signal reception, we use the simplest possible model for the modulated carrier (2.28), the simultaneous transmission (and reception) of two sinewaves, each with the same transmitted power but with slightly different frequencies, the difference between the two frequencies corresponding to the bandwidth of a real information-bearing signal such as that shown in Fig. 2.15(a). Using this simple model to determine the way the multipath-fading channel affects these two signals provides an approximation as to whether real signals arrive distorted by the channel or not, as to whether multipath components are resolvable at the receiver, as well as the effect of terminal motion on the received signal. This last effect, in particular, extends the fading rate discussion of the previous section. Specifically, consider two signals 1 and 2, signal 1 having carrier frequency f_1, signal 2 being transmitted at a somewhat higher frequency f_2. The difference $\Delta f = f_2 - f_1$ will be considered equivalent to a bandwidth B of a real signal. This case is portrayed in the spectral sketch of Fig. 2.15(b). (When we say the frequencies are "slightly" different, we imply the wavelengths of each are very nearly the same. One can readily show that $\Delta\lambda = \lambda_2 - \lambda_1 \ll \lambda_1$ (or λ_2) then implies as well that $\Delta f \ll f_1$ (or f_2).)

Assume now that each of the two transmitted signals is separately scattered over the propagation path, arriving at the mobile receiver as the sum of a large number of multipath rays. For generality in the model, we allow each of the two received scattered signals to have a different number of multipath rays, as well as allowing each of the two received signal groups to have different propagation delays. Extending the approach used in the previous section, in which we had (2.38) incorporating the Doppler shift due to terminal motion, we may write the received signal in the following form

$$S_1(t) = \sum_{k=1}^{L} a_k \cos[\omega_1(t - t_1 - \tau_k) + \omega_k t + \theta_k] \tag{2.43a}$$

and

$$S_2(t) = \sum_{l=1}^{M} a_l \cos[\omega_2(t - t_2 - \tau_l) + \omega_l t + \theta_l] \tag{2.43b}$$

As previously (see (2.17) and (2.24)), $t_j, j = 1$ and 2, represents the large-scale delay of the respective multiple rays; τ_k and τ_l, respectively, represent the additional or incremental random delays of each of the rays in a group; $\omega_k = 2\pi \cos\beta_k v/\lambda$ and $\omega_l = 2\pi \cos\beta_l v/\lambda$; θ_k and θ_l are additional random angles, assumed uniformly distributed from 0 to 2π. The numbers M and L represent the number of rays received for each signal transmitted. β_k and β_l represent the angles of arrival of each ray with respect to the horizontal ground, as in Fig. 2.13. Setting the large-scale delays t_1 and t_2 equal to one another, we get the model of a modulated carrier of bandwidth Δf, with the incremental random delays incurred by each frequency term being used, as we shall see, to determine the frequency selectivity of the wireless channel. Retaining the large-scale delays enables us to determine the relation between the mobile velocity and these delays.

To continue we must define a model for the incremental delays in (2.43a, 2.43b). A commonly used model is to assume these delays are exponentially distributed with some average value τ_{av}. The probability of larger incremental delays thus reduces exponentially. The probability density function $f_\tau(\tau)$ of the incremental delay is thus given by

$$f_\tau(\tau) = \frac{1}{\tau_{av}} e^{-\frac{\tau}{\tau_{av}}}, \quad \tau \geq 0 \tag{2.44}$$

This function is sketched in Fig. 2.16(a). Its second moment $E(\tau^2)$ is readily shown to be $2\tau_{av}^2$, so that the variance $\sigma_\tau^2 \equiv E(\tau - \tau_{av})^2 = \tau_{av}^2$. The standard deviation $\sigma_\tau = \tau_{av}$ of this incremental delay distribution is defined to be the *delay spread*. The total delay of each of the received signals is a random variable $t_i + \tau$ and appears as shown in Fig. 2.16(b).

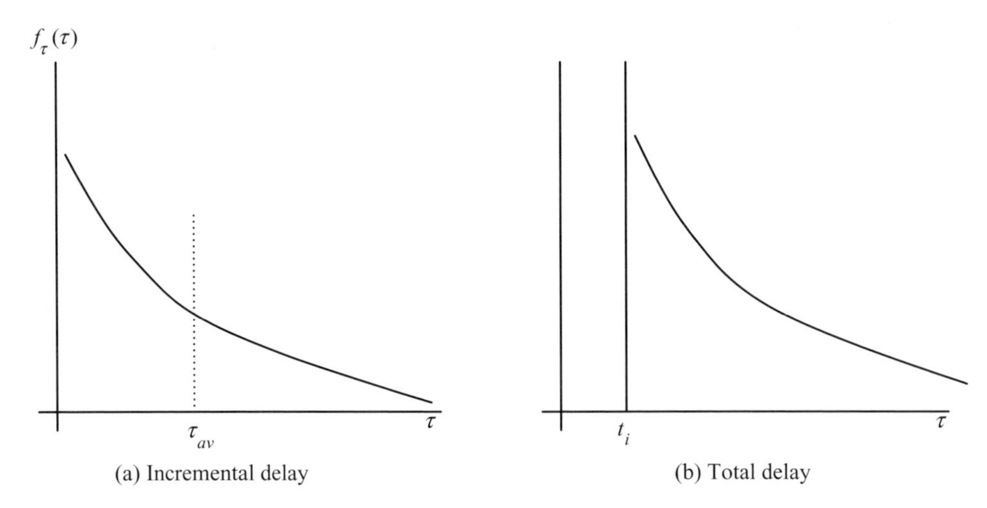

Figure 2.16 Exponential delay distribution

Typical values of the delay spread for macrocellular systems range from 0.2 μsec in rural areas to 0.5 μsec in suburban areas, to from 3 to 8 μsec in urban areas, where more obstacles and consequently greater multipath are expected to be encountered (Yacoub, 1993; Steele, 1992; Rappaport, 2002). Urban microcellular systems have delay spreads of less than 2 μsec, while (indoor) picocellular systems have delay spreads of 50–300 nsec.

Equations (2.43a), (2.43b) and (2.44) may be used to determine the combined effect of receiver mobility and fading on information-bearing modulated signal transmission. To do this we rewrite (2.43a) and (2.43b) in the envelope/phase form of (2.17b). However, we choose to write them in the form $S_i(t) = a_i \cos(\omega_i t + \theta_i), i = 1, 2$, with the propagation delays t_1 and t_2 as well as the carrier frequencies ω_1 and ω_2 multiplying them incorporated in their respective amplitude and phase terms. The random variables a_i and θ_i are, in addition, functions of the incremental delay distribution and the Doppler frequency. A measure of the way in which the two received signals, $S_1(t)$ and $S_2(t)$, compare with one another, representing the effect of the fading channel on a single modulated signal, is given by their envelope correlation function $E(a_1 a_2)$, obtained by averaging over all the random variables appearing in (2.43a) and (2.43b), the a_ks and a_ls, the β_ks and β_ls, the incremental delay distribution, and the random phases θ_k and θ_l. More specifically, the normalized envelope correlation function is more directly useful for our purpose. This function is defined as $[E(a_1a_2) - E(a_1)E(a_2)]/\sigma_1\sigma_2$, where

$$\sigma_i^2 \equiv E[a_i - E(a_i)]^2, \quad i = 1, 2$$

is the variance of envelope a_i and $E(a_i)$ is its average value. We shall henceforth label this normalized correlation function ρ_a. We thus have

$$\rho_a \equiv [E(a_1a_2) - E(a_1)E(a_2)]/\sigma_1\sigma_2 \tag{2.45}$$

Yacoub, M. D. 1993. *Foundations of Mobile Radio Engineering*, Boca Raton, FL, CRC Press.
Rappaport, T. S. 2002. *Wireless Communications, Principles and Practice*, 2nd edn, Upper Saddle River, NJ, Prentice-Hall.

This correlation between two random variables is similar to the autocorrelation function discussed earlier, which, as we indicated, provides a measure of the "connectedness" between two instantiations of a random process spaced an interval τ apart. (We could have normalized the autocorrelation function as well.) As in the autocorrelation case, the two random variables here are said to be uncorrelated if $E(a_1 a_2) = E(a_1)E(a_2)$. We then have $\rho_a = 0$. The two rvs are said to be completely correlated when $\rho_a = 1$, the maximum value of ρ_a. In the specific case under discussion here of the two received signals, ρ_a close to 1 implies the fading channel is non-distorting, i.e., all spectral components of a modulated signal fade in the same manner. When ρ_a is close to zero, two cases arise, as we shall see: the channel can be distorting the original information-bearing signal or, if the bandwidth (the frequency separation in our model) is large enough, distinct rays or "echoes" may appear, allowing them to be separately resolved, actually providing a possible improvement in performance if the individual echoes can be appropriately recombined! Such a technique, involving a *RAKE receiver*, is discussed briefly at the end of this chapter. We shall see in later chapters that CDMA wireless systems incorporate such a procedure combining the separate received rays to improve system performance.

The calculation of ρ_a may be carried out in this case of two signals separated in frequency, assuming Rayleigh fading and exponentially distributed incremental delay. One then finds, after tedious calculations (Yacoub, 1993: 132–138), that ρ_a is given, approximately, by the following relatively simple expression

$$\rho_a(\Delta\tau, \Delta f) \approx \frac{J_0^2(\omega_m \Delta\tau)}{1 + (\Delta\omega\tau_{av})^2} \tag{2.46}$$

$J_0(x)$ is the Bessel function of the first kind and of zeroth order; $\Delta\tau = t_2 - t_1$, the delay in arrival of the two signals (we have arbitrarily chosen signal 2 as arriving later); $\Delta\omega = \omega_2 - \omega_1 = 2\pi(f_2 - f_1) = 2\pi\Delta f = 2\pi B$, B being, as we recall, the transmission bandwidth, or, in the model adopted here, the difference in the two carrier frequencies; τ_{av} is the delay spread defined earlier; and $\omega_m = 2\pi f_m = 2\pi v/\lambda$, the maximum Doppler shift, in radians/sec, as noted earlier. Figure 2.17 provides a sketch of the normalized correlation function ρ_a as a function of $\Delta\omega\tau_{av}$ for various values of $\omega_m\Delta\tau$. Note that, as either $\Delta\omega\tau_{av}$ or $\omega_m\Delta\tau$ increase, the correlation between the two signals goes to zero, implying a distorting channel. We now interpret these results.

Consider first the case in which $\Delta\tau = t_2 - t_1 = 0$. The two signals thus arrive at the same time in a large-scale sense. Recall that these two signals in this case represent two frequency components of the same signal. It is the case in which we are modeling the effect of multipath propagation on an information-bearing signal of bandwidth $B = \Delta f$, enabling us to determine the condition for frequency-selective fading, the focus of this section. Since $J_0(0) = 1$, we have

$$\rho_a \equiv \rho_a(0, \Delta f) = 1/[1 + (\Delta\omega\tau_{av})^2] \tag{2.47}$$

Let ρ_a be less than or equal to 0.5. The two frequency terms are then effectively uncorrelated, leading to *frequency-selective fading*. For this value of ρ_a, $\Delta\omega\tau_{av} \geq 1$. But recall that $\Delta\omega = 2\pi(f_2 - f_1) = 2\pi\Delta f \equiv 2\pi B$, since we started this discussion by noting that the frequency difference of the two signals in this model was equivalent to an information-bearing

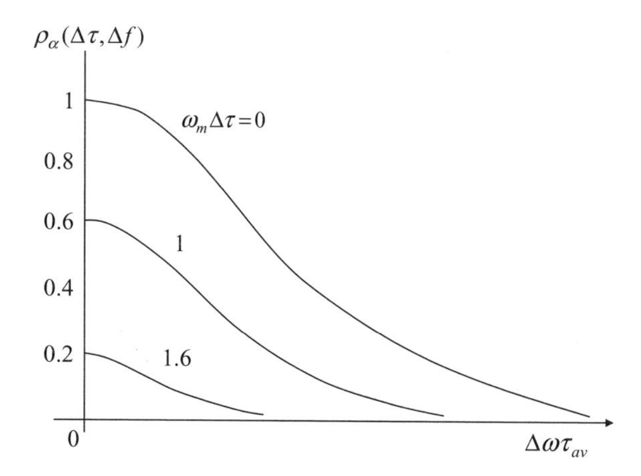

Figure 2.17 Normalized correlation function

signal of bandwidth B Hz. The specific value $\Delta f = B = 1/2\pi\tau_{av}$, for which we have $\rho_a(0, \Delta f) = 0.5$, is defined to be the *coherence bandwidth*. This quantity is a measure of the frequency separation (equivalent bandwidth B of a true information-bearing signal) beyond which frequency-selective fading occurs. Note, however, that one can also write $\tau_{av} > 1/2\pi B$ as the case for which frequency-selective fading occurs, giving rise to inter-symbol interference as noted earlier. The most common digital "symbol" to which this discussion applies is the binary digit or bit. We shall see in Chapter 5 that multi-level digital symbols may be transmitted as well, each symbol corresponding to a number of bits. One then talks of the symbol interval T, in units of time, and inter-symbol interference occurs when successive symbols overlap in time. We shall show in Chapter 5 that the symbol interval T and the digital signal transmission bandwidth B are inverse to one another; more specifically, we have $BT \approx 1$. (This relation comes from Fourier analyis and should be familiar to anyone with some prior background in digital communications or communication theory.) We thus have as the two equivalent conditions for frequency-selective fading, giving rise to inter-symbol interference

$$T < 2\pi\tau_{av} \tag{2.48a}$$

or

$$\tau_{av} > 1/2\pi B \tag{2.48b}$$

Digital symbol intervals T smaller than 5 or 6 times the delay spread τ_{av} thus give rise to frequency-selective fading and consequent inter-symbol interference. This result agrees with intuition.

Consider the following examples:

Example 1

The digital data rate is 10 kbps, with transmission in binary form. The binary interval T is then 0.1 msec and inter-symbol interference would be expected to occur in an environment with a delay spread of 0.02 msec (20 μsec) or more. As noted above, such a large delay spread is unlikely in most fading environments. Most fading media then result in flat or nonfrequency-selective fading for this rate of data transmission.

Example 2

Now let the data rate increase to 200 kbps. The symbol interval decreases correspondingly to 0.005 msec or 5 μsec, with a delay spread of 1 μsec or more, resulting in frequency-selective fading. As noted above, such delay spreads would be encountered in urban environments, as well as some suburban environments. We shall see in later chapters that the GSM cellular system transmits information at a rate of about 271 kbps over a 200 kHz transmission bandwidth.

Example 3

Consider now a data rate of 54 Mbps. As we shall see in Chapter 12, some high bit rate wireless local-area networks transmit at this rate. Frequency-selective fading, with inter-symbol interference resulting, would thus be expected to be a problem since such fading corresponds to delay spreads of 3 nsec or more. This potential problem has been addressed in two ways: the symbol interval is purposely increased by a factor of 6, using the QAM modulation technique discussed in Chapter 5; orthogonal frequency-division multiplexing (OFDM), also discussed in Chapter 5, is used to provide much narrower frequency transmission bands, resulting in flat fading over each band.

This discussion of frequency-selective fading has focused on the case in which inter-symbol interference occurs. Now consider an extreme case for which the signal bandwidth B is large enough so that the delay spread is much greater than $1/2\pi B$. Distinct signal echoes may now appear, each corresponding to different multipath rays, making them individually resolvable. As noted earlier, this phenomenon can be exploited to improve signal detectability in high-bandwidth systems. As an example, consider a system with a transmission bandwidth B of 1.25 MHz. This is approximately the bandwidth of IS-95, the second-generation CDMA cellular wireless standard to be discussed in detail later in this book, as well as the bandwidth of cdma2000, one of the third-generation systems discussed in Chapter 10. Then multipath echoes appearing much greater than $1/2\pi(1.25)$ μsec ≈ 0.13 μsec apart will be resolvable. The individual echoes can be separately detected and appropriately processed to improve the system performance. We discuss this type of processing, called a RAKE receiver, at the end of the next section, as already noted. As a second example, the bandwidth of the digital AMPS system IS-136 discussed in Chapter 8 is 30 kHz. This means that multipath rays or echoes from the same transmitted signal would have to arrive more than $1/2\pi(30)$ msec ≈ 5 μsec apart to be resolvable, a very unlikely situation. (This corresponds to more than a kilometer difference in propagation path. Recall from our discussion of delay spread above, that this quantity can range from fractions of a μsec in rural areas to 3 μsec in urban areas.) Individual multipath echoes thus cannot be resolved in narrowband systems such as IS-136. Channel equalization can be used effectively in such systems to help mitigate the impact of frequency-selective fading. We discuss channel equalization briefly in the next section as well.

We return now to (2.46) and consider the parameters in the numerator involving the Bessel function. We focus on the case $\rho_a(\Delta\tau, 0)$ which provides information on dependence of the normalized correlation function on the Doppler shift and the large-scale difference in time delay. The Bessel function $J_0(\beta)$ is plotted in Fig. 2.18. Note that it is an oscillating function of the parameter β, decreasing in amplitude after each oscillation,

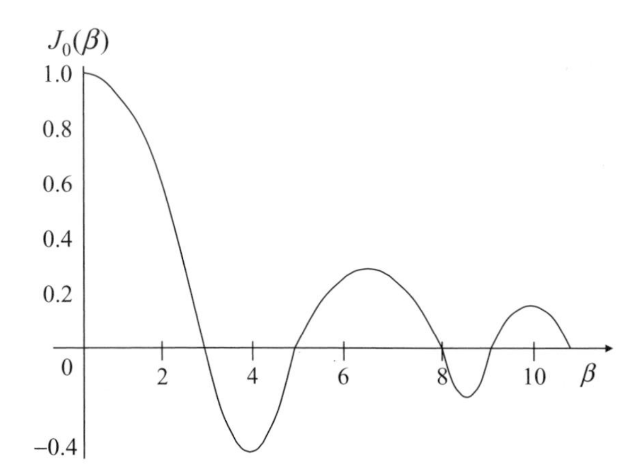

Figure 2.18 Bessel function J_0 (β)

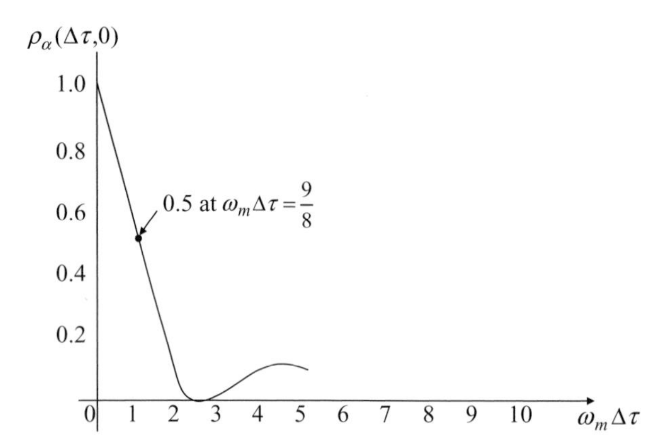

Figure 2.19 Normalized correlation function ρ_α $(\Delta\tau, 0)$

approaching zero for large β. From (2.46) we have $\rho_a(\Delta\tau, 0) = J_0^2\,(\omega_m\Delta\tau)$. This function is sketched in Fig. 2.19. Note that the correlation decreases rapidly, approaching zero for the product $\omega_m\Delta\tau$ large enough. The interpretation here, analogous to the case above involving $\Delta\omega\tau_{av}$ or the equivalent $B\tau_{av}$, is that the different components of the transmitted signal experience different fading conditions due to mobile velocity if they arrive sufficiently "far enough apart" in time. Specifically, the *coherence time* of the fading channel, by analogy with coherence bandwidth, is defined to be the time difference $\Delta\tau$ for which ρ_a $(\Delta\tau,0) = 0.5$. This is just the value $\Delta\tau = 9/8\omega_m = 9/16\pi f_m = 0.18/f_m$. Consider the example used earlier: a 100 km/hr vehicle communicating at a radio frequency of 1 GHz. Then $f_m = v/\lambda = 90$ Hz, and the coherence time is $\Delta\tau = 0.18/f_m = 2$ msec. Say a digital signal is being transmitted over the wireless system. Individual symbols transmitted more than $\Delta\tau = 2$ msec apart will experience different fading conditions, resulting in signal distortion. Alternatively put, if the signal symbol interval

$$T < \Delta\tau = 1/5.6 f_m$$

there will be no distortion. In this example, if $T < 2$ msec, or more than 500 symbols/sec are being transmitted, there is no distortion. In the special case of binary transmission, with T then the binary interval, there is no distortion if the rate of transmission is greater than 500 bps.

We summarize this discussion of possible signal distortion by recapitulating the two kinds of distortion considered in the material above. One has to do with *time-dispersion* (inter-symbol interference) and *frequency-selective fading* encountered when the signal bandwidth exceeds the coherence bandwidth $1/2\pi\tau_{av}$, with τ_{av} the delay spread. We noted as well that, if the bandwidth is large enough, multipath echoes may be resolved, leading to the possibility of enhancing signal detectability despite the distortion. The other has to do with the effect of receiver mobility. If the signal changes rapidly enough in comparison with the coherence time $0.18/f_m$, f_m the Doppler frequency v/λ, then there is no distortion. The Doppler effect is said to lead to *frequency-dispersion* and *time-selective fading*. Note the duality of these two effects occurring in a fading environment. An interesting, detailed, discussion of both of these effects appears in Steele (1992: Chap. 2). The first case of time-dispersion and frequency-selective fading is due to multipath propagation with delay spread. If the signal bandwidth is large enough compared with the reciprocal of the delay spread of its rays taking different paths, frequency-selective fading with resultant distortion and inter-symbol interference is encountered. The second case involving frequency-dispersion and time-selective fading is encountered when the channel changes its characteristics during the transmission of a signal. The change in channel characteristics is directly proportional to the receiver mobility. If a signal itself changes rapidly enough with respect to the reciprocal of the Doppler spread $f_m = v/\lambda$, distortion will not result. (Note specifically that as a signal changes more rapidly, its bandwidth will increase (Schwartz, 1990). Hence a minimum bandwidth is required to eliminate frequency-dispersion and time-selective fading, but, if the signal bandwidth increases too much, frequency-selective fading will be encountered due to delay spread.)

Conditions for these dual effects may be summarized in the following equations:

Frequency-selective fading/time dispersion

$$B > \text{Coherence bandwidth} = 1/2\pi\tau_{av} \tag{2.49}$$

τ_{av} the delay spread, B the signal bandwidth.

Time-selective fading/frequency dispersion

$$T > \text{Coherence time} = 9/16\pi f_m = 0.18/f_m \tag{2.50}$$

f_m the maximum Doppler frequency, v/λ, T the signal symbol interval.

2.6 Fading mitigation techniques

The previous sections describing the transmission of signals through a wireless channel have shown that channel to be a harsh one. Signal fading is ever-present and has to be reckoned with; inter-symbol interference may be encountered with digital signals whose symbol interval is comparable with, or less than, the delay spread. Ever-present noise will further corrupt a transmitted signal. Signal interference, discussed in detail in the next chapter and others following, may result in an error in detecting a given transmitted signal. We note that digital signaling requires the transmission in any one symbol interval of one of a number of possible signals. The simplest case is that of binary transmission, with one of two possible binary digits transmitted every binary symbol interval. In Chapter 5, we discuss the transmission of QAM signals, with one of $M \geq 2$ possible signals transmitted every symbol interval. The object at a receiver is then to determine which one of these signals was transmitted each symbol interval. A signal in a deep fade is much more prone to be received in error due to these effects.

Various techniques have been introduced to overcome the combined effects of fading, noise, and signal interference; these effects, if not dealt with, possibly result in signal decision errors. Interleaving of digital signals to spread them out in time is commonly used to reduce the effect of fast fading and possible bursts of noise. This technique will be described by example in the discussion of second- and third-generation cellular systems in Chapters 8 and 10, respectively, and, quite briefly, in the discussion of high-speed wireless local-area networks (WLANs) in Chapter 12. A technique called orthogonal frequency-division multiplexing (OFDM) is used to transmit sequences of digital signals in parallel, reducing their bandwidth requirements and, hence, converting the channel to one producing flat fading, with consequent reduction of inter-symbol interference. This technique is discussed in Chapter 5, with its application to high-speed WLANs described in Chapter 12. Coding techniques are commonly used to detect or correct digital-signal errors arising during signal transmission; coding procedures are discussed in Chapter 7.

In this section we discuss three other techniques used to mitigate against the effect of fading on signals. These are, first, the use of equalizers, designed to overcome, or at least reduce, inter-symbol interference; second, diversity techniques, with multiple copies of a given signal transmitted and/or received, the appropriate combination of which at the receiver provides an improvement in system performance; third, the RAKE receiver, which provides a special case of diversity reception appropriate for use with very wideband wireless systems, such as those using CDMA, code-division multiple access, technology, deployed in both the second- and third-generation systems discussed in later chapters. It is to be noted that all of these techniques have a long history of usage, predating their application to mobile wireless systems. The possibility of using diversity techniques to counter short-term fading encountered with high-frequency (hf) radio transmission involving reflections from the ionosphere was discovered in 1927, with a radio receiving system incorporating diversity developed soon thereafter by RCA in 1931 (see Schwartz *et al.*, 1966 for specific references). Comparative analysis of some of the diversity schemes to be described later in this section was carried out in the 1950s (Brenner, 1959). The RAKE receiver was introduced as a means of combatting multipath-fading effects in the 1950s as well (Price and Green, 1958). Equalizers were first developed in the mid 1960s, and incorporated in telephone modems at the time, to overcome inter-symbol interference arising in digital telephone transmission (Lucky, 1965).

Brenner, D. G. 1959. "Linear diversity combining techniques," *Proc. IRE*, 47, 6 (June), 1075–1102.
Price, R. and P. E. Green, Jr. 1958. "A communication technique for multipath channels," Proc. IRE, 46, 555–570.
Lucky, R. W. 1965. "Automatic equalization for digital communication," *Bell System Technical Journal*, 44 (April), 558–570.

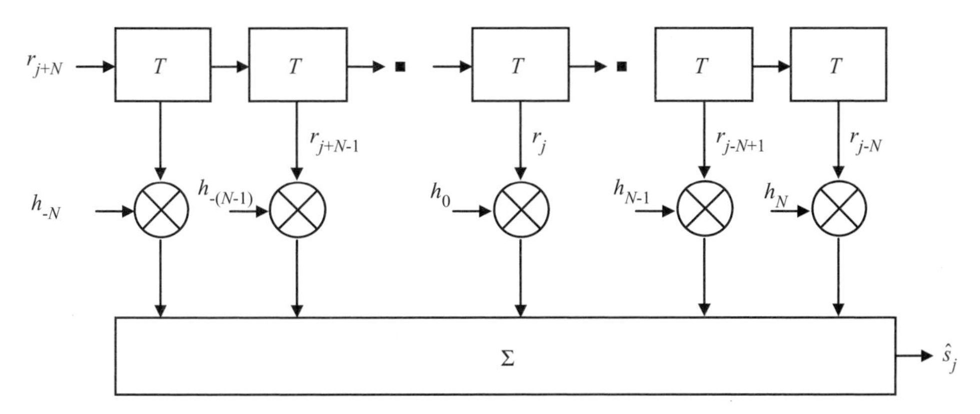

Figure 2.20 Transversal filter equalizer

Equalization techniques

As noted above, the purpose of an equalizer is to eliminate inter-symbol interference if possible, or at least to reduce it. To eliminate the inter-symbol interference over a channel such as the wireless one, it is necessary to estimate and model the channel. Equalization is carried out at baseband, the received distorted signal first being demodulated down to that at the information-bearing range of frequencies. All discussion following therefore considers baseband operation only. This includes the estimation and modeling of the channel, now considered to be the baseband equivalent of the actual channel through which the modulated carrier signal is transmitted. Estimation of the channel is usually carried out by transmitting a prescribed digital test sequence and, knowing the sequence, adjusting or adapting the equalizer to minimize some desired performance objective. Figure 2.20 shows the form of a commonly used equalizer. This device consists of a linear non-recursive or transversal filter, $2N + 1$ taps long, as shown in Fig. 2.20. We focus on this device in this paragraph. The assumption made is that a digital signal transmitted over one symbol interval is spread by the channel over a length of $2N + 1$ symbol intervals. An example might be the digital signal shown in Fig. 2.21(a). There a signal transmitted during one symbol interval is shown at the receiver as rising to a peak value some time later, allowing for propagation time over the channel, and then decaying in time afterwards. A second example in Fig. 2.21(b) shows the signal received decaying in time. The received signal in either case is sampled at the center of each symbol interval to form the input to the equalizer. These sample values are so indicated in Fig. 2.21. In practice, a sequence of input samples appears at the equalizer input. These input samples, taken every T-sec symbol interval, are denoted by r_j at the jth such interval. Note, from Fig. 2.20, that these sample values are each passed through $2N + 1$ T-sec delay elements, the output of each delay is multiplied by a tap coefficient, and the sum of all the resultant tap outputs used to form the estimate $\hat{s}_j$ of the transmitted baseband signal s_j. The equalizer output, corresponding to a sample received at symbol interval j may then be written

$$\hat{s}_j = \sum_{n=-N}^{N} h_n r_{j-n} \tag{2.51}$$

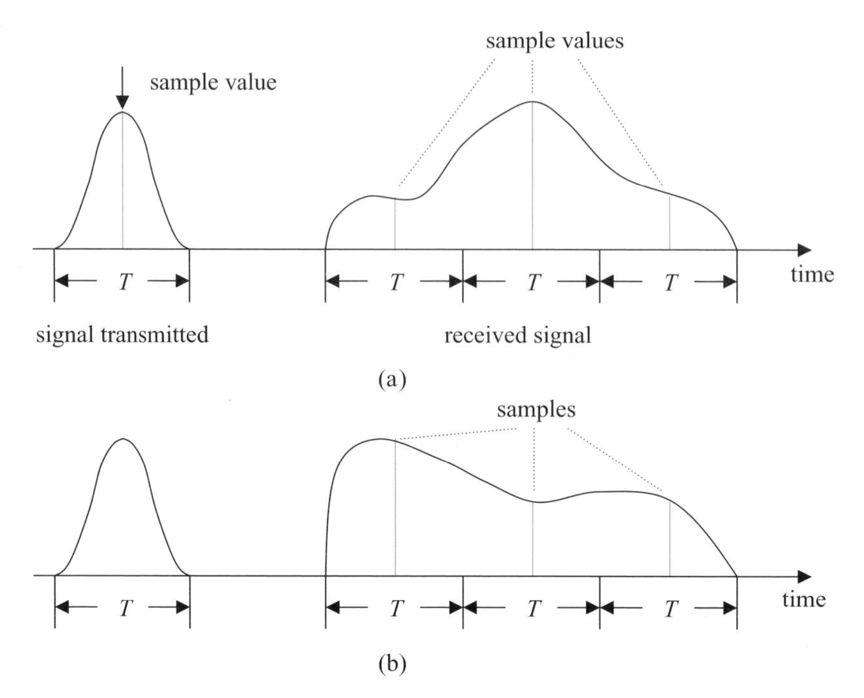

Figure 2.21 Examples of channel response

Note that this output is precisely in the form of a digital convolution similar to the convolution discussed earlier in (2.30)–(2.32), as would be expected since the device of Fig. 2.20 is a digital filter.

The problem now is to identify the $2N + 1$ tap coefficients, h_n, $-N \leq n \leq N$. One common technique is to first transmit a known training sequence K bits long. Pseudo-random sequences discussed in detail in Chapter 6 are often used for this purpose. With both the transmitted and estimated sequence of bits known at the receiver, the tap coefficients may be adapted or adjusted to minimize the difference between the transmitted and estimated signal values, using some performance objective. As the channel characteristics change with time, the equalizer tap gains must be adapted accordingly, with training sequences transmitted to carry out this function. The assumption made here is that the training can be done rapidly in comparison with changes in the channel characteristics. As noted a number of times, these changes are most often due to wireless terminal motion. One simple performance objective used to determine the equalizer tap gains in accordance with the K-bit training sequence is to minimize the mean-squared (ms) difference between the transmitted signal sequence s_j and the estimate $\hat{s}_j$, $1 \leq j \leq K$.

Min. Mean-Squared Objective

$$\text{Find } h_n,\ -N \leq n \leq N,\ \text{such that } \sum_{j=1}^{K} (s_j - \hat{s}_j)^2 \text{ is minimum} \tag{2.52}$$

The desired minimization indicated in (2.52) is readily carried out by taking the derivative of (2.52) with respect to the $2N + 1$ filter coefficients h_n, after replacing $\hat{s}_j$ by its representation in terms of the received signals samples r_j, using (2.51), and setting each term to zero. The following $2N + 1$ equations result

$$\sum_{j=1}^{K} s_j r_{j-l} = \sum_{j=1}^{K} \sum_{n=-N}^{N} h_n r_{j-n} r_{j-l} \qquad -N \leq l \leq N \tag{2.53}$$

These $2N + 1$ equations must now be solved simultaneously to find the desired tap coefficients h_n, $-N \leq n \leq N$. This minimization looks formidable, but may be put into much less forbidding form by defining the following two coefficients

$$R_{l,n} \equiv \sum_{j=1}^{K} r_{j-n} r_{j-l} \qquad -N \leq l, n \leq N \tag{2.54}$$

$$g_l \equiv \sum_{j=1}^{K} s_j r_{j-l} \qquad -N \leq l \leq N \tag{2.55}$$

Equation (2.53) then takes on the simpler looking form

$$g_l = \sum_{n=-N}^{N} h_n R_{l,n} \qquad -N \leq l \leq N \tag{2.53a}$$

The coefficient $R_{l,n}$ defined by (2.54) is referred to as a *sample correlation coefficient* and does, in fact, have a close connection to the autocorrelation function discussed earlier and defined by (2.35). The sample coefficient is, in fact, one way of determining the autocorrelation function in practice, replacing the ensemble average of (2.35) by the sample average of (2.54) (Schwartz and Shaw, 1975). One has to assume, of course, that the system for which the autocorrelation is being estimated does not change over the measurement period, and that the K sample values are large enough in number to provide a good estimate of the autocorrelation (Schwartz and Shaw, 1975).

The $2N + 1$ equations of (2.53a) may be readily put into vector form, further simplifying the form of the analysis to find the equalizer coefficients. Defining $2N + 1$-element vectors $\boldsymbol{g}$ and $\boldsymbol{h}$ elements of which are g_l and h_n, respectively, and then a $2N + 1$ by $2N + 1$ matrix $\mathbf{R}$, with elements $R_{l,n}$, one readily shows that the vector form of (2.53a) is given by

$$\boldsymbol{g} = \mathbf{R}\boldsymbol{h} \tag{2.56}$$

The vector solution for the $2N + 1$ values of the filter coefficients is then given by

$$\boldsymbol{h} = \mathbf{R}^{-1}\boldsymbol{g} \tag{2.57}$$

The simultaneous solution of the $2N + 1$ equations of (2.53), (2.56), and (2.57) can be simplified considerably if the number of samples K is large and a pseudo-random sequence of approximately equal numbers of positive and negative binary digits is chosen as the training sequence. We let their sample values s_j be $+1$ or -1. Further, we shall see in

Schwartz, M. and L. Shaw. 1975. *Signal Processing: Discrete Spectral Analysis, Detection, and Estimation*, New York, McGraw-Hill.

discussing these sequences in Chapter 6 that they have the property that successive bits are nearly independent of one another. We further assume that errors in transmission occur independently and randomly from bit to bit. The successive received bits r_j are then nearly independent of one another as well. Under these conditions, it can be seen that the correlation coefficient $R_{l,n}$ defined by (2.54) is approximately 0 for $l \neq n$, and equals the value K for $l = n$. Alternatively put, the matrix $\mathbf{R}$ is very nearly diagonal. A good approximation to the filter coefficients is then

$$h_n \approx g_n/K = \sum_{j=1}^{K} s_j r_{j-n}/K \qquad -N \leq n \leq N \tag{2.58}$$

The minimum mean-squared criterion used above to find the coefficients of the transversal filter of Fig. 2.20 can also be shown to be equivalent to a maximum-likelihood estimate of those coefficients (Schwartz and Shaw, 1975). Maximum-likelihood estimation is used as well in carrying out the Viterbi algorithm described in Chapter 7 in the context of decoding convolutionally encoded signals. An equalizer based on the Viterbi algorithm and designed for the GSM cellular system is described in Steele (1992).

Criteria other than the minimum mean-squared objective or the equivalent maximum-likelihood estimation procedure may be used to design an equalizer to eliminate or reduce inter-symbol interference. The zero-forcing equalizer, discussed at length in Mark and Zhuang (2003), is one such device. As the name indicates, the objective here is to have the equalizer output $\hat{s}_j$ be equal to the transmitted signal s_j at the jth sample value, while forcing all the sampled inter-symbol values created by the channel response to s_j equal to zero. In the linear z-transform domain, this procedure corresponds to setting the overall transfer function $H(z)$, the product of the channel transfer function and the equalizer transfer function, equal to 1. As noted in Mark and Zhuang (2003), this procedure corresponds to having the impulse response of the equalized system equal to 1 at symbol sample time 0 and 0 at all other symbol interval sample times. This result is only possible for $2N + 1$ symbol intervals, the span of the equalizer, and leads, as in the case of the minimum mean-squared equalizer, to the required solution of $2N + 1$ equations. Other types of equalizers are described in Mark and Zhuang (2003) as well.

Diversity reception

The second method we discuss in this section that is used to mitigate the effect of fading on the transmission of digital signals is that of diversity reception: multiple, independent samples of a digital signal are transmitted and/or received at each symbol interval, and are used to reduce the probability of an error in detecting the signal transmitted that interval (Schwartz *et al.*, 1966). This technique, long used in radio transmission, as noted earlier, is an example of using multiple statistical samples to improve detection or estimation of a corrupted signal. Multiple signal samples may be obtained by use of multiple antennas, a technique often referred to as *space diversity*, by transmission over multiple frequency

---

Mark, J. W. and W. Zhuang. 2003. *Wireless Communications and Networking*, Upper Saddle River, NJ, Pearson Education Inc.

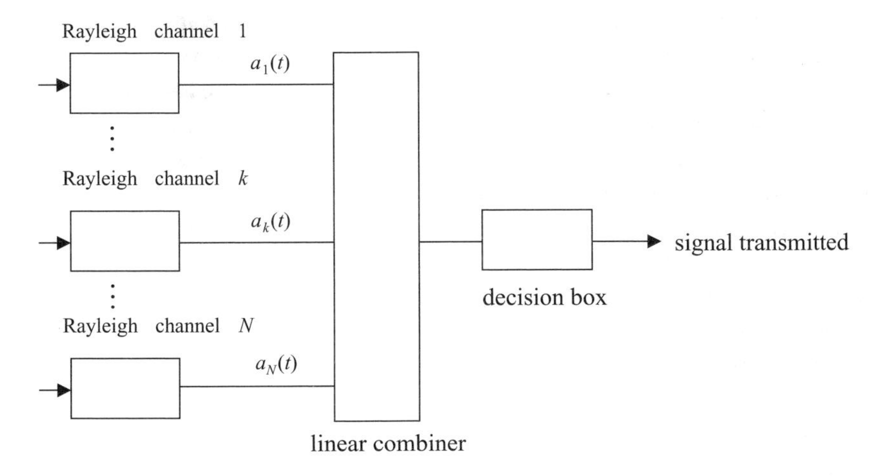

Figure 2.22 Model of N diversity linear combining

channels, referred to as *frequency diversity*, or by reception of multiple samples in time of a given signal, referred to as *time diversity*. Other examples of diversity techniques include angle diversity and polarization diversity. In the former case, directional receiving antennas would be used, pointed in different directions, and receiving independently scattered signals from those directions. In the latter case, a signal would be transmitted simultaneously on each of the two orthogonal polarization directions. Multiple antennas could be located at the receiver, at the transmitter, or at both. We recall that, in the case of Rayleigh fading, fading characteristics change over a distance of the order of a wavelength. Antennas spaced even half a wavelength apart would thus be expected to provide close to independent signal samples. A more general technique of overcoming fading and other signal disturbances, involving the use of multiple antennas at both the transmitter and receiver, and termed generically MIMO reception (multiple in/multiple out) has been received with a great deal of excitement in recent years for its potential in providing large improvements in signal reception. A brief discussion of this technique, with references, appears in Chapter 6.

Consider the general form of diversity reception. Say N independent diversity samples are made available each symbol interval. Whichever diversity technique is actually used, all can be modeled by assuming N independently fading Rayleigh channels are used to transmit the signal, and that the baseband output of each channel is made available to form the diversity sample at each interval. The outputs of the N Rayleigh channels are combined linearly to provide the diversity improvement, with the output of the linear combiner fed to a decision device used to determine the digital signal transmitted each symbol interval. This process of using a diversity technique to improve wireless communication is shown schematically in Fig. 2.22. We use the form of (2.30) to represent the baseband output of a Rayleigh channel, generalized to the kth diversity channel by using the form $a_k(t)$ in place of $a(t)$. For simplicity, we assume the baseband signal to be recovered is binary, $\pm s(t)$. After appropriate combining of the N diversity channels, the decision device then makes the decision, each binary interval, as to whether the signal transmitted was plus or minus.

There are three generic ways of linearly combining the N diversity signals (Brenner, 1959; Schwartz *et al.*, 1966): selection combining, equal-gain combining, and maximal-ratio combining. Selection combining is the simplest operation. The combiner selects as the signal to be used in the decision process the one among the N diversity channels with the largest power.[1] With equal-gain combining, the outputs of the N channels are added together after removing any phase differences on the channels. Maximal-ratio combining, the most complex among the three techniques, weights the output of each diversity channel by the amplitude of that channel before adding them together. Phase differences have to be removed in this case as well. We show later in this section that maximal-ratio combining is the best performing of the three combining techniques. Its improvement in performance over selection combining turns out to be relatively small, however, for small order N of diversity. Performance may be measured in a number of ways. In the next two chapters, as well as elsewhere in this book, we focus on the signal-to-interference ratio, SIR, as a measure of the performance of a mobile wireless system. In particular, we show in later chapters that a well-performing system is one for which the SIR exceeds a specified threshold. We demonstrate diversity improvement in performance due to selection combining in this sense below. The SIR determines the bit error performance in interference-limited digital wireless systems as well. (This last phrase means that interference dominates the performance in contrast to the ever-present noise.) We discuss the specific improvement in bit error performance due to dual diversity (two antennas, for example) in Section 6.4 of Chapter 6.

We now focus on *selection combining*, the simplest of the diversity combining techniques, and demonstrate how even a limited amount of diversity (small N) can provide a significant improvement in performance. Recall from (2.23) that, for a Rayleigh channel, the instantaneous power follows an exponential distribution with average value p, the local-mean power. We assume, for simplicity, that the local-mean power is the same for all N Rayleigh channels. (See Schwartz *et al.*, 1966 for a parallel discussion of selection combining with unequal local-mean powers: the discussion there uses SNR rather than power, however.) From (2.23), the probability distribution of the instantaneous power, or the probability the instantaneous power of a Rayleigh channel will be less than some threshold γ is just $(1 - e^{-\gamma/p})$. Now consider N such channels. The probability that N independent Rayleigh channels will have instantaneous powers less than γ is then just the product of these probabilities, and the probability that at least one of the channels has an instantaneous power greater than a threshold γ is given by

$$\Pr\,[\text{at least one channel } P > \gamma] = 1 - (1 - e^{-\gamma/p})^N \tag{2.59}$$

[1] Most authors use the signal-to-noise ratio SNR on the kth diversity branch as the input to the combiner. In fading environments the signal-to-interference ratio SIR, introduced in the next chapter, is often used in place of the SNR, particularly when the wireless channel is dominated by interference. Since the "noise power" or "interference power" on each diversity branch is often modeled as being the same, it makes no difference in our analysis whether SNR, SIR, or simply power is used. We choose to focus on power at this point for simplicity. We include interference and noise in the discussion of maximal-ratio combining following.

The probability that none of the channels will exceed the threshold, sometimes called the *outage rate* (Schwartz *et al.*, 1966), is given by

$$\Pr\,[\text{all channel } Ps < \gamma] = (1 - e^{-\gamma/p})^N \tag{2.60}$$

From (2.60) it can be noted that the outage rate decreases rapidly with the order of diversity. Conversely, the probability that the instantaneous power of at least one diversity branch exceeds a given threshold, a desirable quality of a wireless system, rapidly approaches 1 with the amount of diversity. A limited amount of diversity can thus provide significant improvement in overcoming short-term fading. Consider some examples:

Example 1

Dual selection diversity is to be used. Compare its threshold performance with that of no diversity when the threshold is 10 (10 dB) and the local-mean power is 100 (20 dB). We have $\gamma/p = 0.1$ here, with $e^{-0.1} = 0.905$. This is the probability that one channel will have a power greater than the threshold. The probability that the channel power will drop below the threshold is then 0.095. The probability that at least one of two channels will have a power greater than the threshold is $1 - (0.095)^2 = 0.991$. The probability that the output of the dual-diversity selection combiner will drop below the threshold is $0.095^2 = 0.009$. This is an order of magnitude less than the single-channel case with no diversity. Each added diversity branch reduces the probability of dropping below the threshold by a factor of $0.095 \approx 0.1$. (*Note:* dBs or decibels relate to power ratios. Following is the definition of dB

$$\text{No. of dB} \equiv 10 \log_{10} \text{(ratio of powers)}$$

Implicit in the numbers above is that the power is being compared with a power of 1. A ratio of 2 is equivalent to 3 dB; a ratio of 10 to 10 dB, etc.)

Example 2

Consider a case now for which the local-mean power is comparable with the threshold. Specifically, take $\gamma/p = 1$. The probability that the instantaneous power, with diversity not used, exceeds the threshold is $e^{-1} = 0.368$. The outage rate, the probability the power drops below the threshold, is 0.632, a relatively high value. The system is thus not performing well in this case. Dual diversity reduces the probability the combiner output power drops below the threshold to $0.368^2 = 0.135$. Four-fold selection diversity reduces the probability the output power drops below the threshold to $0.135^2 = 0.018$; the probability at least one of the four branches will have an instantaneous power exceeding the threshold is then 0.98, a reasonable value.

We now move on to *maximal-ratio combining*. Here we generalize the form of the input to the combiner by indicating that interference and noise appear additively on each of the N Rayleigh channels as well. In particular, the output of the kth channel is given by $a_k(t) + n_k(t)$, with $n_k(t)$ representing the interference and noise. As part of the combining, the kth diversity channel output is multiplied by a gain factor g_k, all signal inputs are brought into phase with one another, and the sum of signal plus interference applied to the decision box of Fig. 2.22. The signal plus interference sum is sampled each symbol interval to determine which digital signal was transmitted. (Recall from (2.30) that the signal term $a_k(t)$ is a complex term whose magnitude is $|a_k(t)|$, while its phase angle has some value $\theta_k(t)$. For maximal-ratio combining, it is assumed that the demodulator can estimate each angle $\theta_k(t)$ and cancel it out, so that all signals appear in phase with one another.) The maximal-ratio combiner output at the sampling time thus has two components, a signal portion given by $s = \sum_{k=1}^{N} g_k a_k$, with a_k the value of $a_k(t)$ at the sampling time, and an interference plus noise portion given by $\sum_{k=1}^{N} g_k n'_k$, with n'_k the sampled value of the interference/noise term at the combiner output coming from the kth diversity channel. The question now is how to pick the gain factors, g_k, $1 \leq k \leq N$. The method selected for maximal-ratio combining is to choose the gain factors to maximize the signal-to-interference power ratio SIR at the sampling time at the combiner output. (Note again that most authors use SNR, the ratio of signal to noise powers. We choose the acronym SIR since interference is the deleterious quantity most often encountered in mobile wireless systems. As used here, however, both interference and noise are included in the definition of the term.) The signal power at the combiner out is proportional to $|s|^2$. To determine the interference plus noise power, we assume the interference on each branch is independent of all the others. The individual interference plus noise powers thus add up at the combiner output, giving the term $I_o = \sum_{k=1}^{N} \left|g_k{}^2\right| \left|n'_k\right|^2$. The SIR at the sampling time is proportional to the ratio $|s|^2/I_o$. Maximal-ratio combining then refers to the procedure for which the individual gains g_k are chosen to maximize the SIR:

Maximal-ratio combining
Choose g_k, all k, to maximize

$$\mathrm{SIR} \equiv |s|^2 \Big/ I_o = \left|\sum_{k=1}^{N} g_k a_k\right|^2 \Big/ \sum_{k=1}^{N} |g_k|^2 \left|n'_k\right|^2 \tag{2.61}$$

This ratio is maximized by applying Schwartz's inequality for complex numbers. This inequality is given by

$$\left|\sum_{k=1}^{N} c_k^* d_k\right|^2 \leq \left[\sum_{k=1}^{N} |c_k|^2\right]\left[\sum_{k=1}^{N} |d_k|^2\right] \tag{2.62}$$

Equality is obtained if $d_k = K c_k$, K an arbitrary constant. The asterisk * represents complex conjugate.

To put (2.61) in the form to which (2.62) may be applied, we let $c_k = a_k^* / |n'_k|$ and $d_k = g_k\, |n'_k|$. We then have, from (2.62) (Schwartz *et al.*, 1966)

$$SIR \leq \sum_{k=1}^{N} |a_k|^2/|n'_k|^2 \tag{2.63}$$

Equality is obtained when

$$g_k = K a_k^*/|n'_k|^2 \tag{2.64}$$

for the kth diversity branch, $k = 1$ to N. The resultant SIR is then equal to the sum of the SIRs, since $|a_k|^2/|n'_k|^2$ is just the SIR on the kth diversity branch. The interpretation is simple and intuitive: choose a high gain factor for a branch whose signal amplitude is large compared with the interference/noise power; reduce the gain for those branches for which the signal is low compared with the interference/noise. This type of optimization forms what is called a *matched filter*. Such a filter occurs in communication theory in which the presence of a signal is to be detected in the presence of noise. The optimum filter, in the sense of maximizing the ratio of signal power to noise power at the filter output, is a matched filter. More generally, this optimum receiving system is an example of a *correlation detector*, in which one of the multiple signals actually transmitted is to be detected (Schwartz, 1990). Note that the problem in the communication theory case is analogous to the one here. The communication theory optimization problem and the matched filter solution were first developed in the context of radar signal detection. An alternate interpretation of such a filter involves considerations in the frequency domain. The interpretation there says that one should design a filter that enhances the signal power in frequency bands where the signal power is larger than the noise and that reduces the signal power in bands where the noise is large compared with the signal (Schwartz, 1990). (More exactly, one should talk of signal and noise spectral densities.)

Comparative performance studies have been made of maximal-ratio combining, equal-gain combining (all diversity gains are the same, hence the diversity channel outputs are simply added together rather than weighting each by a gain factor), and selection combining (Schwartz *et al.*, 1966). Interestingly, equal-gain combining is almost as good as maximal-ratio combining, the difference in performance increasing as the number of diversity channels increases, but still being less than 2 dB poorer for eight-fold diversity. Selection combining is also relatively close to maximal-ratio combining for small orders of diversity, as noted earlier: for dual diversity the difference in performance is about 1 dB. The difference increases with the number of channels combined: it is 2 dB for three-fold combining, just under 4 dB for four-fold combining, and is 6 dB for eight-fold combining.

As noted, maximal-ratio combining requires knowledge of both the phase and amplitude of the signal on each diversity channel. A study of the tolerance to variations in each of these indicates that the performance will deteriorate no more than 1 dB for a phase error of 37.5 degrees, while an error in estimating the signal amplitude by 0.5 dB will result in a performance error of no more than 1 dB (Schwartz *et al.*, 1966).

RAKE receiver

It was noted earlier, in the introduction to this section, that the RAKE receiver can provide significant improvement in the performance of very wideband wireless systems such as the code-division multiple access, CDMA, systems discussed in later chapters. Consider a digital wireless system operating at a high symbol rate, high enough so that the delay spread over a fading channel is greater than the symbol period. This is the case described by (2.48), and results, as noted earlier, in frequency-selective fading. As the symbol interval is reduced, increasing the signal bandwidth correspondingly, individual components of the multipath signal, each arriving at a different delay, may be separately distinguished. If the differential delays as well as relative phases and amplitudes of the individual multipath components can be estimated accurately, the different received rays may be shifted in time, compensating for the differential delays. Maximal-ratio combining can then be applied to the time-compensated arriving rays. This "RAKE" receiver solution for combining separately arriving rays of a signal transmitted over a fading channel gets its name from the fact that the individual rays, after time compensation and combining, resemble the fingers of a rake or fork. Figure 2.22 provides an example of this "rake-like" structure. As noted earlier, such a technique of time diversity was first discussed in Price and Green (1958). (A detailed discussion appears in Schwartz *et al.*, 1966; and Rappaport, 2002 as well.)

In the case of the CDMA systems to be described later in this book, individual binary digits are each multiplied by a much higher bit rate pseudo-random binary stream, resulting in a much smaller symbol interval, called the CDMA "chip." Such systems are also known as *spread spectrum* systems, since the multiplication by the pseudo-random stream increases the transmitted bandwidth, i.e., spreads the frequency spectrum. Users of such a system are individually assigned a different pseudo-random sequence known to both the transmitter and receiver. This sequence then provides the "code" to which the CDMA acronym refers. The coded sequence then modulates the carrier and is transmitted over the wireless channel. Assuming that the channel introduces multipath fading with delay spread greater than the chip interval, the individual rays appear separately at the receiver. The receiving system can rapidly scan through different delay values of the received rays, searching for the sequence corresponding to the one transmitted. Once a number of such differentially delayed received signals are identified, delay compensation can be carried out and maximal-ratio combining used to recover the transmitted sequence. In practice only a few of the earliest arriving rays, normally the strongest in power, are used to carry out the RAKE processing. We shall provide an example, in Chapter 6, of the reduction possible in bit error performance of a CDMA system using the RAKE receiver.

Problems

2.1 Explain, in your own words, the distinction between *average path loss, shadow fading,* and *multipath fading*. How are they related to one another?

2.2 Using Table 2.1, plot $P_{R,db} - P_{T,db}$ for Orlando as a function of distance d, in meters, with $0 < d < 200$ m. Assume transmitter and receiver antenna gains are both 1.

2.3 Determine the shadow-fading parameter σ for each of the four measured curves of Figs. 2.4 and 2.5, and compare. *Hint:* First calculate the average value of each curve and then the root-mean-squared value about these averages.

2.4 The average power received at mobiles 100 m from a base station is 1 mW. Log-normal, shadow, fading is experienced at that distance.

(a) What is the probability that the received power at a mobile at that distance from the base station will exceed 1 mW? Be less than 1 mW?

(b) The log-normal standard deviation σ is 6 dB. An acceptable received signal is 10 mW or higher. What is the probability that a mobile will have an acceptable signal? Repeat for $\sigma = 10$ dB. Repeat both cases for an acceptable received signal of 6 mW. *Note:* Solutions here require the integration of the gaussian function. Most mathematical software packages contain the means to do this. Most books on probability and statistics have tables of the error function used for just this purpose. The error function is defined in Chapter 3 of this text. See (3.12).

2.5 **(a)** Fill in the details of the derivation of the two-ray average received power result given by (2.13a).

(b) Superimpose a $1/d^4$ curve on the measured curves of Figs. 2.7 and 2.8, and compare with the measured curves.

(c) Do the results of Fig. 2.8 validate (2.13a)? Explain.

2.6 (a) Verify, as indicated in the text, that, for the Rayleigh-distributed random variable α in (2.15), σ_r^2 must equal 1/2.

(b) Derive (2.17a) from (2.17) and show that x and y are zero-mean random variables, each with variance σ_R^2 as defined.

(c) Starting with the Rayleigh distribution (2.21) for the received signal envelope a, show the instantaneous received power P_R obeys the exponential distribution of (2.23).

2.7 (a) Show, following the hints provided in the text, that the Ricean distribution (2.25) approaches a gaussian distribution centered about A for $A^2/2\sigma_R^2 \gg 1$.

(b) Verify that the instantaneous received power distribution in the case of a direct ray is given by (2.27). Show that, as the K factor gets smaller (the direct line-of-sight ray decreases relative to the scattered signal terms), the fading distribution of (2.27) approaches an exponential distribution.

2.8 As will be seen throughout the text, simulation is commonly used to determine the performance, as well as verify analysis, of cellular systems. Most critical here is the simulation of fading conditions. This problem provides an introduction to the simulation of Rayleigh fading.

(a) Consider a sequence of n random numbers x_i, $j = 1$ to n, uniformly distributed from 0 to 1. (Pseudo-random number generators are often available in mathematical software packages.) Let $x = (b/n)\Sigma_{j=1}^{n} (x_j - 1/2)$. Show x approximates a gaussian random variable of zero average value and variance $\sigma^2 = b^2/12n$. Repeat for another set of n (independent) uniformly distributed random numbers, calling the sum obtained in this case y. Using x and y, generate a Rayleigh-distributed random variable. Comparing with (2.21), what is the Rayleigh parameter σ_R^2 in this case? *Hint*: Consider the derivation of (2.21) starting with (2.19) and the discussion in the text following.

(b) A different method of obtaining the Rayleigh distribution directly from a uniformly distributed random variable x is to write the expression

$$a = \sqrt{-2\sigma^2 \log_e x}$$

Show the variable a is Rayleigh distributed. How would you now use a sequence of uniformly distributed random numbers to generate a Rayleigh distribution?

(c) Choosing various values for n, generate the Rayleigh distribution using the two methods of (a) and (b), and compare both with each other and with a plot of the Rayleigh distribution. What is the effect on these results of varying n?

(d) Show how you would extend the method of (a) to generate the Ricean distribution. Generate and plot this distribution using the pseudo-random numbers generated in (c) for various values of the Ricean K-factor.

2.9 Consider the average fade duration equation (2.39). Take the case of a vehicle moving at a speed of 100 km/hr. The system frequency of operation is 1 GHz. Say the ratio $\rho = 1$. Show the average fade duration is 8 msec, as noted in the text. Now let the received signal amplitude be 0.3 of the rms value. Show the average fade duration is now 1 msec.

2.10 **(a)** Summarize, in your own words, the discussion in the text on time dispersion and frequency-selective fading.

(b) Consider several cases: a delay spread of 0.5 μsec, one of 1 μsec, and a third one of 6 μsec. Determine whether individual multipath rays are resolvable for the two transmission bandwidths, 1.25 MHz used in IS-95 and cdma2000, and 5 MHz used in WCDMA. (See Chapter 10.)

2.11 Indicate the condition for flat fading for each of the following data rates:

8 kbps, 40 kbps, 100 kbps, 6 Mbps

Indicate which, if any, radio environments would result in flat fading for each of these data rates.

2.12 **(a)** Consider the transversal filter equalizer of Fig. 2.20. A training sequence of K binary digits is used to determine the $2N + 1$ tap gains, as described in the text. Show that, under a minimum mean-squared performance objective, the optimum choice of tap gains is given by (2.53).

(b) Show the vector form of (2.53) is given by (2.56), with the solution given by (2.57).

2.13 **(a)** Work out a simple example of the transversal filter equalizer: Say the equalizer has three taps to be found using the minimum mean-squared performance objective. Choose a set of $K = 10$ arbitrarily chosen binary digits as the training sequence and then let some of these digits be received in "error," i.e., some are converted to the opposite polarity. Find the "best" set of taps in this case. Try to choose the training sequence so that there are equal numbers of +1 and –1 digits. Compare the tap coefficients with those found using the approximation of (2.58).

(b) Repeat this example for a different set of transmitted digits and errors in reception.

(c) Choose a larger example of a transversal filter equalizer and repeat (a) and (b), comparing with the results obtained there.

2.14 **(a)** Plot the improvement in performance obtained with the use of dual selection diversity as the ratio of local-mean power to the threshold varies. Use at least the following cases: (1) the local-mean power 20 times the threshold; (2) local-mean power 10 times the threshold; (3) local-mean power equal to the threshold; (4) local-mean power 0.1 of the threshold. *Note:* Performance may be defined as outage rate or the probability that at least one of the channels has an instantaneous power greater than the threshold.

(b) Repeat (a) for 4- and 8-fold diversity and compare all three orders of diversity.

2.15 Show the optimum maximal-ratio combining gain for the kth diversity branch is given by (2.64). Explain the statement that the SIR is then the sum of the SIRs, summed over the N diversity branches.

2.16 **(a)** Explain how equal-gain combining differs from maximal-ratio combining. In particular, write an expression for the SIR in the case of equal-gain combining. *Hint*: How would this expression compare with (2.61)?

(b) Why would you expect the performance of diversity schemes to be ranked in the order maximal-ratio combining best, equal-gain next best, selection diversity last?

2.17 **(a)** Explain the operation of the RAKE receiver in your own words.

(b) Two third-generation CDMA systems are discussed in Chapter 10. The first, W-CDMA, uses a chip rate of 3.84 Mcps (million chips per second), with a corresponding chip duration of 0.26 μsec; the second system, cdma2000, uses a chip rate of 1.2288 Mcps, with a chip duration of 0.81 μsec. (This is the same chip rate used by the second-generation CDMA system IS-95 discussed in Chapters 6 and 8.) Explain the statements made in Chapter 10 that RAKE receivers can be used to provide multipath time-diversity for paths differing in time by at least those two chip durations, respectively. Which system potentially provides better RAKE performance?

Chapter 3

Cellular concept and channel allocation

3.1 Channel reuse: introduction of cells

In the previous chapter we focused on propagation effects over the radio link between a base station and a mobile terminal. We noted there that most modern wireless systems are organized into geographic cells, each controlled by a base station. (Among the exceptions, as noted there, are small-area systems such as local-area wireless networks and personal-area networks, to be discussed in Chapter 12. *Ad hoc* and wireless sensor networks, not discussed in this book, are other examples of networks not necessarily involving the cellular concept.) We were thus implicitly discussing the propagation effects encountered in a single cell. In this chapter we introduce the cellular concept, showing how the use of cells can increase the capacity of a wireless system, allowing more users to communicate simultaneously.

The number of simultaneous calls a mobile wireless system can accommodate is essentially determined by the total spectral allocation for that system and the bandwidth required for transmitting signals used in handling a call. This is no different than the case for other radio applications such as broadcast radio, AM or FM, or broadcast TV. As an example, consider the first-generation analog mobile system in the US, the AMPS system. Here 25 MHz of spectrum is made available for each direction of transmission in the 800–900 MHz radio band. Specifically, as noted in Chapter 1, in making radio spectrum allocations for mobile radio communication, the Federal Communications Commission, the FCC, in 1981 assigned the 824–849 MHz band to uplink or reverse channel communication, mobile to base station (BS); the 869–894 MHz range was assigned to downlink or forward communication, BS to mobile. Frequency modulation (FM) was the modulation type adopted for these first-generation systems, with the 25 MHz band in each direction broken into 30 kHz-wide signal channels, each accommodating one call. There are thus 832 analog signal channels made available in each direction. Second- and third-generation digital systems use TDMA (time-division multiple access) and CDMA (code-division multiple access) techniques over the same 25 MHz frequency bands in each direction to provide increased capacity, hence allowing more simultaneous calls to be transmitted. The concepts of TDMA and CDMA, as well as detailed descriptions of the second- and

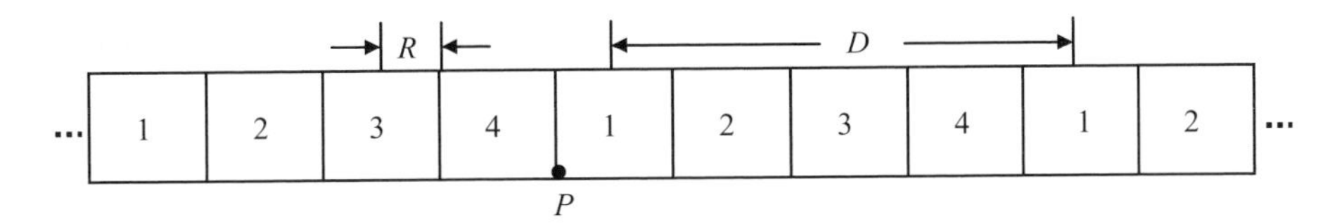

Figure 3.1 One-dimensional cellular array

third-generation systems, will be discussed in later chapters. But they too are limited in their communication capability by the total spectral allocation made available. These numbers, whether 832 analog channels or some multiple thereof using digital technology, are obviously insufficient to handle massive numbers of users, particularly in urban/suburban areas. Breaking a region into a number of geographically distinct areas called cells and reusing the frequencies in these cells allows the number of communication channels to be increased. This concept of *cells*, geographic clusters, each controlled by a base station, to be used to improve the communication capacity of a wireless system, was first proposed as early as 1947 by Bell Laboratories in the US, with a detailed proposal for a "High-Capacity Mobile Telephone System" incorporating the cellular concept submitted by Bell Laboratories to the FCC in 1971 (Jakes, 1974). The first AMPS system was deployed in Chicago in 1983.

The basic idea with cellular systems, as noted above, is to *reuse* channels in different cells, increasing the capacity. But a problem arises: the same frequency assignments cannot be made in adjacent cells because of *inter-channel interference*. The assignments must be spaced far enough apart geographically to keep interference to tolerable levels. Channel reuse is thus not as efficient as might initially be hoped, but the use of a large number of cells does provide an overall gain in system capacity, i.e., the ability to handle simultaneous numbers of calls. In particular, if cells can be reduced in size, more of them can be added in a given geographical area, increasing the overall capacity as a result. The recent trend to the use of much smaller microcells is a step in this direction.

To explain the improvement in system capacity made possible by the introduction of cells, consider first a simple one-dimensional example. (This could serve as a model of a highway-based cellular system, as an example.) Focus on the first-generation analog AMPS system as an example. Say the overall band of 832 channels is first divided into four groups of 208 channels each. Call these groups 1, 2, 3, 4 and locate them as shown in Fig. 3.1. There are thus three cells separating cells with the same set of (interfering) frequencies. We call this *4-cell reuse*. Given N such cells in a system, 208 N channels are made available, compared with the original 832 possible when no cellular structure is invoked. With the number N of cells large enough, a significant increase in system capacity is made possible over the original 832 channels. However, had only two cells been required to separate cells using the same band of frequencies (3-cell reuse) a system with N cells would have resulted in $277N$ usable channels, an even larger improvement in capacity. Which assignment strategy should be used? Clearly this depends on the interference tolerable, since spacing same-band cells three cells apart results in less interference than spacing them two cells apart. This tolerable interference is typically quantified by

Jakes, W. C. ed. 1974. *Microwave Mobile Communications*, AT&T, 1995 edition, New York, IEEE Press.

calculating the *signal-to-interference ratio* SIR, often called the carrier-to-interference ratio CIR. This ratio is defined to be the ratio of the desired *average* signal power at a receiver to the total *average* interference power. (This ratio is comparable with the signal-to-noise ratio used as a performance measure in non-mobile communication systems – Schwartz, 1990.) The SIR should then be greater than a specified threshold for proper signal operation. For example, in the first-generation AMPS system, designed for voice calls, the desired performance threshold, in decibels (dB) is SIR_{dB} equal to 18 dB. ($SIR_{dB} = 10 \log_{10} SIR$.) This number has been found appropriate to provide acceptable voice quality in studies of mobile calls. For the second-generation digital AMPS system (D-AMPS or IS-54/136) discussed in later chapters, a threshold of 14 dB is deemed suitable. For the GSM system, also described in later chapters, a system first proposed for European use and since adopted worldwide, a range of 7–12 dB, depending on the study done, is suggested as the appropriate threshold. We shall see, in a later chapter, that the probability of error in a digital system depends on the choice of this threshold as well.

3.2 SIR calculations, one-dimensional case

We emphasized above that the SIR is calculated on an *average* power basis. Hence it should be clear that calculations of this quantity focus on the distance-dependent part of the received-power equation, equation (2.4) in Chapter 2. The effect of both shadow (log-normal) fading and multipath Rayleigh fading is ignored. More specifically, we shall assume that $g(d)$ in (2.4) is of the form kd^{-n}, with n generally taken to have the values 3 or 4, the latter derived using an ideal two-ray model, as shown in equation (2.13a). We first calculate SIR for a one-dimensional example and then consider the more common two-dimensional case. In the one-dimensional example, let D be the spacing between interfering cells, as shown in Fig. 3.1. The half width (center to each edge) of each cell is taken to have a value R, as shown in Fig. 3.1. We focus on the downlink power received at a mobile, located, in the worst case, at the edge of a cell. This location is indicated by the point P in Fig. 3.1. (Similar calculations can be carried out for the uplink case. We shall, in fact, carry out such calculations explicitly in studying CDMA systems.) Say each base station, located at the center of its cell, transmits with the same average power P_T. (This then implies some sort of power control system about which more will be said later.) The average received power a distance d m from a base station is then $P_T d^{-n}$, $n = 3$ or 4, say. The signal-to-interference ratio at the mobile at point P is then given by

$$SIR = \frac{R^{-n}}{\sum P_{\text{int}}} \tag{3.1}$$

where the sum in the denominator is taken over all interfering base stations and P_{int} represents the *normalized* interfering power. Note that normalized average powers only need be calculated because of the assumption that all base stations transmit at the same average power. It is clear that, theoretically at least, all base stations transmitting at the same frequency will interfere with the home base station's transmission. We shall show by example, however, that only a relatively small number of nearby interferers need be considered, because of the rapidly decreasing received power as the distance d increases.

Schwartz, M. 1990. *Information Transmission, Modulation, and Noise*, 4th edn, New York, McGraw-Hill.

Table 3.1 *SIR calculations, one-dimensional case, single-tier interferers*

Reuse factor	SIR(dB)	
	$n = 3$	$n = 4$
3 cells	19.6	27
4 cells	23	32

In particular, consider the first tier of interferers only, the two interfering base stations closest to the mobile at point P. From Fig. 3.1, assuming all base stations are located at the center of their respective cells, these two base stations are located at distances $(D - R)$ and $(D + R)$, respectively, from the mobile in question. The corresponding SIR is then given by

$$SIR = \frac{R^{-n}}{(D - R)^{-n} + (D + R)^{-n}} \tag{3.2}$$

For 3-cell reuse, $D = 6R$; for 4-cell reuse, $D = 8R$. The results of the calculation, using (3.2), appear in Table 3.1. (Recall again that, by definition, SIR in dB is $10 \log_{10} SIR$.)

As expected, as the distance variation parameter n increases, the interference decreases and the SIR increases. As the reuse factor increases, the interference decreases as well. Now include the second tier of interfering cells. The distances from the centers of these cells, where the base stations are located, to the mobile at point P are given by the values $(2D - R)$ and $(2D + R)$ respectively. The expression for the SIR with these two added interferers included is then given by

$$SIR = \frac{1}{\left(\frac{D}{R} - 1\right)^{-n} + \left(\frac{D}{R} + 1\right)^{-n} + \left(\frac{2D}{R} - 1\right)^{-n} + \left(\frac{2D}{R} + 1\right)^{-n}} \tag{3.3}$$

Note that we have divided through by the cell halfwidth R to normalize this equation. How does the inclusion of the second tier of interferers affect the entries in Table 3.1? For 3-cell reuse with $D/R = 6$, and $n = 3$, one finds $SIR_{dB} = 19.2$ dB. This compares with 19.6 dB when first-tier interferers only are included. It is thus apparent that with the parameter n large enough (3 in this case), the distance dependence becomes weak enough that only a small number of interferers (first-tier cells or at most second-tier cells in addition in this case) need be considered in calculating the SIR. In the next section we turn to the more complex case of two-dimensional cell arrays.

3.3 Two-dimensional cell clusters and SIR

A common model of cellular structures in the two-dimensional case is to consider all cells to be hexagonal in shape and all of the same size. Note that the only three geometric shapes that can be used to tessellate an infinite two-dimensional space are the hexagon, the square, and the triangle. The hexagonal shape is useful in considering base station antennas that

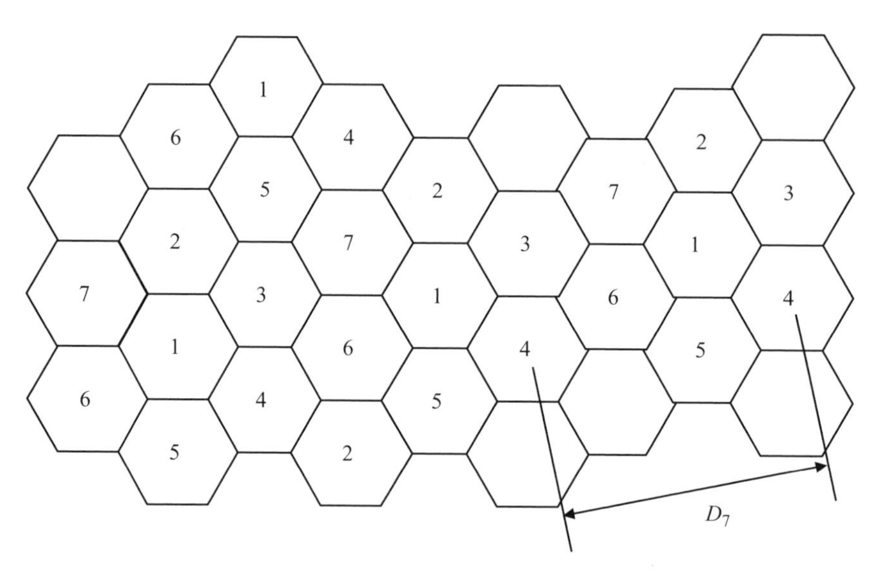

Figure 3.2 Hexagonal cell structure, two dimensions

radiate uniformly in all directions or that cover specific sectors. It provides a fairly good approximation to a circle and has, as we shall see, simple geometric properties that enable SIR calculations to be readily carried out. In a real system, cells have complex shapes, depending on antenna directivity and location, propagation conditions, terrain topography, etc. Software packages are available that take realistic terrain and propagation effects into account. As noted in the previous chapter, however, we are principally interested in this book in providing simple, basic principles for understanding wireless systems and developing useful models for analyzing them. A uniform hexagonal structure falls into this category. Hexagonal geometry will be used in subsequent chapters as well. (We shall occasionally use circular-cell geometry, however, to carry out calculations where this simplifies the analysis. We then compare results with those obtained using the hexagonal model.)

We focus therefore on a two-dimensional structure with hexagonally shaped cells, an example of which appears in Fig. 3.2. It turns out, as shall be shown shortly, that only certain clusters or groups of cells may be accommodated with reuse, among which the total number of available channels is to be divided. An example of a 7-reuse scheme appears in Fig. 3.2. The numbers 1–7 indicate the seven different groups of channels into which the available spectrum is divided. Cells with the same number use the same set of channels and are thus potential interferers of one another. The spacing between closest interfering cells is shown as having a value D_7. More generally, it is found that clusters contain C cells, with C an integer given by the expression

$$C = i^2 + j^2 + ij \tag{3.4}$$

(i, j) the integers $0, 1, 2, \ldots$. Examples of such reuse clusters appear in Table 3.2. We shall demonstrate the validity of (3.4) shortly. Common examples of reuse clusters adopted in practice include those for $C = 3, 7$ (as in Fig. 3.2), and 12. As examples, the AMPS system

Table 3.2 *Reuse clusters, hexagonal cells*

C	i	j
1	0	0
3	1	1
4	2	0
7	2	1
9	3	0
12	2	2
13	3	1
16	4	0
19	3	2
21	4	1
27	3	3

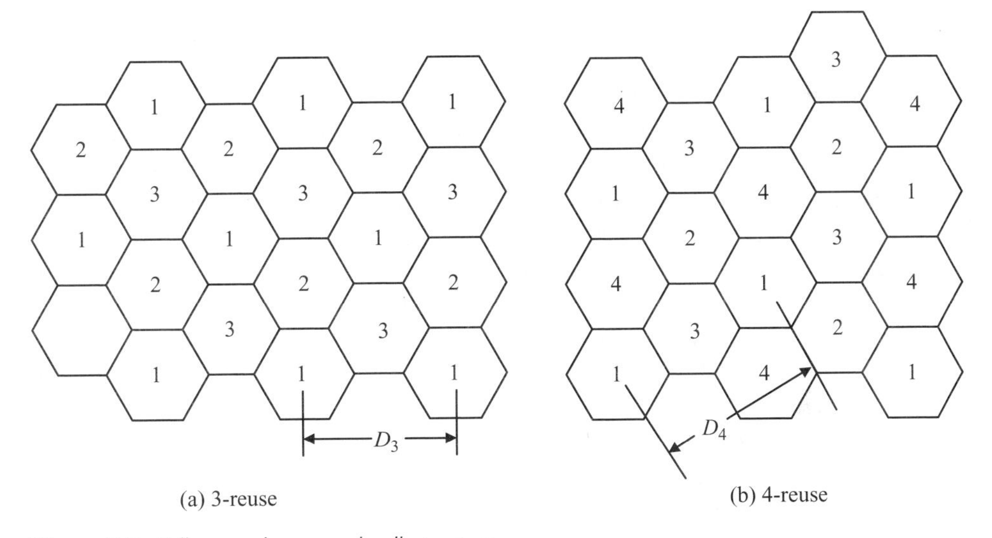

Figure 3.3 Cell reuse, hexagonal cell structure

uses 7; GSM uses 3 or 4. Figure. 3.3(a) shows a frequency reuse pattern of 3; Fig. 3.3(b) shows one for 4. The distance between closest interfering cells is shown as equal to D_3 and D_4, respectively, in the two cases.

Consider a typical hexagonal cell as shown in Fig. 3.4. The distance from the center of the cell to any vertex, the radius of the hexagon, is taken to be a value R. Each edge has a length R as well. The distance across the cell, between parallel edges, is then $\sqrt{3}\,R$, as shown in the figure. The area of the cell is readily calculated to be $6(R/2{*}\sqrt{3}\,R/2) = 3\sqrt{3}R^2/2 = 2.6\,R^2$. Note that a circle circumscribing the hexagon, with the same radius R, has an area $\pi R^2 = 3.14\ R^2$. A circle covering the interior of the hexagon, of radius $\sqrt{3}\,R/2$, has as its area $3/4(\pi R^2) = 2.36\ R^2$. The two circular areas serve as upper and lower bounds of the hexagonal area, respectively. As noted above, calculations involving

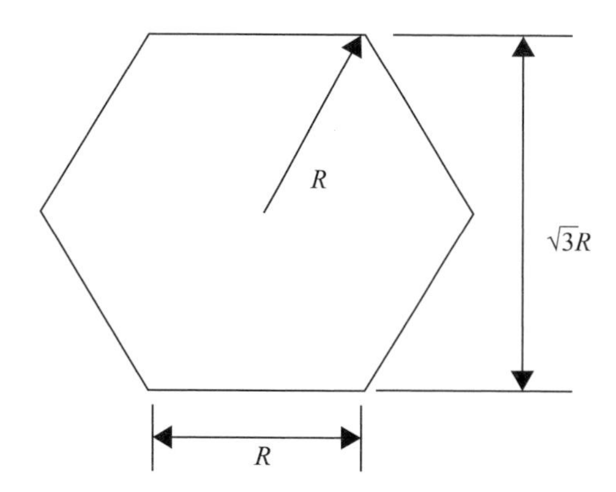

Figure 3.4 Hexagon

circles rather than hexagons are sometimes used in subsequent chapters to simplify the analysis.

Given the hexagonal radius R, the following values of the center-to-center spacing between the closest interfering cells, as indicated in Figs. 3.2 and 3.3, are readily established: $D_3 = 3R = \sqrt{3 \times 3}\,R$; $D_4 = 2\sqrt{3}R = 3.46R = \sqrt{3 \times 4}\,R$; $D_7 = \sqrt{3 \times 7}\,R = 4.58R$. In general, for cluster sizes C, as given by (3.4), we shall show that $D_C/R = \sqrt{3C}$. Hence as the cell cluster increases, the spacing between interfering cells increases, reducing the interference. But the channel assignment per cell decreases as well, as already noted, reducing the effectiveness of the channel reuse, the reason for introducing a cellular structure. Consider the AMPS and IS-54/136 examples again. Say N cells constitute a geographic area. The number of frequency channels available for this area is then $832N/C$. The number per cell is $832/C$. For a 7-reuse system the number per cell is $832/7 \cong 118$ channels. For a 3-reuse system, the number is $832/3 \cong 277$ channels per cell, as already noted in the earlier one-dimensional example of the previous chapter.

We have yet to establish the validity of the "magic" numbers given by (3.4) and $D_C/R = \sqrt{3C}$. Consider the latter condition first. We adopt an approach used in Yacoub (1993). Note from Fig. 3.2 that the cluster consisting of seven cells may itself be approximated by a hexagon of the form shown more generally in Fig. 3.5. This larger hexagon has as its distance between "edges" just the distance D_7 indicated in Fig. 3.2. In the general case this would be D_C, the distance between closest interferers, as indicated in Fig. 3.5. A hexagon with this distance between edges has a radius $R_C = D_C/\sqrt{3}$. Its area is then $A_C = 3\sqrt{3}R_C^2/2 = \sqrt{3}D_C^2/2$. The number of cells within a cluster is given by the ratio of the cluster area to the cell area. The cell area $a = 3\sqrt{3}\,R^2/2$. Taking the ratio of areas, we find $C = A_C/a = {D_C}^2/3R^2$, from which we get $D_C/R = \sqrt{3C}$.

Now consider (3.4). The specific form of this expression is a property of the hexagonal tessellation of the two-dimensional space. To demonstrate this property, consider the set of hexagons of radius R shown in Fig. 3.6. The location of the centers of the hexagons may be specified by drawing two axes labeled u and v as shown. Axis v is chosen to be vertical; axis u is chosen to be at an angle of 30° ($\pi/6$ rad) with respect to the horizontal. Starting at an arbitrary hexagon whose center is taken to be (0, 0), axes u and v intersect

---

Yacoub, M. D. 1993. *Foundations of Mobile Radio Engineering*, Boca Raton, FL, CRC Press.

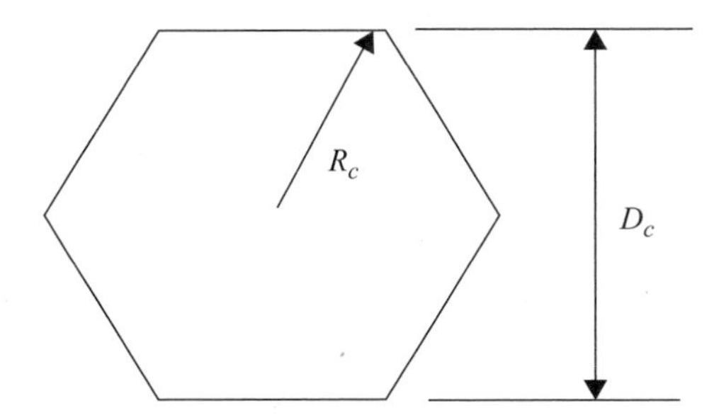

Figure 3.5 Larger hexagon

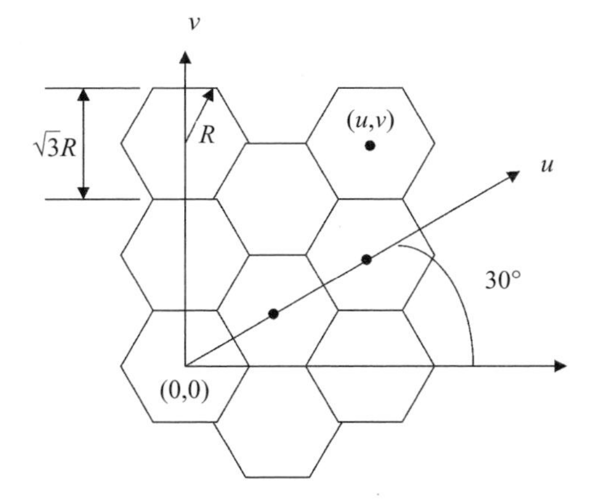

Figure 3.6 Hexagonal space tessellation

the centers of adjacent hexagons as indicated in the figure. Consider the distance D now to the center of a hexagon whose location is at point (u,v). From the Pythagorian Theorem this is given by

$$D^2 = (v + u \sin 30°)^2 + (u \cos 30°)^2 = u^2 + v^2 + uv \tag{3.5}$$

But note that u and v each increment by multiples of $\sqrt{3}\,R$, the distance between edges of a hexagon. Thus $u = i\,\sqrt{3}\,R$ and $v = j\sqrt{3}\,R$, i and $j = 0, 1, 2, \ldots$. We thus have, from (3.5)

$$D^2 = 3R^2(i^2 + j^2 + ij) \tag{3.5a}$$

Combining this equation with the relation $C = D_C^2/3R^2$, we get (3.4).

Given this introduction to two-dimensional cell clusters, hexagonal geometry, and the spacing of first-tier interferers, we are in a position to calculate the two-dimensional SIR, just as was done in the previous section for the one-dimensional case. Note first that, given any cell, there will be six first-tier interferers located about it. This is apparent from the cluster constructions of Figs. 3.2 and 3.3. These six interferers are drawn schematically in Fig. 3.7. The actual location of these interferers depends on the cluster size C. Assume, for simplicity, that that they fall along the six hexagonal axes spaced 60° apart as shown in Fig. 3.7. For the worst-case SIR calculation, let the mobile be placed at point P, a corner of the cell, as shown in Fig. 3.7. Its distance from its own base station at the center of the cell is R. Its distance from the closest interferer, the base station at another cell at

Table 3.3 *Approximate SIR calculations [using (3.6)]*

C	D/R	SIR		SIR_{dB}	
		$n=3$	$n=4$	$n=3$	$n=4$
3	3	3.4	8.5	5.3	9.3
7	4.58	14.3	62.5	11.6	18
12	6	30.3	167	14.8	22.2

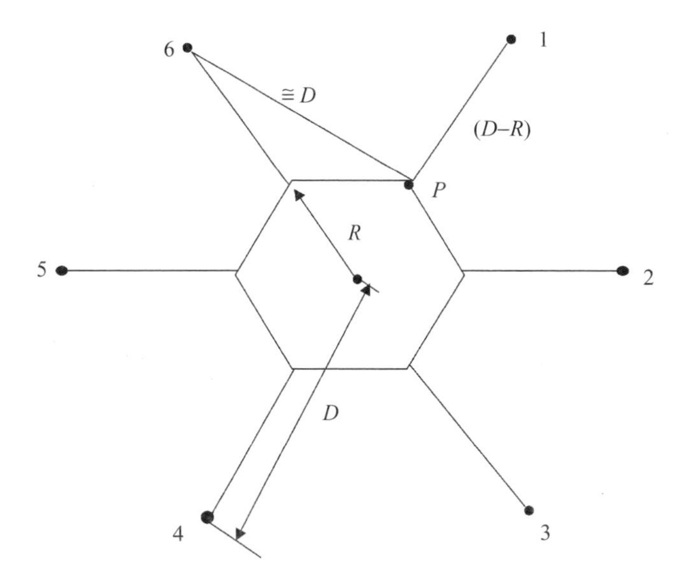

Figure 3.7 Six first-tier interferers, approximate location

point 1, is then approximately $(D - R)$. Its distance from the first-tier interfering base station furthest away, at point 4, is approximately $(D + R)$. The other four interfering base stations are varying distances away. (By symmetry, the distances from the mobile at point P to points 2 and 6 are the same, as are the distances to points 3 and 5.) But recall that with $D_C/R = \sqrt{3C}$, $D/R \geq 3$, $C \geq 3$. Then as a good approximation, the four interferers at points 2, 6, 3, 5 are about a distance D away from the mobile. An approximation to the SIR is then given by

$$SIR \approx \frac{1}{\left[\left(\frac{D}{R} - 1\right)^{-n} + \left(\frac{D}{R} + 1\right)^{-n} + 4\left(\frac{D}{R}\right)^{-n}\right]} \tag{3.6}$$

The results of using (3.6) appear in Table 3.3. Details are left to the reader.

How valid are the approximations used in obtaining (3.6) and the entries of Table 3.3? G. L. Stüber (1996: 468) has used as an approximation two interferers each at distances of $(D - R)$, D, and $(D + R)$. He obtains more conservative results: for $C = 3$, his SIRs for $n = 3$ and 4, respectively, are 2.8 and 6.1 compared with our values of 3.4 and 8.5 (Table 3.3). For $C = 7$, $n = 3$ and 4, respectively, he gets 13.2 (11.2 dB) and 55.6 (17.5 dB)

Stüber, G. L. 1996. *Principles of Mobile Communications*, Boston, MA, Kluwer Academic Publishers.

compared with our 14.3 (11.6 dB) and 62.5 (18 dB), a difference of only 0.5 dB. Rappaport (1996: 39) uses another approximation: two interferers at a distance of $(D - R)$, and one each at $(D - R/2)$, D, $(D + R/2)$, $(D + R)$. These results are yet more conservative, resulting in an SIR for $C = 3$ of 2.6 for $n = 3$ and 5.8 for $n = 4$. But note that, in dB measure, they do not differ very much from the results of Stüber. As C increases, increasing D/R, the differences between the various approximations become negligible.

The upshot of these results is that with a required SIR_{dB} of 18 dB, IS-54/136 (digital AMPS or D-AMPS) would have to use a 7-reuse scheme. In practice $C = 7$ is used, but three antennas per cell, each covering a 120° sector, are required to attain the 18 dB performance level, with the area of interference further reduced by a factor of three. (One third as many interferers, or two first-tier interferers, per sector now appear. Details are left to the reader.) The GSM system requires an SIR of 7 dB, with the interference reduced by introducing frequency hopping, a technique in which the transmission frequencies are periodically changed in a prescribed, pseudo-random manner. From Table 3.3 this system can then use a value of C as small as 3, assuming a propagation dropoff factor of $n = 4$.

3.4 Traffic handling capacity: Erlang performance and cell sizing

With the appropriate cell cluster size or reuse parameter C determined by the SIR required for a given system, the number of channels available per cell is immediately established. For example, the first-generation AMPS provides $832/C$ channels per cell, as already noted. With $C = 7$, as determined in the previous section, the system can provide 118 channels per cell. We will see in later chapters that the second-generation digital AMPS system IS-54/136 provides three times as many channels per cell. In this system, three users are accommodated per frequency channel assignment, using time slot, TDMA, technology to be described later. Given the number of channels available per cell, we can now determine the geographical extent or size of cells. This determination is based on a consideration of the traffic expected in a given region and the system performance, other than SIR, that is to be provided to a typical user.

In this introductory section on system performance we focus exclusively on the *probability of call blocking* as a measure of performance or user satisfaction. We shall see later in this book that other measures of performance must be considered as well. These include the probability of dropping a call on handoff, and, for third-generation wireless systems based on packet-switching technology, such measures as the packet loss probability and packet delay. The call blocking probability has, historically, been the principal measure of performance in telephone systems worldwide. Simply put, it describes the chance that a user, attempting to place a call, receives a busy signal. This measure depends, as might be expected, on the number of channels available to handle simultaneous calls and the traffic expected to utilize the system. In the mobile wireless domain the calculation of the call blocking probability in a given cell depends on the number of channels available, which we have already discussed, and the traffic expected to use that cell. With the call blocking probability specified (this is variously taken to be a number somewhere between 1% and

---

Rappaport, T. S. 1996. *Wireless Communications, Principles and Practice*, Upper Saddle River, NJ, Prentice-Hall PTR.

5%), a limit must be put on the amount of traffic expected to use that cell. This can be done by properly adjusting the cell size, as we shall see. (It is to be noted, parenthetically, that one normally talks of the amount of traffic at some given time, usually taken as the "busy hour" during a typical week.) In the material following, we first define quantitatively what is meant by "amount of traffic," or traffic intensity, or traffic load, using a simple statistical model, then indicate how the blocking probability is related to it.

Traffic intensity is commonly defined to be the product of the average number of call attempts per unit time, times the average call length. Let the symbol λ represent the average number of call attempts per unit time and $1/\mu$ be the average call length. The traffic intensity or load A is then λ/μ, in units of *Erlangs.* (A. K. Erlang was a great Danish engineer who, in the early 1900s, was the first to put traffic engineering on a firm quantitative basis.) As an example, say there is an average of 50 call attempts per minute, with each call, if accepted, lasting 3 minutes. The traffic load is then 150 Erlangs. The usual statistical model is to assume, in addition, that the pattern of call attempts or call arrivals obeys a Poisson distribution, with average rate of arrival λ. Call lengths are assumed in this simple model to be exponentially distributed, with average value $1/\mu$. A simple analysis then shows that, with N channels available, the call blocking probability P_B is given by the so-called *Erlang-B* formula

$$P_B = \frac{A^N/N!}{\sum_{n=0}^{N} A^n/n!} \tag{3.7}$$

(Details of the derivation appear in Chapter 9. A more thorough discussion arising from traffic models developed for wire telephony appears in Schwartz (1987). It turns out that this formula is independent of the actual call arrival–call length statistics. Its derivation is most readily carried out using the Poisson/exponential assumptions for call arrival and call length statistics, respectively. This is what is done in the derivation in Chapter 9.) A table or plot of P_B vs A indicates that $N \approx A$ for $P_B = 10\%$. $N > A$ channels are required to have $P_B = 5\%$ or less. (Detailed tables of the Erlang-B function appear in Lee, 1993.) As an example, consider the traffic load $A = 150$ Erlangs calculated above. For $P_B = 5\%$, $N = 153$ channels are found to be required. For a blocking probability of 1%, 169 channels are required. Small increases in the number of channels made available reduce P_B substantially. Consider the converse now. Say the number of channels is fixed. How does the blocking probability vary with Erlang load? Let $N = 100$ channels, to be concrete. To keep P_B at 1% or less, a load of no more than 84 Erlangs can be allowed. If A were to increase by 13%, to 95 Erlangs, the blocking probability would jump to 5%, a five-fold increase! At low blocking probabilities, the blocking probability is very sensitive to changes in load.

Consider now the significance of these numbers in relation to mobile usage of a given channel. Say a call lasts 200 seconds on the average. Say a typical user makes a call every 15 minutes, on the average. If 100 channels are available and a blocking probability of

---

Schwartz, M. 1987. *Telecommunication Networks: Protocols, Modeling, and Analysis*, Reading, MA, Addison-Wesley.

Lee, W. C. Y. 1993. *Mobile Communications Design Fundamentals*, New York, Wiley Interscience.

1% is desired, 378 users can be accommodated. (How is this number obtained?) If the average call duration increases to 400 seconds, only 189 users can be accommodated. If the typical call lasts 200 seconds, on the average, but a mobile user attempts a call every ten minutes, on the average, the number of users allowed in the cell is now 250. So cell phone or terminal usage clearly plays a role in determining the number of mobiles that may be accommodated.

Now consider the impact of these calculations on the sizing of a cell. To be concrete, we focus first on an AMPS example. Recall that with a total of 832 frequency channels and $C = 7$ reuse, 118 channels are available per cell. To keep the blocking probability P_B to less than 1%, the Erlang load allowed turns out to be, using (3.7) or tables or plots based on that equation, $A = 101$ Erlangs. Say a typical user makes 200-second-long calls once every 15 minutes on the average, then the total call arrival rate is $\lambda = 0.5$ call/sec, on the average, and $15 \times 60/2 = 450$ users may be accommodated in each cell. Assume now that these users are uniformly distributed over the cell. Since the area of a hexagonal cell of radius R is $3\sqrt{3}\,R^2/2 = 2.6\,R^2$, the allowable user density is $173/R^2$ mobiles/unit of area. Consider first a rural region. Say the density of mobile terminals is two terminals per km^2. Then the cell radius should be set at about $R = 9.3$ km or 5.6 miles. Say now a suburban region has a mobile terminal density of 100 mobiles/km^2. The corresponding cell radius drops to 1.3 km. In an urban region with a mobile density of 1000 mobiles/km^2, the cell radius would reduce further to 0.26 km or about 800 ft. Similar calculations can be carried out for the second-generation, digital version of AMPS, D-AMPS or IS-54/136. As already noted, this system can handle three times as much traffic as AMPS, since three time slots in TDMA operation are made available per frequency channel. Cells for this system could be designed to have radii $\sqrt{3} = 1.7$ times as large as those for AMPS. The Pan-European second-generation mobile system GSM has 124 frequency channels, of 200 kHz bandwidth each, defined in a 25 Mhz total assigned bandwidth. As we shall see later in this book, eight TDMA time slots can be individually assigned as user channels per frequency channel. This results in a total of 992 channels. With a reuse parameter of $C = 3$, 330 channels per cell become available. If reuse of $C = 4$ is adopted instead, 248 channels become available. Calculations similar to those carried out for the AMPS and D-AMPS cases can then be carried out for the GSM system to determine cell radii, assuming the hexagonal cell model. The reader is encouraged to carry out these calculations.

Somewhat more accurate calculations may be carried out that take into account the fact that mobiles do not necessarily cover an area uniformly, but, particularly in the case of an urban environment, are constrained to follow streets which cover only part of the area. As an example, consider the case of a uniform street grid, such as in Manhattan, New York City (Jakes, 1974: Ch. 7). Say streets are spaced w units apart and carry two-way traffic. Let S be the spacing of mobile terminals. One can then find the number of mobiles in a cell of radius R. A simple calculation shows that this number M is given by

$$M = \alpha R^2/wS \tag{3.8}$$

with $\alpha = 10.4$ for a hexagon. (Consider a cell of area A. Break it into a two-dimensional, "Manhattan-street" grid, with streets spaced a distance w apart, as shown in Fig. 3.8. The total street length L is then given approximately by the number of small squares of area

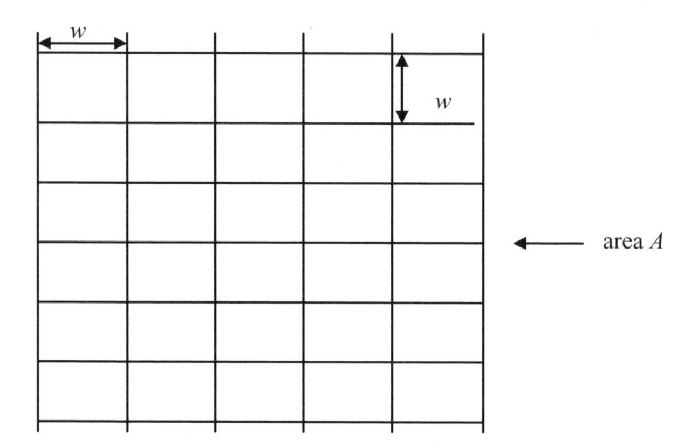

Figure 3.8 Manhattan street grid

w^2 each in the cell times the length 4 w of each square. Thus $L \cong 4\ w\ A/w^2 = 4\ A/w$. For a hexagon, $A = 2.6\ R^2$. Then $L \cong 10.4\ R^2/w$. Spacing mobiles a distance S apart, we then find $M = L/S = 10.4\ R^2/wS$. Other geometrical shapes give slightly different values of α.)

As an example, let w be 10 m, while the spacing S between mobiles is 30 m. Let each mobile terminal attempt a call every 15 minutes, on the average, each call lasting an average of 200 seconds. The user call statistics are thus chosen to be the same as those used in the example above. The Erlang load per mobile user is therefore $200/15 \times 60 = 0.22$ Erlang. The total load is 0.22 M. Consider the same AMPS example used above: with reuse factor $C = 7$, 118 channels per cell are available. Then a total load of 101 Erlangs only may be accommodated if the desired call blocking probability is to be kept at 1%. Equating this number to 0.22 M, we find the radius of a cell in this environment to be $R = 400$ m $= 0.4$ km.

3.5 Probabilistic signal calculations

To this point we have focused on average power considerations. We calculated, in both Section 3.1 and Section 3.2, the average SIR, using this quantity as a measure of cellular system performance. We neglected completely the phenomenon of fading discussed in Chapter 2 and the fact that both the signal power and the interference power might vary statistically within the confines of a cell. A more appropriate measure of performance in a cell might be the probability that the signal power for either a mobile (downlink) or a base station (uplink) was greater than any of the interfering signal powers measured at that point. Alternatively, one could calculate the probability that the ratio of actual signal power received at a mobile terminal to the interference power, the measured SIR, rather than the average SIR, was greater than a specified threshold. Such calculations have been carried out in the literature, and we will have occasion later in this book to study related probabilistic performance measures. These calculations can become quite complex, however, and, at this point in our discussion, we choose instead to focus on a simpler problem: the probability that the received signal power at a mobile exceeds a specified threshold. Using this calculated quantity we can, in turn, determine the fraction of a cell's area that will be expected to receive acceptable service. We use the model of shadow or log-normal fading introduced in Chapter 2 to carry out the desired calculation. We follow the approach introduced in Jakes (1974: 125–128).

Recall that in Chapter 2 we noted that large-scale signal fading experienced over distances of many wavelengths could be modeled as a log-normal process: the received signal in dB varied randomly about the average signal, obeying a gaussian distribution. Figure 2.3 portrays a typical plot of received signal power in dB as a function of transmitter–receiver distance, with power points shown scattered about the average power. These scattered points represent the random fluctuations of the power about the average value. The log-normal distribution of the power is used to model the fact that, due to large-scale fading, actual measurements of power typically scatter about the average power at any distance d from a transmitter. Specifically, repeating (2.4) here, we have the received signal power P_R given by

$$P_R = \alpha^2 10^{x/10} g(d) P_T G_T G_R \tag{3.9}$$

α^2 is the small-scale Rayleigh or Ricean-fading term due to multipath that we will neglect in this discussion. The term $g(d)$ represents the distance-dependent d^{-n} term on which we focused in calculating the SIR in this chapter. The other terms are, of course, the transmitted power and antenna gains at the transmitter and receiver, respectively. It is the function $10^{x/10}$ on which we focus here.

Thus, consider the received power measured in dB. Then we have, neglecting the small-scale fading term, i.e., setting $\alpha^2 = 1$, $P_{R,dB} = x + P_{R,av,dB}$, where $P_{R,av,dB} = 10\log_{10}(g(d)P_T G_T G_R)$. To simplify the notation, let $p \equiv P_{R,dB}$ and $p_{av} \equiv P_{R,av,dB}$. Then the randomly varying received signal power p, measured in dB, is gaussian distributed about its average value p_{av}, also measured in dB. Its probability distribution $f(p)$ is given by, repeating (2.9)

$$f(p) = \frac{e^{-(p-p_{av})^2/2\sigma^2}}{\sqrt{2\sigma^2}} \tag{3.10}$$

The standard deviation σ takes on values in the range 6–10 dB, as noted in Chapter 2. How do we now use this gaussian distributed model for the received signal power, measured in dB? Say the threshold for acceptable received signal power at a mobile is p_0 dB. Then, given p_{av} and σ, both in dB, we can calculate the probability the received signal power is above an acceptable level. This probability, $P[p \geq p_0]$ is, of course, given by the following integral

$$\begin{aligned} P[p \geq p_0] &= \int_{p_0}^{\infty} \frac{e^{-(p-p_{av})^2}}{\sqrt{2\pi\sigma^2}} dp \\ &= \frac{1}{2} - \frac{1}{2}\text{erf}\left(\frac{p_0 - p_{av}}{\sqrt{2}\sigma}\right) \end{aligned} \tag{3.11}$$

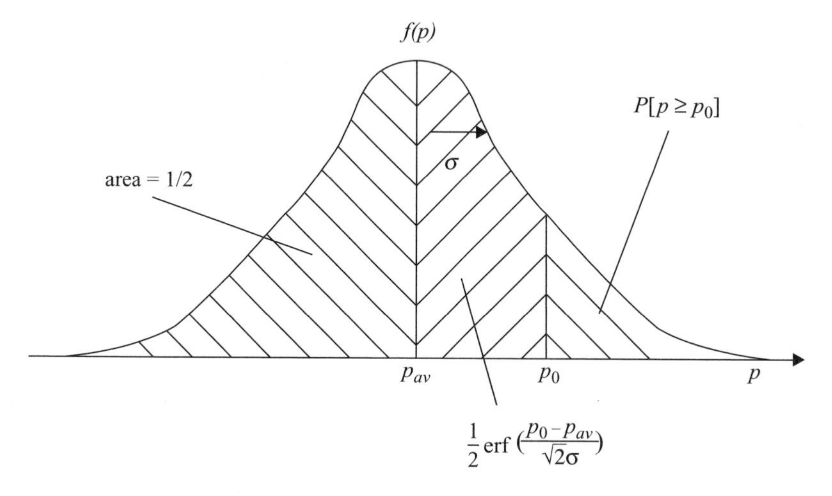

Figure 3.9 Probability received signal (dB) is greater than a threshold

where the error function erf (x) is defined as

$$\text{erf}\,(x) \equiv \frac{2}{\sqrt{\pi}} \int_0^x e^{-x^2} dx \tag{3.12}$$

The error function is tabulated in many books on probability and statistics, and is available as well in most mathematics software packages. Figure. 3.9 displays the error function as well as (3.11) pictorially. Note that (3.11) also represents the probability that a specified power received p_0 is exceeded, given the average received power p_{av} and the standard deviation σ about that value.

As an example, let $p_0 = p_{av}$. Then there is a 50% chance the received signal power is acceptable. Now say the threshold p_0 is 10 dB above p_{av}. For example, the average received power might be 1 mw, while the threshold is set at 10 mw. Let the standard deviation σ of the shadow-fading characteristic be 10 dB as well. Then the probability the threshold power is exceeded is $1/2 - 1/2\ \text{erf}(1/\sqrt{2}) = 0.16$. The probability that the received power is greater than 10 dB *below* the average value p_{av}, again given $\sigma = 10$ dB, is 0.84.

Now consider another application of modeling the shadow-fading effect as log normal (Jakes, 1974). Let the acceptable power level within a cell be p_0 as above. Say this level is set at the average power level at a distance R from the base station. The question now is, what is the fraction of service area within this radius that has acceptable service, i.e., that has a received power greater than or equal to p_0? Note that without shadow fading all of the area within that radius would receive satisfactory service, since the power increases as $(r/R)^{-n}$, $r < R$. Consider a circle of radius R drawn about the base station at the center of the cell. Assume omnidirectional antennas for simplicity. (Clearly, directional antennas with correspondingly higher gain would improve the situation.) Call the fraction of useful service area F_u. Then

$$F_u = \frac{1}{\pi R^2} \int_{over\,\pi R^2} P(p \geq p_0)\, dA \tag{3.13}$$

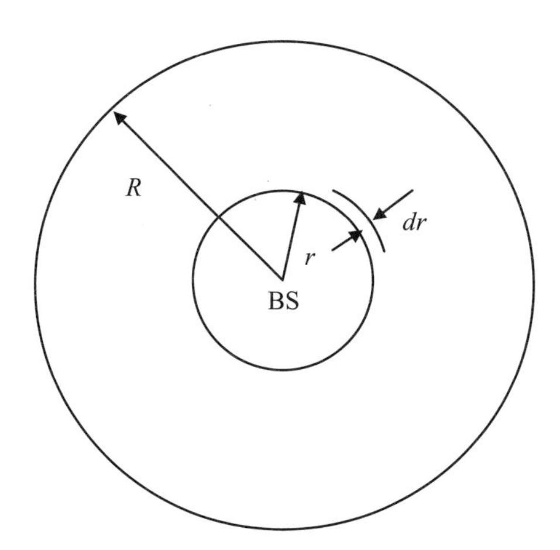

Figure 3.10 Calculation of useful service area

with $dA = 2\pi r dr$. The geometry involved in the calculation of (3.13) is sketched in Fig. 3.10. Since the average received power varies as r^{-n}, p_{av}, the average received power expressed in dB, may be written as $p_{av} = p_0 - 10n\log_{10} r/R$, since we have taken $p_0 = p_{av}$ at $r = R$. Inserting this expression into (3.11) we have

$$P[p \geq p_0] = \frac{1}{2} - \frac{1}{2}\mathrm{erf}\left(\frac{10\log_{10} r/R}{\sqrt{2}\sigma}\right) \tag{3.14}$$

(We assume, as an obvious approximation, that the r^{-n} variation may be taken all the way in to $r = 0$.) Equation (3.13) is then given by

$$F_u = \frac{1}{\pi R^2}\int_0^R 2\pi R\left[\frac{1}{2} - \frac{1}{2}\mathrm{erf}\left(\frac{10n\log_{10} r/R}{\sqrt{2}\sigma}\right)\right]dr \tag{3.15}$$

Simplifying (3.15) by converting $\log_{10}$ to $\log_e$, defining a parameter $b \equiv 10n\log_{10} e/\sqrt{2}\sigma$, defining a variable x by $e^x \equiv (r/R)^b$ or $x = b\log_e(r/R)$, and then replacing the integration variable r by x, we get

$$F_u = \frac{1}{2} + \frac{1}{b}\int_0^\infty e^{-\frac{2x}{b}}\mathrm{erf}(x)\,dx \tag{3.15a}$$

The integration in (3.15a) may be carried out directly by replacing erf(x) by its defining integral (3.12). The resultant double integral may then be partially evaluated by the usual technique of interchanging the order of integration. The final result obtained, after replacing the parameter b by its actual value noted above, is then given by (3.15b) below

$$F_u = \frac{1}{2} + \frac{1}{2}e^{\left(\frac{\sigma}{3n}\right)^2}\left[1 - \mathrm{erf}\left(\frac{\sigma}{3n}\right)\right] \tag{3.15b}$$

(In obtaining the very simple dependence on $\sigma/3n$, we have approximated $10\log_{10} e/\sqrt{2} = 3.13$ by the value 3.)

As an example of the use of this expression, say $n = 3$ (i.e., r^{-3} power variation with distance) and $\sigma = 9$ dB. Then we find $F_u = 0.71$. This says that, in this case, 71% of the area within a radius R of the base station has a signal above the threshold, defined as the average power at R, when half the locations at R (along the circumference) have a received signal above the threshold. If $n = 4$ and $\sigma = 6$ dB, as another example, $\sigma/3n = 0.5$. Then $F_u = 0.83$ (Jakes, 1974: Fig. 2.5–1, p. 127). From the simple form of (3.15b), with dependence on σ/n, it is apparent that if n increases (σ decreases), F_u increases. This variation is as to be expected: with increasing n the received power increases more rapidly as one moves into the base station, increasing the fractional area whose signal is likely to be above a desired threshold. Similarly, decreasing σ, a measure of the shadow-fading variation about the average received power, increases the useful service area coverage.

More generally, let the threshold of acceptable received signal power p_0 be different from the average power $p_{av}(R)$ at a distance R from the base station. Then one finds (Jakes, 1974), defining a parameter $a \equiv [p_0 - p_{av}(R)]/\sqrt{2}\sigma$, that F_u is now given by

$$F_u = \frac{1}{2}\left\{1 - \operatorname{erf}(a) + \exp\left(\frac{1 - 2ab}{b^2}\right)\left[1 - \operatorname{erf}\left(\frac{1 - ab}{b}\right)\right]\right\} \tag{3.16}$$

Here $b = 3n/\sigma$ is the same parameter defined previously.

As the threshold p_0 drops below the average value $p_{av}(R)$, i.e., the parameter a becomes negative, F_u increases. For the probability $P[p_{av}(R) > p_0]$ increases, and, in turn, a higher portion of the area within a radius R about the base station has a received signal power greater than the threshold. There is thus an increased useful service area. As an example, say the threshold p_0 is reduced, so that $P[p_{av}(R) > p_0] = 0.7$, rather than the previous 0.5. Let $n = 3$ and $\sigma = 9$ dB again, as in the example cited above, so that $b = 1$. Then (Jakes, 1974: Fig. 2.5–1) the fraction of the area within a radius R from the base station increases to about 0.84 from the previous value of 0.71.

Problems

3.1 **(a)** Verify the entries in Table 3.1.

(b) Compare with the effect of including second-tier interferers as well. Verify that $SIR_{dB} = 19.6$ dB for the case of $D/R = 6$ and $n = 3$, as indicated in the text.

3.2 **(a)** Consider a hexagonal cell of radius R, as shown in Fig. 3.4. Verify the area of the cell is $3\sqrt{3}R^2/2 = 2.6\,R^2$. Show that, for reuse cluster size C, the center-to-center spacing between the closest interfering cells is given by $D_c/R = \sqrt{3C}$.

(b) Referring to Fig. 3.6, fill in the details of the derivation of (3.4).

3.3 **(a)** Verify the entries of Table 3.3.

(b) Consider the entries in Table 3.3 for the 7-reuse system. Compare these numbers with those obtained using the two other approximations suggested in the text. Can you propose another approximation for the location of the interferers?

3.4 Calculate the worst-case uplink SIR assuming the co-channel interference is caused only by the closest interfering mobiles in radio cells a distance $D = 3.46\ R$ away from the cell. Assume the simplest path-loss model $g(d) = 1/d^4$. Repeat for $n = 3$.

3.5 Calculate the Erlang loads on a system for the following cases:

1 An average call lasts 200 seconds; there are 100 call attempts per minute.

2 There are 400 mobile users in a particular cell. Each user makes a call attempt every 15 minutes, on the average. Each call lasts an average of 3 minutes.

3 The number of users in 2 is increased to 500; each user makes a call attempt every 20 minutes, on the average. Repeat if the average call length doubles to 6 minutes.

3.6 Consider a mobile system supporting 100 channels per cell. A call blocking probability of 1% is desired. Mobile users typically use their cell phones once per 10 minutes, on the average, their calls lasting an average of 10 minutes. Say the system is concentrated in an urban area with a density of 500 cell phones per km^2. Calculate the required cell radius if a hexagonal topology is assumed. Repeat if the mobile users "stay on the line" for 4 minutes, on the average. What happens to the cell size if, in addition, they start using their phones more often?

3.7 The worst-case uplink SIR is to be calculated for a $C = 3$ reuse system, for which path loss may be modeled by

$$\begin{aligned} g(d) &= d^{-1.3} & 0 \le d \le 150\,\text{m} \\ &= 150\,\text{m}^{-1.3}(d/150\,\text{m})^{-3.5} & 150\,\text{m} \le d \end{aligned}$$

(See (2.6a) in Chapter 2.)

Do this for two cases: cell radius = 1000 m and cell radius = 100 m.

3.8 Refer to Section 3.4 of the text. The statement is made that IS-136 or D-AMPS systems capable of handling three times the traffic of the analog AMPS system can cover cells with radii $\sqrt{3} = 1.7$ times as large. Explain. What assumptions are implicit in this statement?

3.9 Consider a GSM system with a reuse parameter $C = 3$. For this system, 330 channels per cell become available. Assume that the allowable Erlang traffic load is about the same number. (The reader is encouraged to consult tables or graphs of the Erlang-B formula to determine the resultant call blocking probability.) Assuming a hexagonal cell model, calculate the allowable cell radius for a suburban region with a mobile density of 200 mobiles/km^2. Repeat for a system with a reuse parameter of 4. How do these results change for an urban region with a mobile density of 1000 mobiles/km^2? In all cases assume a typical user makes a call once per 15 minutes on the average, the call lasting, on the average, 200 sec.

3.10 Refer to tables of the error function defined by (3.12). Equation (3.15b) represents the fraction of acceptable service area within a cell assuming shadow fading with log-normal variance σ^2 and power-variation parameter n. Show this fraction is 0.71 for $\sigma = 9$ dB and $n = 3$. Show the fraction increases to 0.83 for $\sigma = 6$ dB and $n = 4$. Try various other combinations of σ and n, and repeat.

Chapter 4

Dynamic channel allocation and power control

In Chapter 3 we discussed channel allocation in a cellular environment with reuse constraints built in to keep interference to a tolerable level. The reuse constraints reduce the number of channels that can be allocated to each cell, reducing thereby the improvement in system traffic capacity expected through the use of the cell concept. Various strategies have been proposed and/or adopted to obtain further improvement in system performance. The use of directional antennas to reduce the number of interfering signals with which a desired signal has to contend is one possibility (Stüber, 2001). This procedure has, in fact, been adopted in current cellular systems. Another approach is that of reducing cell size, hence gaining more cells in a geographic region, with consequent increase in capacity, with a given reuse characteristic included. Cellular systems are, in fact, moving toward a hierarchy of cell sizes: larger cells, called *macrocells*; *microcells*, to be used, ideally, in the more crowded urban environments where the traffic load does dictate reducing the cell size (base station antennas and their attendant transmitter–receiver systems may be located at the reduced height of street lamp posts); and *picocells*, to be used principally in indoor cellular systems. One problem with the microcell approach is that handoffs, to be discussed in Chapter 9, become more frequent, increasing the need to handle handoff calls more effectively. Macrocell overlays on microcellular systems have been proposed as well to improve traffic-handling capability (Stüber, 2001).

Other methods used to improve system performance include *dynamic channel allocation* (DCA) strategies for reducing call blocking, hence increasing traffic throughput, as well as power control for reducing interference. (DCA is sometimes referred to as *dynamic channel assignment*.) We discuss both of these techniques in this chapter. Some simulation studies have shown that the combination of DCA and power control can provide substantially greater improvements in system performance than the use of either technique alone (Lozano and Cox, 1999).

Stüber, G. L. 2001. *Principles of Mobile Communications*, 2nd edn, Boston, MA, Kluwer Academic.

Lozano, A. and D. C. Cox. 1999. "Integrated dynamic channel assignment and power control in TDMA wireless communication systems," *IEEE Journal on Selected Areas in Communications*, 17, 11 (November), 2031–2040.

We provide an introduction to DCA in Section 4.1 following, focusing on one allocation strategy only to simplify the discussion. Simple versions of DCA were adopted in a number of so-called *low-tier* wireless systems. These systems, such as DECT, the Digital European Cordless Telephone system and the Japanese Personal Handyphone System (PHS), are low-power wireless systems designed to handle pedestrian traffic principally, as contrasted to the *high-tier* second-generation mobile systems we describe later in this book (Cox, 1995; Goodman, 1997; Rappaport, 1996). Third-generation mobile systems discussed in Chapter 10, as well as mobile systems being considered for future use, may be expected to use DCA techniques to a greater extent. Power control in mobile wireless systems is discussed in Section 4.2. We provide some examples of power control algorithms, again to motivate the subject and to provide concrete examples of the control procedures involved. We have already shown the need to control transmission power to maintain a specified SIR in Chapter 3. We shall find, in discussing CDMA systems in later chapters, that power control is an absolute requirement in those systems.

4.1 Dynamic channel allocation

The allocation of channels to cells described in the previous chapter has implicitly assumed *fixed channel allocation* (FCA): groups of channels are assigned permanently to given cells following a prescribed reuse pattern. (Recall from the previous chapter that, with the reuse strategy described there, the total available set of channels is divided into C groups, each group then being assigned to a specific cell in a cluster of C cells.) DCA refers to a variety of techniques in which channels may be allocated to users in a cell in accordance with varying traffic demands. The techniques could range from one in which no permanent assignments are made, all channels being kept in a pool, to be assigned to a cell as the need arises, to techniques in which channels are nominally assigned to cells, but with borrowing by other cells of currently unused channels allowed, if the need arises. Reuse constraints still have to be respected, however. Borrowing strategies can thus become complex as the reuse region changes. The possibility of using DCA to improve system capacity was recognized in early studies of cellular systems (Cox and Reudink, 1972, 1973). Many algorithms have been proposed for DCA in the years following. Each has its particular attributes, whether it be simplicity of use or improved system performance under overload traffic conditions. Overload refers to traffic intensity greater than the traffic load for which the system has nominally been designed. This could be due to a larger number of users than expected or to increased usage (users calling more often or making longer calls) or both. Recall from the previous chapter that we determined the nominal traffic load to be the one providing the desired cell blocking probability, given the number of channels assigned to a cell. As the load increases beyond this value, overloading the

Cox, D. C. 1995. "Wireless personal communications: what is it?," *IEEE Personal Communications*, 2, 2, 20–35.
Goodman, D. J. 1997. *Wireless Personal Communication Systems*, Reading, MA, Addison-Wesley.
Rappaport, T. S. 1996. *Wireless Communications, Principles and Practice*, Upper Saddle River, NJ, Prentice-Hall PTR.
Cox, D. C. and D. O. Reudink. 1972. "Dynamic channel assignment in two-dimensional large-scale mobile communication systems," *Bell System Technical Journal*, 51, 1611–1672.
Cox, D. C. and D. O. Reudink. 1973. "Increasing channel occupancy in large-scale mobile radio systems: dynamic channel assignment," *IEEE Transactions on Vehicular Technology*, VT-22, 218–222.

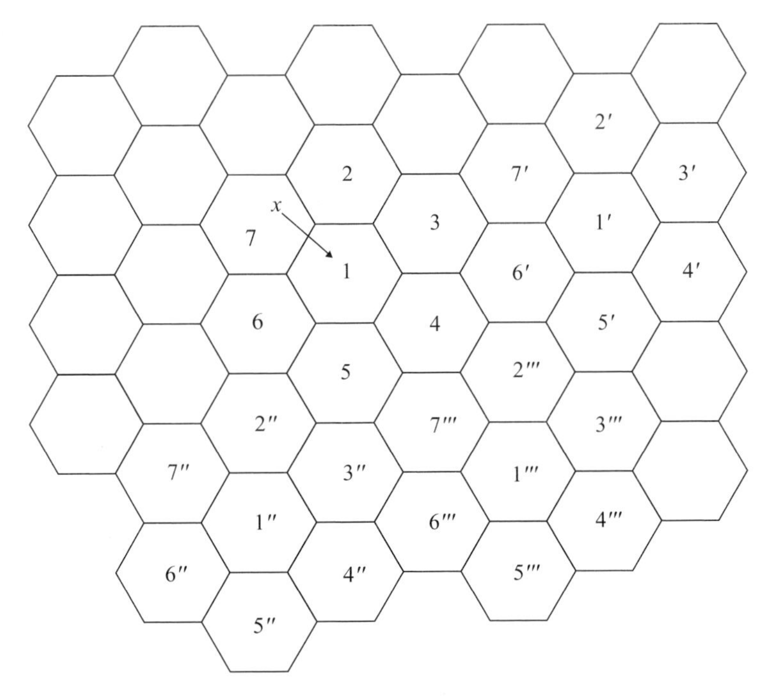

Figure 4.1 Channel locking

system more, the call blocking probability increases. A channel allocation scheme is said to be better than another in a given load region if its blocking-probability load curve is below the other's curve.

An overview of DCA, with qualitative comparisons made between different algorithms proposed, appears in Katzela and Naghshineh (1996). As noted above, DCA introduces flexibility to the process of channel assignment. Different DCA schemes differ in the channel assignment rules. It is to be noted that all DCA schemes out-perform FCA at low and moderate traffic conditions. At very heavy loads, however, FCA is found to perform better since, at these volumes of traffic, relatively few channels are normally free for assignment to a new call. In addition, so-called channel locking, to be described below, exacerbates the situation, leading to performance degradation. Hybrid schemes have been proposed which convert to FCA at high loads.

In implementing DCA, channels can be borrowed by a cell, when needed, from neighboring cells. But with the borrowing of channels, other cells are locked out from using these channels because of reuse constraints. Borrowing in DCA thus gives rise to channel locking. At moderate traffic loads, this poses no problem and DCA serves to accommodate fluctuations in traffic from cell to cell. In heavy traffic, however, locking in turn causes cells to borrow even more, ultimately reducing the performance rather than improving it. Channel locking can be described by reference to Fig. 4.1. Say channels are assigned to clusters of seven cells as shown. Channels may not be reused by neighboring cells, but

Katzela, I. and M. Naghshineh. 1996. "Channel assignment schemes: a comprehensive survey," *IEEE Personal Communications*, 3, 3 (June), 10–31.

may be used one step further away. Consider a DCA scheme in which channel x in cell 7 is borrowed by cell 1 as shown. Then channel x is locked out from all cells surrounding cell 1, as well as cells $7'$ and $7'''$. It cannot be borrowed by cells surrounding cells $7'$ and $7'''$ as well. This is the channel-locking problem, due to reuse constraints, that, as noted above, degrades the performance of DCA at high loads.

But note that this simple borrowing strategy is too stringent. It is possible to introduce *directional locking* to alleviate some of this problem. Note with this same example, referring again to Fig. 4.1, that cells on the far side of cell $7'$, away from cell 1, such as cells $2'$ and $1'$, could borrow channel x from cell $7'$ without violating the reuse condition. The use of directional locking does in fact improve the DCA performance.

As noted above, a variety of DCA techniques have been proposed in the literature (Katzela and Naghshineh, 1996). These include strategies of preferred cells from which to borrow; different channel assignment rules; assignment of some channels to individual cells, with the rest pooled to be assigned freely when needed; using different borrowing rules in the inner and outer portions of cell, etc. Rather than study each of these exhaustively, however, in this introductory treatment we choose to focus on one DCA strategy that has been shown, by both analysis and simulation, to provide performance superior to a number of other algorithms proposed in the literature. The interested reader can then peruse the literature for other proposed DCA strategies. The scheme we describe here, the *BDCL* DCA scheme proposed by Zhang and Yum (1989), is a scheme for which analysis is possible, in contrast to simulation alone, making it easier to explain the operation of the algorithm. The acronym BDCL stands for *b*orrowing with *d*irectional *c*hannel *l*ocking, so that the scheme uses directional locking described above to obtain performance improvement. The complete scheme is defined by three specific features: specification of lock directions, channel ordering, and immediate channel reallocation. Under *channel ordering*, channels assigned to a cell are numbered. The lowest-numbered channel has the highest priority to be assigned to the next local call, i.e., to a call associated with this particular cell. The highest-numbered channel, on the other hand, has the highest priority to be borrowed by neighboring cells. Local calls are thus assigned channels in order from the lowest-numbered channel; channels to be borrowed start with the highest number and work their way down. *Immediate channel reallocation* is a strategy designed to pack channels used by local calls closely together; borrowed calls are packed together as well. This feature incorporates four parts: 1. Packing of local-call ordering – when a call on a channel terminates and there is a call ongoing on a higher channel, the latter is moved to the (lower) channel that has been released. 2. When a local call using a locally assigned (nominal) channel terminates and there is another local call ongoing, using a borrowed channel, the latter call is moved to the nominal channel and the borrowed channel is released. 3. Packing of borrowed channels – when a borrowed channel call terminates, releasing a borrowed channel, and there is a borrowed-channel call ongoing using a lower-numbered channel, the latter is switched to the higher (released) channel 4. When a channel is unlocked by termination of a call in an interfering cell, any call on a

---

Zhang, M. and T.-S. P. Yum. 1989. "Comparisons of channel assignment strategies in cellular telephone systems," *IEEE Transactions on Vehicular Technology*, 38, 4 (November), 211–218.

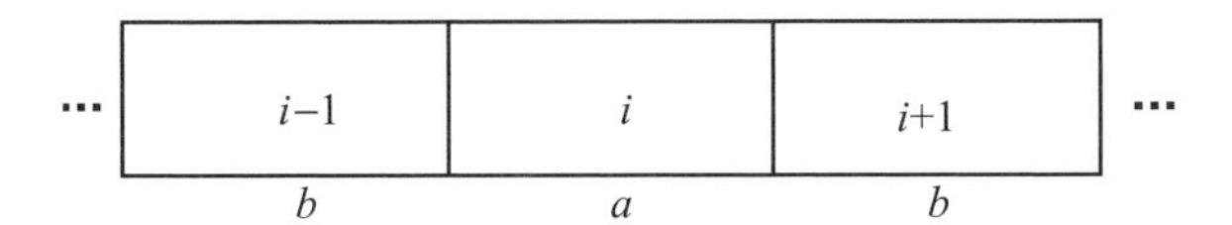

Figure 4.2 Channel assignment in one-dimensional cellular system

borrowed channel or a higher-order channel is switched to this channel. A performance study shows that this scheme performs well compared with FCA at moderate loads and provides an improvement, as expected, to a scheme not using directional locking (Zhang and Yum, 1989).

We focus, in the paragraphs following, on a one-dimensional version of this scheme and show how one can carry out an approximate performance analysis. We follow the approach of Yeung and Yum (1995). Say there are a total of $2m$ channels available in the system and assign them m at a time to alternate cells. Call the first set of m channels the a set; the second set is called the b set. Fig. 4.2 shows the assignment to cells $i-1$, i, and $i+1$ of the complete one-dimensional set. Say all m channels in cell i are occupied. Under FCA, if a new call arrives it will be blocked. The blocking probability is then given by equation (3.7) in Chapter 3, with $N=m$. With the DCA strategy, i may borrow a channel from $i-1$ or from $i+1$, if available. If it borrows a channel from $i-1$ $(i+1)$, however, the corresponding channel in $i+1$ $(i-1)$ is then locked. Under this DCA strategy with directional locking, blocking of calls at cell i occurs if all $2m$ channels covering this three-cell system are in use. For i can then not borrow a channel from a neighbor, even if available, since that channel would be in use in the other neighboring cell. This comment is expanded on further below.

The approximate analysis of this scheme and a comparison with FCA is made possible by the observation by Yeung and Yum that the three cells in Fig. 4.2 (hence any three neighboring cells in the one-dimensional system) may be decoupled from the system and treated in isolation. The subsequent analysis of the three-cell system is then shown to lead to a (tight) upper bound on the blocking-probability performance.

This observation is developed as follows (Yeung and Yum, 1995). Say a typical cell i has lent one or more channels to its neighbors. It is then readily shown that it either will not, or cannot, borrow from its neighbors. In particular, say cell i has lent k channels, numbered $m-k+1$ to m, to cell $i+1$. (Recall that cell is nominal set of channels a is numbered from 1 to m.) The implication then is that cell $i+1$ already had m calls in progress, i.e., its full complement of m assigned channels occupied. Cell $i+1$ thus has $m+k$ calls in progress. Say a call arrives at cell i. There are two possibilities: 1. The number of calls ongoing in i is less than $m-k$, in which case i can handle the call directly. 2. Cell i has $m-k$ calls in progress. In this case, there are $2m$ channels occupied in cells i and $i+1$. Therefore cell i cannot borrow a channel from cell $i-1$ and the new call is blocked. As a result of this analysis, borrowing of channels does not propagate. Hence the blocking-probability calculation of the three-cell group may be carried out by decoupling the group

Yeung, K. L. and T.-S. P. Yum. 1995. "Cell group decoupling analysis of a dynamic allocation strategy in linear microcell radio systems," *IEEE Transactions on Communications*, 43, 2/3/4 (February–April), 1289–1292.

from the rest of the one-dimensional network and treating it in isolation. Yeung and Yum then show the analysis of the blocking-probability performance of the three-cell group in isolation leads to a tight upper bound on the blocking probability. We follow their approach and work out some simple examples, determining blocking probability as a function of load. These results can then be compared with the equivalent FCA performance, showing, as noted earlier, that DCA does provide a performance improvement at low to moderate loads, performing more poorly than FCA at high loads, however.

The blocking-probability analysis proceeds as follows: we focus on the three-cell group of Fig. 4.2. Assume again that there are $2m$ channels available in the one-dimensional system. m of these are assigned to cell i; the other m channels are assigned to neighboring cells $i-1$ and $i+1$. Because of borrowing, cell i can, however, be using up to $2m$ channels. If all $2m$ channels are assigned to it, both cells $i-1$ and $i+1$ have no channels available because of locking. In the other extreme case, $i-1$ and $i+1$ can each have $2m$ channels under use at the same time, through borrowing, since reuse is obeyed. Cell i will then have no channels available for use. (Even though we focus on the three-cell group for analysis, the two cells $i-1$ and $i+1$ can each borrow from its left-, and right-hand neighbor, respectively.) Cell i will thus have 0 channels available when both $i-1$ and $i+1$ have $2m$ calls ongoing (hence $2m$ channels occupied) at the same time. But these channels duplicate one another. It is thus apparent that the channel *numbers* occupied by this three-cell group can never exceed the maximum value of $2m$. A call arriving at i will be blocked if and only if all $2m$ channels assigned to the group are occupied. Let x_1 be the number of channels occupied (carrying ongoing calls) in cell i, x_2 the number of channels occupied in cell $i+1$, and x_3 the number occupied in cell $i-1$. We also call these numbers the states of each cell. The triplet (x_1, x_2, x_3) represents the vector of states (number of channels occupied) of the three-cell group. The set of possible states is then represented by the three-dimensional diagram of Fig. 4.3, which takes into account the possibility of channel borrowing, with directional locking incorporated. Note, in particular, that the plane cut at $x_1 = x$ has the rectangular shape shown with the four boundary points given by $(x, 0, 2m-x)$; $(x, 0, 0)$; $(x, 2m-x, 0)$; and $(x, 2m-x, 2m-x)$. These points are so labeled in the figure.

How many possible states are there in total? We count these by varying x_1 from $2m$, the maximum point shown in Fig. 4.3, with one state only possible here, as shown, to its minimum value of 0. At $x_1 = 2m-1$, there are $2^2 = 4$ states; at $x_1 = 2m-2$, there are $3^2 = 9$ states. Continuing to $x_1 = 0$, at which there are $(2m+1)^2$ states, the total count of possible states is given by

$$\sum_{k=1}^{2m+1} k^2 = (2m+1)(2m+2)[2(2m+1)+1]/6$$

For example, let $m = 2$. The total number of channels available is $2m = 4$, and the total number of possible states is 55. For $m = 1$, with $2m = 2$ channels available in the system, there is a total of 14 such states. Of these possible numbers of states, some are blocking states, i.e., states for which a channel is not available to handle an additional call request in a particular cell, either at the cell itself, or through borrowing from neighboring cells.

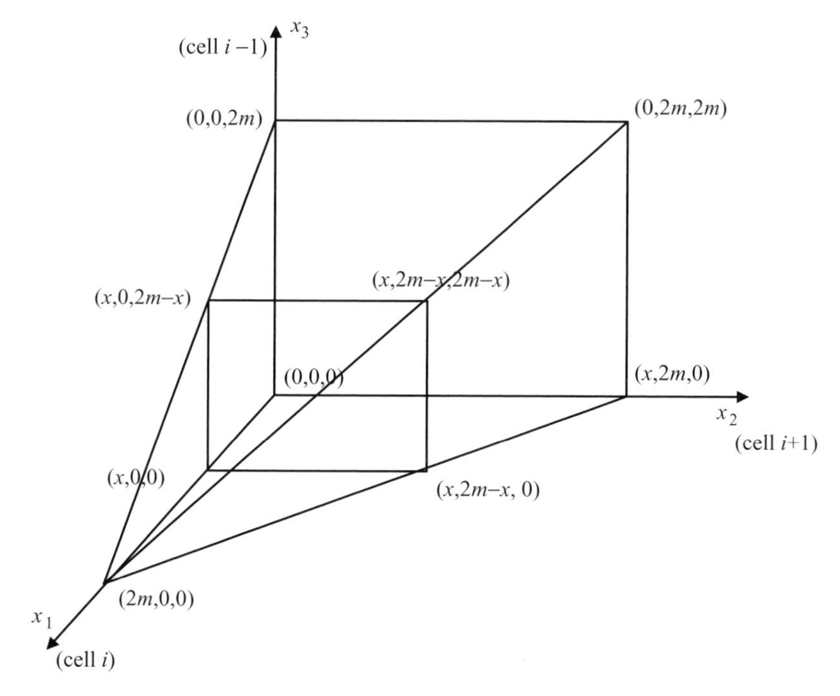

Figure 4.3 Possible states, one-dimensional BDCL strategy (from Yeung and Yum, 1995)

For example, say a call request arrives at cell i. This call can only be accepted if cell i has one or more channels free, or, if none is available in cell i, one can be borrowed from either cell $i-1$ or cell $i+1$. If no channel is available, the call is blocked. Blocking occurs when all $2m$ channels in the three-cell group are in use, as noted earlier.

Consider first the simplest case of two channels available in the one-dimensional system. Hence $m = 1$ channel is initially assigned to each cell. There are 14 states possible in this case, as has already been noted. Of these 14 states, we can readily show that nine are blocking states. Call the two channels available 1 and 2. Let channel 1 be assigned to cell i, channel 2 to cells $i-1$ and $i+1$. (Channels 1 and 2 then correspond, respectively, to sets a and b in Fig. 4.2.) Clearly, if each cell has one call active, any new call arriving at cell i will be blocked. This corresponds to the vector state (1, 1, 1). The other blocking states may be determined by considering, in order, $x_1 = 0$, 1, and 2, its maximum value, and finding the corresponding values of x_2 and x_3 that result in blocking. The nine states found are tabulated in Table 4.1, with a brief note appending each. The specific channels occupied in each cell are indicated in parentheses.

Summing over the probability of being in each of these states, we find the blocking probability B_i of cell i. More generally, it can be shown that, for a total of $2m$ channels, the blocking probability of cell i for the one-dimensional BDCL algorithm is given by Yeung and Yum (1995)

$$B_i = \sum_{x_1+x_3 \leq 2m} P[x_1, 2m - x_1, x_3] + \sum_{x_1+x_2 < 2m} P[x_1, x_2, 2m - x_1] \qquad (4.1)$$

Table 4.1 *Blocking states, cell i, BDCL algorithm, m = 1 channel per cell*

x_1, cell i	x_2, cell $i+1$	x_3, cell $i-1$	
0	0	2(1,2)	Here cell $i-1$ has borrowed one channel, 1, from cell i. Even though cell $i+1$ has no active call using its channel 2, cell i cannot use it.
0	2(1,2)	0	Same comment as above, by symmetry.
0	2(1,2)	2(1,2)	$i-1$ and $i+1$ can each have $2m = 2$ calls in progress since this doesn't violate reuse.
0	1(2)	2(1,2)	cell $i-1$ has borrowed ch. 1 from cell i.
0	2(1,2)	1(1)	Same comment as above, by symmetry.
1(1)	0	1(2)	Cell i cannot borrow $i+1$'s inactive ch. 2.
1(1)	1(2)	0	Same as above, by symmetry.
1(1)	1(2)	1(2)	Discussed in text.
2(1,2)	0	0	Cell i has borrowed ch. 2 from one of its neighbors, locking the use of that channel by the other neighbor.

Here $P[x_1, x_2, x_3]$ is the probability that the system is in state (x_1, x_2, x_3). (We drop the designation *vector state* for simplicity, and use just *state*. It should be apparent from the context as to whether we are referring to the state of one cell or to all three cells.) It is left for the reader to show that, with $m = 1$, the states over which (4.1) is summed are precisely those indicated in Table 4.1.

How does one now calculate the probability $P[x_1, x_2, x_3]$ to determine the blocking probability B_i from (4.1)? Assume, as in the fixed channel allocation (FCA) case considered in Chapter 3, that call requests arrive at cell i at a Poisson rate λ_i, with an average exponentially distributed holding time $1/\mu_i$. The Erlang load at cell i is then given by $A_i = \lambda_i/\mu_i$. Similarly, let the call requests at cells $i+1$ and $i-1$ arrive, respectively, at Poisson rates λ_{i+1} and λ_{i-1}. The call holding times at cells $i+1$ and $i-1$ are similarly considered to be exponentially distributed with average values $1/\mu_{i+1}$ and $1/\mu_{i-1}$, respectively. We will be assuming the homogeneous traffic case, so will henceforth take all the call holding times to have the same average value $1/\mu$. The Poisson arrival and exponential holding time assumptions define a set of transitions between the various states of Fig. 4.3. For example, say the three-cell system i, $i+1$, and $i-1$ is at state (x_1, x_2, x_3). Say a call arrives (a request is generated) at cell i. Then, if the call request can be accepted, the state moves to state $(x_1 + 1, x_2, x_3)$. If a call arrives at cell $i+1$, the state moves to state $(x_1, x_2 + 1, x_3)$. On the other hand, if a call completes at cell i, the state changes to state $(x_1 - 1, x_2, x_3)$. We have indicated above that the average rate of call arrivals (requests) at cell i is just λ_i. The rate of completion of a call is given by μ. Using these call arrival and call completion rates, one may construct a state-transition diagram showing the rate at which various states are traversed. Such a diagram for the states traversed in the case of the three decoupled cells i, $i+1$, and $i-1$, appears as Fig. 4.4. One can use this diagram to calculate the steady-state probability $P[x_1, x_2, x_3]$ of being in state (x_1, x_2, x_3), and, hence,

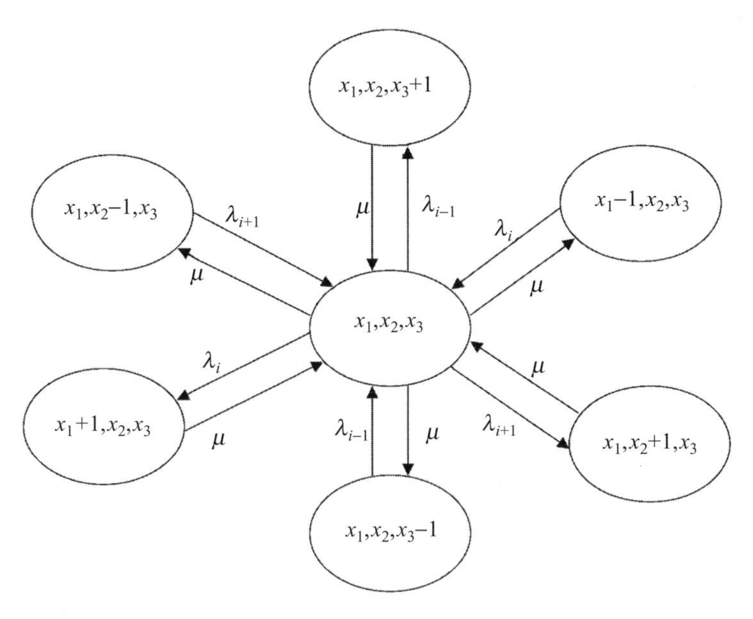

Figure 4.4 State-transition diagram, three decoupled cells (from Yeung and Yum, 1995: Fig. 3)

using (4.1), the blocking probability B_i at cell i. Note, however, that the call arrival rates λ_{i+1} and λ_{i-1} at the cells neighboring cell i are actually smaller in the real, non-decoupled system than shown in Fig. 4.4, since cell i can use certain channels which appear blocked in the decoupled system. As an example, say channel k of cell $i+1$ is lent to cell $i+2$, the non-interfering co-channel of cell i. Hence cell $i+1$ cannot use this channel, but cell i *can* use it. This reduces the actual blocking probability at cell i. B_i as calculated using (4.1) is thus an *upper* (conservative) bound on the actual blocking probability.

With the Poisson call arrival– exponential holding time assumptions made in describing the call arrival traffic in our model of calls in the cellular environment (this is the same model used universally for telephone traffic), the state-transition diagram of Fig. 4.4, and Fig. 4.3, on which it is based, describes a special case of transitions in a Markov chain, a subject which has been well-studied for many years. In particular, it can be shown (Yeung and Yum, 1995) that, in this case, the desired steady-state probability $P[x_1, x_2, x_3]$ is of the *product-form* type (Schwartz, 1987)

$$P[x_1, x_2, x_3] = G^{-1} \prod_{j=1}^{3} A_j^{x_j} / x_j!, \quad A_j \equiv \lambda_j / \mu_j \tag{4.2}$$

The term *product-form* is used to indicate the fact that the probability appears as the product of a number of terms, in this case three. Thus, even though the state occupancies x_1, x_2, x_3 of the three cells i, $i-1$, and $i+1$ are clearly coupled, the probability of state occupancy acts as if they are independent. They are not truly independent, however, since they are coupled in (4.2) by the term G^{-1}, required to ensure the probability is normalized

Schwartz, M. 1987. *Telecommunication Networks: Protocols, Modeling, and Analysis*, Reading, MA, Addison-Wesley.

to 1. Thus, the quantity G is obtained by normalizing $P[x_1, x_2, x_3]$ over all the possible states

$$\sum_{allstates} P[x_1, x_2, x_3] = 1 \tag{4.3}$$

Hence it is clear that the quantity G in (4.2) is given by

$$G = \sum_{allstates} \left(\prod_{j=1}^{3} A_j^{x_j} / x_j! \right) \tag{4.4}$$

To show the distinction between the product-form solution applicable here and truly independent probabilities, consider the case of fixed channel allocation (FCA) for the same three cells. In this case, we have, explicitly

$$P[x_1, x_2, x_3] = P(x_1)P(x_2)P(x_3) \tag{4.5}$$

with

$$P(x_j) = \frac{A_j^{x_j}}{x_j!} \Bigg/ \sum_{x_j=1}^{m} \frac{A_j^{x_j}}{x_j!} \tag{4.6}$$

Equation (4.6) is readily obtained from elementary queueing theory for the Poisson arrival–exponential holding time model (Schwartz, 1987). It is readily shown, by summing over all values of x_j, that (4.6) is properly normalized to 1. Note that, with $x_j = m$, one obtains the Erlang-B blocking formula stated as equation (3.7) in Chapter 3. We shall use (3.7) shortly in comparing DCA with FCA. (Further discussion of Poisson arrivals and the Erlang formula appears in Chapter 9.)

Given (4.2) and (4.4), one can now calculate the blocking probability B_i of cell i for this one-dimensional case of the BDCL dynamic channel allocation scheme on which we have focused in this chapter. (Recall again our comment, however, that the use of (4.2) and (4.3) provides an upper bound on B_i.) Given B_i, the probability of blocking B of the entire N-cell system is given by the weighted sum

$$B = \sum_{i=1}^{N} \lambda_i B_i \Bigg/ \sum_{i=1}^{N} \lambda_i \tag{4.7}$$

In the homogeneous case with which we will be dealing, i.e., the uniform traffic case, with $\lambda_i = \lambda$, and $\mu_i = \mu$, $B = B_i$, as given by (4.1). We consider here the two special cases, $m = 1$ ($2m = 2$ channels are thus available in the system) and $m = 2$ ($2m = 4$). (Much larger examples are presented in the reference on which the material presented here is based – Yeung and Yum, 1995). For the homogeneous case, with the average traffic arrival rate λ the same in each cell, we have the Erlang load A_j of (4.2) just equal to A. Consider first the case $m = 1$ channel allotted to each cell. It is then readily shown, using (4.2) in (4.1), that the blocking probability B is given by the following expression

$$B = \frac{\frac{7}{2}A^2 + 2A^3 + \frac{A^4}{4}}{1 + 3A + \frac{9}{2}A^2 + 2A^3 + \frac{A^4}{4}} \tag{4.8}$$

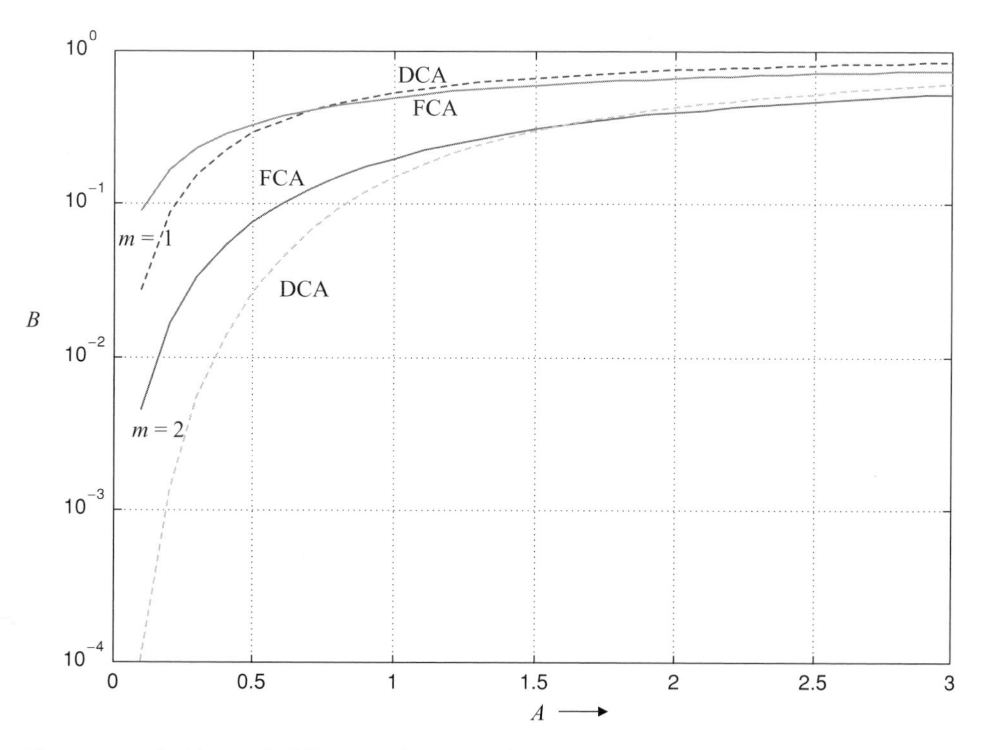

Figure 4.5 Blocking probability, one-dimensional BDCL scheme, compared with FCS

(The numerator of (4.8) is obtained by summing the numerator product terms of (4.2) over the nine blocking states of Table 4.1 and simplifying the resultant sum. The denominator corresponds to calculating G in (4.2): the sum of product terms over the same nine blocking states plus the five non-blocking states.) Equation (4.8) is plotted as a function of the Erlang load A in Fig. 4.5. Plotted on the same figure, for comparison, is the blocking-probability performance curve for the equivalent 2-channel ($m = 1$) FCA scheme, with one channel of the two assigned alternately to neighboring cells. The blocking probability in this case of $m = 1$ channel per cell is given simply by

$$B = A/(1 + A) \tag{4.9}$$

using (3.7), or (4.6) with $x_j = m = 1$. Note that the DCA strategy in this simplest possible case does provide some blocking-probability improvement over FCA at loads below $A = 0.75$. Above this value of load, FCA provides slightly better performance, as expected. Also plotted on Fig. 4.5 are the comparative performance curves for DCA and FCA, for the case of $m = 2$ channels per cell, or a total of four channels for the system. Details of the DCA calculation, using (4.1) to (4.4), are left to the reader. Recall from our discussion of Fig. 4.3 earlier that there are 55 active states to consider in this second example of the linear one-dimensional BDCL strategy. Of these 55 states, 25 are found to be blocking states; the other 30 states are non-blocking. An expression similar to that of (4.8) can be calculated for the blocking probability in this case, again containing the sum of the product-form terms over the blocking states, obtained from (4.2), in the numerator, while the sum of all the product-form terms, taken over all 55 states, appears in the denominator. The resultant performance curve in this case, plotting the blocking probability B vs the load A, appears in Fig. 4.5. The equivalent performance curve for the $m = 2$ FCA case is plotted as well. This curve is a plot of the expression

$$B = A^2/2/(1 + A + A^2/2) \tag{4.10}$$

again obtained using (3.7) or (4.6) with $x_j = m = 2$. Note that in this case the DCA algorithm provides considerable improvement over the equivalent FCA scheme for Erlang loads less than $A = 2$. Above this value of load FCA provides better performance. These results are in agreement with our comment earlier that DCA schemes are found to provide performance improvement over FCA in low to moderate traffic cases. In the high traffic cases, FCA is found to perform better.

Note also, from these two simple cases considered in Fig. 4.5, that the range of loads over which the DCA advantage is manifested increases with increasing number of channels. This is substantiated by the example calculated in Yeung and Yum (1995), with $2m = 20$ channels considered. The nominal load assumed in that paper is $\lambda = 100$ calls/hr. With an average call holding time of $1/\mu = 3$ min., the Erlang load is $A = 5$ Erlangs. At this value of load, FCA provides a blocking probability of 0.12. The BDCL strategy is found to provide a blocking probability about an order of magnitude less. The difference between these two schemes remains about the same for load increases up to 50% above the nominal load. The performance of the two schemes would presumably converge, with FCA eventually taking over, at much higher traffic loads. Note also, as expected, that, with 20 channels available in this system, much higher loads can be accommodated than in the almost trivial cases we have considered here, with just two and four channels available.

We have focused here, as noted earlier, on one particular DCA algorithm only for simplicity and to keep from bogging down in detailed explanations of various DCA strategies. The discussion, using this example of a DCA algorithm, demonstrated as well how the use of DCA improves the blocking performance over the FCA strategy in the low-to-moderate traffic region. The questions arise, however, as to how general the performance improvement is, and as to other improvements possible using DCA algorithms. The performance improvement over FCA in low-to-moderate traffic cases has been demonstrated through the use of simulation for a variety of DCA algorithms. In particular, substantial improvements in the system performance of GSM through the proposed introduction of DCA have been demonstrated through simulation (Priscoll *et al.*, 1997; Le Bris and Robison, 1999). We recall from Chapter 1 that GSM is one of the three second-generation cellular systems successfully adopted worldwide. GSM will be described in detail in Chapters 6 and 8. The simulation studies appearing in Priscoll *et al.* (1997) indicate that the introduction of DCA could increase the number of users accommodated by a GSM network by as much as 40–50% over FCA for blocking probabilities in the range of 1%. The studies

Priscoll, F. Della *et al.* 1997. "Application of dynamic channel allocation strategies to the GSM cellular network," *IEEE Journal on Selected Areas in Communications*, 15, 8 (October), 1558–1567.

summarized in Le Bris and Robison (1999) focus on the error rate experienced by users in a Rayleigh fading environment with a traffic load of 70%, i.e., 70% of the channels busy. Their results indicate that the use of a DCA technique improves the error rate by 25% in the uplink direction and by 15% downlink. (The concept of error rate is summarized briefly in Chapter 6.)

Other examples of the improvement due to the introduction of DCA appear in Katzela and Naghshineh (1996) and in the references cited there. Simulation has generally been used to demonstrate the improvement possible. More generally, however, the performance advantages of DCA have been demonstrated by the use of a bounding technique, studying the performance of an idealized form of DCA which provides close to optimum blocking-probability performance. This idealized version of DCA has been labeled a "maximum packing" policy and its performance evaluated by Everitt and Manfield (1989) using Monte-Carlo simulation and analytically, for some simple cellular structures, by Raymond (1991).

With the maximum packing strategy a newly arriving call is always assigned a channel so long as one is available, if necessary by rearranging the channel assignments of ongoing calls. The only stipulation is that reuse constraints must be satisfied. Everitt and Manfield have used as their model for comparing maximum packing with FCA a 350-channel, 49-cell hexagonal structure with 7-cell reuse. In the case of FCA 50 channels are thus made available for each cell. They then find that the maximum packing strategy, i.e., DCA, provides an order of magnitude reduction in blocking probability over FCA when the traffic load is such as to produce blocking probabilities of 2% and 5% using FCA. This result agrees with the result cited above for the (smaller) system analyzed in Yeung and Yum (1995). More specifically, in their example, FCA can accommodate a traffic load of 40.3 Erlangs/cell, on the average, if the blocking-probability is to be kept to 2%, while the average traffic load rises to 44.5 Erlangs/cell allowable if the acceptable blocking-probability is increased to 5%. For these same traffic loads the blocking probabilities using maximum packing are a magnitude lower. Alternatively, the maximum packing strategy or DCA can accommodate 47.3 Erlangs/cell, on the average, for the 2% blocking-probability case and 50.7 Erlangs/cell, on the average, for 5% blocking. The allowable traffic is thus increased by 15% or more through the use of maximum packing.

Everitt and Manfield have also studied the effect of traffic variability on the comparative performance of the maximum packing strategy and FCA. They do this by modeling the traffic in each cell as a normally distributed random variable with the average traffic given by the numbers above, and with the variance increasing to model increasing variability. The results obtained indicate that DCA blocking performance is much less sensitive to traffic variability than in the FCA case. Raymond's analytical performance results, using a linear cellular model, are similar for the average traffic case. He also finds a blocking-probability

---

Le Bris, L. and W. Robison. 1999. "Dynamic channel assignment in GSM networks," Proc. VTC'99, IEEE Vehicular Technology Conference, Amsterdam, September, 2339–2342.

Everitt, D. and D. Manfield. 1989. "Performance analysis of cellular mobile communication systems with dynamic channel assignment," *IEEE Journal on Selected Areas in Communications*, 7, 8 (October), 1172–1180.

Raymond, P.-A. 1991. "Performance analysis of cellular networks," *IEEE Transactions on Communications*, 37, 12 (December), 1787–1793.

reduction of a magnitude or so through the use of maximum packing/DCA for a nominal traffic load resulting in a blocking probability, using FCA, of 1–2%. This improvement in blocking-probability performance reduces as the average traffic increases, both DCA and FCA providing about the same performance at triple the average load, for which the blocking probability has risen to about 40% or so. This value clearly corresponds to the high-traffic case. Both studies of this idealized version of DCA thus corroborate results found for specific examples of DCA schemes, including the BDCL algorithm discussed in this section: DCA provides significant performance improvement at nominal traffic loads, those resulting in 1%–5% blocking probability when FCA is used. The effect of traffic variability at these loads is reduced significantly as well with the use of DCA. The performance improvement reduces with load, however, and, at high loads, disappears, FCA taking over at high-enough loads.

4.2 Power control

In introducing the cellular concept in mobile wireless systems, we noted that this concept was motivated by the need to increase the capacity of these systems. Thus, system channels may be reused, allowing more mobile users to access the system, providing cells using the same channels are not "too close" together. We specified the "closeness" by indicating that the total interference power from co-channel cells must be no more than a specified fraction of the signal power within a given cell. We quantified this concept by introducing the signal-to-interference ratio SIR on a given channel as a measure of performance. Given a specified requirement for the SIR and the number of channels actually available, one can then determine the number of mobile users that may be accommodated within any given cell (and hence the entire mobile system). This is accomplished by specifying, as a performance objective, the blocking probability expected to be experienced by a typical user. The use of DCA, as described in the previous section, enables the system, under moderate traffic loads, to accommodate more mobile users for a given number of channels than is possible under FCA.

Critical to all of this is the maintenance of an appropriate SIR. This requires monitoring and controlling mobile power to ensure that the SIR in a given cell does not drop below a specified threshold. All mobile cellular systems incorporate means for monitoring mobile uplink power with base stations sending power control messages to mobile units in their cells. We shall, in fact, see when discussing CDMA systems in Chapters 6, 8, and 10, that tight power control is absolutely essential for the proper working of those systems. In this section we describe and compare two *distributed power control* strategies proposed in the literature, whereby transmitter powers are individually adjusted so that the SIR at all receivers is as large as possible. We shall show that the resultant SIR turns out to be the *same* at all receivers. These algorithms and variations of them proposed in the literature apply principally to cellular systems requiring a reuse constraint, as discussed in Chapter 3. Such systems are exemplified by the second-generation TDMA/FDMA (time-division multiple access/frequency-division multiple access) systems to be discussed in Chapters 6 and 8 (D-AMPS or IS-136 and GSM). (The concepts of TDMA and FDMA are discussed in Chapter 6.) The second-generation CDMA (code-division multiple access) system IS-95 to be discussed in Chapters 6 and 8, as well as the third-generation CDMA systems discussed in Chapter 10, do not involve a reuse constraint: the same frequency may be reused in each cell. Coding is then relied on to handle interfering signals, as we shall see in Chapter 6. The two algorithms discussed in this section, and the examples we use to compare them, thus do not specifically apply to CDMA systems. We therefore follow the detailed discussion of these algorithms, and a modification leading to a universal form of these and related algorithms, by a brief description of a related algorithm used in CDMA systems to control transmitter power. We show that this algorithm does have a commonality to the TDMA/FDMA algorithms discussed first.

For simplicity in describing the various algorithms, we focus on uplink power control. We thus assume the transmitters to be located at the mobiles and the receivers at the base stations. The SIR is thus measured at each of the base stations, for a given channel. This situation is, in fact, the more critical one for CDMA systems in which interfering mobiles will all be using the same transmission frequency. It will become clear, however, as we describe the algorithms, that they are equally applicable to downlink power control, ensuring in that case that the SIR measured at a given mobile terminal is as large as possible.

Recall from Sections 3.2 and 3.3 that the uplink SIR is the ratio of the (desired) power received at a given base station from a mobile using the channel in question to the sum of all the (interfering) powers received at that base station from all other mobiles in the system using that same channel. Focusing on average power only, as we did in the earlier analysis in Chapter 3, we can generalize that analysis by writing the SIR_i in cell i as the following ratio

$$SIR_i \equiv \frac{P_i/d_{ii}^n}{\sum\limits_{j=1, j\neq i}^{M} P_j/d_{ij}^n} \tag{4.11}$$

Here, as shown in Fig. 4.6, d_{ii} is the distance from the mobile in cell i using the channel in question to its base station, transmitting at power P_i, while d_{ij} is the distance from an interfering mobile in cell j to the base station in cell i. P_j represents the transmitting power of that mobile. The parameter n, ranging typically in value from 2 to 4, is, of course, the propagation exponent discussed in Chapter 2. There are assumed to be some number M of interfering mobiles using the same channel, each located in a different cell. Equation (4.11) may be rewritten in a somewhat more general form as

$$SIR_i = \frac{P_i G_{ii}}{\sum\limits_{j=1}^{M} P_j G_{ij} - P_i G_{ii}} = \frac{P_i}{\sum\limits_{j=1}^{M} P_j (G_{ij}/G_{ii}) - P_i} \tag{4.12}$$

The term G_{ij} is the so-called gain factor (actually an attenuation!) between transmitter j (mobile j in the uplink case) and receiver i (base station i in this case). In the average power case under discussion here, we, of course, have $G_{ij} \equiv 1/d_{ij}^n$. The objective with power control is to keep each SIR in the system above a required threshold

$$SIR_i \equiv \gamma_i \geq \gamma_0, \; 1 \leq i \leq M \tag{4.13}$$

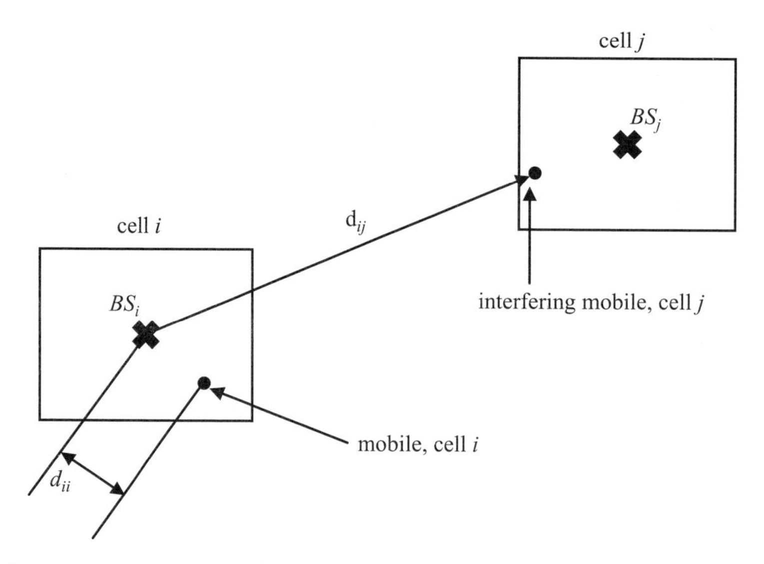

Figure 4.6 Distances involved in calculating SIR_i

To simplify the notation, we shall henceforth use the symbol γ_i as indicated in (4.13) to represent SIR_i.

A number of papers have appeared on the subject of power control in the sense indicated in (4.13) of keeping each SIR above a threshold value. We shall cite only a few. The reader can then check the references in each to come up with a comprehensive list. In particular, decentralized algorithms were first applied to power control for satellite systems. Aein (1973) and Meyerhoff (1974), which appeared in the 1970s, are among the earliest papers describing such algorithms. Zander and his group in Sweden have pioneered in developing power control algorithms for cellular wireless systems. We focus here first on a distributed power control algorithm, the *distributed balancing algorithm* or DBA, proposed by Zander (1992). The objective here is to adjust the transmitted power values, P_i, $1 \leq i \leq M$, in a distributed fashion, each of the M mobiles working in conjunction with its base station, to attain the maximum achievable SIR. This quantity is defined as the maximum over all the mobile powers of the minimum SIR. Using the symbol γ to represent SIR, as noted above, we have, as the maximum achievable SIR

$$\gamma^* = \max_{allpowers} \min \gamma_i \qquad 1 \leq i \leq M \tag{4.13a}$$

The DBA algorithm due to Zander, as well as the one to be described following this discussion, both achieve the same, maximum achievable, SIR γ^* for *all M* mobiles using the same channel. The price to be paid in the use of this algorithm, as we shall see, is that

Aein, J. M. 1973. "Power balancing in systems employing frequency reuse," *Comsat Technology Review*, 3, 2 (Fall).

Meyerhoff, H. J. 1974. "Method for computing the optimum power balance in multibeam satellites," *Comsat Technology Review*, 4, 1 (Spring).

Zander, J. 1992. "Distributed cochannel interference control in cellular radio systems," *IEEE Transactions on Vehicular Technology*, 41, 3 (August), 304–311.

a limited amount of communication between each mobile and its base station, as well as between the M base stations, is necessary. (A modification of this and related algorithms, due to Foschini and Miljanic (1993), to be discussed later, eliminates this problem.) The DBA algorithm is a stepwise iterative algorithm. It requires as input the complete set of gains, $G_{ij} \equiv 1/d_{ij}$, $1 \leq i, j \leq M$. It assumes as well that these gains do not change during the operation of the algorithm. This implies that rapid convergence to the required SIR value γ^* is necessary. This algorithm turns out, unfortunately, to take relatively long to converge. The second algorithm to be considered, following, which is a modified version of this one, converges much more rapidly.

Both algorithms require stepwise iteration, starting from some initial set of power values, each mobile in synchronism adjusting its power iteratively. The two algorithms differ in the adjustment strategy. Consider the nth iteration at mobile i. To simplify the notation, let $r_{ij} \equiv G_{ij}/G_{ii}$, i.e., the gain factor G_{ij} normalized to G_{ii}. The DBA algorithm makes the following power adjustment

$$P_i^{(n)} = c^{(n-1)} P_i^{(n-1)} \left[1 + \frac{1}{\gamma_i^{(n-1)}} \right] \tag{4.14}$$

where

$$\gamma_i^{(n-1)} = P_i^{(n-1)} \bigg/ \left(\sum_{j=1}^{M} P_j^{(n-1)} r_{ij} - P_i^{(n-1)} \right) \tag{4.15}$$

The constant $c^{(n-1)}$ must be greater than zero and the same value is to be selected by all the mobiles. Then, simplifying (4.14) with the use of (4.15), we have

$$P_i^{(n)} = c^{(n-1)} \sum_{j=1}^{M} P_j^{(n-1)} r_{ij} \tag{4.16}$$

The power iteration described by (4.16), as carried out simultaneously by each mobile using the channel in question, may be written much more simply in vector form. Thus, define $\boldsymbol{P}^{(n)}$ as the M-element vector whose ith entry is $P_i^{(n)}$. Define $\mathbf{R}$ as the $M \times M$ matrix with r_{ij} as its (i, j) entry. Note that $\mathbf{R}$ is a non-negative matrix, with unit diagonal entries and off-diagonal entries greater than zero but less than one, since $r_{ii} \equiv G_{ii}/G_{ii} = 1$ and $r_{ij} \equiv G_{ij}/G_{ii} < 1$, $i \neq j$. The DBA algorithm of (4.16) is then written in vector form quite simply as

$$\boldsymbol{P}^{(n)} = c^{(n-1)} \mathbf{R} \boldsymbol{P}^{(n-1)} \tag{4.17}$$

Since $\mathbf{R}$ is a non-negative matrix with unit diagonal entries, it may readily be shown, using the Perron–Frobenius theorem for non-negative matrices (see, for example, Noble and Daniel, 1988) that its largest eigenvalue $z^* > 1$, with a multiplicity of 1, while the other $(M - 1)$ eigenvalues all satisfy $|z_j| < z^*$. (Recall that $\mathbf{R}$ is an $M \times M$ matrix, hence

---

Foschini, G. J. and Z. Miljanic. 1993. "A simple distributed autonomous power control algorithm and its convergence," *IEEE Transactions on Vehicular Technology*, 42, 4 (November), 641–646.
Noble, B. and J. H. Daniel. 1988. *Applied Linear Algebra*, 2nd edn, Englewood Cliffs, NJ, Prentice-Hall.

has M eigenvalues.) In particular, we shall now show that the DBA algorithm of (4.17) has the SIR at *each* cell converging to the same limiting value γ^* given by (4.18)

$$\lim_{n\to\infty} \gamma_i^{(n)} = \gamma^* = 1/(z^* - 1) \tag{4.18}$$

It is thus clear that, since one would like the SIR γ^* to be as large as possible, one wants the eigenvalue z^* as close to 1 as possible. However, as we shall see shortly, the closer z^* is to 1, the slower the convergence of the algorithm. This point motivates the need for the second algorithm we shall discuss, which converges to the same value of SIR γ^*, but much more rapidly.

Following is the proof that the DBA algorithm, applied synchronously by all M mobiles that use the same channel and hence may be expected to interfere with one another, has γ_i converging in all mobiles to the same desired value γ^*: Let each mobile independently choose an initial transmitting power value. This establishes an initial power vector $\boldsymbol{P}^{(0)}$. Iterating (4.17) n times, one then has

$$\boldsymbol{P}^{(n)} = \mathbf{R}^n \left[\prod_{j=0}^{n-1} c^{(j)}\right] \boldsymbol{P}^{(0)} \tag{4.19}$$

We noted above that the $M \times M$ matrix $\mathbf{R}$ has M eigenvalues z_j, $1 \le j \le M$. Each has its corresponding eigenvector ϕ_j. These M eigenvectors represent an orthogonal set that can be used as a basis for a unique expansion of the initial power vector $\boldsymbol{P}^{(0)}$

$$\boldsymbol{P}^{(0)} = \sum_i a_i \phi_i \tag{4.20}$$

The expansion coefficients a_j appearing in (4.20) are given by the usual respective dot products

$$a_j = (\boldsymbol{P}^{(0)}, \phi_j)/(\phi_j, \phi_j) \tag{4.21}$$

(Recall that if $\boldsymbol{x}$ and $\boldsymbol{y}$ are each M-component vectors, $(\boldsymbol{x}, \boldsymbol{y}) \equiv \sum_j x_j y_j$.)

Using (4.20), $\mathbf{R}^n\ \boldsymbol{P}^{(0)}$ in (4.19) may be replaced by $\sum_j a_j \mathbf{R}^n \phi_j$. But note that, by definition of eigenvalues and eigenvectors, $\mathbf{R}\phi_j = z_j \phi_j$. Hence $\mathbf{R}^n \phi_j = z_j^n \phi_j$, and $\mathbf{R}^n \boldsymbol{P}^{(0)} = \sum_j a_j z_j^n \phi_j$. Without loss of generality, let z_1 be $z^* > 1$, the largest of the M eigenvalues. Then for the iteration variable n large enough, $(z^*)^n$ must dominate all the terms in the sum over j, and $\sum_j a_j z_j^n \phi_j \to a_1 (z^*)^n \boldsymbol{P}^*$, where we have used the symbol $\boldsymbol{P}^*$ to represent ϕ_1, the eigenvector corresponding to $z_1 \equiv z^*$. Hence $\lim_{n\to\infty} \mathbf{R}^n \boldsymbol{P}^{(0)} = a_1 (z^*)^n \boldsymbol{P}^*$, and we get, from (4.19)

$$\lim_{n\to\infty} \boldsymbol{P}^{(n)} = \boldsymbol{P}^* a_1 \lim_{n\to\infty} (z^*)^n \prod_{j=0}^{n-1} c^{(j)} \tag{4.22}$$

The object now is to choose the constant $c^{(j)}$ on the jth iteration so that the limit (4.22) is bounded, i.e., to keep the limiting powers at the M mobiles from getting too large. One possibility is to simply let $c^{(j)} = 1/\max\ (\boldsymbol{P}^{(j)})$, where by max $(\boldsymbol{P}^{(j)})$ we mean the maximum among the M transmitted powers on the jth iteration. Note, however, that this choice requires all mobiles carrying out the algorithm to synchronously pass their respective power values on the jth iteration to each other. Alternatively, each base station

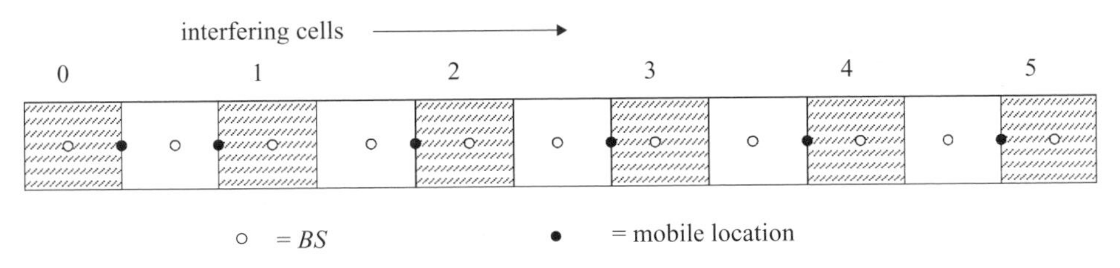

Figure 4.7 11-cell linear cellular system

must pass its mobile's power value on to all other base stations, and thence on to the mobile in each cell.

Recapping the results of the analysis thus far, we have shown that the ith component of the power vector $\boldsymbol{P}^{(n)}$ at the nth iteration of the DBA algorithm, i.e., the power transmitted by the mobile in cell i, approaches the following expression for n "large enough":

$$P_i^{(n)} = c^{(n-1)} P_i^* a_1 (z^*)^n \prod_{j=0}^{n-2} c^{(j)} \tag{4.23}$$

But recall that $P_i^{(n)}$ may also be written in the form of (4.14). Replacing $P_i^{(n-1)}$ in (4.14) by its equivalent (4.23), with $n-1$ used in place of n, we also get

$$P_i^{(n)} = c^{(n-1)} \left[1 + \frac{1}{\gamma_i^{(n-1)}} \right] P_i^* a_1 (z^*)^{n-1} \prod_{j=0}^{n-2} c^{(j)} \tag{4.24}$$

Comparing (4.23) and (4.24), it is clear that, taking the limit in n, one gets finally

$$\lim_{n\to\infty} \gamma_i^{(n-1)} \equiv \gamma^* = 1/(z^* - 1) \tag{4.25}$$

This expression is precisely that of (4.18), the common value of SIR to which we indicated the SIRs of all M mobiles would converge.

We now consider two one-dimensional examples of the application of this algorithm. The first example is the 11-cell linear cellular system shown in Fig. 4.7. We assume channels can be reused every two cells. We focus on the six cells indicated by cross hatching. The base stations are indicated by open circles; the $M = 6$ potentially interfering mobiles, each assigned the same channel, are indicated by the black dots. The mobile locations have all been chosen to lie at the cell edges, corresponding to a worst-case interference situation. (Note that the distances d_{ij} and hence the corresponding SIRs depend critically on the location of the various mobiles with respect to the base stations. As the mobiles move, the distances change. This is the reason, noted above, why it is important to have a rapidly converging power control algorithm.) Each cell is defined to have a width of one unit. (One can pick the width to be D units, in some distance measure, but since we deal with ratios of powers of distance in calculating the elements $r_{ij} = G_{ij}/G_{ii} = d_{ii}^n/d_{ij}^n$ of the matrix $\mathbf{R}$, the specific value of D cancels out.) It is left to

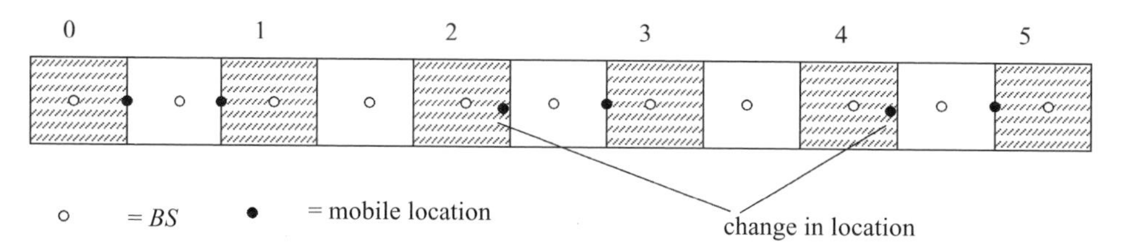

Figure 4.8 Symmetrical case of mobile locations

the reader to show that the matrix **R** for this example appears as follows in (4.26)

$$\mathbf{R} = \begin{bmatrix} 1 & \frac{1}{3^4} & \frac{1}{7^4} & \frac{1}{11^4} & \frac{1}{15^4} & \frac{1}{19^4} \\ \frac{1}{3^4} & 1 & \frac{1}{3^4} & \frac{1}{7^4} & \frac{1}{11^4} & \frac{1}{15^4} \\ \frac{1}{7^4} & \frac{1}{5^4} & 1 & \frac{1}{3^4} & \frac{1}{7^4} & \frac{1}{11^4} \\ \frac{1}{11^4} & \frac{1}{9^4} & \frac{1}{5^4} & 1 & \frac{1}{3^4} & \frac{1}{7^4} \\ \frac{1}{15^4} & \frac{1}{13^4} & \frac{1}{9^4} & \frac{1}{5^4} & 1 & \frac{1}{3^4} \\ \frac{1}{19^4} & \frac{1}{17^4} & \frac{1}{13^4} & \frac{1}{9^4} & \frac{1}{5^4} & 1 \end{bmatrix} \tag{4.26}$$

The corresponding six eigenvalues are then readily found, using any mathematics software, to be given by 1.0137, 1.006, 1, 0.999, 0.994, 0.987. The maximum eigenvalue is thus $z^* = 1.0137$ and the SIR to which all mobile SIRs converge is $\gamma^* = 1/(z^* - 1) = 73.3$ or, in dB, 18.7 dB.

For the second example, we simply switch two mobile locations, creating the symmetrical 11-cell case of Fig. 4.8. Note that this slight change shows the effect of mobility on power control. The **R** matrix in this case is correspondingly symmetrical. It is then readily shown that the six eigenvalues of matrix **R** are 1.0140, 1.012, 1.011, 0.988, 0.988, and 0.987. Then $z^* = 1.0140$ and $\gamma^* = 71.6$ or 18.5 dB. Note that there is a slight change in the SIR because of the mobility introduced. More significant changes in mobile location would introduce correspondingly greater changes in the SIR.

Since the SIR to which the power control algorithm converges is mobile position dependent due to changes in the distance or gain parameters, it is important that the algorithm converge rapidly, as noted earlier. (Recall our basic assumption that the gain parameters, or distances on which they are based, were constant.) However, it was also noted earlier that the DBA algorithm on which we have been focusing converges slowly. The reason is clear from the examples just described: the eigenvalues are quite closely spaced together. But in the proof of convergence we made the assumption that, for iteration *n large enough*, $z_1^n \gg z_i^n, i \neq 1$. But for z_i close to z_1, n has to be very large for this condition to hold true. Note that in both examples the eigenvalues were very closely spaced, so that convergence is expected to be slow. It is readily shown from simple matrix analysis that this algorithm will always converge slowly: the eigenvalues of the matrix **R** will always cluster around the value 1. This observation is demonstrated as follows. The trace Tr(**R**) of a matrix **R**, defined to be the sum of the diagonal terms, is also known to be given by the sum of the eigenvalues. We thus have

$$\text{Tr}(\mathbf{R}) \equiv \sum_{i=1}^{M} r_{ii} = \sum_{i=1}^{M} z_i \tag{4.27}$$

In our case, with diagonal entries r_{ii} all equal to 1, Tr(**R**) $= M$. If we desire a high SIR γ^* we must have $z^* - 1 = z_1 - 1 = 1/\gamma^* \ll 1$. Hence $z^* = 1 + \varepsilon$, $\varepsilon \ll 1$. (Note in our two examples we have $\varepsilon = .0137$ and 0.014, respectively.) Since Tr(**R**) $= M$, we have, from (4.27)

$$\sum_{j=2}^{M} z_j = M - z^* \approx M - 1 \tag{4.28}$$

Hence large M implies *all* z_js are close to 1 and are closely spaced. Slow convergence is thus expected. In fact, referring to our two examples, even with M only equal to 6, note how closely spaced the six eigenvalues are. Consider the ratio of the two largest eigenvalues in our first example, that of Fig. 4.7. Here we have $z_2/z_1 = 1.006/1.0137 = 0.992$. Then for $n = 20$ iterations, z_2^n/z_1^n is only 0.85, rather than $\ll 1$, as assumed in our derivation of the convergence of the algorithm. At $n = 60$, z_2^n/z_1^n is still only 0.52. The other eigenvalues, which are relatively close to $z^* = z_1$ as well, also continue to enter into the convergence until n is extemely large.

What if we have both z_1 and z_2 less than 1, however? Then convergence is much more rapid. This motivates the second algorithm, for which we shall see the eigenvalues are all exactly 1 less than the values found here; i.e., calling them λ_i, $1 \leq i \leq M$, we shall show that $\lambda_i = z_i - 1$. In our first example of Fig. 4.7, then, we have $(\lambda_2/\lambda_1)^n = (0.006/0.0137)^n = 0.44^n$. For $n = 8$, this number is 0.0015, and it is clear that convergence is already close. For the second example of Fig. 4.8, we have $(\lambda_2/\lambda_1)^n = 0.86^n$ and convergence is expected to be slower, but clearly still much more rapid than for the DBA algorithm. This second, modified algorithm, called simply the *distributed power control* or DPC algorithm, is due to Grandhi, Vijayan, and Goodman (1994). It is based on the Meyerhoff power balance algorithm proposed for satellite systems (Meyerhoff, 1974). Instead of using the power iteration equation (4.14) of the DBA algorithm, it adopts simply the iteration

$$P_i^{(n)} = c^{(n-1)} P_i^{(n-1)} / \gamma_i^{(n-1)} \tag{4.29}$$

In matrix-vector form this may be written as

$$\boldsymbol{P}^{(n)} = c^{(n-1)} \mathbf{A} \boldsymbol{P}^{(n-1)} \tag{4.30}$$

with **A** an $M \times M$ matrix whose elements $a_{ij} \equiv G_{ij}/G_{ii}$, $i \neq j$; $a_{ii} = 0$. The matrix **A** is thus precisely the same as the matrix **R** in the DBA algorithm, except that it is reduced by the identity matrix **I**: $\mathbf{A} = \mathbf{R} - \mathbf{I}$. Note that the M eigenvalues λ_j, $1 \leq j \leq M$ of matrix **A** are given simply by $\lambda_j = z_j - 1$, with the z_js the eigenvalues of the matrix **R** discussed above. It is readily shown, using the same approach used in proving convergence of the DBA

Grandhi, S. A. *et al.* 1994. "Distributed power control in cellular radio systems," *IEEE Transactions on Communications*, 42, 2/3/4 (February–April), 226–228.

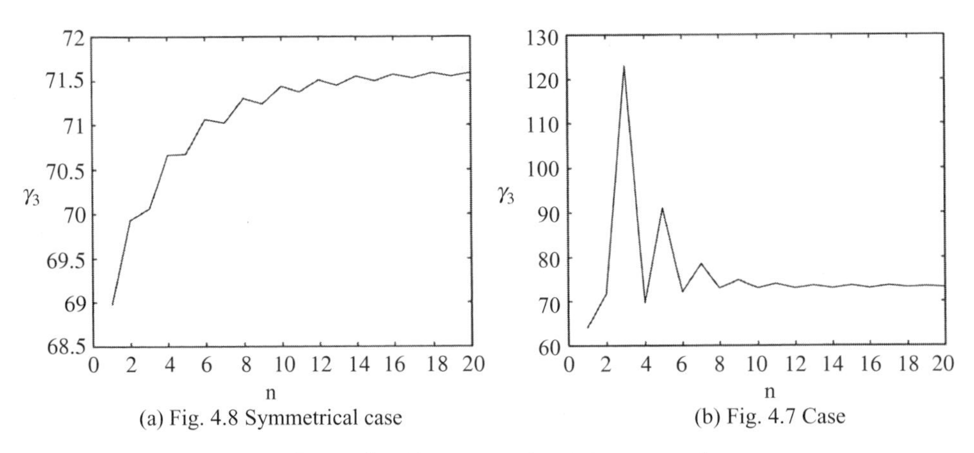

(a) Fig. 4.8 Symmetrical case (b) Fig. 4.7 Case

Figure 4.9 Convergence of DPC algorithm, example of Figs. 4.7 and 4.8

algorithm, that the DPC algorithm converges to the same SIR γ^* as the DBA algorithm, but given this time by $\gamma^* = 1/\lambda^*$, with $\lambda^* = z^* - 1$ the largest eigenvalue of matrix **A**. The convergence rate is now governed by λ_2/λ^*, a much smaller number than z_2/z^*, which governed the convergence rate in the DBA algorithm. This point has already been made above.

Consider as examples the same two cases of the 11-cell linear network discussed previously, as described by Figs. 4.7 and 4.8. Fig. 4.9 plots the convergence of $\gamma_3 = \text{SIR}_3$, the SIR of cell 3 in these two cases. Note that the SIR of cell 3 for case 1, shown in Fig. 4.9(b), does converge rapidly, within eight iterations, as predicted above, albeit with some initial wild oscillations! The corresponding SIR for the symmetrical case of Fig. 4.8 converges more slowly, within about 20 iterations, again as predicted above, but with fewer and much smaller oscillations. Grandhi *et al.* (1994) have carried out similar simulations for a 50-cell one-dimensional system. Each channel in their system is reused every three cells, resulting, of course, in a much higher SIR. They have chosen three representative mobile positions, rather than the two we have selected for our examples. They find that, at most, five iterations are required for convergence with the resultant minimum SIR ranging from 31.8 to 34 dB. The corresponding convergence rate for the DBA algorithm is much slower, with the DBA algorithm still far from convergence after 50 iterations in the three cases considered.

We have focused, thus far, in this discussion of distributed power control, on convergence of the SIR. What are the resultant transmitter powers at each mobile after convergence has taken place? Consider our 11-cell examples. Recall again that six mobiles are assigned the same channel, assuming a reuse of two. In case 1, that of Fig. 4.7, the relative powers at each mobile, relative to the one with the highest power, are found to be given by the following list:

mobile →	0	1	2	3	4	5
relative power →	0.86	1	0.214	0.052	0.016	0.004

Note that the algorithm acts to reduce the powers to very small values in mobiles 3, 4, 5. Consider case 2, however, with two of the six mobiles moved from their positions in case 1. The corresponding relative powers are now listed below:

mobile →	0	1	2	3	4	5
relative power →	0.64	0.71	1	1	0.71	0.64

This result is to be expected on grounds of symmetry, with cells to the left of cell 2 and to the right of cell 3 behaving the same way.

Further studies in distributed power control for cellular systems have focused on eliminating some of the problems associated with implementing the two algorithms described in this section. These problems include the effect of variations in the gain coefficients, assumed constant here, as mobiles move; the need for some communication between base stations in carrying out the algorithms; the need for synchronous implementation of the algorithms. A modification of the DPC algorithm proposed by Foschini and Miljanic (1993) removes the need for communication; the modified algorithm relies on local measurements only, and the authors prove convergence for this algorithm. Mitra has shown as well that the modified DPC algorithm converges with asynchronous implementation, so that mobiles do not have to be assumed to carry out the algorithm iterations at the same time (Mitra, 1994).

The Foschini–Miljanic modification of the DPC algorithm replaces (4.29) by the equation

$$P_i^{(n)} = \gamma P_i^{(n-1)} / \gamma_i^{(n-1)} \tag{4.31}$$

The parameter γ replacing $c^{(n-1)}$ in (4.29) is a "target" SIR which the mobiles all aspire to attain. This target SIR must be defined by $\gamma \leq \gamma^*$. All mobile iterations are then locally carried out, with no inter-mobile communication required, as is the case in using $c^{(n-1)}$. An interesting variation of (4.31) is obtained by rewriting that iterative expression in decibel (dB) form. Recalling that $P_{dB} \equiv 10\log_{10}P$, we have, immediately

$$P_{i\ dB}^{(n)} = P_{i\ \ dB}^{(n-1)} + \left(\gamma_{dB} - \gamma_{i\ \ dB}^{(n-1)}\right) \tag{4.31a}$$

The interpretation of this expression is quite interesting. It says that the difference between the target SIR expressed in dB and the SIR measured by a mobile in dB is added to the previous estimate of transmit power to form the new estimate. Blom and Gunnarson (1998) have shown that most distributed power control algorithms proposed in the literature turn out to be of the form of (4.31a) if dB formulations are used. The unification of these algorithms is obtained by multiplying the difference term in (4.31a) by a suitably chosen parameter which may vary with iteration.

It was noted at the beginning of this section that the algorithms discussed here are appropriate for TDMA/FDMA mobile systems in which reuse constraints play a role. The one-dimensional examples on which we focused did, in fact, assume two-cell reuse. We also noted at the beginning of this section that a form of this algorithm was used as the uplink (mobile transmitter) power control algorithm in the IS-95 CDMA system

---

Mitra, D. 1994. "An asynchronous distributed algorithm for power control in cellular systems," in *Wireless and Mobile Communications*, ed. J. M. Holtzman, Boston, MA, Kluwer/Academic.

Blom, J. and F. Gunnarson. 1998. "Power control in cellular systems, Linköping studies in science and technology," Thesis No. 706, Department of Electrical Engineering, Linköpings Universitet, Linköping, Sweden.

to be described in Chapters 6 and 8. The objective in the CDMA system is to ensure that the SIR as measured at a base station due to each mobile in its cell is the same, independent of where the mobile is located within the cell. We shall see in Chapter 6 that this condition results in the best CDMA performance. To attain this objective, a two-part power control procedure is used (Viterbi, 1995). There is a coarse, "open-loop," power control used in which each mobile adjusts its transmitter power using its own automatic-gain control measurement of its received downlink power from the base station. But this is only a coarse adjustment since the downlink and uplink propagation conditions may differ considerably. The implication here is that the uplink and downlink gain parameters G_{ij} and G_{ji} as defined in (4.12) are the same. Clearly, this can be a coarse approximation only. CDMA systems must incorporate a "closed-loop" power control scheme as well to attain the desired equality of all received SIRs at the base station. The "closed-loop" power control strategy adopted for IS-95 by which the base station controls the transmit power of each mobile in its cell is of the "bang-bang" type: a mobile is commanded by the base station to increase its transmitter power by a fixed dB amount if the estimated received SIR at the base station due to that mobile is below a specified threshold; it is commanded to decrease its transmitter power by a fixed dB amount if the estimated received SIR is above the threshold (Viterbi, 1995). Specifically, we have, using the same notation adopted heretofore in this section

$$P_i^{(n)}{}_{dB} = P_i^{(n-1)}{}_{dB} \pm \Delta \tag{4.32}$$

where Δ is a fixed increment in dB (for example, one dB), and the + value is used if the estimated $SIR_i < \gamma$, while the – value is used if $SIR_i > \gamma$. But note, by comparing (4.32) and (4.31a), that the two power control procedures are quite similar. In fact, the two expressions are identical if the difference term of (4.31a) is quantized to one bit. CDMA and TDMA/FDMA power control strategies are thus closely related. The basic difference lies in the manner in which SIR_i is measured or estimated. Note that the CDMA algorithm requires just one bit of transmission from the base station to each mobile after the estimation of each received SIR is carried out.

Problems

4.1 Refer to Fig. 4.1. Say there are ten channels assigned per cell.

(a) How many channels are there in the system?

(b) The DCA algorithm BDCL described in the text is to be implemented for this system: Make a channel table numbering the channels from 1 to 10. Apply channel ordering as described in the text, assigning local calls priority from 1 up, assigning channels to be borrowed by neighboring cells priority from 10 down. Focus on one cell. Work out different cases of the application of this algorithm, with different numbers of calls in progress, both local, and those using channels borrowed from neighboring cells. Demonstrate channel locking for this example, and then show how directional locking can help alleviate this

Viterbi, A. J. 1995. *CDMA, Principles of Spread Spectrum Communication*, Reading, MA, Addison-Wesley.

problem. Include examples that demonstrate the features of immediate channel reallocation described in the text.

(c) Compare this algorithm with another one, either taken from the literature, or one you have invented yourself.

4.2 Problem 4.1 above is to be repeated for the one-dimensional case of Fig. 4.2.

(a) How many channels are there in the system if ten channels are assigned per cell?

(b) Show how the BDCL algorithm is implemented for this system, with different numbers of calls in progress.

(c) Consider the case of ten calls in progress in cell i. Say cell $i-1$ has seven calls in progress. A mobile user in cell i now attempts to make a call. The base station for that cell borrows a channel from cell $i-1$. Applying the BDCL algorithm, are any channels now locked? Which one(s)? Show, by example, that borrowing of channels does not propagate, as stated in the text.

4.3 Calculate the blocking probability for the one-dimensional BDCL algorithm for the case of $2m = 2$ channels in the system. Assume homogeneous traffic conditions. Show that (4.8) results. Plot this probability vs the traffic intensity A in Erlangs and compare with the FCA result.

4.4 Consider the case of $2m = 4$ channels in a one-dimensional BDCL system.

(a) Show there is a total of 55 possible states.

(b) Indicate some of the 25 blocking states. Explain, in words, why they are blocking states.

(c) Calculate and plot the blocking probability as a function of the Erlang load A, assuming homogeneous traffic conditions. Compare with the plot of Fig. 4.5. Plot the FCA blocking probability as well and compare with the DCA result. Using the results of both this problem and problem 4.3, discuss the effect of increasing the number of channels on both the FCA and DCA algorithms.

4.5 Refer to Figs. 4.7 and 4.8. The DBA power algorithm is to be applied to each of these examples.

(a) Find the **R** matrices in each case. In particular, show that (4.26) is the **R** matrix for the case of Fig. 4.7.

(b) Find the six eigenvalues of **R** in each case and show they agree with the values appearing in the text.

(c) Show $\text{Tr}\,(\mathbf{R}) = M = \sum_i z_i$ in each case.

(d) Run the DBA algorithm for each of the cases and show the algorithm converges to the two values of SIR, 18.7 dB and 18.5 dB, indicated in the text. Plot the SIR at a number of base stations as a function of iteration n for at least 20 iterations. Comment on the results.

4.6 This problem is concerned with the analysis of the DPC algorithm, as defined by (4.29).

(a) Starting with (4.29) show the matrix–vector form of this algorithm may be written in the form of (4.30), with the matrix **A** defined as indicated in the text.

(b) Relate the eigenvalues of the **A** matrix to those of the matrix **R** defined for the DBA algorithm.

(c) Prove the DPC algorithm converges to the same SIR as the DBA algorithm. Show why this algorithm is expected to converge more rapidly than the DBA algorithm.

4.7 Repeat problem 4.5 for the DPC algorithm: Find the **A** matrices for the two cases of Figs. 4.7 and 4.8. Find the eigenvalues in each case. Compare with those of the DBA algorithm. Run and plot the algorithm in each case for at least 20 iterations and compare with the DBA algorithm.

4.8 The DBA and DPC distributed power control algorithms are to be compared for a one-dimensional cellular network of your choice, consisting of at least 13 cells.

(a) Carry out the iterative analysis for both the two-reuse and three-reuse cases, plotting the resultant SIRs as a function of iteration parameter at a number of base stations. Try to choose worst-case mobile locations.

(b) Compare the SIR to which the DPC algorithm converges with the largest eigenvalue of the **R** matrix of the DBA algorithm.

(c) Find the relative powers at each of the mobile terminals and comment on the results.

4.9 Consider the Foschini–Miljanic algorithm defined by (4.31).

(a) Compare this algorithm with the DPC algorithm defined by (4.29). Show the two algorithms converge to the same SIR.

(b) Show why no inter-mobile communication is needed in carrying out this algorithm.

(c) Apply this algorithm to the 11-cell linear system of Fig. 4.7 and compare with the results shown in the text.

4.10 The object of this problem is to provide an example of the application of the fixed-increment power control algorithm of (4.32). Start with a cellular network of your own choice. (A simple choice would be a one-dimensional network such as those used as examples in the text.) Let each potentially interfering mobile choose an initial transmit power. The base station for each mobile estimates its SIR and then commands the mobile to reduce or increase its transmit power by a fixed increment, as discussed in the text. This procedure is then iterated until convergence is reached. Carry out this algorithm for various values of the increment. Discuss your results.

Chapter 5

Modulation techniques

We have stressed thus far in this book basic system concepts in mobile cellular systems. We have discussed the propagation environment and modeling thereof, the use of geographically assigned cells to increase capacity by channel reuse, and, finally, in the last chapter, the use of dynamic channel allocation and power control to improve system performance. Implicit in all of this was the assumption that users communicate over assigned *channels*. Depending on the system under study, a channel can be a frequency band, a time slot in a frequency band, or a unique code, a sequence of defined binary symbols. In each of these cases, communication is carried out using a specified carrier frequency, with various modulation techniques used to transmit the desired information, whether it be voice, data, or other types of information-bearing signals, over this carrier. In this chapter we describe various modulation techniques proposed or adopted to utilize the channels most effectively.

Modulation or variation of the frequency of a carrier is universally used in communication systems to bring the information to be transmitted up to a desired operating or carrier frequency. In the case of mobile wireless systems, the resultant modulated radio signal is then transmitted uplink, from mobile to base station, or downlink, from base station to mobile, as the case may be. The transmitted power discussed in previous chapters was the power in this modulated carrier signal. First-generation cellular systems, such as the AMPS system mentioned in Chapter 1, used analog FM as the modulation scheme. Second-generation systems, such as digital AMPS or D-AMPS (also referred to as IS-54/136), IS-95, or GSM, introduced in Chapter 6 following, and then described in detail in Chapter 8, use digital modulation schemes. Third-generation systems discussed in Chapter 10 and the personal communication networks described in Chapter 12 use digital modulation techniques as well. Consider second-generation systems as examples. In the USA, as noted in Chapter 1, mobile cellular systems have been assigned the frequency band of 824–849 MHz for uplink communications and 869–894 MHz for downlink transmissions. Both AMPS and D-AMPS slot the 25 MHz-wide bands available in each direction into 832 30 kHz channels. As we shall see, D-AMPS further uses time slots in each 30 kHz band to increase the capacity of the system by a factor of 3–6. The modulation technique chosen for D-AMPS is $\pi/4$-DQPSK, to be described shortly. IS-95 in the US uses the same uplink and downlink frequency bands, but channels, codes in this case, occupy much wider 1.25 MHz frequency bands. It uses QPSK for downlink communications and OQPSK for uplink transmission. Both modulation techniques will also be described in this chapter.

In Europe, the bands assigned to, and used by, GSM are 890–915 MHz in the uplink direction and 935–960 MHz in the downlink direction. The 25 MHz bands available in each direction are further broken into 124 frequency channels of 200 kHz each. Each frequency channel in turn is accessed by up to eight users, using time slot assignments. The modulation technique adopted, to be described later in this chapter, is called 0.3 GMSK.

Additional wireless communication channels for cellular communications have been made available in the US in the 1.85–1.99 GHz band, the so-called PCS band, and bands ranging from 1.71 to 1.9 GHz in Europe. For simplicity in providing examples in this chapter, we focus on the modulation techniques used in the 800–900 MHz bands.

The discussion in this chapter will focus on the digital modulation techniques noted above. Thus, DQPSK, GMSK, QPSK, and OQPSK will be described in some detail after first providing some introductory material on digital modulation. It is to be noted, however, that multiple criteria are involved in the specific choice of a modulation technique to be used. Examples of such criteria include bandwidth efficiency, power efficiency, cost and complexity, and performance in fading channels. Note that the battery-operated mobile or subscriber terminal, in particular, must be reasonably inexpensive, small in size, and parsimonious in its use of power. Operation in a fading environment means that constant-amplitude modulation techniques with nonlinear amplifiers used are favored. This is one reason for the original choice of FM for the first-generation analog systems.

Our approach in this chapter will be to first describe the simplest, basic digital modulation techniques. These include on–off keying (OOK) or amplitude-shift keying (ASK), phase-shift keying (PSK), and frequency-shift keying (FSK). We will then touch briefly on quadrature amplitude modulation (QAM), used, incidentally, in wireline modems to increase the transmission bit rate over a bandwidth-limited channel, such as the telephone access line from home or office to a telephone exchange or central office. Quadrature phase-shift keying (QPSK) is a special case of QAM. We will go on to discuss enhanced digital modulation techniques such as OQPSK or MSK, and GMSK, chosen to work relatively well in the constrained bandwidth, fading environment of wireless cellular systems.

We conclude this chapter with an introduction to orthogonal frequency-division multiplexing (OFDM). This scheme, which goes back historically at least to the 1960s, is being considered for use in advanced cellular systems. It has already been applied to high-speed wireless LANs and to high-speed DSL modems developed for use over telephone copper wire access lines to the home. We discuss its application to wireless LANs as part of the IEEE 802.11g and a standards in Chapter 12.

5.1 Introduction to digital modulation techniques

Consider an unmodulated sinusoidal carrier $A \cos \omega_0 t$ operating continuously at a carrier frequency of f_0 Hz, or $\omega_0 = 2\pi f_0$ radians/sec. Its amplitude, frequency, and phase may

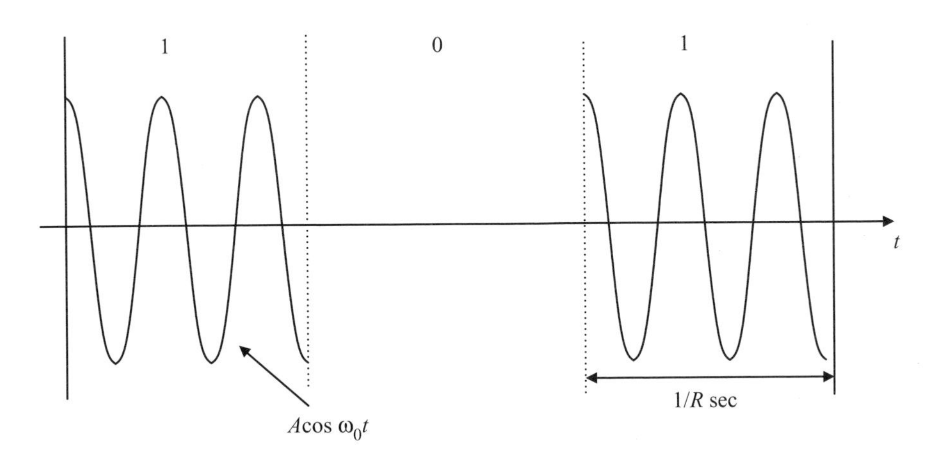

Figure 5.1 OOK signal corresponding to binary 1, 0, 1

individually be varied or modulated in accordance with an information-bearing signal to be transmitted to provide, in the case of the first two, the well-known AM and FM high-frequency signals used in (analog) radio broadcast systems with which we are all familiar. The simplest digital equivalents are called, respectively, as noted above, on-off keyed or amplitude-shift keyed transmission, OOK or ASK, frequency-shift keyed transmission, FSK, and phase-shift keyed transmission, PSK. In this digital case, successive binary digits (bits), 0 or 1, are used to vary or modulate the amplitude, frequency, and phase, respectively, of the unmodulated carrier. The binary, information-bearing, digits are also referred to as comprising the *baseband* signaling sequence. In the simplest case, that of OOK transmission, a binary 0 will turn the carrier off; $A \cos \omega_0 t$ will be transmitted whenever a binary 1 appears in the baseband signaling sequence. Say the binary sequence is being transmitted at a bit rate of R bits/sec. Then each binary symbol or bit lasts $1/R$ sec. An example of an OOK signal corresponding to the three successive bits 1, 0, 1 appears in Fig. 5.1. Note that OOK transmission is not suitable for digital mobile systems because of the variation in amplitude of the signals transmitted. Recall that we indicated earlier that constant-amplitude signals are preferred because of the fading environment, which, by definition, introduces unwanted random amplitude variations. We discuss OOK transmission here to provide a more complete picture of digital communication.

In actuality, the instantaneous change in amplitude shown occurring in Fig. 5.1, as the OOK signal turns off and then comes back on again, cannot occur in the real world. Such abrupt transitions would require infinite bandwidth to be reproduced. In practice, the signals actually transmitted are shaped or pre-filtered to provide a more gradual transition between a 0 and a 1, or vice-versa. The resultant transmitted signal corresponding to a baseband binary 1 may be written $Ah(t) \cos \omega_0 t$. The low-frequency time function $h(t)$ represents a signal-shaping function. This signal shaping is designed to allow the transmitted OOK sequence to fit into a prescribed system bandwidth with little or no distortion. We shall discuss signal shaping briefly in the next section. The rf carrier transmission bandwidth, the frequency spread of the modulated transmitted signal about the carrier, is readily shown to be given by $2B$, with B the baseband bandwidth, the bandwidth required

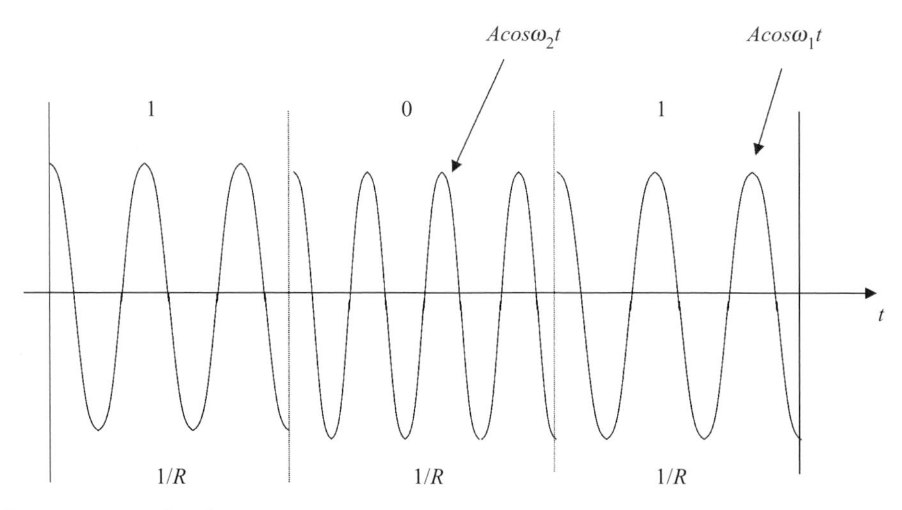

Figure 5.2 FSK signal

to transmit the baseband binary sequence of 1s and 0s (Schwartz, 1990). We shall see, in the next section, that the baseband bandwidth, in Hz, varies between $R/2$ and R for the type of shaping discussed there. Hence the transmitted signal bandwidth varies between R and $2R$ Hz. As an example, if a 10 kbps digital information signal is to be transmitted over the wireless radio channel, the baseband bandwidth ranges from 5 kHz to 10 kHz, depending on the shaping used. The radio channel bandwidth would range from 10 to 20 kHz. Since OOK transmission, even with signal shaping, results in amplitude variation and is hence undesirable for mobile communication, we look to other types of digital modulation to provide the desired digital signaling. Most implementations in practice use some variation of phase-shift keying or PSK. However, we continue our discussion by first considering frequency-shift keying or FSK, the binary version of FM. This brief discussion of FSK will turn out to be useful in our discussion later of more exotic forms of digital communication.

In the case of FSK, the carrier frequency shifts between two different frequencies $f_1 = f_0 - \Delta f$ and $f_2 = f_0 + \Delta f$, depending on whether a 0 or a 1 is being transmitted. The term Δf is called the *frequency deviation* about the average carrier frequency f_0. In radian frequencies we have $\omega_1 = \omega_0 - \Delta\omega$, $\omega_2 = \omega_0 + \Delta\omega$, each radian frequency being 2π times the corresponding frequency in Hz. In particular, let $A \cos \omega_1 t$ correspond to a 1 transmitted and $A \cos \omega_2 t$ to a 0 transmitted in a binary interval $1/R$ seconds long. (Strictly speaking, each carrier term must be multiplied by the shaping term $h(t)$, but we ignore that term in the interests of simplicity.) An example appears in Fig. 5.2. It is then found that the transmitted signal bandwidth is given approximately by $2\Delta f + 2B$, where the baseband bandwidth B again varies between $R/2$ and R, depending on the shaping used (Schwartz, 1990). The FSK transmission bandwidth can thus be substantially more than the bandwidth required for OOK transmission. The same result is obtained for analog

Schwartz, M. 1990. *Information Transmission, Modulation, and Noise*, 4th edn, New York, McGraw-Hill.

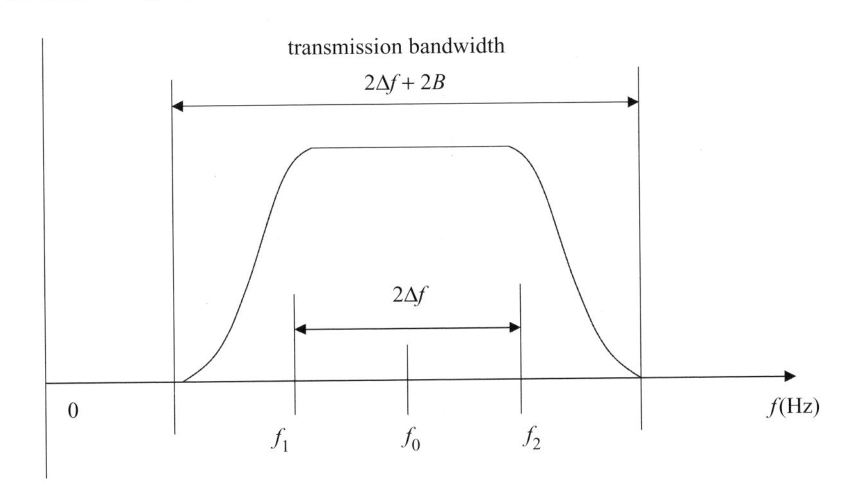

Figure 5.3 FSK spectrum

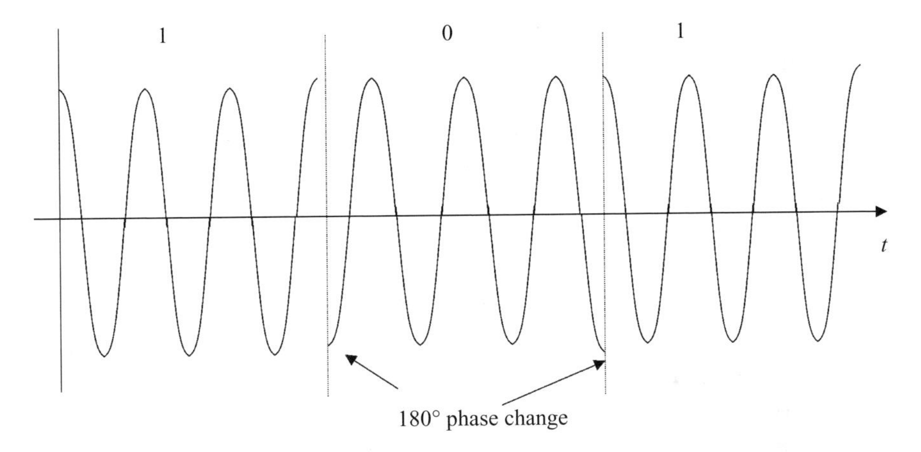

Figure 5.4 PSK signal

FM. A simple pictorial example of the FSK spectrum, indicating the carrier frequencies and the transmission bandwidth, appears in Fig. 5.3.

Consider now phase-shift keying, or PSK, the last of the three basic ways of digitally modulating a carrier in conformance with the information-bearing binary signal sequence. In this case, as an example, we might have the carrier term $A \cos \omega_0 t$ transmitted when a binary 1 appears in the baseband binary sequence; $-A \cos \omega_0 t = A \cos (\omega_0 t + \pi)$ when a 0 appears. (We have again left shaping factors out to simplify the discussion.) An example appears in Fig. 5.4. PSK transmission turns out to be the best scheme to use in the presence of noise (Schwartz, 1990). It requires a phase reference at the receiver, however, in order for accurate, error-free, detection of the transmitted binary signal sequence to be carried out. Note that both FSK and PSK provide ostensibly constant-amplitude transmission, a desirable property in the presence of signal fading with its attendant random variation of amplitude, as already noted. However, note from Fig. 5.4 that abrupt *phase* changes occur when the signal switches from a 1 to a 0 and vice-versa. (The effect is exaggerated in

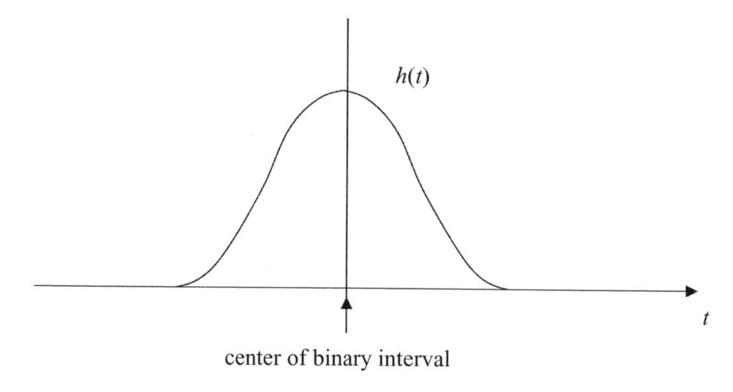

Figure 5.5 Typical shaping function

Fig. 5.4 by not showing the signal shaping that is required.) These abrupt changes in phase (π radians or 180° in the case of Fig. 5.4) will be modified somewhat and smoothed during transmission, since abrupt changes imply infinite bandwidth in the signal spectrum, a physical impossibility. However, the phase changes do result in an undesirable increase in signal transmission bandwidth and a resultant undesirable variation in the transmitted signal amplitude, because of the limited bandwidth encountered during transmission. Steps thus have to be taken to reduce these potentially large, abrupt phase changes. We shall, in fact, discuss modifications to the modulation technique that accomplish this. Before we do this, however, we digress somewhat in the next section to discuss signal shaping in a quantitative way. This is important since the concept of shaping plays a key role in the enhanced modulation techniques appropriate to mobile wireless communications that we will be describing later.

5.2 Signal shaping

We have noted in the previous section that shaping must be used in digital communication schemes to keep the bandwidth of the resultant modulated transmission signal within the prescribed system bandwidth. We pursue this concept of shaping in this section, focusing on a class of shaping functions that have been used for many years in designing digital communication systems. Other types of shaping functions will be encountered as well in the specific modulation schemes designed for mobile systems that we will be discussing in the next section. The discussion in this section should provide the necessary insight to understand signal shaping in general.

Recall that the shaping of a high-frequency radio carrier signal at frequency f_0 was indicated by writing the carrier signal in the form $h(t) \cos \omega_0 t$. The time-varying function $h(t)$ is the shaping function. To keep the modulated signal smooth, particularly during bit transitions from a 1 to a 0 and vice-versa, it is clear that the function $h(t)$ must have a form such as that shown in Fig. 5.5. It should be maximum at the center of the binary interval $1/R$ and then drop off smoothly on either side of the maximum. One can then reproduce the original binary symbol, whether a 1 or a 0, by sampling the value at the center. Note, as an example, that if such a function were to multiply the OOK sinewave of Fig. 5.1, it would reduce considerably the abrupt amplitude changes encountered in

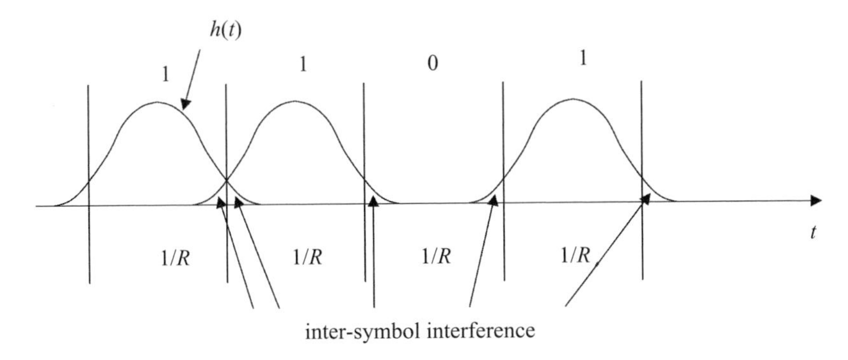

Figure 5.6 Binary sequence

switching from a 1 transmitted to a 0 and vice-versa. Rather than focus on the modulated high-frequency signals of Figs. 5.1, 5.2, and 5.4, and the effect of multiplying them by $h(t)$, however, it is simpler to consider shaping of the binary baseband signal sequence of 1s and 0s. The multiplication by $\cos \omega_0 t$ then serves to shift the shaped baseband spectrum up to a spectrum centered at the carrier frequency f_0. Shaping can thus be carried out either at baseband, as we shall assume, for conceptual simplicity, or directly at the carrier frequency. The result is the same.

Thus, consider a baseband sequence of 1s and 0s, corresponding to a bit rate of R bits/sec (bps), each bit multiplied by $h(t)$, as shown in Fig. 5.6. The 1s are each represented by the appearance of $h(t)$. The 0s give rise to a gap in the time sequence. It is well known from Fourier analysis that as a time function is narrowed, its spectral bandwidth increases correspondingly. As the function broadens in time, its bandwidth decreases. The width of a pulse and its bandwidth are inverse to one another (Schwartz, 1990). Thus if the width of $h(t)$ in Fig. 5.6 is reduced, the bandwidth of the baseband signal sequence is increased; if the width of $h(t)$ increases, the corresponding bandwidth decreases. As the width of $h(t)$ increases however, the pulses begin to intrude into the adjacent binary intervals. This gives rise to *inter-symbol interference*, as shown in Fig. 5.6. Initially there is no problem. For it is always possible to sample a binary pulse at its center and determine whether it is a 1 or 0. But eventually the inter-symbol interference becomes large enough to affect the value of the binary signal at the center of its particular interval, and mistakes such as confusing a 1 for a 0 and vice versa may occur. This is particularly true if noise is present, interfering with the signal transmitted. There is thus a tradeoff between inter-symbol interference and the bandwidth needed to transmit a sequence of pulses. The same observation applies to the modulated binary sequence obtained by multiplying by $\cos \omega_0 t$ and shifting up to the carrier frequency f_0. We shall in fact see this tradeoff occurring in discussing specific shaping functions in the next section, as used in digital mobile communications systems.

To clarify this tradeoff between bandwidth and inter-symbol interference more specifically, we discuss now a specific type of shaping, adopted in many digital communication systems. Other examples used in current, second-generation digital mobile systems will be provided in the next section, as has already been noted. This type of shaping is termed *sinusoidal rolloff* shaping. To introduce this type of shaping, we consider first a special case, *raised-cosine* shaping. The term *raised-cosine* is used because the Fourier Transform or

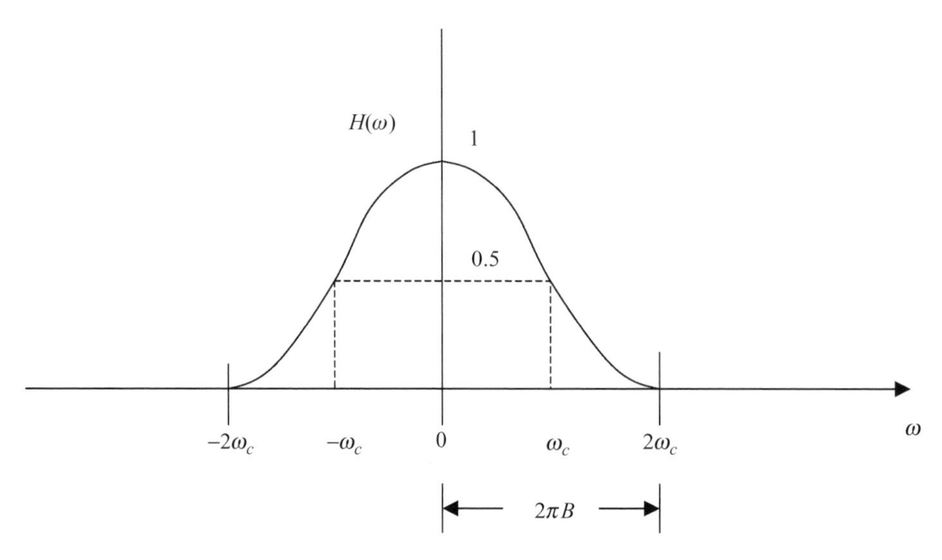

Figure 5.7 Raised-cosine spectrum

spectrum, $H(\omega)$, of the shaping function $h(t)$ is of a raised-cosine form, as presented below in (5.1)

$$
\begin{aligned}
H(\omega) &= \frac{1}{2}\left(1 + \cos\frac{\pi\omega}{2\omega_c}\right) \quad |\omega| \leq 2\omega_c \\
&= 0 \quad \text{elsewhere}
\end{aligned}
\tag{5.1}
$$

This function is sketched in Fig. 5.7. This type of shaping and sinusoidal rolloff shaping in general are special cases of Nyquist-shaping functions, named after the great AT&T researcher H. Nyquist, who first described these shaping functions in the context of telegraphy in the 1920s (Nyquist, 1928; Schwartz, 1990). Note that the raised-cosine spectrum of (5.1) and Fig. 5.7 has $2\omega_c$ as its (baseband) bandwidth, in units of radians/sec, with ω_c an adjustable parameter to be discussed further below. The bandwidth B in units of Hz is then given by $B = \omega_c/\pi$, as indicated in Fig. 5.7. The high-frequency or rf bandwidth using OOK or PSK transmission would then be double this value, as already noted. Defining a variable $\Delta\omega \equiv \omega - \omega_c$, with $\Delta\omega$ measuring the variation away from the variable ω, as shown in Fig. 5.8, one obtains, as given in (5.2) below, an alternate way of writing (5.1), leading directly to the more general case of sinusoidal rolloff shaping discussed below

$$
H(\omega) = \frac{1}{2}\left(1 - \sin\frac{\pi}{2}\frac{\Delta\omega}{\omega_c}\right) \quad -\omega_c \leq \Delta\omega \leq \omega_c \tag{5.2}
$$

Taking the inverse transform of (5.1) or (5.2), it may be shown that the impulse function or shaping function $h(t)$ in time is given by

$$
h(t) = \frac{\omega_c}{\pi}\frac{\sin\omega_c t}{\omega_c t}\frac{\cos\omega_c t}{1 - (2\omega_c t/\pi)^2} \tag{5.3}
$$

Nyquist, H. 1928. "Certain topics in telegraph transmission theory," *Transactions of the AIEE*, 47 (April), 617–644.

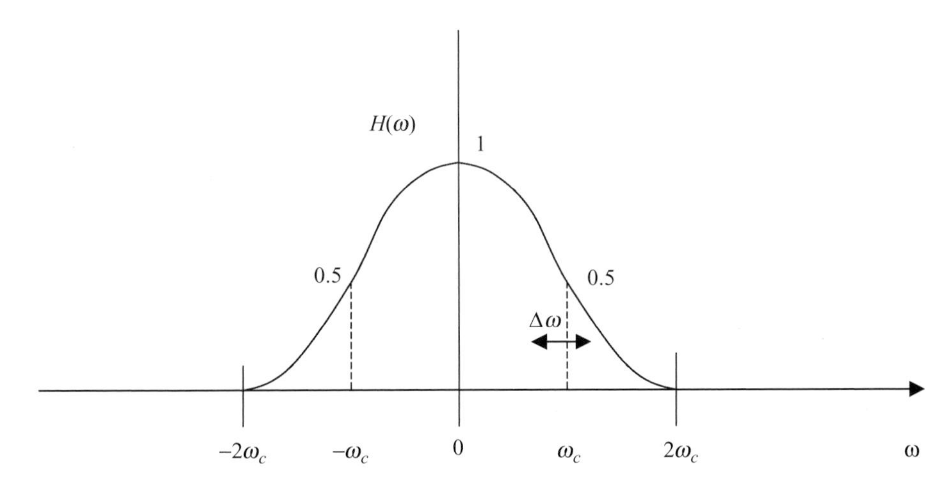

Figure 5.8 Alternate variable representation

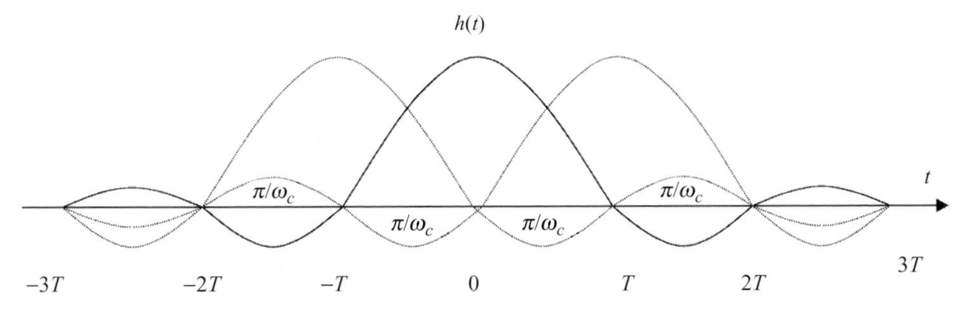

Figure 5.9 Impulse response for spectrum of Figs. 5.7, 5.8

This function is shown sketched in Fig. 5.9. Note that this function has the property that it goes through 0 at intervals spaced π/ω_c sec apart. Binary digits spaced that distance apart and centered at the 0 points of the other digits will thus not interfere with one another. This is indicated by the added pulses indicated by dashed lines and spaced π/ω_c sec on either side of the main pulse shown in Fig. 5.9. In practice, some pulse time jitter will always occur, resulting in some inter-symbol interference. But note from (5.3) that the tails of these binary pulses drop off as $1/t^3$ for t large enough, so that any inter-symbol interference can be reduced to a tolerable level. Selecting the binary interval $1/R$ to be just this spacing between zeros of $h(t)$, we have $1/R = \pi/\omega_c$. A binary transmission rate R thus requires a baseband bandwidth, in Hz, of $B = \omega_c/\pi = R$. As an example, binary transmission at the rate of 14.4 kbps requires a baseband bandwidth of 14.4 kHz if raised-cosine shaping is used. The corresponding high-frequency bandwidth, centered about the carrier frequency, would be 28.8 kHz for OOK or PSK transmission. Doubling the binary transmission rate doubles the bandwidth required to transmit these two types of digital high-frequency transmission.

Consider now the more general case of sinusoidal rolloff shaping. This is used to control the inter-symbol interference, while keeping the bandwidth required for transmission as low as possible. As noted above, raised-cosine shaping is a special case. By sinusoidal

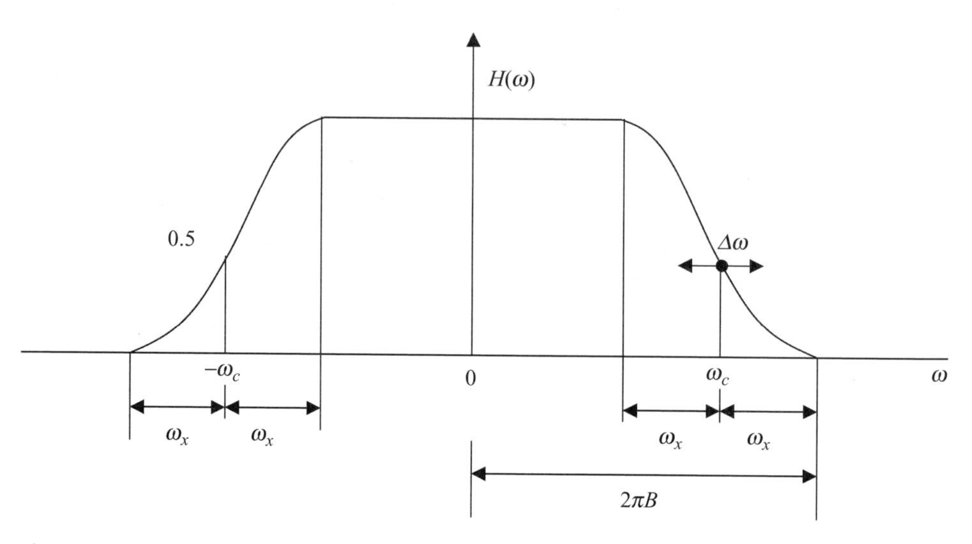

Figure 5.10 Sinusoidal rolloff shaping spectrum

rolloff shaping we mean shaping with the spectrum $H(\omega)$ given by

$$\begin{aligned} H(\omega) &= \frac{1}{2}\left(1 - \sin\frac{\pi}{2}\frac{\Delta\omega}{\omega_x}\right) && |\Delta\omega| \le \omega_x \\ &= 1 && -(2\omega_c - \omega_x) < \Delta\omega < -\omega_x \\ &= 0 && |\Delta\omega| > \omega_x \end{aligned} \tag{5.4}$$

This spectrum is shown sketched in Fig. 5.10. Note that it introduces a parameter $\omega_x \le \omega_c$ in place of the parameter ω_c appearing in the raised-cosine spectrum of (5.2). It thus generalizes the raised-cosine shaping spectrum, as already noted. Letting $\omega_x = \omega_c$ in (5.4), one obtains (5.2). The variable $\Delta\omega$ in (5.4) is again defined to be $\Delta\omega = \omega - \omega_c$. The corresponding pulse shape $h(t)$, the inverse Fourier Transform of $H(\omega)$, may be shown to be given by

$$h(t) = \frac{\omega_c}{\pi}\frac{\sin\omega_c t}{\omega_c t}\frac{\cos\omega_x t}{1 - (2\omega_x t/\pi)^2} \tag{5.5}$$

Note that this function has precisely the same form as that of the raised-cosine impulse response of (5.3) except that the parameter $\omega_x \le \omega_c$ replaces the parameter ω_c in the second part of the expression, as is to be expected. From Fig. 5.10 the radian bandwidth is given by $2\pi B = \omega_c + \omega_x$. In units of Hz we get $B = f_c + f_x$, with the two frequency parameters f_c and f_x obtained by dividing the corresponding radian frequency terms by 2π. The bandwidth B is thus smaller than that required in the raised-cosine case, only approaching that case as f_x approaches f_c. The $\sin\omega_c t/\omega_c t$ term in (5.5) indicates that successive binary digits may still be spaced $1/R = \pi/\omega_c$ sec apart with 0 inter-symbol interference. However, as $f_x = \omega_x/2\pi$ is reduced, reducing the bandwidth B, timing jitter must be more tightly controlled, since the $1/t^2$ dropoff in the denominator of the second term of $h(t)$ is now less rapid than in the raised-cosine case, because the term $(\omega_x t)^2$ is smaller than the equivalent term in (5.3). There is thus a tradeoff between bandwidth and timing jitter control. As f_x is increased, approaching f_c, its value in the raised-cosine case, the possibility of encountering inter-symbol interference due to timing jitter is reduced, but the bandwidth is increased. Note that Fig. 5.9 applies to this more general form of shaping, but with the rate of decrease of the lobes of the function $h(t)$ decreasing as f_x decreases.

The bandwidth $B = f_c + f_x$ required to transmit these pulses is commonly written in a form involving the ratio f_x/f_c, defined to be the rolloff factor r. Note from Fig. 5.10 that this ratio determines the rate at which the shaping spectrum drops from its maximum value of 1 to 0. As f_x, or, equivalently, r, decreases, approaching 0, the spectrum drops more rapidly, reducing the bandwidth B, but, as already noted, increasing the possibility of inter-symbol interference due to timing jitter. Recall that we had the bit rate R, from Fig. 5.7, given by $R = \omega_c/\pi = 2f_c$, since $\omega_c = 2\pi f_c$. This enables us to write the bandwidth B in the form

$$B = f_c + f_x = (R/2)(1 + f_x/f_c) = (R/2)(1 + r) \tag{5.6}$$

Here, as already noted, the rolloff factor $r \equiv f_x/f_c \leq 1$. The transmission or radio bandwidth for OOK or PSK transmission is then $2B = R\,(1+ r)$. The transmission bandwidth in these cases thus ranges from values just above R to $2R$, the value in the raised-cosine case with $r = 1$. Labeling the transmission bandwidth B_T, we have

$$B_T = R(1 + r) \tag{5.6a}$$

In subsequent references to transmission bandwidth, we will usually drop the subscript and simply use the letter B to refer to that bandwidth. There should be no confusion with this change in designation, the discussion always indicating whether it is the baseband or transmission bandwidth to which reference is being made.

Say, as an example, that 14.4 kbps transmission is desired. If sinusoidal rolloff shaping and PSK modulation are used, the required transmission bandwidth is 21.6 kHz if a rolloff factor of 0.5 is used; 28.8 kHz if the rolloff factor is 1. If 28.8 kbps is desired, these bandwidths double. Conversely, given the bandwidth available, one can determine the maximum binary transmission rate for a given rolloff factor. The second-generation cellular wireless system D-AMPS (IS-136), as an example, uses sinusoidal rolloff shaping, with a rolloff factor $r = 0.35$. It slots its overall frequency band assignment into 30 kHz-wide user bands, as noted earlier. If PSK modulation were to be used, the maximum binary transmission rate would be $R = 30{,}000/1.35 = 22.2$ kbps. The modulation scheme actually used, to be described in the next section, is labeled DQPSK. The actual transmssion rate is 48.6 kbps. The Japanese PDC system also uses sinusoidal rolloff shaping, with the rolloff factor $r = 0.5$. The bandwidth available per user is 25 kHz. For PSK modulation, the binary transmission rate would then be 16.7 kbps. This system also uses DQPSK and the actual binary transmission rate is 42 kbps. We shall see, in the next section, how more sophisticated modulation schemes allow substantially higher rates of transmission.

5.3 Modulation in cellular wireless systems

We noted in Section 5.1, on digital modulation schemes, that specific problems arise in dealing with transmission over a bandwidth-limited cellular wireless channel, which has the potential for introducing random amplitude variations due to fading. Modulation schemes must be selected that provide an essentially constant amplitude of transmission. PSK and FSK discussed in Section 5.2 do have this property, but instantaneous phase shifts of up to π radians (180°) that may be incurred in changing from one bit value to another give rise to amplitude changes during transmission due to limited bandwidth. In addition, as noted in Section 5.2, the bit rate possible using PSK over a bandlimited channel of bandwidth B is strictly limited by B and directly proportional to it. We thus look to somewhat more complex digital modulation schemes that produce transmitted signals with limited amplitude and phase variations, from one bit to another, while allowing higher bit rates to be used for a given bandwidth. These types of modulation, examples of which are provided in this section, are based on, and utilize, phase-quadrature transmission.

We thus begin the discussion in this section by focusing on a technique called *quadrature* PSK or QPSK, which allows a system to transmit at double the bit rate for a given bandwidth. Consider transmitting *two* carriers of the same frequency, f_0, at the same time, with one in phase quadrature to the other: $\cos\omega_0 t$ and $\sin\omega_0 t$. Let each have two possible amplitude values, say ± 1, for simplicity. Let each last the same length of time, T sec. Each then corresponds to a PSK signal. Let $a_i = \pm 1$ be the value assigned to the cosine (inphase) carrier of the ith such signal; let $b_i = \pm 1$ be the value assigned to the sine (quadrature) carrier of the same signal. Say each of the two carriers is shaped by a shaping factor $h(t)$. The ith signal $s_i(t)$, lasting an interval T, is then given by the expression

$$s_i(t) = a_i h(t)\cos\omega_0 t + b_i h(t)\sin\omega_0 t \quad 0 \le t \le T \tag{5.7}$$

The shaping factor $h(t)$ is assumed here to be centered about the middle of the interval T. This expression may also, by trigonometry, be written in the following amplitude/phase form

$$s_i(t) = r_i h(t)\cos(\omega_0 t + \theta_i) \quad 0 \le t \le T \tag{5.7a}$$

The four possible values of the amplitude-phase pair (r_i, θ_i) are obviously related one-to-one to the corresponding values of (a_i, b_i). Four possible signals are thus obtained, depending on the values of the combination (a_i, b_i), each one of which requires the same bandwidth for transmission. How are the actual values of a_i and b_i determined? How would transmission be carried out? Let the bit rate be R bps, as previously. Successive *pairs* of bits are stored every T sec, giving rise to four different two-bit sequences. Note then that we must have the two-bit interval $T = 2/R$. Each of these four sequences is mapped into one of the four carrier signals, and that particular signal, lasting T sec, is transmitted. One possible mapping of these signals appears in Table 5.1. A block diagram of the resultant QPSK modulator appears in Fig. 5.11.

If we use sinusoidal rolloff shaping, as in (5.4) or (5.5), the *baseband* bandwidth $B = (1/T)(1 + r)$, with $1/T$ replacing the previous $R/2$. The bandwidth required to transmit a given bit sequence has thus been reduced by a factor of two. Conversely, given a bandwidth B, we can now transmit at twice the bit rate. The transmission or rf bandwidth, as in the case

Table 5.1 *Example of mapping, binary input sequence → QPSK (a_i, b_i)*

Successive binary pairs	a_i	b_i
0 0	−1	−1
0 1	−1	+1
1 0	+1	−1
1 1	+1	+1

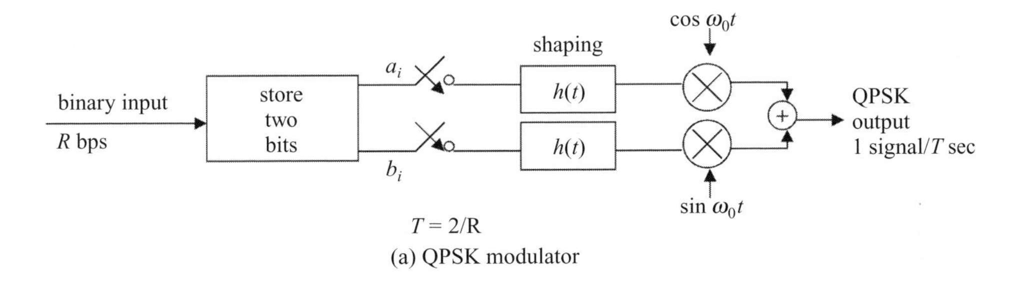

(a) QPSK modulator

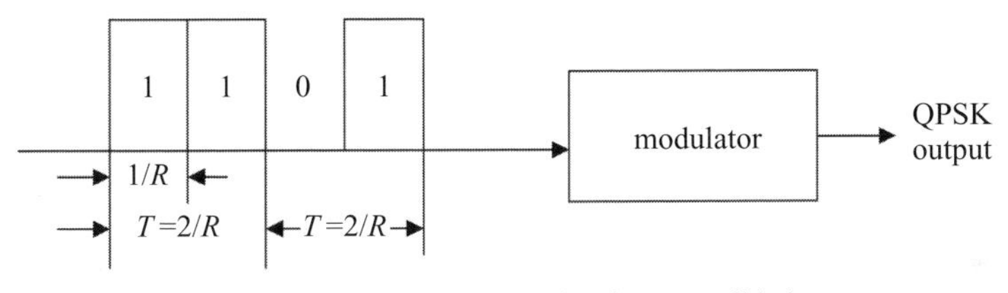

(b) Example of two possible inputs

Figure 5.11 Generation of QPSK signals

of PSK, is double the baseband bandwidth, or $(2/T)(1 + r)$. Demodulation, the process of reconstructing the original binary sequence at a receiver, is readily carried out by noting that the cosine and sine terms in the QPSK signal of (5.7) are orthogonal to one another. Multiplying the QPSK received signal in two parallel operations by cosine and sine terms, respectively, each exactly in phase with the transmitted cosine and sine terms, and then integrating over the T-sec interval (effectively low-pass filtering the two multiplied terms), will extract each term separately (Schwartz, 1990). This operation requires exact phase knowledge of the transmitted signal, precisely the same problem encountered with PSK transmission. An alternative procedure, to be described briefly later in this section, is to use successive signals as phase references for the signals following.

The operation of quadrature modulation may be diagrammed geometrically on a two-dimensional diagram, as shown in Fig. 5.12(a). The horizontal axis represents the coefficient a_i of the inphase, cosine, term of (5.7); the vertical axis represents b_i, the coefficient of the quadrature, sine term. Note that putting in the four possible values of (a_i, b_i),

Table 5.2 $\pi/4$ *DQPSK*

(a_i, b_i)	$\Delta\theta(n) = \theta(n+1) - \theta(n)$
$-1, -1$	$-3\pi/4$
$-1, +1$	$-\pi/4$
$+1, -1$	$\pi/4$
$+1, +1$	$3\pi/4$

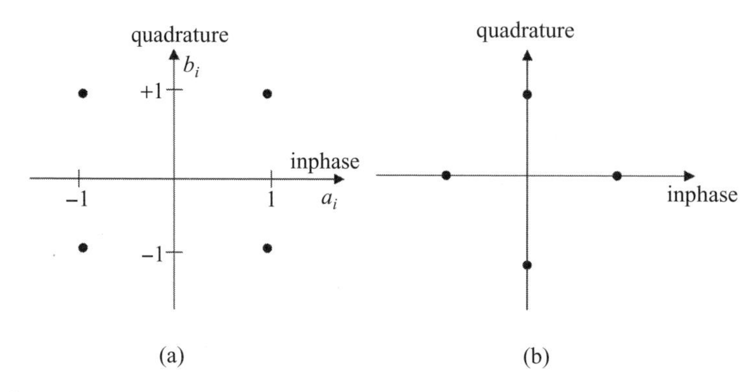

Figure 5.12 QPSK signal constellation

one obtains the four points of operation shown in Fig. 5.12(a). These points correspond directly as well to the four values of (r_i, θ_i) appearing in (5.7a). These points are referred to as the *signal constellation* of the QPSK signals under discussion. Each point may be written as the complex number or phasor $\boldsymbol{a}_i$, the real part of which is a_i, the imaginary part of which is b_i. A little thought will make it clear that an equivalent representation of the QPSK signals would be obtained by rotating the signal constellation of Fig. 5.12(a) 45° or $\pi/4$ radians, producing the constellation shown in Fig. 5.12(b). This constellation corresponds to transmitting one of the four signals $\pm\cos\omega_0 t$, $\pm\sin\omega_0 t$, depending on the particular value of $\boldsymbol{a}_i = (a_i, b_i)$. It thus represents another mapping of the four possible values of (a_i, b_i), the first example of which was provided in Table 5.2. Note that, although the four points in the signal constellation lie on a circle, transitions between these points, and hence signals transmitted, will result in some signal amplitude variation in the vicinity of the transition times.

The T-sec long interval we have introduced, corresponding to the time to store two bits, and the time over which each QPSK signal (one of four possible) is transmitted, is called the *baud* or *symbol* interval, as distinct from the binary interval $1/R$. One refers to a given number of baud, rather than bits/sec, and it is the baud number that determines the bandwidth. In the special case of two-state, binary, transmission the number of baud and the bit rate are the same. For the QPSK case, however, they differ by a factor of 2. Equations (5.6) and (5.6a) must be changed with the symbol or band rate $1/T$ used in place of R. This change has been noted above. For example, say we transmit 28.8 kbps using QPSK transmission. This corresponds to 14.4 *k*baud or 14.4 *k*symbols per second. The

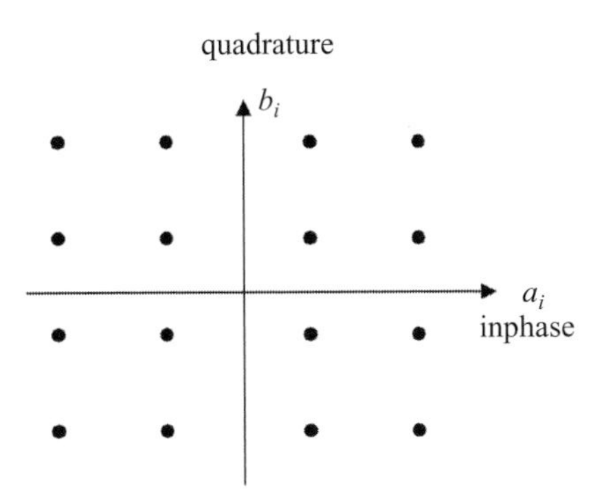

Figure 5.13 Signal constellation 16-QAM

baseband bandwidth is $7.2(1 + r)$ kHz if sinusoidal rolloff shaping is used. The bandwidth required to transmit the radio-frequency QPSK signal is twice that or 14,400 $(1 + r)$ Hz, with r the rolloff factor, as before.

Consider now extending this idea of storing two bits to reduce the bandwidth required for transmission. Say three successive bits are stored, as an example. This results in $2^3 = 8$ possible 3-bit sequences. Corresponding to these sequences, we could transmit eight different signals, all of the same amplitude but differing in phase by $\pi/4$ radians. Alternately, these eight signals could differ in both phase and amplitude. The former case is referred to as 8-PSK signaling. This type of signaling would be represented by a signal constellation similar to that of Fig. 5.12 with eight points symmetrically located on a circle centered around the origin. This 8-PSK signaling scheme has, in fact, been adopted for an enhanced bit-rate version of GSM called *EDGE*. This scheme is discussed further in Chapter 10 on third-generation cellular systems.

The second case noted above, when designing a set of eight signals, each corresponding to one of the 3-bit input sequences, is to have both amplitude and phase differ. The signal representations of equations (5.7) and (5.7a) still apply. The only difference with the QPSK and the 8-PSK cases is that a signal constellation consisting of a mapping of eight possible values of the amplitude pair (a_i, b_i) is obtained, each of which may be defined as a complex number with a_i its real part in the plane of the signal constellation and b_i its imaginary part. More generally now, say k successive bits are stored, resulting in $M = 2^k$ possible k-bit sequences. Each of these sequences can again be mapped into a signal of the form of (5.7). The resultant set of M different signals must then differ in the values of a_i and b_i. This modulation technique is referred to as *quadrature amplitude modulation*, or QAM. These signals correspond to M possible points in a two-dimensional signal constellation. QPSK and 8-PSK are clearly special cases, with the signal points arranged on a circle. A specific example of a 16-point signal constellation appears in Fig. 5.13. In this case, one of 16 possible signals of the form of (5.7) would be transmitted every T-sec interval. Four successive bits would be stored in this case to generate the corresponding 16-QAM signal. The bandwidth required to transmit this sequence of signals is one-fourth that of the equivalent PSK signals, since the length of a given one of the 16 possible signals has been increased by a factor of four. Conversely, given a specified channel bandwidth one can transmit at four times the bit rate. More generally, with k successive bits stored, the

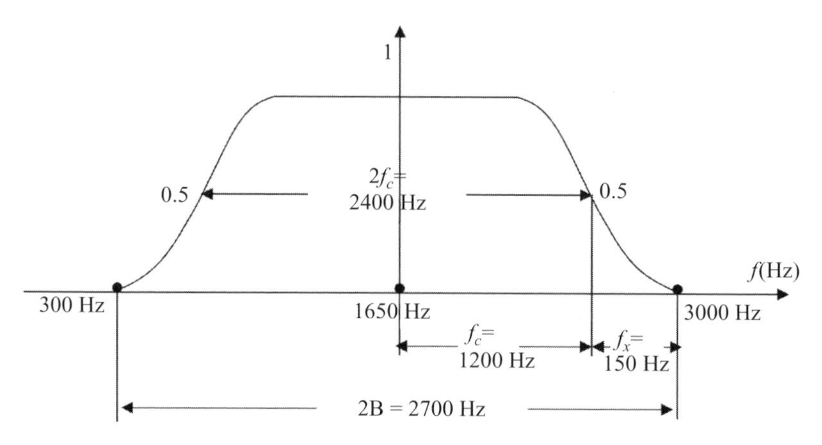

Figure 5.14 Telephone modem bandwidth

baud length T increases by a factor of k, allowing k times the bit rate to be transmitted over a given bandwidth. Aside from the special case of 8-PSK, QAM has not generally been adopted for cellular wireless systems, because it does introduce amplitude variations in the resultant signal set transmitted. It has been adopted for wireless local area networks (LANs), to be discussed in Chapter 12. We also refer to QAM in the next Section 5.4, on orthogonal frequency division multiplexing, or OFDM, a technique adopted, as we shall see, for high-speed data transmission over wireless LANs. It is of interest to discuss the technique in somewhat more depth, however. We do this through its use with telephone modems because of the relative simplicity of that application. In particular, QAM is the basis of operation of the 28.8 kbps telephone-wire modems (from *mo*dulator–*dem*odulator) that were in use prior to the introduction of the 56 kbps modems commonly used at this time. (The 56 kbps modems use a version of OFDM, but it would be going too far afield to focus on that application of OFDM.)

We introduce QAM for telephone modems with the following simple example. Say the copper telephone access line to a home or business is capable of transmitting signals over the frequency range 300–3000 Hz. QAM is to be used to carry binary information over this channel. Hence the modem is to be of the QAM type. A carrier frequency f_0 is selected at the center of this band, at 1650 Hz. This band is then equivalent to the much-higher rf bands used in wireless communications. A diagram appears in Fig. 5.14. 12.5% rolloff shaping is used, i.e., $r = 0.125$, so that the frequency f_c, referring to Fig. 5.14, is 1200 Hz, while the frequency f_x is 150 Hz. The transmission bandwidth is then 2700 Hz. Recall from Fig. 5.9 that, in the binary communication case, we had $1/R = \pi/\omega_c = 1/2f_c$ Hz, or $R = 2f_c$. In the case of QAM transmission, the symbol rate $1/T = R/k$, in units of baud, plays the same role as R in the binary case. We thus have $1/T = 2f_c$ baud. In particular, in the telephone example of Fig. 5.14, $1/T = 2400$ baud. Consider two cases now. In case 1, $k = 4$ bits are stored, giving rise to $2^4 = 16$ possible QAM signals, as shown by (5.7). The actual transmitter or modulator would, as an example, be of the form of Fig. 5.11(a), except that four successive bits are stored, rather than the two shown in that figure. A possible signal constellation is the one shown in Fig. 5.13. The bit rate achievable with this scheme is then $R = 4/T = 9600$ bps. For case 2, let 12 successive bits be stored, so

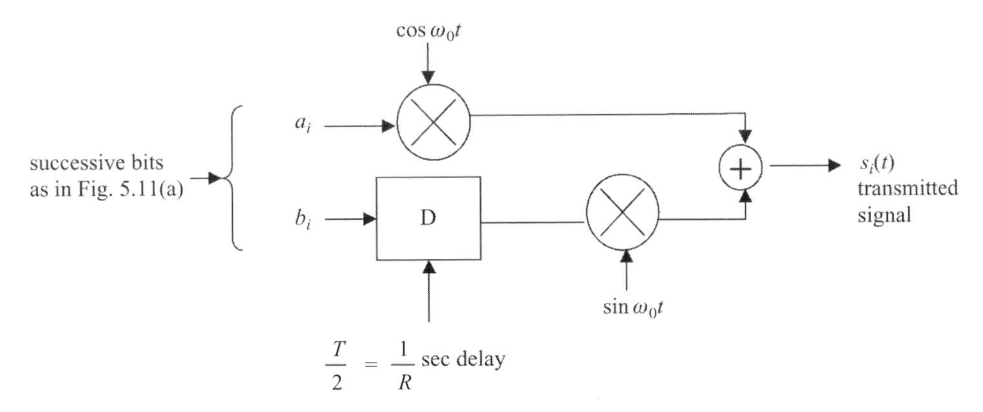

Figure 5.15 OQPSK block diagram; T = baud or symbol interval

that, in each interval $T = 1/2400$ sec long, one of $2^{12} = 4096$ possible QAM signals is transmitted, depending on the particular sequence of 12 bits entering the modulator. The signal constellation would have 4096 points. The bit rate possible, for the same symbol rate of 2400 baud over the telephone line with transmission bandwidth of 2700 Hz, is now 28.8 kbps. This is, of course, the modem bit rate used with PCs before the advent of the 56 kbps modem, as noted above.

After this brief aside, we return to our primary subject of interest in this section, the digital cellular mobile domain. (We return briefly to QAM in discussing OFDM in the next section, as noted earlier.) We focus on QPSK, which has no amplitude variation, as is apparent from the signal constellations of Fig. 5.12. As noted earlier, this modulation scheme doubles the bit rate possible over the band-limited wireless channel, a desirable feature in this environment. There is still the problem of an abrupt 180° (π rad.) phase shift occurring during the switch from one QPSK signal to another at intervals of T sec, a problem we noted existed with PSK, if both a_i and b_i change sign. This is apparent from an examination of the two constellations of Fig. 5.12. Note that the phase changes by 90° ($\pi/2$ rad.) if only one of these coefficients changes. A scheme developed to reduce the maximum phase shift, and hence keep the amplitude quasi-constant during transmission over the band-limited wireless channel, is called *offset QPSK* (OQPSK) or, in a modified form, *minimum-shift keying* (MSK). This scheme serves as the basis for modulation schemes selected for some of the current cellular wireless systems (Pasupathy, 1979). In OQPSK the modulation by one of the two (quadrature) carriers is offset in time by one bit. ($T/2 = 1/R$ sec.) A block diagram appears in Fig. 5.15. The system to the left of the two input lines in the figure is the same as that appearing in Fig. 5.11(a).

MSK is similar, except that shaping or multiplying by the sine function $\cos \pi t/T$ is first carried out, before carrying out the multiplying (modulation) and delay functions shown in Fig. 5.15. This is shown schematically in the block diagram of Fig. 5.16. Note that the acronym "lpf" shown there represents the shaping by $\cos \pi t/T$. It stands for low-pass filtering (GMSK, to be described later, uses gaussian shaping). The system to the left of

---

Pasupathy, S. 1979. "Minimum-shift keying: a spectrally efficient modulation," *Communications Magazine*, 17, 4 (July), 14–22.

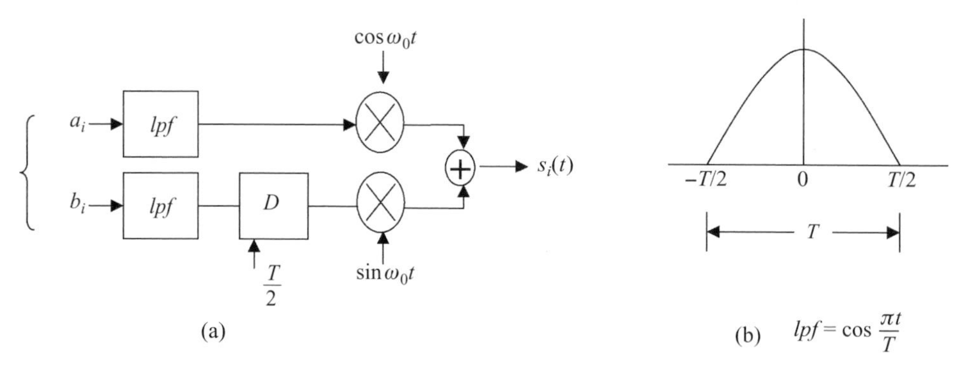

Figure 5.16 MSK modulation

the two input lines is again the same as that appearing in Fig. 5.11(a). The (a_i, b_i) pair again represents the mapping of two successive information bits into one of four possible pairs, an example of which appeared in Table 5.1.

The MSK waveform may therefore be written in the form

$$s_i(t) = a_i \cos\frac{\pi t}{T} \cos\omega_0 t + b_i \sin\frac{\pi t}{T} \sin\omega_0 t \quad 0 \leq t \leq T \tag{5.8}$$

The MSK signal is thus similar to the QPSK signal of (5.7), except that the $\sin \pi t/T$ shaping function multiplying the quadrature carrier term is in phase quadrature to the $\cos \pi t/T$ shaping function used to multiply the inphase carrier term. The $\sin \pi t/T$ term arises simply because of the $T/2 = 1/R$ delay following the lpf shaping in the quadrature section of the modulator. Note that the coefficients a_i and b_i change every $T = 2/R$ sec. By trigonometry, (5.8) may be rewritten in the following equivalent form

$$s_i(t) = \cos\left(\omega_0 t - \frac{a_i b_i \pi t}{T} + \theta\right)$$
$$\theta = 0, a_i = 1; \theta = \pi, a_i = -1 \tag{5.9}$$

Note from this form of the MSK signal s_i(t) that the signal has the desired constant amplitude. Because of the $T/2$ delay separating the influence of a_i and b_i (Fig. 5.16), only one of these coefficients can change at a time. The maximum phase change possible is thus $\pi/2$ radians, rather than the maximum phase change of π radians in the QPSK case.

But why the *minimum* shift keying designation? This is explained by rewriting the expression for the MSK signal in yet another form

$$s_i(t) = \cos(\omega_0 t \pm \Delta\omega t + \theta)$$
$$\Delta\omega = \pi/T, \quad \Delta f = 1/2T = R/4, \quad a_i b_i = \mp 1 \tag{5.10}$$

This form is representative of frequency-shift keying (FSK) (see Section 5.1), with frequency deviation Δf, in Hz, and frequency spacing, in Hz, given by $2\Delta f = 1/T = R/2$, R the information bit rate. It is in this context that the *minimum* shift keying appellation arises. To demonstrate this, consider the recovery of the information bit sequence at the receiver, at the other end of the wireless channel. We have not discussed this so-called *detection* process to this point, except to note, in the discussion of PSK in Section 5.1,

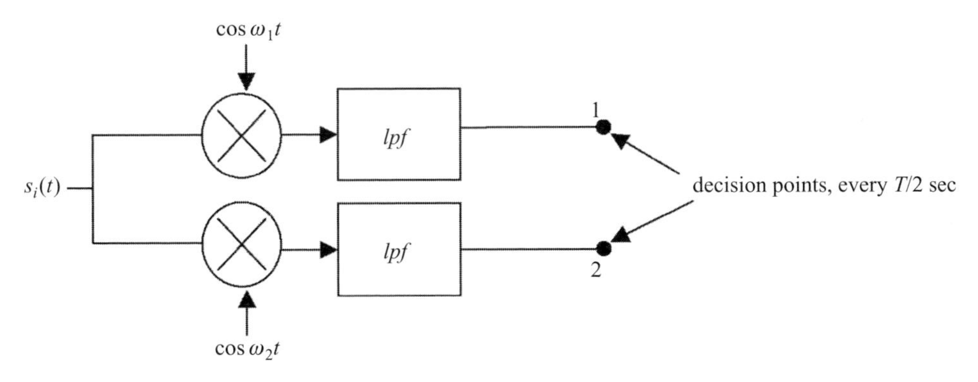

Figure 5.17 Detecting FSK signals; used to detect MSK signal

that a phase reference was required in that case. To simplify the discussion, let $\omega_1 = \omega_0 - \Delta\omega$ in (5.10), the case for which $a_ib_i = +1$, and $\omega_2 = \omega_0 + \Delta\omega$, the case for which $a_ib_i = -1$. For the time being, let $\Delta\omega$ be arbitrary. Fig. 5.17 shows a synchronous FSK detector that is applicable as well to MSK detection. The term *synchronous* detector means that both the phase and frequency of the two locally generated sinewave signals multiplying the incoming received signal $s_i(t)$ in each of the two branches shown in Fig. 5.17 are identical to, or locked to, those of the corresponding transmitted signals. The box labeled "lpf" is a low-pass filter, whose frequency response is considerably below that of the high-frequency MSK signals (or FSK signals in general). We shall approximate this device by an integrator, providing the simplest type of low-pass smoothing operation. The decision points are points at each of which the (low-pass) signal is sampled every $T/2$ sec. A decision is made at these sampling times as to the value (polarity in this case) of a_i or b_i, alternating one after the other, as generated at the transmitter. Thus, having decided on the value of a_i, for example, we use this value to determine the transmitted value of b_i $T/2$ sec later; then, given b_i, use this to determine the value of a_{i+1}, and repeat the process.

We now show how these decisions are made. Say a_i, as an example, has been found to be $+1$. In determining b_i, by sampling the two decision points $T/2$ sec later, consider each of the two possibilities. If $b_i = +1$, $a_ib_i = 1$, implying, from (5.10), that $\cos(\omega_0 - \Delta\omega)t = \cos\omega_1 t$ was transmitted. The value of the low-pass signal sample at the upper position 1 in Fig. 5.17 is then closely approximated by the integral over $T/2$ sec (the lpf output) $2/T[\int_0^{T/2} \cos^2\omega_1 t dt]$. It is left to the reader to show that, with $\omega_1 T \gg 1$, a reasonable assumption (why?), this is very nearly $1/2$. Consider the lower decision point 2 at the same sampling time. We now show that, under a simple condition, its value is zero. In particular, the value of the signal sample at decision point 2 is closely approximated by

$$\frac{2}{T}\int_0^{T/2} \cos\omega_1 t \cos\omega_2 t dt = \frac{1}{T}\int_0^{T/2} [\cos(\omega_1 + \omega_2)t + \cos 2\Delta\omega t]dt$$
$$\cong \sin\Delta\omega T/2\Delta\omega T$$

Here we have applied simple trigonometry to replace the product of cosine functions by the sum and difference angles, have used the fact that $\omega_2 - \omega_1 = 2\Delta\omega$, and have assumed

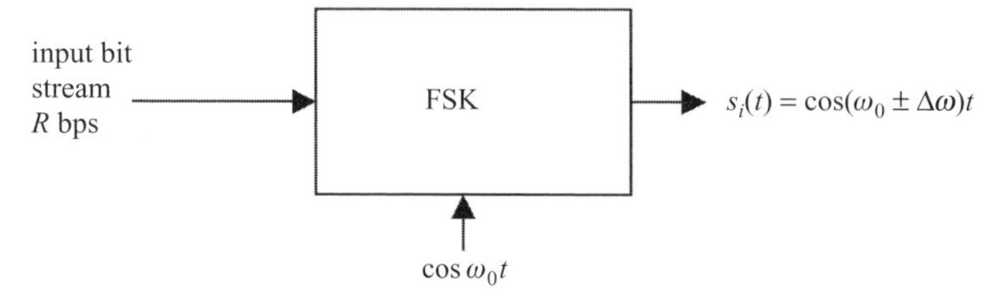

(a) FSK equivalent of MSK

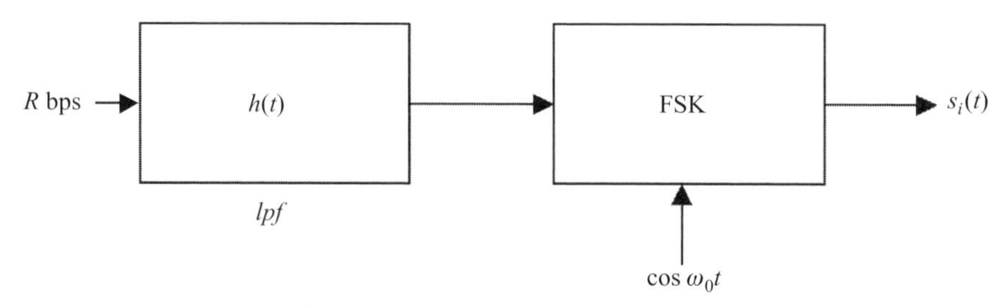

(b) $h(t)$: Gaussian shaping–gaussian MSK

Figure 5.18 Gaussian MSK modulation

that $(\omega_1 + \omega_2)\, T \gg 1$. We now choose $\Delta\omega T$ such that $\cos \Delta\omega T = 0$. With this choice of $\Delta\omega T$ this analysis indicates that, given a_i to have been found to be $+1$, a value of $b_i = +1$ would produce a positive signal at receiver decision point 1 and no output at point 2. It is left for the reader to show, using a similar analysis, that $b_i = -1$ would produce a signal at point 2 and no signal at point 1. Similarly, if $a_i = -1$, $b_i = +1$ produces an output at point 2 and nothing at point 1, while $b_i = -1$ produces the reverse. Unique decisions at the receiver as to the value of b_i can thus be made, given a_i known. Continuing the argument, it is again left to the reader to show that with b_i known, a_i can be uniquely determined. From the sequence of a_is and b_is, the information-bearing bit stream at the transmitter can in turn be reconstructed at the receiver. But note that having the decisions correctly made requires having a non-zero signal sample appear at one of the decision points only. The condition for this to happen is the one noted above: $\sin \Delta\omega T = 0$. The *smallest* frequency deviation $\Delta\omega$ satisfying this condition is $\Delta\omega T = \pi$, or $\Delta f = \Delta\omega/2\pi = 1/2T = R/4$, just the MSK condition of (5.10). To summarize, MSK modulation, looked at as an FSK modulation scheme, uses the *smallest* possible frequency shift that ensures correct detection, at the receiver, of the transmitted signal stream. (This is not to say that the signals are always correctly received as transmitted. Noise introduced in the system, fading effects, timing jitter leading to inter-symbol interference, and interference from other transmitters, may occasionally cause errors in signal detection to occur, a well-designed system not withstanding.)

Now consider *gaussian* MSK, or GMSK, used in the GSM mobile wireless system. This modulation scheme extends MSK to include gaussian shaping. Figure 5.18 demonstrates the process. In Fig. 5.18(a) we show the FSK equivalent of MSK, as given by (5.10) above. In GMSK, a low-pass gaussian shaping filter $h(t)$ precedes the FSK modulator, as shown

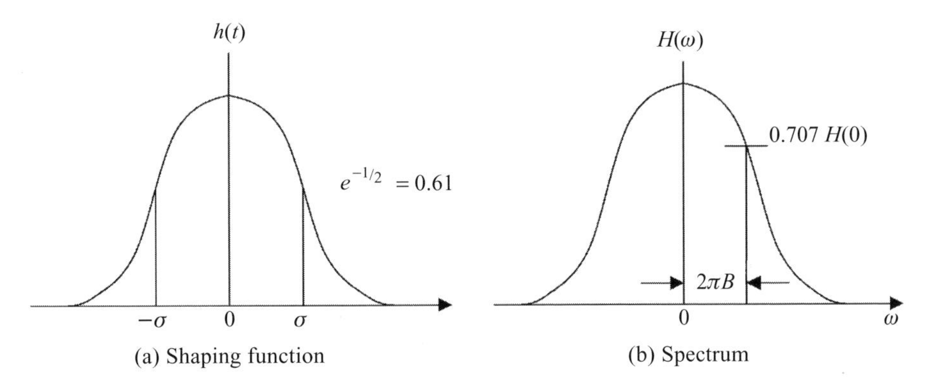

Figure 5.19 Gaussian shaping

in Fig. 5.18(b). This shaping filter has the specific gaussian form given in (5.11) below, hence its name

$$h(t) = e^{-t^2/2\sigma^2} \tag{5.11}$$

This shaping is to be applied to each information bit input, as shown in Fig. 5.18(b). Note that the parameter σ controls the width of the resultant shaped binary pulse entering the modulator. Increasing σ increases the spread of the pulse into adjacent bit intervals, increasing the inter-symbol interference. Unlike the case of sinusoidal rolloff shaping, where the shaping function is specifically chosen to attain zero inter-symbol interference, except for the effect of timing jitter, here *controlled* inter-symbol interference is introduced. The reason is that by increasing the shaped binary pulse width, the transmission bandwidth is thereby reduced: this represents the well-known inverse relationship between pulse width and bandwidth (Schwartz, 1990). This inverse relationship is demonstrated in this specific case by calculating the Fourier Transform of $h(t)$, its spectrum. It is readily shown that this transform, $H(\omega)$, is also gaussian shaped and is given by (5.12) following

$$H(\omega) = Ce^{-\sigma^2\omega^2/2} \tag{5.12}$$

C a constant. The 3 dB bandwidth, i.e., the bandwidth in Hz at which $H(\omega)$ is $1/\sqrt{2} = 0.707$ of its peak value, is readily shown to be given by $B = 0.133/\sigma$. Note the inverse relation between pulse width σ and bandwidth B. This shaping function $h(t)$ and its transform $H(\omega)$ are sketched in Fig. 5.19.

An alternate repesentation of the shaping function $h(t)$ may be written, replacing σ by its equivalent in terms of the bandwidth B

$$h(t) = e^{-t^2/2\sigma^2} = e^{-(2\pi B)^2 t^2/2 \ln_e 2} \tag{5.11a}$$

Note that as $\sigma \to 0$ or $B \to \infty$, the shaping effectively disappears and we recover the original MSK signal at the modulator output.

The effect of the shaping function on the input to the FSK modulator of Fig. 5.18(b) is found by considering an unshaped square-topped binary pulse, representing a binary 1, say, passing through the (low-pass) shaper before being acted on by the modulator. This pulse, of width $1/R = T/2 \equiv D$, excites the gaussian filter of Fig. 5.18(b). The desired output $g(t)$, which is, in turn, the input to the FSK (MSK) modulator, is found by convolving the two time functions. The resultant time function is found to be given by (5.13), suppressing a multiplicative constant

$$g(t) = \text{erfc}[2\pi B(t - D)/\sqrt{\ln 2}] - \text{erfc}[2\pi B(t + D)/\sqrt{\ln 2}] \tag{5.13}$$

The complementary error function, erfc(x), appearing in (5.13) is well-known as the integral of the gaussian function

$$\text{erfc}(x) = \frac{2}{\sqrt{\pi}} \int_x^{\infty} e^{-y^2} dy \tag{5.14}$$

The effect of the gaussian shaping may be determined by plotting the shaped time function $g(t)$ as a function of the normalized bandwidth $B_N \equiv BD= BT/2 = B/R$ (Steele, 1992: 535). As B_N decreases, reducing the bandwidth with the binary input rate R fixed, $g(t)$ spreads more into adjacent binary intervals, increasing the inter-symbol interference. The resultant spectrum, however, the Fourier Transform of $g(t)$, or, equivalently, the effect of multiplying $H(\omega)$ of (5.12) by the $\sin x/x$ spectrum of the square binary input pulse, turns out to have reduced spectral lodes as compared with the MSK spectrum (Steele, 1992: 540, 541). GMSK thus effectively provides a reduced transmission bandwidth as compared with plain MSK, at the cost of inter-symbol interference. A good compromise value of B_N turns out to be $B_N = 0.3$. This value provides a substantial reduction in transmission bandwidth with not too much inter-symbol interference introduced. This is the value that has, in fact, been adopted for the GMSK modulation used in GSM systems.

The designers of IS-136 (previously known as IS-54) or D-AMPS, the second-generation North American TDMA cellular system, chose for its modulation scheme $\pi/4$-DQPSK (*differential* QPSK). This scheme uses *eight* different phase positions, with constant amplitude, as its signal constellation. The four different values of the (a_i, b_i) pair discussed previously are assigned to corresponding *differential changes* in the constellation phase positions (Gibson, 1996: Ch. 33). Given a current point in the constellation at which the system is operating, say at the end of the T-sec interval n, there are four different points to which the system can move, depending on which of the four values of (a_i, b_i) apears in the next T-sec interval, $n + 1$. Table 5.2 provides the necessary mapping of (a_i, b_i) to the differential phase changes, $\Delta\theta(n) = \theta(n + 1) - \theta(n)$. A diagram of the eight-point constellation, with an example of the differential phase changes indicated, appears in Fig. 5.20. In this example, the system phase point at the end of interval n is taken to be $3\pi/4$, as shown in the figure. The pair $(a_i, b_i) = (-1,-1)$ arising in the next T-sec interval would then change the phase by $-3\pi/4$ rad., rotating the phase point to $\theta(n + 1) = 0$, as shown in the figure. The three other possible rotations, corresponding to the three shown in Table 5.2, are indicated in Fig. 5.20. The transmitted signal is then of the constant-amplitude phase shift type, similar to the signal indicated earlier in (5.7a), using one of the eight possible phase positions. Transitions between signal points will result in amplitude variations in the successive signals transmitted, just as in the case

Steele, R. ed. 1992. *Mobile Radio Communications*, London, Pentech Press; New York, IEEE Press.
Gibson, J. D. ed. 1996. *The Mobile Communication Handbook*, Boca Raton, FL, CRC Press/IEEE Press.

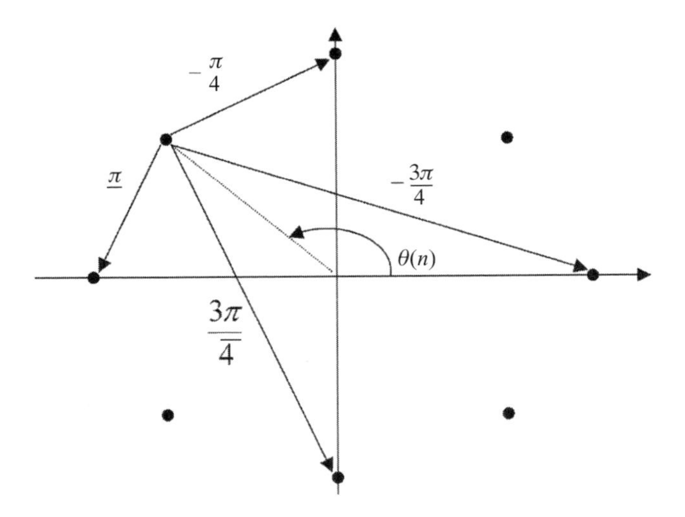

Figure 5.20 $\pi/4$ DQPSK

of QPSK. These variations, just as those for 8-PSK, will be reduced as compared with QPSK, however.

How would detection at the receiver be accomplished in this case? It is clear that phase detection is necessary, since the information to be reproduced is carried in the phase of the received signal. It was noted earlier that the receiver must generally be locked in precisely to the phase and frequency of the transmitted signal in the case of phase-shift keyed and FSK-type modulation. Such techniques then require the receiver to correctly track the absolute phase of the transmitted signal for correct detection to take place. Such absolute phase information is not needed, however, in using this type of differential phase modulation scheme. *Differential* phase shift decisions only need be made, each T-sec interval, to correctly determine the (a_i, b_i) pair generated at the transmitter, from which, in turn, the corresponding binary information sequence may be reconstructed. In essence, the transmitted signals, one after the other, provide the necessary phase reference for the signal following. This can simplify the phase detection process considerably. Provision must be made, however, for detecting and correcting an erroneous phase decision; otherwise the phase error could perpetuate indefinitely!

5.4 Orthogonal frequency-division multiplexing (OFDM)

We conclude this chapter with an introduction to OFDM. As noted in the introduction to this chapter, OFDM is currently used in high-speed DSL modems over copper-based telephone access lines. It has also been standardized as part of the IEEE standards 802.11g and 802.11a for high bit rate, 54 Mbps, data transmission over wireless LANs (WLANs). Those applications are described as part of our WLAN discussion in Chapter 12. OFDM has been proposed as well for advanced cellular systems.

The application of OFDM to wireless systems is due to the increasing need for higher bit rate, higher bandwidth data transmission over radio-based communication systems. We noted, in Chapter 2, that, as the transmission bandwidth increases, frequency-selective

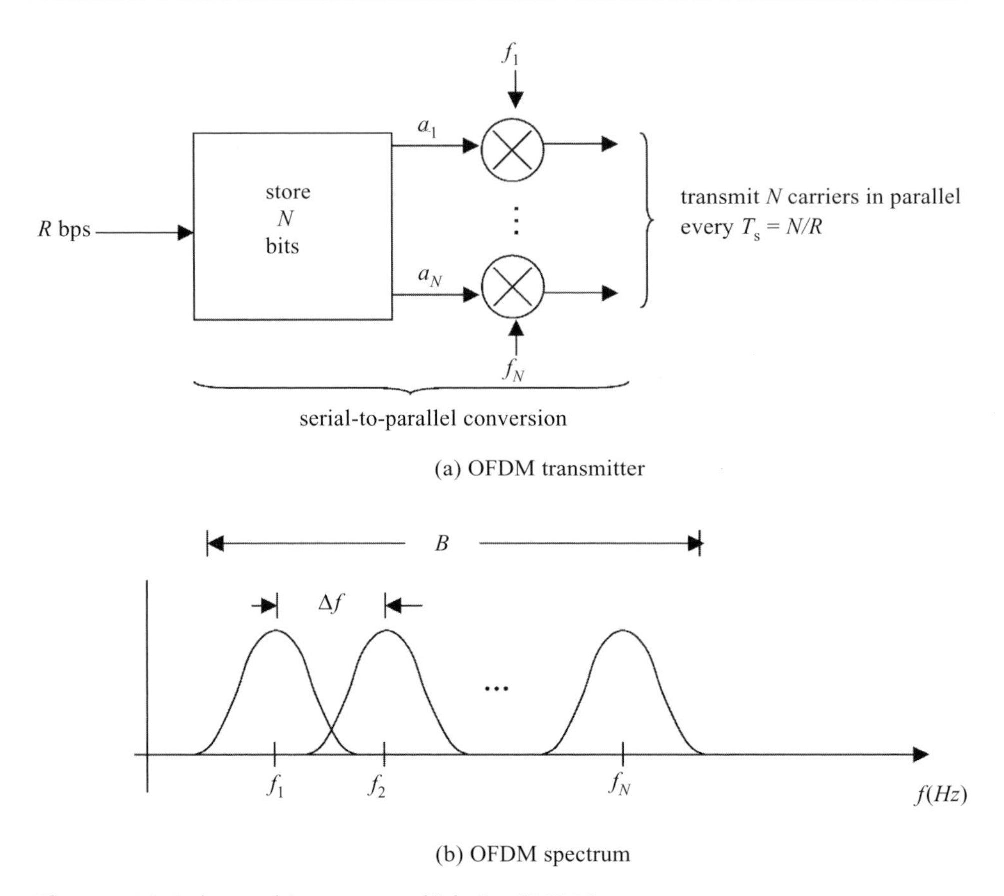

Figure 5.21 Orthogonal-frequency multiplexing (OFDM)

fading and consequent signal distortion are encountered. In the case of digital transmission, particularly, inter-symbol interference occurs, with successive digital symbols overlapping into adjacent symbol intervals. OFDM mitigates this effect by essentially dividing the signal transmission spectrum into narrow segments and transmitting signals in parallel over each of these segments. If the bandwidth of each of these frequency-spectrum segments is narrow enough, flat or non-frequency-selective fading will be encountered and the signal transmitted over each segment will be received non-distorted. It is this virtue that has resulted in OFDM taking on special significance for the wireless application.

Specifically, consider the case of binary digits transmitted at a rate of R bps. The bandwidth B required to transmit these bits is, from (5.6a), $R(1 + r)$, with r the Nyquist rolloff factor, or the order of R Hz. Now consider a sequence of N of these bits stored for an interval $T_S = N/R$. We call this interval the OFDM symbol interval. Serial-to-parallel conversion is then carried out, with each of the N bits stored used to separately modulate a carrier. All N modulated-carrier signals are then transmitted simultaneously over the T_S-long interval. This procedure constitutes the essence of OFDM. Figure 5.21(a) portrays this simple OFDM generation process. The parameters a_k, $1 \leq k \leq N$, represent the successive bits stored, while the frequencies f_k, $1 \leq k \leq N$, represent the N carrier frequencies transmitted in parallel. Figure 5.21(b) shows the resultant OFDM spectrum. To make the various carrier frequencies orthogonal to each other, it suffices to have the spacing Δf between carriers equal to $1/T_S$. We thus have $B = N\Delta f$, B the transmission bandwidth as shown in Fig. 5.21(b). Note that Δf is effectively the bandwidth of each of the N parallel frequency channels as well.

It is thus clear that this signal-spreading process has reduced the transmission bandwidth of each of the signals transmitted in parallel by the factor of N. With N large enough, flat fading, rather than frequency-selective fading, occurs on each of the frequency channels used, thus overcoming any frequency-selective fading incurred without this serial-to-parallel process. (Note that this process of storing a sequence of bits and then transmitting N carriers in parallel differs from QAM in which one carrier only is used.) What value of N or, alternatively, Δf, is desirable? Recall from Chapter 2 that flat fading occurs if the transmission bandwidth B and delay spread τ_{av} are related by

$$\tau_{av} < 1/2\pi B \tag{5.15}$$

Hence we must have $\Delta f < 1/2\pi\tau_{av}$. In equivalent terms of time, using $\Delta f = 1/T_S$, we must have

$$T_S = N/R \gg \tau_{av} \tag{5.16}$$

Simply stated, then, to avoid inter-symbol interference the OFDM symbol interval must be much larger than the delay spread. This is what one expects intuitively.

Consider some examples. As the first example, let the transmission bandwidth be 1 MHz, with 800 kbps data transmission attempted over this channel. (A Nyquist rolloff factor of 0.25 is assumed here.) A delay spread of 1 μsec would result in frequency-selective fading, with intersymbol interference encountered. Using OFDM with ten carriers spaced 100 kHz apart, the inter-symbol interference would be mitigated, with flat fading encountered over each of the ten parallel 100 kHz-wide channels. (These OFDM carriers are often referred to as subcarriers.)

As the second example, consider the case of a 20 MHz-wide channel, the one adopted for the high data rate IEEE 802.11g and 802.11a wireless LAN standards. Without the use of OFDM, delay spreads greater than 10 nsec would result in frequency-selective fading. As noted in Chapter 2, however, delay spreads encountered in various environments, including those of microcells and indoor environments, are much larger than this value. A technique such as OFDM is thus required in going to very high bandwidth systems. A carrier separation Δf, and hence subcarrier bandwidth, of 312.5 kHz has, in fact, been adopted for the 20 MHz-wide wireless LAN systems, as we shall see in Chapter 12. With an OFDM subcarrier bandwidth of 312.5 kHz, inter-symbol interference is incurred with delay spreads greater than 0.5 μsec, a substantial improvement over the case without OFDM introduced.

The implementation of OFDM using multiple carriers transmitted in parallel can, however, produce a problem in implementation. We now show, by analysis, that this problem can be avoided by the use of the discrete Fourier Transform technique. Consider the N parallel output signals of Fig. 5.21 transmitted over a T_S-long symbol interval. Call the sum of these signals, the total signal transmitted, $v(t)$. Define the kth carrier frequency f_k as $f_c + k\Delta f, 0 \le k \le N - 1$. In effect, then, we are simply redefining the carrier nomenclature, the lowest of the N parallel subcarriers written as f_c, the rest all spaced intervals of Δf

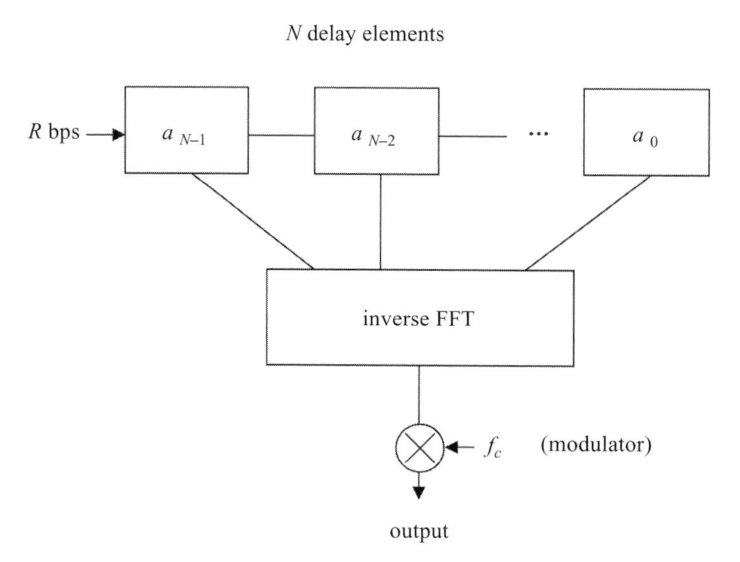

Figure 5.22 Equivalent generation of OFDM signal

above that value. Writing the kth carrier term as $\cos 2\pi(f_c + k\Delta f)t$, $v(t)$ may be written as

$$\begin{aligned} v(t) &= \text{Re}\left[\sum_{k=0}^{N-1} a_k e^{j2\pi(f_c+k\Delta f)t}\right] \\ &= \text{Re}\left[e^{j2\pi f_c t}\sum_{k=0}^{N-1} a_k e^{j2\pi k\Delta f t}\right] \\ &= \text{Re}\left[e^{j2\pi f_c t} a(t)\right] \end{aligned} \tag{5.17}$$

with $a(t) \equiv \sum_{k=0}^{N-1} a_k e^{j2\pi k\Delta f t}$. Here Re[] represents the "Real part of."

We now sample $a(t)$ at intervals T_S/N apart, i.e., at the rate of R samples per second. We thus have, in place of $a(t)$, the sampled function $a(n)$, with t replaced by nT_S/N, $n = 0 \ldots N-1$. But, recalling that $\Delta f T_S = 1$, we can write $a(n)$ as

$$a(n) = \sum_{k=0}^{N-1} a_k e^{j2\pi kn/N} \qquad n = 0 \ldots N-1 \tag{5.18}$$

We now note that (5.18) is exactly in the form of the inverse discrete Fourier Transform (Oppenheim and Willsky, 1997), and may readily be evaluated using Fast Fourier Transform (FFT) techniques. In place of the parallel transmission of N orthogonal carriers as represented diagrammatically by Fig. 5.21 and expressed mathematically by (5.17), the process of OFDM transmission has been replaced by an FFT calculation, with transmission at the single carrier frequency f_c following! This equivalent process is diagrammed in Fig. 5.22. As we shall see in Chapter 12, this is precisely the procedure followed in the standards adopted for high-speed wireless LANs. At the receiver, the reverse process is carried out each symbol interval T_S after demodulation of the received modulated carrier signal: the discrete Fourier Transform is calculated, from which the N coefficients a_k,

Oppenheim, A. V. and A. S. Willsky. 1997. *Signals and Systems*, 2nd edn, Upper Saddle River, NJ, Prentice-Hall.

$k = 0 \ldots N - 1$ are recovered and parallel-to-serial conversion used to generate the desired output bit stream. It is to be noted that N chosen as multiples of 2 is helpful in carrying out the FFT calculations.

This process of generating the OFDM signal from binary input samples using the inverse discrete Fourier Transform may be generalized to an operation on QAM signal samples, allowing higher bit rate data signals to be transmitted over a specified bandwidth. In particular, say QAM is first used to reduce the bandwidth required to transmit a given input bit stream. To be specific, let $K < N$ successive binary digits be stored, generating one of 2^K possible QAM signals. As noted earlier in our discussion of QAM, that scheme gives rise to a signal constellation such as that shown in Fig. 5.13. We noted there that each point in the constellation, representing one of the 2^K signals to be transmitted, may be represented as a complex number. Specifically, let the kth such number be labeled $\boldsymbol{a}_k$. Carrying out QAM generation using each successive group of K binary digits, and then storing the resultant N successive complex numbers, $\boldsymbol{a}_k$, $k = 0 \ldots N - 1$, over the T_S OFDM symbol interval, one may use serial-to-parallel conversion to transmit these N complex numbers using a different subcarrier for each. We have thus obtained a more general form of OFDM. But this process of first generating a QAM signal before applying OFDM corresponds to generalizing (5.17) to incorporate the complex numbers $\boldsymbol{a}_k$ in place of the binary coefficients a_k used there. The inverse discrete Fourier Transform technique is again obtained as a method of carrying out the OFDM operation by again sampling the resultant equation, but, this time, by sampling at the QAM intervals K/R units of time apart. One now obtains the inverse discrete Fourier Transform of (5.18), but modified to include the more general case of incorporating the complex coefficients $\boldsymbol{a}_k$ in place of the a_ks. This modified and more general form appears in (5.19) below

$$a(n) = \sum_{k=0}^{N-1} \boldsymbol{a}_k e^{j2\pi kn/N} \tag{5.19}$$

Fast Fourier techniques may now be carried out to calculate the inverse discrete Fourier Transform (IFFT) operation of (5.19). The transmitted signal $v(t)$ is again precisely that of (5.17), but with the more general form of (5.19) used in place of (5.18). This more general OFDM system incorporating QAM operations before carrying out serial-to-parallel conversion is diagrammed in Fig. 5.23. The I and Q signals shown at the output of the IFFT box represent the inphase and quadrature components of the complex IFFT operation.

We conclude this section with some examples of the OFDM system incorporating QAM operations.

Example 1

Let the transmission bandwidth be 1 MHz. Say a rolloff factor of $r = 0.25$ is used. Then, as noted above, data traffic may be transmitted at a rate of 800 kbps. Frequency-selective fading with consequent inter-symbol interference will be incurred for delay spreads of about 0.2 μsec or more. Let OFDM with $N = 16$ equally spaced carriers be used. The subcarrier spacing Δf is 67 kHz and the OFDM symbol interval is $T_S = 1/\Delta f = 15$ μsec. Delay spreads of 3 μsec or less can be encountered without incurring inter-symbol interference. If 16-QAM is now utilized for the OFDM system, with four successive bits stored to generate one of 16 possible points in the complex-number QAM constellation, traffic at the rate of 3.2 Mbps can be accommodated for the same delay-spread numbers.

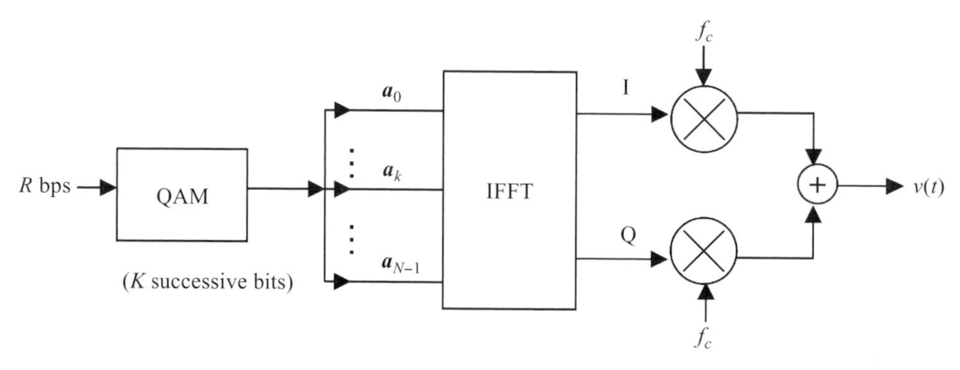

Figure 5.23 OFDM output, QAM incorporated

> Example 2
>
> For the same 1 MHz example, say it is desired to transmit at a 4.8 Mbps data rate, but with no inter-symbol interference for delay spreads up to 25 μsec. Then 64-QAM must first be used to reduce the symbol rate to 800 kbaud that the channel allows. To accommodate a delay spread of 25 μsec, we choose an OFDM symbol interval T_S at least five times the maximum delay spread or 125 μsec. Choosing $N = 128$ subcarriers means a subcarrier spacing of $\Delta f = 7.8$ kHz, with a symbol interval $T_S = 1/\Delta f = 128$ μsec.

As already noted, OFDM will be further discussed in Chapter 12 in the treatment there of the IEEE 802.11g and 802.11 high bit rate WLAN standards.

Problems

5.1 **(a)** Why are constant-envelope modulation techniques preferred for use on radio channels for mobile communications?

(b) Explain, on the basis of (a), why FM, rather than AM, was adopted for use with the first-generation AMPS.

(c) Again referring to (a), why do digital mobile wireless systems avoid using some form of QAM requiring multiple amplitude levels, despite the ability thereby to attain higher bit rates over the channels provided?

(d) Why are bandwidth and power efficiency important considerations in the design of mobile wireless systems?

5.2 **(a)** QPSK is to be used to transmit data over a wireless link operating at a carrier frequency of 871.5 MHz. The transmission bandwidth available at that frequency is 1.25 MHz. Sinusoidal rolloff shaping is to be used. Calculate the data rate that can be used for the following rolloff factors: $r = 0.25, 0.5, 1$. What would the comparable rates be if PSK were to be used? Would the results differ at a different carrier frequency? Explain.

(b) Repeat (a) for a bandwidth of 30 kHz centered about the same carrier frequency. Again, would the results differ at a different carrier frequency?

5.3 **(a)** Choose an input sequence of ten or more binary digits to be applied to a QPSK modulator as in Fig. 5.11. Map them to the five appropriate QPSK signal pairs using Table 5.1. Sketch the corresponding output QPSK signals using the signal representation of (5.7). Ignore the shaping function for simplicity. Let the carrier frequency be some multiple of $1/T$. Note the times at which π-radian and $\pi/2$-radian phase shifts occur. Correlate these with the input bit pairs.

(b) Repeat (a) for the OQPSK modulator of Fig. 5.15. What is the maximum phase shift in this case?

5.4 **(a)** Focus on QPSK. Explain, using a specified sequence of binary digits, the effect on phase transitions of changes in the sign of first, one, and, then, both of the coefficients a_i and b_i appearing in (5.7).

(b) Show, for the same sequence as in (a), how OQPSK differs from QPSK. What is it about this result that makes OQPSK preferable to QPSK for mobile communication?

(c) Compare MSK with QPSK for the same sequence as in (a).

5.5 **(a)** MSK modulation is defined by (5.8) or by Fig. 5.16. Show that (5.8) may be expressed in the equivalent forms of (5.9) and (5.10).

(b) Explain, by example, the statement in the text, that only one of the coefficients a_i and b_i can change at a time with MSK. Why is the maximum phase change then only $\pi/2$ radians?

5.6 Consider the MSK modulator shown in Fig. 5.16. Sketch the output MSK signal for a sequence of at least ten input bits, arbitrarily chosen. Demonstrate that the signal does have a constant envelope. Do any phase discontinuities appear at the bit transition points? Explain.

5.7 **(a)** The analysis of MSK detection following (5.10) shows that if $f = R/4$, as in (5.10), successive values of a_i and b_i will be uniquely recovered using the FSK detector of Fig. 5.17. Carry out the analysis, filling in all of the details left out, and demonstrate the validity of the argument used.

(b) Continuing the discussion in (a) of the FSK detector of Fig. 5.17, show, as suggested in the text, that for all possible values of a_i (b_i) correctly established, the detector will, an interval $T/2$ later, uniquely and correctly produce the value of b_i (a_i) that may have been transmitted.

5.8 Equation (5.13) represents the effect of shaping a binary pulse of width D by a gaussian filter whose impulse response (characteristic in time) is given by (5.11) and frequency spectrum by (5.12). Plot (5.13) as a function of time for various values of $B_N = BD$, including specifically the value 0.3, as well as values above and below that value. Show the effect of reducing bandwidth B is to introduce more inter-symbol interference, as indicated in the text. You will need to use a software package, or a table, of the complementary error function erfc(x) defined by (5.14).

5.9 Consider $\pi/4$ DQPSK as described in the text. The initial phase of the system is set to start at zero. Let the incoming bit stream be 0 0 1 0 1 1 1 0 0 1 1 1, ordered from left to right. Show the points on the constellation diagram of Fig. 5.20 through which the modulator moves as these bits are processed, one after another.

5.10 A transmission bandwidth of 2 MHz is available. Nyquist rolloff shaping is used in transmitting data.

(a) Find the bit rates that may be transmitted over this channel using PSK for rolloff factors of (1) 0.2, (2) 0.25, and (3) 0.5.

(b) Nyquist rolloff shaping of 0.25 is used. It is desired to transmit at a rate of 6.4 Mbps over this channel. Show how this may be done. Repeat for 9.6 Mbps.

5.11 The channel of problem 5.10 is to be used for wireless data transmission at the rate of 6.4 Mbps.

(a) Estimate the maximum delay spread that may be accommodated.

(b) It is desired to accommodate a delay spread of 8 μsec. OFDM is to be used. Find an acceptable number of subcarriers, the bandwidth of each, and the resultant OFDM symbol interval. Draw a diagram of the resultant spectrum.

(c) Draw a block diagram of the OFDM transmitter of (b) if discrete Fourier Transform processing is used for its implementation.

Chapter 6

Multiple access techniques: FDMA, TDMA, CDMA; system capacity comparisons

We used the term "channel" rather abstractly in our discussion of FCA and DCA in Chapters 3 and 4. In this chapter we make the concept more concrete and provide examples of different types of "channels" used in current cellular systems. The word *channel* refers to a system resource allocated to a given mobile user enabling that user to communicate with the network with tolerable interference from other users. Channels are thus implicitly *orthogonal* to one another. The most common types of channels adopted for cellular systems are frequency channels, time slots within frequency bands, and distinct codes. These three different ways of providing access by multiple users to a cellular system are termed, respectively, *frequency-division multiple access* or FDMA; *time-division multiple access* (TDMA); and *code-division multiple access* (CDMA). We describe these different multiple access techniques in this chapter, using the three most widely deployed second-generation digital cellular systems as examples. All three of these systems utilize FDMA as well. Two of the systems, GSM and D-AMPS or IS-136, are TDMA-based systems; the third system, IS-95, uses CDMA. Since FDMA underlies all of the cellular systems to be discussed in this book, including the third-generation systems discussed in Chapter 10, we describe the FDMA concept briefly first. We then devote separate sections to TDMA and CDMA systems, ending the chapter with a comparison of their "channel capacities," or the number of users each multiple-access scheme can accommodate per cell in a specified frequency band.

Consider, therefore, first, the concept of FDMA. This concept is very simply explained. This scheme, in its simplest and pure form, divides a given frequency band into frequency channels, allocating each to a different system user or mobile terminal. This is precisely the strategy adopted for first-generation analog mobile systems. This channelization of a given frequency spectrum was, of course, historically first used for broadcast radio and then for broadcast TV. In the broadcast case, each station is assigned a given channel or range of frequencies over which to broadcast, using, for example, amplitude modulation (AM) or frequency modulation (FM), with sufficient guard band between channels to ensure limited interference between them. Radio and TV stations sufficiently far apart geographically can then reuse the same channels without suffering from interference from one another. First-generation, analog, cellular systems operate on the same principle. The North American AMPS system, for example, as noted in Chapter 1, was allocated the 869–894 MHz range of frequencies for downlink or forward transmission, base station (BS) to mobile (MS). Uplink or reverse communication is carried out using the 824–849 MHz band. These 25 MHz bands are, in turn, broken into 832 30 kHz channels, a pair of which, one channel in each direction, is assigned to one mobile user. Such an assignment of frequency channels is sometimes referred to as a frequency-orthogonal assignment, since it enables multiple users to simultaneously use the system. The AMPS system is thus an example of a pure FDMA system. The pairing of channels to provide two-way communication in either direction, uplink or downlink, is termed *frequency-division duplex* or FDD. The composite scheme, with separate frequency channels assigned for each direction of communication, is thus labeled FDMA/FDD. Note how this differs from the broadcast model in which one-way communication only is invoked, and in which all recipients of the broadcast information share the same channel.

In the sections following, we discuss the concepts, as noted above, of TDMA and CDMA, using as examples GSM, D-AMPS, and IS-95. These second-generation systems are further discussed in more detail in Chapter 8. (Third-generation digital cellular systems using both time-division and code-division multiple access techniques are discussed in detail in Chapter 10.) Specifically, in Section 6.1 following, we discuss TDMA systems, using GSM and D-AMPS as examples. In Section 6.2 we provide an introduction to CDMA. Sections 6.3 to 6.5 are then devoted to determining the capacity of CDMA systems, in terms of the number of users a given cell can accommodate. Using the performance results of Section 6.5, we are then able to compare the capacities of GSM, D-AMPS, and the IS-95 CDMA system in Section 6.6.

6.1 Time-division multiple access (TDMA)

TDMA systems, exemplified by the pan-European GSM system and the North American IS-136 or D-AMPS system, gain capacity over the first-generation FDMA systems by assigning multiple users to one frequency channel. These second-generation TDMA systems are implicitly circuit-switched systems, as exemplified by the modern digitally switched telephone system. Digital signals sent out on a given frequency channel or band are transmitted in specified time slots ("circuits") in a repetitive frame structure operating at the carrier frequency assigned to that channel. Each user is assigned one or more of these time slots per frame, and keeps the time slot as long as needed, i.e., until a "call" is completed. This is why the word circuit is used. The digital signals themselves are transmitted using one of the modulation schemes (PSK or a version of QPSK) discussed earlier in Chapter 5. An example of such a repetitive frame structure containing N slots per frame appears in Fig. 6.1. Provision has to be made, of course, to distinguish both the beginning and end of a frame, as well as to delineate the slots within a frame. As noted above, users are each assigned one or more slots in a frame. These second-generation systems were designed principally to carry voice calls, just as is the case with digital wire-based telephone systems, although the transmission of data over these systems has begun to play an increasingly significant role. We discuss the transmission of voice calls over these systems in detail in Chapter 8. The third-generation systems described in Chapter 10 have

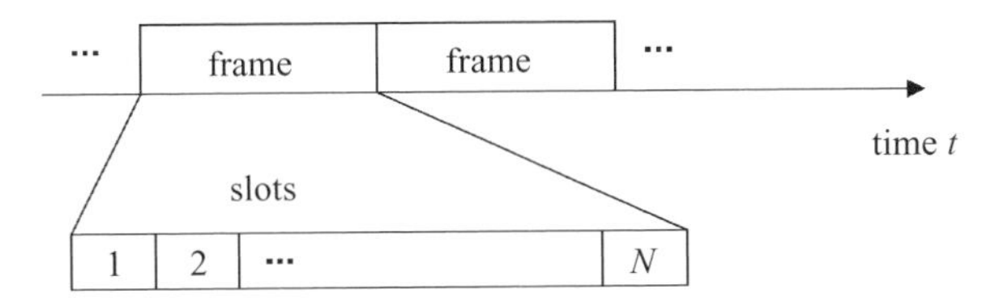

Figure 6.1 TDMA slot structure

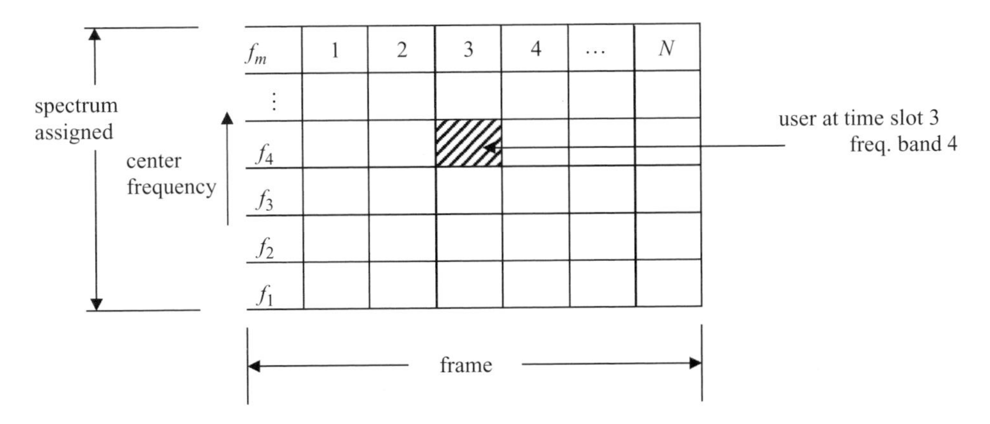

Figure 6.2 FDMA/TDMA channel assignment

been designed for the transmission of data traffic and use packet-switching technology, the technology used for the Internet, for that purpose.

These second-generation TDMA systems are effectively FDMA/TDMA systems, since the entire frequency spectrum allocated to a given system is, first, split into smaller frequency channels or bands, each centered about a specified carrier frequency. Each such band is then time-slotted following a specified frame structure, to accommodate multiple users. The entire FDMA/TDMA scheme may be diagrammed as shown in Fig. 6.2. A "channel" then corresponds to one or more time slots within a given frequency band. In this figure there appear M frequency bands, each using a different carrier frequency, and N time slots per frame, assigned to a given carrier. An example of a user channel corresponding to a specified time slot assigned to one of the frequency bands appears in the figure. There are thus potentially NM channels in this system. We shall see shortly that the North American D-AMPS or IS-136 mobile cellular system normally assigns each user two time slots per frame. This effectively doubles the bit rate possible per user over assigning one slot per frame only (twice as many bits are transmitted per frame), but reduces the number of possible simultaneous users correspondingly. In the GSM system one slot only per frame is assigned per user, but users may hop among the different frequency channels to reduce the chance of getting caught in a deep fade. (Control must of course be exerted to ensure only one user at a time occupies any given channel.)

Consider GSM as a specific example of a FDMA/TDMA system. As already noted earlier in this book, in Europe this system has been allocated two 25 MHz spectral bands,

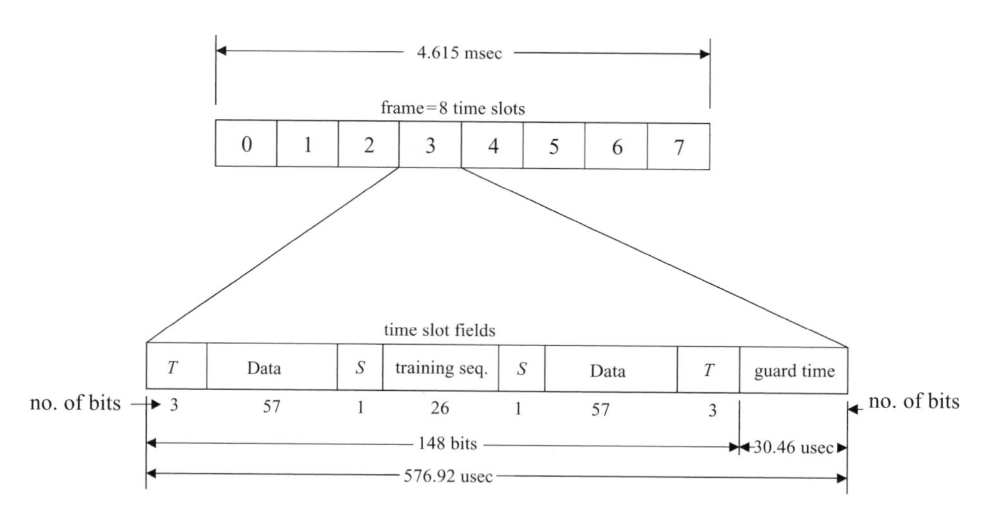

Figure 6.3 GSM frame structure

890–915 MHz uplink, in the reverse, MS to BS direction, and 935–960 MHz downlink, in the forward, BS to MS direction. (GSM has also been adapted to the 2 GHz band in Europe and North America, but we focus here only on the 900 MHz system.) Each 25 MHz band is, in turn, split into 200 kHz-wide bands. One such band is used as a guard band, so there are effectively 124 frequency assignments allocated in each direction. GMSK transmission is used as the modulation technique, as noted in Chapter 5. Each 200 kHz frequency band carries an 8-slot TDMA frame structure, frames repeating at intervals of 4.615 msec. As noted above, each user is assigned one time slot per frame. There are thus up to $124 \times 8 = 992$ channels available per GSM system. This number corresponds to the number *NM* appearing in Fig. 6.2. (A multiframe structure is also imposed on the system. This structure will be described in our much-more detailed discussion of GSM later in Chapter 8.)

The GSM frame structure is diagrammed in Fig. 6.3. Note that each time slot is 576.92 μsec long and contains 148 bits. Slots are separated by a guard time of 30.46 μsec, equivalent to 8.25 bits. Of the 148 bits per slot, 114 are data bits, placed in two groups of 57 bits each. Three bits (called T fields) each define the beginning and end of a slot. A 26 bit training sequence provides necessary time synchronization, and two 1-bit flag bits complete the slot. The system thus transmits at a bit rate of 154.25 bits/576.92 μsec = 270.833 kbps. This resultant bit transmission corresponds to a channel usage of 1.35 bits/Hz over the 200 kHz-wide frequency channel. Recall that in Chapter 5 we indicated that GSM used 0.3 GMSK transmission. The 0.3 factor indicated that $B/R = 0.3$, B being the low-pass 3 dB bandwidth and R the transmission bit rate. The *3 dB transmission bandwidth* is thus $0.6R = 162.5$ kHz. The 200 kHz channel bandwidth is the bandwidth between points approximately 10 dB down from the center of the band (Steele, 1992: 540).

The user data rate is a bit more complicated to calculate. Each user gets to transmit 114 bits per frame. However, as will be shown in discussing GSM in detail in Chapter 8,

Steele, R. ed. 1992. *Mobile Radio Communications*, London, Pentech Press; New York, IEEE Press.

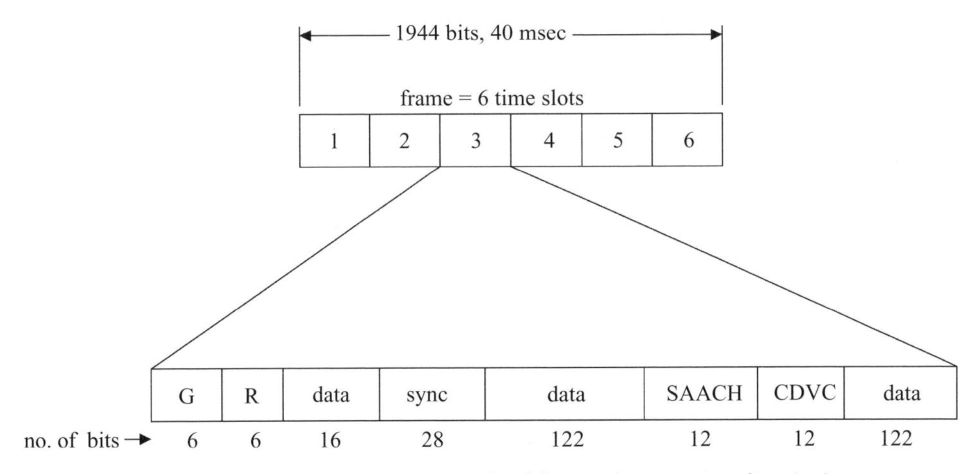

Figure 6.4 IS-136 (D-AMPS) frame structure (mobile => base station direction)

data channels occupy 24 slots of every 26 (multiframe) frames. The other two time slots are used to carry control information. The user bit rate made available is thus 114/4.615 msec × 24/26 = 22.8 kbps. These bits are, however, transmitted at the 270.833 kbps rate. The 0.3 GMSK modulation scheme adopted for GSM plus the use of frequency hopping enable GSM to operate at an SIR as low as 6.5 dB. A cellular reuse factor of four is thus made possible, rather than the 7-reuse factor required for IS-136. One-fourth of the 124 frequency assignments is therefore allocated per cell, so that 992/4 = 248 channels are made available per cell. This number then corresponds to the GSM system capacity and will be used later in comparing the capacities of the three second-generation systems on which we focus in this book.

The IS-136 D-AMPS cellular system uses a completely different set of FDMA/TDMA channels. As already noted earlier in this book, IS-136 uses the same frequency allocation plan as the original analog AMPS system: the 25 MHz spectrum allocation in each direction of transmission is split into 832 30 kHz frequency bands. In the case of IS-136, however, as noted in Chapter 5, DQPSK is used as the basic modulation scheme. Each band is, in turn, used in TDMA fashion, with repetitive frames carrying six time slots each. The frame-slot structure for the uplink, reverse direction appears in Fig. 6.4. Note that frames are 40 msec long, hence repeat at a rate of 25 frames per second. They carry 1944 bits each. The transmission rate is thus 1944 × 25 = 48.6 kbps. Two types of users can be accommodated in the 6-slot TDMA frame structure: full-rate and half-rate users. Full-rate users are allocated two time slots per frame each; hence three such users may be accommodated per 30 kHz frequency channel. Six half-rate users, each assigned one time slot per frame and each transmitting at half the bit rate of the full-rate user, may be accommodated per frequency channel. Consider the full-rate users. They can send 260 data bits per slot for a total of 520 bits per frame. This thus provides an information bit rate of 520 bits/frame × 25 frames/sec = 13 kbps, sent at the transmission rate of 48.6 kbps. For speech transmission this bit rate handles coded compressed speech delivered from a speech coder running at 7.95 kbps. Normal 64 kbps speech is thus compressed by a factor of about 8:1. Details of how speech (voice) is handled in this system appear in Chapter 8.

The different fields in the expanded time slot structure shown in Fig. 6.4 will be explained in more detail as well in Chapter 8. It suffices at this point to note that G is a guard time, R is called the ramp-up time, the Sync field is a specified 28-bit sequence providing synchronization, the SAACH field provides a control channel, and CDVC, standing for "coded digital verification control code," is an 8-bit number plus four code bits that keeps the base station aware that a user is still active: the BS transmits this number to the mobile; the mobile replies with the same number. If the mobile does not send this number, the slot is relinquished.

This brief introduction to GSM and D-AMPS concludes our discussion at this point of time-division multiple access, TDMA. In the following sections of this chapter we focus on *code-division multiple access* or CDMA. We return briefly to both GSM and IS-136 later in this chapter, in comparing the user capacity of these systems with that of CDMA.

6.2 Code-division multiple access (CDMA)

Code-division multiple access (CDMA) is the third type of multiple access technique used in cellular systems. It is the access technique used in the second-generation system IS-95, which, in North America, uses the same assigned frequency spectrum as AMPS and IS-136 (D-AMPS). More wide-band versions of CDMA have been adopted for third-generation cellular systems. Details of these more advanced systems appear later in this book, in Chapter 10. We note that CDMA is a scheme based on spread-spectrum technology invented and developed many years ago principally for military communication systems.

Orthogonality of users in the CDMA case is accomplished by assigning each user a distinct digital *code*. Codes are selected so as to be "orthogonal" to one another in the following sense: say a code i is made up of a specified sequence of l bits $\{x_{ik}\}$, $k = 1 \ldots l$, where the binary digit $x_{ik} = \pm 1$. Let this sequence of bits be repesented by the l-element vector $\mathbf{c}_i$. Orthogonality then means the dot product $\mathbf{c}_i \bullet \mathbf{c}_j = \Sigma_k x_{ik}x_{jk} = 0, j \neq i$. Multiple users can thus transmit simultaneously if each is assigned a different code. A receiver for code i can then uniquely reproduce is signal by carrying out the dot product operation. An output should only appear if code i is present. Because of the orthogonality, all other signals will be rejected. In practice, as we shall see, a user code is generated by a shift-register. The resulting code is an example of a *pseudo-random sequence* (Viterbi, 1995). The use of this pseudo-random sequence in generating a CDMA signal is indicated in Fig. 6.5. PSK modulation is shown being used in this figure. Other types of modulation schemes could be used as well. (OQPSK and QPSK are used in IS-95, as will be noted in Chapter 8.)

The pseudo-random sequence modulates, or multiplies, each bit in the information bit stream as shown in Fig. 6.5. Each bit in the pseudo-random sequence, called a *chip*, is of length T_c, chosen to be much shorter than the information bit length $1/R$: $T_c \ll 1/R$. The effect of multiplying information bits by the pseudo-random sequence is to convert the binary information stream to a noise-like sequence of much wider spectral width. This accounts for the term *spread spectrum communications*, used interchangeably here with the acronym CDMA. The bandwidth W of the resultant wider-band binary sequence

---

Viterbi, A. J. 1995. *CDMA, Principles of Spread Spectrum Communication*, Reading, MA, Addison-Wesley.

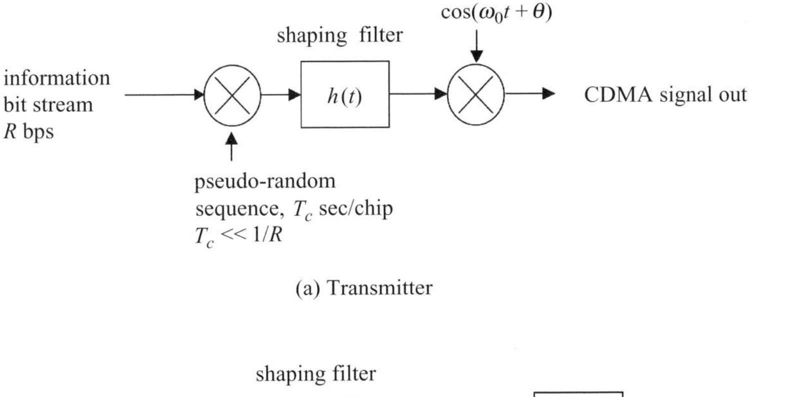

(a) Transmitter

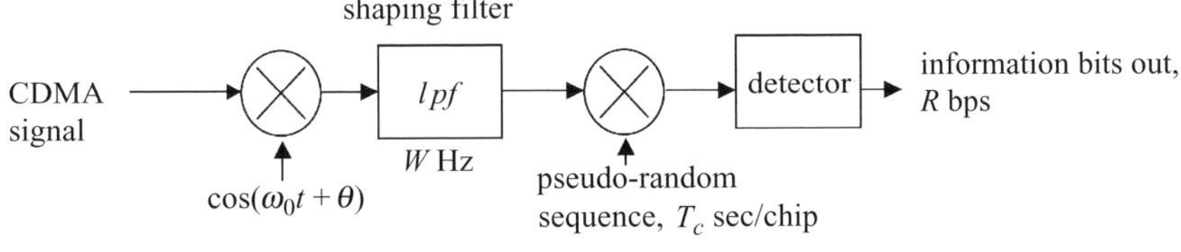

(b) Receiver

Figure 6.5 CDMA system-transmitter and receiver

is approximately $1/T_c$. Hence the bandwidth of the original information bit stream is effectively multiplied by W/R, called the *spreading gain* of the system. We shall see shortly that, the larger the spreading gain, the more effective the performance of the CDMA system. (Note that, in practice, mod-2 addition is used rather than multiplication of the binary signal by the pseudo-random sequence. The two are equivalent. See Chapter 8.)

Multiplying by the same pseudo-random sequence at the receiver, as shown in Fig. 6.5(b), reproduces the original information bit stream. Other coded signals using the same carrier frequency are rejected because of the orthogonality of the codes. The use of CDMA makes universal frequency reuse possible: the same user code may be used in *any* cell of a given system to which a mobile may roam. This simplifies the problem of assigning a channel in any new cell to which a mobile hands off. It creates a problem, however, with interference since *all* users in *all* cells using the same frequency contribute to the interference at a given base station or mobile (depending on the direction of transmission). Despite this increased interference power, the net result of spreading is to improve the detectability of a given signal, as will be demonstrated later in this section.

The word *pseudo-random* implies that the binary sequence or user code multiplying the information bit stream has random or noise-like properties. Such a sequence is thus often labeled *pseudo-noise* (PN). This sequence is deterministic since it is generated using a linear shift register. In particular, a *maximum length* shift register may be shown to have the desired noise-like properties (Viterbi, 1995). Say the shift register is of length n. Its mth binary output a_m is then given by the recursive relation

$$a_m = \sum_{i=1}^{n} c_i a_{m-i}, \qquad c_i = 0, 1; c_n = 1 \tag{6.1}$$

Note that the binary output here is 1 or 0. The 0s would have to be converted to -1s to ensure orthogonal codes are used.

A maximum length shift register is defined to be one whose period $P = 2^n - 1$ for all nonzero initial vectors. It may then be shown (Viterbi, 1995) that such a shift register has the following noise-like properties (Pr[] stands for probability of []):

1 $\Pr[0] = 1/2(1 - 1/P)$; $\Pr[1] = 1/2(1 + 1/P)$

For $n \geq 10$, then, $\Pr[0] = \Pr[1] = 1/2$ to within a factor of $1/P \leq 10^{-3}$. The relative occurrence of 0s and 1s is thus equally likely to within $1/P$, the expected property of a random sequence.

2 The relative frequency of occurrence of any sequence (run) of 0s and 1s l bits long is $1/2^l$, $l \leq n - 1$; $1/2^{n-1}$, $l = n$. As an example, one should expect that, in a random sequence of 0s and 1s, 1/2 of all run lengths are of length 1; 1/4 are of length 2; 1/8 are of length 3, etc. The maximum length shift register has this property.

3 Shifting a random sequence by a number of elements results in an equal number of agreements and disagreements with the original sequence. Again it is found (Viterbi, 1995) that the maximum length shift register has this property to within a factor of $1/P$. This property implies that two users having codes shifted by even one chip will still be separately detectable to within $1/P$.

This last property, in particular, is used to quantify the effectiveness of the CDMA orthogonal code structure. To be specific, consider the kth chip of the code of user j. Its binary value, as noted above, is written as $x_{jk} = \pm 1$. Say it has the value of the code element of user i displaced by l chips: $x_{jk} = x_{i,k+l}$. Then it may be shown (Viterbi, 1995) that

$$\frac{1}{P}\sum_{k=1}^{P} x_{ik}x_{i,k+l} = -\frac{1}{P}, l > 0; \quad = 1, l = 0 \tag{6.2}$$

The receiver multiplication, chip by chip, shown in Fig. 6.5(b), carries out precisely this *correlation* operation. Hence, based on (6.2), the signal-to-interference (SIR) detectability is $P = 2^n - 1$. This can be made as large as desired by having the shift register length n as large as desired. The price paid is that this implies decreasing the chip length T_c, to pack more pseudo-random bits into a given information bit $1/R$ sec long. The transmission bandwidth goes up accordingly. This potential SIR detectability indicated by (6.2) assumes the codes of various users are aligned or synchronized. This may not be the case, however, with uplink transmissions from multiple mobiles. This lack of code alignment and hence code orthogonality does lead to user interference. With many such interfering users, the interference can be considered noise-like. The discussion in sections following makes this assumption.

We now focus on the performance attainable with CDMA. By this we mean simply the capacity of the system, the number of users a given system may accommodate. In the next section, 6.3, we take a simple, first-cut approach to this determination of CDMA capacity by focusing on interference from one cell only. This helps as well in solidifying our understanding of CDMA. In Section 6.4 we pause to present the results of some probability of bit error calculations in a fading environment needed in studying CDMA capacity performance. In Section 6.5 we return to CDMA interference determination, taking the interference from all users of a system into account. The final result of the determination of capacity is then used to compare the capacity of the CDMA-based IS-95 system with that of the TDMA-based GSM and IS-136. A discerning reader following the presentation of these next sections will note that the analysis throughout is based on the same idealized models of system geometry, propagation effects, and fading phenomena we have adopted throughout this book. These models can be very useful in providing much-needed insight into the environment in which mobile cellular communications is carried out, as well as providing some measure of the performance attainable with these systems. Caution is required, however, in using the results of these models to definitively compare the relative performance of competing systems. The performance of complex systems such as the mobile cellular systems under study here under realistic working conditions might differ markedly from that calculated here and, in fact, do vary with location and system implementation, as well as other factors.

6.3 CDMA capacity: single-cell case

As noted above, in this section we show how one determines the capacity of a CDMA system, for the simplest case of interference due to mobile users within a single cell only. Say there are K users within the cell. Hence each user has $K - 1$ interfering users. We focus on the uplink, mobile to BS, direction since this is the one in which communication is most difficult to control. All users in this cell are assumed to be power controlled to have the *same* power P_R as received at the BS. Power control is *critical* to the performance of CDMA systems. Otherwise, close-in users would have a built-in advantage. Examples of such power control techniques were discussed earlier in Chapter 4.

If the number of users transmitting is large enough, each using its own pseudo-random code, it may be assumed that the composite, interfering behavior approaches that of white gaussian noise. The basic problem then becomes that of detecting a signal in the presence of noise. This problem has been studied for many years. In particular, it is well-known from communication theory that the detectability of a (binary) signal in the presence of white gaussian noise (noise whose probability density function is gaussian or Normal, and whose frequency spectrum is flat) depends on the ratio of received signal bit energy E_b to noise spectral density N_0 (Schwartz, 1990; Wozencroft and Jacobs, 1965). By *detectability* we mean the ability to distinguish correctly between either of two binary signals (1 or 0, or 1 or -1, as the case may be), to within a specified probability of error. In the case of mobile wireless communication, noise is usually small or even negligible. It is the noise-like *interference* power that dominates. This is the power due to the $K - 1$ interfering users, given by $(K - 1)*P_R$. We thus use the signal bit energy to interference noise power spectral density as a measure of signal detectability.

Since the bandwidth of the wideband coded CDMA signals is taken to be a value W Hz $\gg 1/R$, R the information bit rate, as noted in the previous Section 6.2, the power

---

Schwartz, M. 1990. *Information Transmission, Modulation, and Noise*, 4th edn, New York, McGraw-Hill.
Wozencroft, J. M. and I. M. Jacobs. 1965. *Principles of Communication Engineering*, New York, Wiley.

spectral density I_0 of the interference, found by dividing the interference power by the spectral bandwidth, is given by $(K-1)P_R/W$. The received signal bit energy is $E_b = P_R/R$. The ratio E_b/I_0, comparable with the signal bit energy-to-noise spectral density E_b/N_0 determining the detectability of signals in the usual communication theory context, is thus given by

$$\frac{E_b}{I_0} = \frac{P_R/R}{(K-1)P_R/W} = \frac{W/R}{(K-1)} \tag{6.3}$$

The larger the spreading gain W/R, the more users K can be accommodated. Given the desired value of E_b/I_0, based on an acceptable value of bit error probability, and the value of spreading gain W/R possible in a specified system, the number of users that may be accommodated in a cell, neglecting at this point interference from other cells, is, from (6.3), readily determined to be

$$K = (W/R)/(E_b/I_0) + 1 \tag{6.4}$$

As an example, say $E_b/I_0 = 5$ (or 7 dB), with the information bit rate $R = 10$ kbps and the transmission bandwidth $W = 1.25$ MHz (the values selected for IS-95). Then $K =$ 26 users/cell may be accommodated within this 1.25 MHz bandwidth. The total number of system users in the 25 MHz spectrum allocated to cellular systems in North America would then be $20 \times 26 = 520$ users. Note that frequency reuse plays no role here in a CDMA system because users are individually assigned pseudo-random codes and *all* frequencies are used in *all* cells; this is the essence of universal frequency reuse made possible through the use of spread spectrum CDMA technology.

In Section 6.5 we improve this simple single-cell interfering user model considerably by incorporating signal propagation effects, shadow fading, and interference encountered from *all* system users. The single-cell model will be found to be overly optimistic, as might be expected. The more comprehensive analysis of Section 6.5 enables us to compare CDMA and TDMA system capacities, at least to within the confines of the idealized models we use. But before going on to the improved model, we discuss the E_b/I_0 requirement for signal detectability. Why, in particular, choose $E_b/I_0 = 5$, as done in the example above? Where does this number come from? In the next section, 6.4, we therefore summarize some bit error probability performance numbers obtained from communication theory comparing the detectability of binary signals in the presence of gaussian noise for various modulation schemes. We then cite some numbers extending the performance evaluation to the case of rapidly fading Rayleigh channels, of the kind discussed earlier in Chapter 2. For those readers with prior background in communication theory, these results may serve as a review of work they may have previously encountered. Readers with no prior background in communication theory will have to accept the performance results on faith, or go to the references cited for details on the analyses used.

6.4 An aside: probability of bit error considerations

We begin this section on bit error probability and its connection to mobile binary signal detectability by first summarizing some pertinent comparative performance results from communication theory of the detectability of signals in gaussian noise for various modulation schemes such as those discussed in Chapter 5.

For example, it can be shown that PSK is the best-performing binary modulation scheme using the simplest model of a communication channel, one in which additive white gaussian noise, noise with gaussian statistics that is asumed *added* to the signal during transmission, is the only deterrent to signal detectability (Schwartz, 1990; Wozencroft and Jacobs, 1965). By *performance* one means the bit error probability attainable, the probability that a binary digit will be received in error. The *best-performing* scheme is then the one providing the smallest error probability. The probability of bit error P_e for PSK in the presence of additive white gaussian noise is readily found to be given, in terms of the complementary error function, by the following expression (Schwartz, 1990: 475)

$$P_e = \frac{1}{2}\text{erfc}\sqrt{\frac{E_b}{N_0}} \tag{6.5}$$

Here E_b is the received signal bit energy and N_0 is the noise spectral density, both as defined previously. The complementary error function erfc$x$ is defined by the following integral

$$\text{erfc}x \equiv \frac{2}{\sqrt{\pi}}\int_x^\infty e^{-y^2}dy \cong \frac{e^{-x^2}}{x\sqrt{\pi}}, \quad x > 3 \tag{6.6}$$

From (6.5) and (6.6) then, we get the following expression for the PSK bit error probability for small values of the probability

$$P_e \cong \frac{1}{2}\frac{e^{-E_b/N_0}}{\sqrt{\pi E_b/N_0}} \tag{6.7}$$

The probability of error thus varies inversely as the exponential of E_b/N_0. Say, as an example, that it is desired to have $P_e = 10^{-5}$. One bit in 10^5 transmitted is thus received in error, on the average. From (6.7) we find that the required E_b/N_0, in dB measure, is 9.6 dB. Reducing E_b/N_0 by half, or 3 dB, to 6.6 dB, one finds $P_e = 10^{-3}$, a two orders of magnitude increase, because of the exponential variation with E_b/N_0! The error probability performance is thus critically dependent on the received signal bit energy to noise spectral density ratio E_b/N_0.

If one now goes to FSK modulation instead, using the same additive white noise channel, it is found that a 3-dB loss is incurred: the required values of E_b/N_0 are now twice as much, or 3 dB larger. For fixed noise spectral density this says the required signal bit energy or power must be doubled to obtain the same error probability as PSK. As examples, if $P_e = 10^{-5}$ is desired, using FSK modulation, $E_b/N_0 = 12.6$ dB; for $P_e = 10^{-3}$, $E_b/N_0 = 9.6$ dB. These numbers for PSK and FSK require accurate phase synchronization between transmitter and receiver, as noted in Chapter 5. One can detect FSK signals non-coherently, however, on the basis of comparison of the relative signal amplitudes or envelopes at each of the two frequencies transmitted, obviating the need for phase synchronization. (This is clearly impossible with PSK.) The price paid is a loss of about 0.7 dB, i.e., non-coherent FSK requires an increased signal energy or power of 0.7 dB – the required E_b/N_0 increases to 13.3 dB, for example, if $P_e = 10^{-5}$ is desired.

It may also be shown that differential PSK or DPSK requires almost a dB more of signal power than does PSK: $E_b/N_0 = 10.5$ dB for $P_e = 10^{-5}$. (This scheme is similar to DQPSK, discussed in Chapter 5.) These numbers can be improved considerably by coding the binary signals prior to carrying out the carrier modulation. As an example, if rate 1/2 convolutional coding is used, with PSK as the modulation scheme, the required signal energy to noise spectral density E_b/N_0 ranges from 4 to 6 dB at $P_e = 10^{-5}$, depending on the type of coder used, a considerable reduction from the 9.6 dB figure quoted above (Viterbi, 1995; Schwartz, 1990: 693). Coding techniques, including convolutional coding, are discussed in the next chapter, Chapter 7.

Now consider a fading environment. These numbers change considerably (Schwartz *et al.*, 1966). The PSK bit error probability is now found to be given, approximately, by

$$P_e = 1/4(E_b/N_0) \tag{6.8}$$

This inverse linear relation is to be compared with the inverse exponential variation of (6.7) in the non-fading environment. This means that, even for $E_b/N_0 = 10$ dB, the bit error probability is approximately 1/40 or 0.025, a more than three-magnitude increase from the 10^{-5} figure quoted earlier. Both DPSK and FSK turn out to require 3 dB more signal power than the PSK case, the same performance reduction as in the non-fading case. The diversity schemes discussed in Chapter 2 can be shown to recoup some of this severe performance deterioration due to fading (Schwartz *et al.*, 1966). As an example, say FSK is used as the modulation scheme. The use of dual diversity, combining the output of two receiving antennas with the signals received by each assumed fading independently of one another, is then found to improve the bit error probability performance by an order of magnitude! For $E_b/N_0 = 10$ dB, for example, P_e decreases from 0.05 in the non-diversity case to 0.005 using dual diversity. (We ignore here the type of combining scheme that might be used. As discussed in Chapter 2, one might use selection diversity, selecting the largest of the multiple received signals available; simply add the multiple received signals together, i.e., equal-gain combining; or use maximal-ratio combining, with each received signal term weighted by its signal-to-noise ratio. As noted in Chapter 2, this last scheme may be shown to provide the best diversity performance (Schwartz *et al.*, 1966).)

Recall that these diversity techniques for improving system performance in a fading environment have been known for years, predating their possible use in mobile wireless systems. They generally incorporate the use of two or more receiving antennas, as in the example just cited. Quite recently, however, studies have shown that significantly substantial improvements in transmission capacity could be expected by the use of multiple antennas at both the transmitter and receiver. Modern digital signal processing (DSP) techniques make the use of multiple antennas practical. Systems incorporating multiple antennas have been termed *multiple-input multiple-output* (MIMO) systems (Matsumoto *et al.*, 2003). MIMO schemes may be combined with recent improvements in coding

Schwartz, M., W. R. Bennett, and S. Stein. 1966. *Communication Systems and Techniques*, New York, McGraw-Hill; reprinted, IEEE Press, 1996.

Matsumoto, T. *et al.* 2003. "Overview and recent challenges of MIMO systems," *IEEE Vehicular Technology News*, 50, 2 (May), 4–9.

schemes such as turbo coding, to be discussed in Chapter 7 following, the combined systems being referred to as using space-time coding. These recent developments have generated a great deal of interest in their application to practical wireless systems. Space-time coding has been proposed for use with the third-generation CDMA mobile systems discussed in Chapter 10, as well as with future-generation systems yet to be developed. A pioneering MIMO implementation called *BLAST* has been developed at Bell Labs for use with CDMA systems (Foschini, 1996). Simulation studies show that BLAST can potentially provide very high increases in transmission capacity.

Consider now spread spectrum communications specifically, as exemplified by the CDMA systems we have been discussing in this chapter. Recall that, in Chapter 2, we described the special case of a diversity scheme called the RAKE receiver which utilizes the fact that, for very wideband signals, multipath versions of the same signal may arrive at distinctly different times, allowing them to be separately resolved. The RAKE receiver combines these separate multipath signals, resulting in substantial diversity improvement (Price and Green, 1958). But CDMA does, in fact, produce very wideband signals because of the pseudo-random coding, with its high spreading gain. Put another way, the use of pseudo-random coding with chip values much smaller than the digital symbol intervals converts the signal to one in which the delay spread can be much greater than the signal interval, the chip. Calculations then show that dual diversity reception of convolutionally coded CDMA signals using a RAKE receiver requires an E_b/I_0 ratio of 5 (7 dB) to attain a bit error probability $P_e = 10^{-3}$(Gilhausen *et al.*, 1991; Viterbi, 1995). This number represents a significant improvement over the number found above for dual-diversity FSK. (Note that here we revert back to the use of interference spectral density I_0 rather than noise spectral density N_0.) If E_b/I_0 decreases by 1 dB to 6 dB or a value of 4, the bit error probability P_e is found to increase by a magnitude to $= 0.01$! So the specific value of signal energy to interference spectral density is clearly critical to the performance of a given system.

These comments on the effect of Rayleigh-type multipath fading on bit error probability and the use of diversity techniques to mitigate this effect are summarized in Table 6.1. Tabular values applicable to the CDMA scheme discussed in this chapter are in bold. To summarize, this brief discussion of the interplay of the ratio of signal bit energy to noise or interference spectral density and bit error probability indicates that the capacity of a CDMA system, the number of allowable users per cell, depends both on the choice of bit error probability deemed acceptable and the coding/diversity scheme used in overcoming the effect of fading. We shall use the value $E_b/I_0 = 5$ (7 dB) consistently throughout the remainder of this chapter. Other authors have adopted a value of $E_b/I_0 = 4$ (6 dB), which provides, as noted above, a bit error probability of 10^{-2}. Note that second-generation cellular systems were originally designed for, and are still used primarily for, voice traffic. Digital voice is relatively robust with respect to error probability. That is why $P_e = 10^{-2}$

Foschini, G. J. 1996. "Layered space-time architecture for wireless communication in a fading environment when using multiple antennas," *Bell Laboratories Technology Journal*, 1, 2, 41–59.
Price, R. and P. E. Green, Jr. 1958. "A communication technique for multipath channels," *Proc. IRE*, 46, 555–570.
Gilhausen, K. H. *et al.* 1991. "On the capacity of a cellular CDMA system," *IEEE Transactions on Vehicular Technology*, 40, 2 (May), 303–312.

Table 6.1 *E_b/I_0 vs probability of bit error, different communication channels*

Modulation type and channel		Required E_b/I_0 (dB)			
	$P_e \rightarrow$	0.01	0.005	0.001	10^{-5}
PSK					
no fading, no coding		4		6.6	9.6
with convolutional coding					4–6
fading channel		14	17	24	
with dual diversity RAKE and convolutional coding		6		7	
DPSK					
no fading					10.5
fading channel		17	20	27	
FSK					
no fading		7		9.6	12.6
fading channel		17	20	27	
with dual diversity			10		

or 10^{-3} may provide acceptable performance. These numbers would be unacceptable for the transmission of data.

6.5 CDMA capacity calculations: CDMA compared with TDMA

In Section 6.3 we determined the number of allowable users per cell for a CDMA system, using a simple model incorporating single-cell cell users only, and not including propagation effects as well as shadow fading. In this section we generalize the calculations to include these effects. In particular, we include the effect of interfering users from throughout the system region, not just from the same cell, as well as the effect of shadow fading, using the model described in Chapter 2. (The effect of rapid multipath fading is assumed taken care of with dual-diversity RAKE receivers and is included in the value of E_b/I_0 chosen, using the numbers discussed in Section 6.4.) We again focus on the uplink, mobile to BS, direction. As noted in Section 6.3, the operation of a CDMA system depends critically on the power control. We assume all mobiles have their power controlled by downlink signals from the base station with which they communicate, so that the uplink signals, as received at any base station, all have the same average power value P_R. We recall again that the power control techniques described in Chapter 4 are designed to achieve this condition. In addition, shadow fading is assumed to vary slowly enough so that the BS power control can handle changes in received power induced by this type of fading.

We focus initially on *hard handoff*: a mobile communicates with one BS in a given cell only. When it crosses a border to another cell, it immediately switches, or hands off, to the new BS. The analysis of handoff strategies is carried out in Chapter 9. CDMA allows *soft handoff*, however. In this case, a mobile may be in communication with two or more

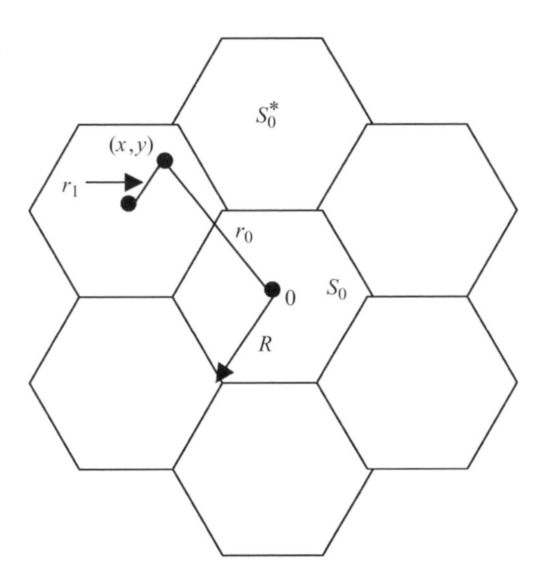

Figure 6.6 Calculation of CDMA outside interference power

base stations. It will then be assigned the one to which the propagation loss is the least. This turns out to reduce the total interference power and increase the system capacity, the number of simultaneous users allowed per cell. We will provide the numbers obtained through analysis in this case later. The model chosen for analysis assumes hexagonal cells. (We shall see later that analysis based on a circular-cell structure provides essentially the same results. This means the results are, to first order, robust with the cellular geometry assumed.) Homogeneous conditions are assumed throughout: users are taken to be uniformly distributed in each cell, with the same density of users throughout; all received powers are taken to have the same value P_R throughout the system; propagation loss and shadow-fading parameters are the same, from cell to cell, throughout the system. This is thus a very idealized model.

Consider then the hexagonal cell structure of Fig. 6.6. We follow the approach used in Viterbi (1995) in calculating the total interference power at a base station. In particular, because of assumed uniformity throughout the system, focus on a typical cell S_0 with its base station taken to be at its center, the point 0. Interfering mobiles from outside this cell, neglected in the previous analysis of Section 6.3, are thus those in all cells outside S_0, the region denoted as S_0^*. These mobile users, in transmitting signals to their own base stations, also produce interference power at base station BS_0. It is the total interference power in the outside region region S_0^* that one would like to determine. As an example, consider a mobile at point (x, y) in cell 1, as shown. It is in communication with its base station BS_1, but its power is received as interference power at BS_0 as well. As stated above, assume the mobiles are uniformly distributed over a cell, i.e., there is a *uniform* traffic density ρ users/m^2. Since the area of a hexagonal cell of radius R m is $(3\sqrt{3}/2)\,R^2$ (see Chapter 3), the traffic density, given K users per cell, is

$$\rho = 2K/3\sqrt{3}R^2 \tag{6.9}$$

We indicated that each BS requires a received power P_R, assumed to be the same throughout the system. This then determines the power required to be transmitted by a mobile a distance r m from its base station. For recall, from (2.4) in Chapter 2, that the received power may be written, ignoring the rapid fading term

$$P_R = P_T r^{-n} 10^{z/10} \tag{6.10}$$

P_T is the mobile's transmitted power, n is the propagation exponent, usually taken as 4, and z is the gaussian-distributed shadow-fading random variable, centered on the average power, with standard deviation σ, in dB. Its probability density function is thus

$$f(z) = e^{-z^2/2\sigma^2}/\sqrt{2\pi\sigma^2} \tag{6.11}$$

Now focus on the interfering mobile located at point (x, y), at a distance r_1 from its BS in cell 1. Let the shadow-fading random variable in this case be written as z_1. The mobile's transmitted power must be given by

$$P_{T_1} = P_R r_1^n 10^{-z_1/10} \tag{6.12}$$

(Note, by the way, that, since the shadow-fading random variable z is symmetrically distributed about 0, one could equally well write $10^{-z/10}$ in (6.10) and $10^{z/10}$ here.) The (interfering) power received at BS_0 from this mobile, a distance r_0 from BS_0, is then given by

$$P_{T_1} r_0^{-n} 10^{z_0/10} = P_R \left(\frac{r_1}{r_0}\right)^n 10^{(z_0 - z_1)/10}$$

The number of mobiles in a differential area $dA(x, y)$ centered at point (x, y) is $\rho dA(x, y) = 2K dA(x, y)/3\sqrt{3}R^2$. Using this number in place of the single interfering mobile described above, integrating over all S_0^*, and then taking the average or expectation (denoted by the symbol E) over the random variables, one finds the *total* average interference power at base station BS_0 due to mobiles *outside* cell S_0 to be given by the following integral (Viterbi, 1995):

$$\begin{aligned} I_{S_0^*} &= \frac{2K}{3\sqrt{3}R^2} P_R E\left\{ \iint_{S_0{}^*} \left[\left(\frac{r_1}{r_0}\right)^n 10^{(z_1 - z_0)/10} \right] dA(x, y) \right\} \\ &= \frac{2K}{3\sqrt{3}R^2} P_R E\left[10^{(z_0 - z_1)/10}\right] \iint_{S_0{}^*} \left(\frac{r_1(x, y)}{r_0(x, y)}\right)^n dA(x, y) \end{aligned} \tag{6.13}$$

The expression for the outside interference power thus breaks into the product of two terms, one purely geometric, involving an integration over the outside region S_0^*, the other an average over shadow-fading terms. Numerical integration over the purely geometric double integral, including the term $2/3\sqrt{3}R^2$ in the evaluation, provides a value of 0.44 for the case $n = 4$ (Viterbi, 1995). (Closed-form integration using circular-cellular geometry will be shown shortly to provide a very similar value.) Consider now the shadow-fading expression. It will turn out that this term is very significant, producing a relatively large value for the model adopted here. Hence shadow fading adds significantly to the interference power. (The effect of the geometric term dies out relatively quickly for the propagation parameter $n = 4$, so that first-tier cells surrounding the cell S_0 account for most of the interference power. We shall demonstrate this effect directly using the circular-cell model.)

We now focus on evaluation of the shadow-fading expression in (6.13). The analysis to follow was first described in Viterbi *et al.* (1994) and appears as well in Viterbi (1995). The two shadow-fading random variables z_0 and z_1 represent power variations as measured at the two base stations BS_0 and BS_1, respectively. Note, however, that, because the power received at these base stations is due to mobiles transmitting in the vicinity of point (x, y), shadow-fading effects in that region must be incorporated as well. Viterbi *et al.* (1994) and Viterbi (1995) model this by assuming that the shadow-fading random variable (rv) measured at each base station is given by the sum of two random variables. One, common to both shadow-fading terms, represents the effect of shadow fading in the vicinity of the transmitting mobiles at (x, y). The second rv represents random power variations encountered along the propagation path and is assumed independent along the two paths from (x, y) to BS_0 and BS_1. In particular, the assumption made is to write the two shadow-fading random variables z_0 and z_1 in the form

$$z_i = ah + bh_i \qquad i = 0, 1 \quad a^2 + b^2 = 1 \tag{6.14}$$

The rv h represents the shadow fading in the vicinity of the transmitting mobile, hence common to the power received at the two base stations in question, the one, BS_1, to which directed, the other, BS_0, at which it serves as interference. The other rv, h_i, $i = 0$ or 1, represents the added shadow fading due to the (independent) propagation conditions encountered along the two paths. If half of the effect of the shadow fading is due to the common region about the transmitting mobile, the other half to the independence of the shadow-fading terms received at the two base stations, one must have $a^2 = b^2 = 1/2$. Note that, from this discussion, the three different random variables appearing in (6.14), h, h_0, and h_1, must individually be gaussian-distributed and independent. Hence the first and second moments must be related as follows

$$E(z_i) = 0 = E(h) = E(h_i); \quad E\left(z_i^2\right) = \sigma^2 = E(h^2) = E\left(h_i^2\right);$$
$$E(hh_i) = 0 = E(h_0 h_1)$$

Consider now the evaluation of the expectation over the shadow-fading expression appearing in (6.13). Note first, from (6.14), that $(z_0 - z_1) = b(h_0 - h_1)$ is a gaussian rv with zero average value and variance $2b^2\sigma^2$ from the moment relations just tabulated. To simplify the calculation of the expectation in (6.13), define the variable transformation

$$e^y \equiv 10^{(z_0 - z_1)/10} \tag{6.15}$$

Then $y = (z_0 - z_1)/10 * \log_e 10 = 0.23\ (z_0 - z_1)$ is also gaussian, with zero average value and variance $\sigma_y^2 = (0.23)^2 * 2b^2\sigma^2 = 0.053\sigma^2$, if $b^2 = 1/2$, as suggested above. The

Viterbi, A. J. *et al.* 1994. "Soft handoff extends CDMA cell coverage and increases reverse link coverage," *IEEE Journal on Selected Areas in Communications*, 12, 8 (October), 1281–1288.

expectation term in (6.13) is then readily evaluated, as shown, following, in (6.16)

$$E\left[10^{(z_0-z_1)/10}\right] = E(e^y) = \int_{-\infty}^{\infty} e^y \frac{e^{-y^2/2\sigma_y^2}}{\sqrt{2\pi\sigma_y^2}} dy = e^{\sigma_y^2/2} \tag{6.16}$$

As an example, take the common shadow-fading modeling case $\sigma = 8$ dB. (Viterbi (1995) tabulates the results for a number of other cases, including the effect of varying the propagation exponent n. To keep the discussion concise, we focus on this one example only.) Then $\sigma^2 = 64$ and $E(e^y) = 5.42$. Note that this is quite a large value, as indicated earlier in the discussion following (6.13). Shadow fading thus has a large effect on the system capacity. Using this value for the shadow-fading standard deviation and including it with the value of 0.44 found for the geometric double integral in (6.13), we finally get, as the average outside cell interference power

$$I_{S_0^*} = P_R K \cdot 0.44 \cdot 5.42 = 2.38 P_R K \tag{6.17}$$

This interference power term representing interference power from outside a cell must now be added to the in-cell interference power $(K - 1)P_R$ from the $K - 1$ interfering mobiles inside a cell discussed in Section 6.3. Note that this outside interference power is larger than the in-cell interference power, due principally to the effect of shadow fading. The total interference power I, the sum of the two terms, is thus, for this commonly used case of $n = 4$ and $\sigma = 8$ dB, $I = P_R(3.38K - 1)$, and the corresponding interference spectral density, for a W-Hz transmission bandwidth, is now $I_0 = (3.38K - 1)P_R/W$. The ratio E_b/I_0 of signal bit energy to interference spectral density, the term whose value determines the bit error probability, now becomes, for the case under discussion here

$$E_b/I_0 = (W/R)/(3.38K - 1) \tag{6.18}$$

This value of E_b/I_0 is to be compared with that of (6.3) in Section 6.3. Note that the effect of outside cell interference and shadow fading is to reduce E_b/I_0 by a factor of more than three for the same number of users per cell, or, conversely, to keep the ratio to a specified value, the number of users must be reduced by more than three. Specifically, take $E_b/I_0 = 5$ (7 dB), the value we chose previously. Letting W again be 1.25 MHz, the transmission bandwidth chosen for IS-95, and $R = 10$ kbps, the information bit rate also used in the example of Section 6.3, we get $K = 7$ or 8 users per cell. This represents a considerable reduction in capacity from the value of 26 users per cell obtained previously, using in-cell interference only for the calculation. (For the complete 25 MHz band, this means that, on the average, no more than 160 users can be accommodated per cell.) We have, however, neglected other improvements in system capacity made possible by the use of CDMA. These include, among other factors, soft handoffs, mentioned earlier; the possibility of cell sectoring (three antennas covering 120° sectors each can, ideally, reduce the number of interferers by a factor of three); and voice silence detection. These will all be described shortly and their effects on system capacity evaluated. But, first, we pause to check the figure of 0.44 obtained numerically for hexagonal geometry for the double integral of (6.13). As noted earlier, a circular-cell model which we now describe

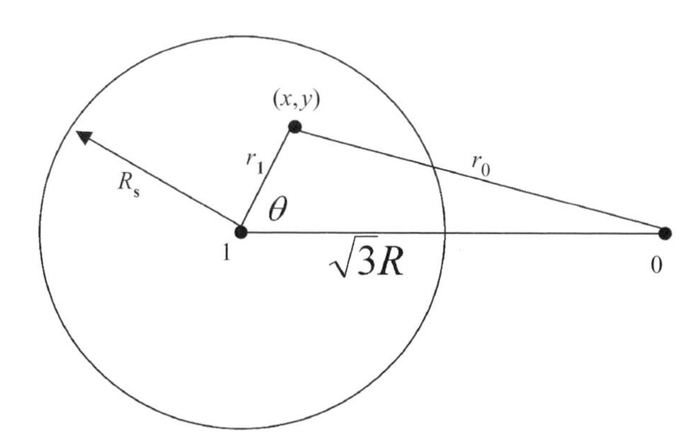

Figure 6.7 Calculation of outside interference power, circular-cell model

gives almost the same result – a value of 0.43! This at least validates the robustness of the calculation with respect to geometry. We follow here the analysis carried out in Kwok and Wang (1995). (This paper carries out an independent calculation of the outside-cell interference, neglecting shadow-fading, for hexagonal geometry as well, obtaining results in agreement with those of Viterbi (1995).)

Consider the circular-cell interferer model indicated in Fig. 6.7. The circular cell shown there is equivalent to the hexagonal interfering cell of Fig. 6.6. Note from Fig. 6.6 that it represents one of six first-tier interfering cells surrounding cell S_0. To make this model equivalent to the hexagonal-cell one, we set the area of the circular cell, of radius R_S, equal to that of the hexagonal cell, of radius R. We thus have $\pi R_S^2 = 3\sqrt{3}R^2/2$, or $R_S = 0.91R$. Note also, from the geometry of Fig. 6.6, that the centers of the two cells, the points 0 and 1, are spaced a distance $\sqrt{3}R$ apart. This is so indicated in Fig. 6.7. We carry out the "outside cell interference" calculation for the first-tier interfering cells only, to show the approach used with the circular-cell model. The interesting thing is that this analysis can be done quite readily, leading to a closed-form expression, rather than requiring numerical evaluation of the double integral of (6.13), as was the case with the hexagonal model. This is done by calculating the total interference power emanating from the first-tier circular cell shown in Fig. 6.7, and then multiplying by six, because of symmetry, to obtain the total first-tier interference power. The use of the circular geometry makes this task relatively simple. Specifically, adopting the same approach as used in writing the expression in (6.13), we assume the K interfering mobiles in the circular cell of Fig. 6.7 are uniformly distributed over the cell area πR_S^2. We again invoke the r^{-n} propagation law, and again require power control such that the power received at any base station from any mobile has the same value P_R. We again first calculate the interfering power received at the base station at point 0 from the mobiles clustered in the differential region dA about the point (x, y) in the interfering cell, but now adopt polar coordinates. Using the polar coordinates indicated in Fig. 6.7, $dA = r_1 dr_1 d\theta$. We then integrate over the entire interfering cell and multiply by six to obtain the total first-tier interference power. It is left for the reader to

Kwok, M.-S. and H.-S. Wang. 1995. "Adjacent-cell interference analysis of reverse link in CDMA cellular radio system," IEEE PIMRC95, Toronto, Canada, September, 446–450.

show that the resultant expression for the first-tier interference, comparable with that of (6.13), with shadow fading neglected, is given by

$$I_{1st,S_0^*} = \frac{6KP_R}{\pi R_S{}^2} \int_0^{R_S}\int_0^{2\pi} \left(\frac{r_1}{r_0}\right)^n r_1\, dr_1\, d\theta \tag{6.19}$$

Note, now, that the variable r_0 in (6.19), the distance from the interfering mobiles in the differential area about the point (x, y) to the base station at point 1 in cell S_0, can be written, from the geometry of Fig. 6.7, in terms of the polar coordinates r_1 and θ

$$\begin{aligned} r_0^2 &= \left(\sqrt{3}R - r_1 \cos\theta\right)^2 + r_1^2 \sin^2\theta \\ &= \left(3R^2 + r_1^2\right) - 2\sqrt{3}r_1 R \cos\theta \end{aligned} \tag{6.20}$$

Replacing r_0 in (6.19) by (6.20), integrating first with respect to θ, and then replacing r_1 in the remaining integral over r_1 by the dummy variable $x = r_1/\sqrt{3}R$, one gets, for the special case $n = 4$

$$I_{1st,S_0^*} = 36KP_R \int_0^{0.525} \frac{x^5(1+x^2)}{(1-x^2)^3} dx \tag{6.19a}$$

It is left for the reader to show that the integral in (6.19a) is readily evaluated, and one obtains for the first-tier interference power

$$I_{1st,S_0^*} = 0.38KP_R \tag{6.19b}$$

This value is to be compared with the hexagonal geometry value $0.44KP_R$ obtained, using numerical integration, of the double integral of (6.13) over all space. (The equivalent first-tier number, using hexagonal geometry, as calculated in Kwok and Wang (1995), turns out to be 0.384!) Continuing for other tiers, Kwok and Wang obtain the value of $0.43KP_R$, as already noted earlier. The circular and hexagonal model results are thus effectively the same, indicating, as again noted above, that this calculation is a robust one with respect to geometry. Note also that the first-tier results are already close to the results obtained by integrating over all space. This is obviously due to the r^{-4} propagation dependence assumed here. Interference effects from cells further away quickly drop to zero as the distance increases. We noted this effect as well in our calculation of SIR in earlier chapters.

We now return to the calculation of the number of users the CDMA system can carry, its user capacity, for a specified value of E_b/I_0, the ratio of the signal bit energy to the interference spectral density. We have shown above how outside-cell interference is calculated, including the effect of shadow fading. We now discuss the improvement in CDMA user capacity made possible by taking advantage of additional effects such as *soft handoff*, *voice silence detection*, and *antenna sectoring*. We have modeled, in the discussion above, the calculation of interference assuming hard handoff. Because of the power control required for CDMA, soft handoff, as described earlier, becomes possible, with a mobile deciding among two or more base stations with which to communicate. Calculations indicate, in particular, that the outside cell interference term decreases from the value of 2.38 shown in (6.17) to 0.77, if a mobile selects the better of two base stations

(Viterbi, 1995). (The term reduces further to 0.57 if the mobile is allowed to select the best one of three base stations (Viterbi, 1995).) This results in a potential increase in user capacity of 3.38/1.77, or a factor of almost two to one (see (6.18)). This factor assumes perfect power control. A 1-dB power control loss factor is typical, however, reducing the potential improvement by 1.25 (Viterbi, 1995). A significant further increase in capacity is made possible by automatic *silence detection*: it is well-known that, in normal speech, speakers tend to alternate between very brief intervals of silence and active speech generation or talk spurts (Schwartz, 1996). A talk spurt lasts, typically, from 0.4 to 1.2 sec, followed by an interval of silence lasting from 0.6 to 1.8 sec. On the average, speech is active about 0.4 of the time. Voice telephone networks have, for years, capitalized on this phenomenon to allow almost 2.5 times the number of users to access the networks as would normally be the case. In the case of a CDMA system carrying voice signals, silence implies that no power is transmitted, automatically reducing the interference power correspondingly. No silence detection equipment is needed, as is the case for voice telephony. We can, therefore, allocate an average reduction of about 2.5 in CDMA interference power due to speech silence detection. (Viterbi (1995) uses a factor of 2.67.) Finally, assume that 120° sector antennas are used at the base station. It turns out that antenna sectoring is required for AMPS and D-AMPS systems, even with a reuse factor of 7. In the case of CDMA this use of antenna sectoring results in a potential reduction of another factor of three in interference power, since interference at any one of the antennas can come only from outside-cell mobiles in the corresponding sector. This factor of three assumes perfect base station antennas, however. Allowing a loss of 1 dB in antenna gain reduces the potential improvement of three by a factor of 1.25, to 2.4.

Putting together the various factors described above, the final result for the capacity of the IS-95 CDMA system in a 1.25 MHz transmission band turns out to be 84 mobiles per cell if two-cell soft handoff is used, and 96 mobiles if three-cell handoff is used. Note again that these results are obtained for the case of a shadow-fading model with $\sigma = 8$ dB, a propagation exponent of $n = 4$, and for a required E_b/I_0 of 7 dB, a value of 5. If E_b/I_0 can be reduced to a value of 4 (6 dB), i.e., the corresponding increase in bit error probability is tolerable, the corresponding capacity values increase to 108 and 120 users per cell. Recall again our initial precautionary comment: these results have been obtained using an idealized model of the cellular system. Note also that these are average capacity values. Some statistical performance measures instead of average quantities were described briefly in Chapters 2 and 4. More details on statistical measures appear in the performance analyses of Chapter 9.

Calculations of the CDMA capacity for values of parameters such as the propagation constant n and the shadow-fading standard deviation, as well as soft handoff effects, other than those chosen here, appear in Viterbi (1995). The CDMA capacity results for the case discussed here are summarized in Table 6.2, where they are compared with corresponding capacity values for the AMPS, D-AMPS or IS-136, and GSM systems for the 1.25 MHz frequency transmission range. The AMPS and D-AMPS results assume a reuse factor of 7; the GSM results, a reuse factor of 4. Details are left to the reader. (Note that in our previous discussion of the capacity of the TDMA systems in Section 6.1 we cited their

Schwartz, M. 1996. *Broadband Integrated Networks*, Englewood Cliffs, NJ, Prentice-Hall.

Table 6.2 *Comparative capacity, mobiles/cell, 1.25 MHz band*

AMPS[1] (analog)	D-AMPS[1] (TDMA)	GSM[2] (TDMA)	IS95[3] (CDMA)
6	18	12–13	84–96

Notes: [1] Reuse factor of 7, 120°-sector antennas.
[2] Reuse factor of 4.
[3] $E_b/I_0 = 7$ dB, $n = 4$, $\sigma = 8$ dB, voice silence detection factor of 2.4, soft handoff of 2 or 3, power control loss of 1 dB, 120°-sector antennas, antenna gain loss = 1 dB.

capacity per cell for the full 25 MHz band. The numbers obtained there must, therefore, be reduced by 20 to obtain the 1.25 MHz frequency range numbers of Table 6.2.) It would thus appear, on the basis of the idealized models used here, that the CDMA system provides substantial average capacity improvement over the TDMA systems. This calculated improvement in performance of CDMA as compared with TDMA has been debated for years. It is to be noted that almost all of this capacity improvement is due to the use of soft handoff combined with automatic voice silence detection. Voice silence detection could be incorporated into TDMA systems, but at the price of some system complexity. Simulation studies of TDMA systems incorporating features such as voice silence detection do, in fact, indicate, that they provide capacity performance comparable to that of CDMA (Gudmundson *et al.*, 1992; Sköld *et al.*, 1995). Despite the controversy over the CDMA calculations assuming idealized conditions, it is still a useful exercise in this author's opinion since it does focus attention on the operation of a CDMA system, and the various features it incorporates. We shall see, in Chapter 10, that CDMA has become the preferred multiple access technique for third-generation cellular systems handling high-speed data as well as voice. A prime reason is its ability to provide flexible handling of both types of information. Spread-spectrum technology has been adopted as well for local-area, personal-area, and *ad hoc* wireless networks, as will be seen in Chapter 12.

Problems

6.1 **(a)** Verify the figure 832 as the capacity (number of users possible) in the analog AMPS system. What is the corresponding figure for D-AMPS (IS-136)?

(b) Show that the European spectral allocation for GSM makes 992 channels available.

(c) Verify the comparative capacities of the AMPS, D-AMPS, and GSM systems shown in Table 6.2.

Gudmundson, B. *et al.* 1992. "A comparison of CDMA and TDMA systems," Proc. IEEE Vehicular Technology Conference, Denver, CO, May, 732–735.
Sköld, J. *et al.* 1995. "Performance and characteristics of GSM-based PCS," Proc. 45th IEEE Vehicular Technology Conference, 743–748.

6.2 **(a)** Explain, in your own words, the distinction between *user information bit rate* and *transmission rate*. In particular, show that the GSM user rate is 22.8 kbps while the transmission rate is 270.833 kbps.

(b) Calculate the corresponding information bit rates and transmission rates for IS-136 (D-AMPS).

6.3 **(a)** Draw a diagram of a CDMA system and identify its major building blocks.

(b) What is the purpose of power control in CDMA systems?

(c) What is meant by *soft handoff*?

(d) Explain why CDMA offers universal reuse of channels, in which each channel can be used in all cells.

6.4 Compare TDMA and CDMA schemes, identifying the advantages of each scheme over the other.

6.5 The CDMA system IS-95 speech encoder operates at a bit rate of 9.6 kbps. Find the spreading gain if the chip rate is 1.2288 Mchips/sec. How many chips per bit does this represent? Superimpose a sketch of a sequence of chips on top of a sequence of bits.

6.6 **(a)** Under the assumptions made in the text, show that the CDMA system IS-95, as used in the United States, has the potential of accommodating 520 users if interference in an isolated cell only is considered.

(b) Show the capacity of IS-95 in the US reduces to the range of values shown in Table 6.2 if intercell interference as well as the other considerations noted in the text are taken into account.

6.7 **(a)** Using (6.5) (or (6.7) where applicable) and (6.8) plot and compare PSK bit error probability as a function of E_b/I_0, the ratio of signal energy per bit to interference power spectral density, in the two cases of non-fading and fading media.

(b) Using any reference on the effect of a fading environment on digital transmission (one example appears in the text), check some of the probability of bit error entries appearing in Table 6.1.

6.8 This problem is devoted to establishing some of the CDMA interference expressions used in the text.

(a) Demonstrate that the total average interference power at a base station due to mobiles outside its own cell is given by (6.13).

(b) Explain why two paths must be considered in evaluating the shadow-fading expression appearing in (6.13).

(c) Follow the discussion in the text of the evaluation of the shadow-fading expression appearing in (6.13) and demonstrate that one obtains (6.16). Evaluate (6.16) for the example considered in the text and show that the average outside cell interference power is given by (6.17). Compare this result with that obtained for other values of the shadow-fading standard deviation σ.

(d) Use the circular-cell interferer model discussed in the text and show that the first-tier interference power is given by (6.19b).

6.9 Verify that the ratio E_b/I_0 with total CDMA interference power included is given by (6.18) for the example chosen in the text. Find the equivalent expression for a number of other choices of the shadow-fading parameter σ. (See problem 6.8(c).)

6.10 **(a)** The text discusses, briefly, a number of features of CDMA such as soft handoff and automatic silence detection that improve the capacity of such systems. Which of these features could be implemented with TDMA systems as well? Explain your conclusions.

(b) Put the effect of these features together, as done in the text, and show how the capacity of 84 mobiles per cell is obtained. Indicate how this number changes as the assumptions made for the various features change as well.

Chapter 7

Coding for error detection and correction

We have noted a number of times that the mobile radio environment is a hostile one. This was particularly spelled out in Chapter 2, in our discussion of fading phenomena in that environment. We shall see, in our discussion of voice-oriented second-generation (2G) cellular systems in the next chapter, Chapter 8, that extensive use is made of coding techniques to mitigate the effect of transmission vagaries such as fading on the information transmitted. By "coding" is meant the purposeful introduction of additional bits in a digital message stream to allow correction and/or detection of bits in the message stream that may have been received in error. In Chapter 10, describing packet-switched data transmission in third-generation (3G) cellular systems, as well as Chapter 12 on wireless local-area networks, we shall see that coding techniques are commonly used as well. Three such techniques have been adopted for wireless systems: block coding for error detection, with error-detection bits added to each message to be transmitted; convolutional coding for error correction; and, in high-speed 3G systems, the use of turbo coding in place of the convolutional coding procedure. (Turbo coding is a relatively recent, and particularly effective, coding technique.) Although block codes are used extensively in telecommunications systems to correct errors, as we shall see in Section 7.1 following, wireless systems in particular use block codes most frequently to detect errors not corrected by such error-correction procedures as convolutional and turbo coding. In this chapter we provide a brief introduction to these techniques to better equip the reader in understanding their use in the systems discussed in Chapters 8 and 10. The discussion in this chapter of block and convolutional coding summarizes and follows quite closely the more detailed introductory treatment appearing in Schwartz (1990). A more comprehensive treatment appears in Michelson and Levesque (1985). Both books contain other references to the literature in the field. We begin the chapter by describing block coding first, then move on to convolutional coding, concluding with a brief tutorial on turbo coding.

Schwartz, M. 1990. *Information Transmission, Modulation, and Noise*, 4th edn, New York, McGraw-Hill.
Michelson, A. M. and A. H. Levesque. 1985. *Error-Control Techniques for Digital Communication*, New York, John Wiley & Sons.

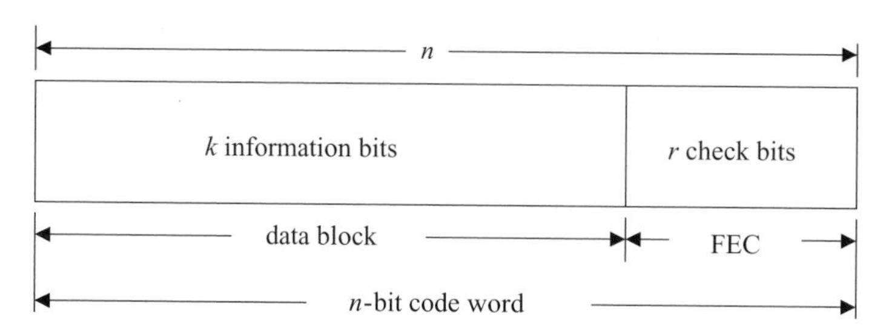

Figure 7.1 Block coding

7.1 Block coding for error correction and detection

Say a block of k bits is used to carry digital information. Block coding consists of adding r bits, formed by linear operations on the k information bits, to the block. These added bits are often referred to as *parity-check bits*. Figure 7.1 portrays this concept pictorially. (Note that the stream of bits is read from left to right, the first bit in time being the left-most bit. We shall be using this convention throughout this chapter, except where otherwise noted.) Depending on the numbers k and r, and on the linear operations used to form the parity-check bits, one may detect or correct errors occurring during transmission of the full block of $n = k + r$ bits. The code is called an (n, k) code, and the resultant coded n-bit block is referred to as a *codeword*. (With k-bit blocks, there are then 2^k possible codewords.)

Consider a simple example. Say $r = 1$ parity check bit equal to the modulo-2 (mod-2) sum of the k information bits is added to the k-bit block. The mod-2 sum of the $n = k + 1$ bits transmitted is then always 0, and the added-parity-check bit is referred to as an even-parity-check bit. A little thought will indicate that it is possible to detect an *odd* number of errors. It is similarly readily shown that choosing the parity-check bit to be the complement of the mod-2 sum of the k information bits also provides a scheme for detecting any odd number of errors. As an example, let the information block be 00110101, with this $k = 8$-bit block again assumed read in from left to right (first 0, then 0, 1, 1, . . .). The mod-2 sum is 0, so that the even-parity-check bit would be 0, and the full 9-bit codeword would be 001101010. The complement of the sum is clearly 1, so that an odd-parity-check bit would be 1. In either case, any odd number of errors would be detectable.

Now consider increasing the number r of parity-check bits. It is apparent that more errors may be detected and some errors corrected as well. The appropriate choice of the parity-check bits, as a linear operation on the information bits, is a fundamental problem in coding theory. Many different codes have been developed, with different properties. There exist codes to detect or correct independently occurring errors, to detect or correct burst errors, to provide error-free or low-error synchronization of binary information, etc. (Burst errors, in which a whole sequence of bits may be corrupted during transmission, might be expected to occur relatively frequently in the wireless fading environment, particularly for slowly moving mobiles "trapped" in a region prone to fading. The occurrence of burst errors can be reduced, however, by storing a number of successive bit blocks and reading the bits out in a different order than generated. This technique of *block interleaving* separates out the bits in any one block, converting burst errors to independently generated ones.

(Such a technique is commonly used in cellular systems, as will be noted in Chapters 8 and 10.) To develop codes with specific properties, a significant amount of structure must be incorporated in them. In particular, we focus in this introductory chapter on coding on so-called *systematic codes*, in which the parity-check bits always appear at the end of the code word, the property assumed above in discussing single-parity-bit error checking. Of this class, we shall, in addition, stress *cyclic codes*, which, as we shall see, are describable in polynomial form, a property that makes their implementation and analysis much simpler to carry out.

Consider now the k-bit data information block $d_1, d_2, \ldots, d_k$, with d_i, $i = 1$ to k, either 0 or 1. We denote this sequence of binary numbers by the vector $\boldsymbol{d}$:

$$\boldsymbol{d} = (d_1, d_2, \ldots, d_k)$$

The corresponding codeword will be denoted by the n-bit vector $\boldsymbol{c}$

$$\boldsymbol{c} = (c_1, c_2, \ldots, c_k, c_{k+1}, \ldots, c_n)$$

A systematic code is one for which $c_1 = d_1, c_2 = d_2, \ldots, c_k = d_k$, i.e., c_1 to c_k represent the first k bits transmitted. The $r = n - k$ parity-check bits $c_{k+1}, c_{k+2}, \ldots, c_n$ are then given by the weighted mod-2 sum of the data bits, as given by (7.1) below

$$\begin{aligned} c_{k+1} &= h_{11}d_1 \oplus h_{12}d_2 \oplus \ldots \oplus h_{1k}d_k \\ c_{k+2} &= h_{21}d_1 \oplus h_{22}d_2 \oplus \ldots \oplus h_{2k}d_k \\ &\vdots \\ c_n &= h_{r1}d_1 \oplus h_{r2}d_2 \oplus \ldots \oplus h_{rk}d_k \end{aligned} \tag{7.1}$$

The h_{ij} terms are 0 or 1, and represent the weights referred to above.

A simple example would be a (7, 3) code ($n = 7$ output bits for every $k = 3$ input bits) with the following choice of the four parity-check bits

$$\begin{aligned} c_4 &= d_1 \oplus d_3 \\ c_5 &= d_1 \oplus d_2 \oplus d_3 \\ c_6 &= d_1 \oplus d_2 \\ c_7 &= d_2 \oplus d_3 \end{aligned} \tag{7.2}$$

The $2^3 = 8$ possible codewords in Table 7.1 then result. (We have purposely inserted a space between the input bits and the four parity-check bits in the column of codewords shown.)

A close look at (7.1), with the (7, 3) code of equation (7.2) just described as an example, indicates that the n-element vector $\boldsymbol{c}$ for a systematic codeword may be written as a $k \times n$-matrix operation on the k-element information data word (or block) $\boldsymbol{d}$

$$\boldsymbol{c} = \boldsymbol{d}\mathbf{G} \tag{7.3}$$

Because $\boldsymbol{c}$ is a systematic codeword (see the example of the (7, 3) codewords), the $k \times n$ matrix $\mathbf{G}$, called the *code-generator matrix*, must have as its first k columns the $k \times k$ identity matrix $\mathbf{I}_k$ while the remaining $r = n - k$ columns represent the transposed array of weight coefficients h_{ij} of (7.1). (Why must the array be transposed in defining $\mathbf{G}$? *Hint*: What if we had chosen to write $\boldsymbol{c} = \mathbf{G}\boldsymbol{d}$?) This r-column array may be represented by

Table 7.1 *Codewords, (7,3) code*

d	c
000	000 0000
001	001 1101
010	010 0111
011	011 1010
100	100 1110
101	101 0011
110	110 1001
111	111 0100

a $k \times r$ matrix we label **P**. We thus have

$$\mathbf{G} = [\mathbf{I}_k \mathbf{P}] \tag{7.4}$$

with

$$\mathbf{P} = \begin{bmatrix} h_{11} & h_{21} & \cdots & h_{r1} \\ h_{12} & h_{22} & \cdots & h_{r2} \\ \vdots & \vdots & \vdots & \vdots \\ h_{1k} & h_{2k} & \cdots & h_{rk} \end{bmatrix} \tag{7.5}$$

Consider the (7, 3) code as an example. Then it is left for the reader to show that

$$\mathbf{P} = \begin{bmatrix} 1 & 1 & 1 & 0 \\ 0 & 1 & 1 & 1 \\ 1 & 1 & 0 & 1 \end{bmatrix} \tag{7.6}$$

and

$$\mathbf{G} = \begin{bmatrix} 1 & 0 & 0 & 1 & 1 & 1 & 0 \\ 0 & 1 & 0 & 0 & 1 & 1 & 1 \\ 0 & 0 & 1 & 1 & 1 & 0 & 1 \end{bmatrix} \tag{7.7}$$

Continuing with the (7, 3) code as an example, look carefully at the eight codewords in Table 7.1 above. Note that the codewords all differ in at least three positions. A single error in any one bit of an 8-bit transmitted codeword should be correctable since the received codeword would be *closer* to that transmitted codeword than to any of the seven other possible codewords, in the sense that all the others would differ in more than one bit position from this received codeword. This (7, 3) code is an example of a *single-error-correcting code*. (Certain numbers of errors will be detectable, although not correctable as well. We discuss error detection later on.) This single-error-correcting capability is obviously due to the fact that the codewords differ in at least 3-bit positions. The difference in the number of bit positions between two codewords is called the *Hamming distance* after the prominent researcher R. W. Hamming, who, while at Bell Laboratories in the 1940s and 1950s, did some of the basic work on coding theory for communication systems. For two errors to be correctable, the Hamming distance d should be 5; for x errors correctable, the Hamming distance should be at least $2x + 1$.

Given the code used, how does one decode the received n-bit codeword? An obvious procedure is to repeat the parity-check calculation at the receiver and compare the resultant parity-check bits with those received. If there is a difference, it is apparent that at least one error has occurred. An error-correction procedure can then be invoked, if the code used has been designed for this purpose. An example of such an error-correction procedure would be a comparison made with entries in a codebook, choosing the codeword most likely to have been transmitted. (This method can become difficult to apply, however, for long codewords, since the list of entries grows exponentially with the data block size k.)

The parity-check procedure at the receiver can be represented as follows: since the n-bit codeword $\boldsymbol{c} = \boldsymbol{d}\mathbf{G}$ is a systematic one, with the matrix $\mathbf{G}$ written in the form of (7.4), it is apparent that $\boldsymbol{c}$ may be written in the form

$$\boldsymbol{c} = [\boldsymbol{d}\ \boldsymbol{d}\mathbf{P}] = [\boldsymbol{d}\ \boldsymbol{c}_P] \tag{7.8}$$

with the symbol $\boldsymbol{c}_P$ representing the $r = n - k$ bit parity-check sequence. Say an n-bit codeword is now received at the receiving system. Its first k bits must represent one of the 2^k possible k-bit data vectors $\boldsymbol{d}$. Using this received vector $\boldsymbol{d}$, the decoder carries out the operation $\boldsymbol{d}\mathbf{P}$ and does a mod-2 comparison of the resultant r-bit vector, bit-by-bit, with the received parity-check vector (the last $r = n - k$ bits received). If the received sequence represents a proper codeword, this comparison must result in a zero value. We write this comparison mathematically as

$$\boldsymbol{d}\mathbf{P} \oplus \boldsymbol{c}_P = 0 \tag{7.9}$$

(It is to be noted that mod-2 subtraction is the same as mod-2 addition.)

An alternate, equivalent, procedure is obtained by noting that (7.9) may be rewritten in vector-matrix form as follows

$$[\boldsymbol{d} \quad \boldsymbol{c}_P]\begin{bmatrix} \mathbf{P} \\ \mathbf{I}_{n-k} \end{bmatrix} = 0 \tag{7.10}$$

The symbol $\mathbf{I}_{n-k}$ represents an identity matrix of order $n - k$. We now define a new $n \times r$ matrix $\mathbf{H}^T$ given by the post-multiplication term of (7.10):

$$\mathbf{H}^T = \begin{bmatrix} \mathbf{P} \\ \mathbf{I}_{n-k} \end{bmatrix} \tag{7.11}$$

Then, from (7.8) and (7.10), we have

$$\boldsymbol{c}\mathbf{H}^T = 0 \tag{7.12}$$

The transpose of $\mathbf{H}^T$ is called the *parity-check matrix* $\mathbf{H}$ and, from (7.11), must be given by

$$\mathbf{H} = [\mathbf{P}^T \quad \mathbf{I}_{n-k}] \tag{7.13}$$

This matrix plays a significant role in studies of error-correcting codes. Note, from (7.5), that it may be written as

$$\mathbf{H} = \begin{bmatrix} h_{11} & h_{12} & \cdots & h_{1k} & 1 & 0 & \cdots & 0 \\ h_{21} & h_{22} & \cdots & h_{2k} & 0 & 1 & \cdots & 0 \\ \vdots & \vdots & \vdots & \vdots & \vdots & \vdots & \vdots & \vdots \\ h_{r1} & h_{r2} & \cdots & h_{rk} & 0 & 0 & \vdots & 1 \end{bmatrix} \tag{7.14}$$

As an example, for the (7, 3) code we have been considering, it is left for the reader to show that

$$\mathbf{H} = \begin{bmatrix} 1 & 0 & 1 & 1 & 0 & 0 & 0 \\ 1 & 1 & 1 & 0 & 1 & 0 & 0 \\ 1 & 1 & 0 & 0 & 0 & 1 & 0 \\ 0 & 1 & 1 & 0 & 0 & 0 & 1 \end{bmatrix} \tag{7.15}$$

(What is the matrix $\mathbf{H}^T$ in this example?)

Now return to the question of error detection and possible correction at a receiver. We indicated that the comparison of (7.9) provides a means of checking for errors in transmission. The use of the matrix $\mathbf{H}^T$ provides another means. Say an error has occurred in one or more of the digits of the code vector $\boldsymbol{c}$. The received vector $\boldsymbol{r}$ may then be written as the mod-2 sum of the transmitted vector $\boldsymbol{c}$ and an n-bit error vector $\boldsymbol{e}$, with a 1 appearing wherever an error occurred

$$\boldsymbol{r} = \boldsymbol{c} \oplus \boldsymbol{e} \tag{7.16}$$

For example, if errors occurred in the first, second, and last bits, we would have $\boldsymbol{e} = (1, 1, 0, 0, \ldots, 1)$. Operating on the received vector $\boldsymbol{r}$ with the matrix $\mathbf{H}^T$, we then get, using (7.12)

$$\boldsymbol{r}\mathbf{H}^T = (\boldsymbol{c} \oplus \boldsymbol{e})\mathbf{H}^T = \boldsymbol{e}\mathbf{H}^T \equiv \boldsymbol{s} \tag{7.17}$$

The r-element vector $\boldsymbol{s}$ is called the *syndrome*. If it is non-zero this is an indication that one or more errors have occurred. (Conversely, a zero $\boldsymbol{s}$ is no guarantee that no errors have occurred, since codes may not necessarily detect all error patterns.)

As a special case, say a single error has occurred in the ith digit of a transmitted codeword, resulting in a 1 in the ith digit of the error vector $\boldsymbol{e}$. From (7.17) and the definition of the matrix $\mathbf{H}^T$, it is readily shown that the syndrome $\boldsymbol{s}$ must be given by the ith row of the $\mathbf{H}^T$ matrix or the ith column of the parity-check matrix $\mathbf{H}$

$$\boldsymbol{s} = (h_{1i}, h_{2i}, \ldots, h_{ri}) \tag{7.18}$$

With all rows of $\mathbf{H}^T$ (columns of $\mathbf{H}$) distinct, the error is therefore not only detectable, but correctable as well. A code satisfying this condition will then be single-error correcting. Consider the (7, 3) code again as an example. Say the information vector consists of the three bits 1 1 0. The 7-bit codeword, from our discussion earlier, is given by 1 1 0 1 0 0 1. Say an error occurs in the third position, changing the 0 to a 1. The error vector $\boldsymbol{e}$ is then (0 0 1 0 0 0 0), and the received vector $\boldsymbol{r}$ has the bit-elements 1 1 1 1 0 0 1. Post-multiplying $\boldsymbol{r}$ by the matrix $\mathbf{H}^T$ for this example (see (7.17)), it is readily shown that the syndrome vector $\boldsymbol{s}$ is (1 1 0 1). Comparing with (7.15), we note that this is just the third column of the parity-check matrix $\mathbf{H}$ or the third row of $\mathbf{H}^T$.

To generalize the discussion a bit, what are the requirements on a code to correct more than one error? The so-called *Hamming bound* (one of a number of such bounds) provides a measure of the number r of parity-check bits required. Say *at least t errors* are to be corrected. The Hamming bound says simply that the number of possible check bit patterns 2^r must be at least equal to the number of ways in which up to t errors can occur. For an (n, k) code, with $r = n - k$, we must therefore have

$$2^r \geq \sum_{i=0}^{t} \binom{n}{i} \tag{7.19}$$

As an example, for a single-error-correcting code, $t = 1$, and we have $2^r \geq n + 1$. If $r = 3$, then n can be no more than seven bits long. The (7, 3) code we have been using as an example fits, of course, into this category. It is left for the reader to show that (7, 4) and (15, 11) codes should also correct one error. But note that the Hamming bound is no more than that. It is a bound that indicates the possibility of finding a code that corrects this number of errors. It is not prescriptive. It provides possible candidates for codes. It does not provide a method for finding the code itself. We shall see later that the cyclic code polynomial formulation does provide a method for establishing the code.

Consider the possibility now of correcting at least two errors. Applying (7.19), one finds that (10, 4), (11, 5), and (15, 8) codes should exist that provide double-error correction. We indicated earlier that the resultant codewords should then have a Hamming distance of 5. For correcting at least three errors, the Hamming distance must be 7. Examples of codes correcting at least three errors would then be (15, 5), (23, 12), and (24, 12) codes. These examples can be checked by using the Hamming bound of (7.19).

There are two problems, however, with attempting to do error correction using (n, k) block codes. Note first that a relatively large number of check bits must be added to correct even small numbers of errors, with the codeword sizes constrained to relatively small values. Large-size information blocks would therefore require large numbers of parity-check bits. This problem may be summed up by noting that the so-called *code efficiency*, defined as the ratio k/n, tends to decrease as the desired error-correction capability increases. The second problem arises out of the first one: for real-time transmission, the transmission rate, and, hence, transmission bandwidth must increase as the number of check bits is increased. For $n = r + k$ bits must now be transmitted in the same time interval as required to transmit the original k-bit information block. This increased bandwidth would result in added noise let into the system. Alternately, if the time interval required to transmit the original information block is kept fixed, bit lengths are reduced by the ratio k/n. This reduces the energy per bit, increasing the probability of error in receiving the codeword. The specific connection between bit energy E_b and bit error probability was noted in Section 6.4 of Chapter 6 in summarizing some important performance equations relating error probability to signal energy and noise spectral density. It was noted there that, in the case of PSK digital transmission, as an example, the bit error probability in the presence of white noise and a non-fading environment was given by $P_e = {}^1/_2 \operatorname{erfc} \sqrt{E_b N_0}$, with erfc the complementary error function, E_b the energy in a bit, and

N_0 the noise-spectral density (see (6.5)). It was shown there, that for small error probability, the error probability decreased exponentially with the critical parameter E_b/N_0. In a fading environment, however, the bit error probability decreased only linearly with E_b/N_0. (Recall our discussion in Chapter 6 in which we indicated that the error probability equations were still fairly valid in the cellular environment with the gaussian-like signal interference used in place of white noise.) As parity-check bits are added, reducing the bit length and hence bit energy correspondingly, the bit error probability increases, counteracting the intended use of coding to reduce errors!

To demonstrate this effect, consider the (7, 3) code single-error-correcting code we have been using as an example throughout this section. Without the use of the code, the probability of error of the uncoded 3-bit block is just the probability that at least one bit in three will be in error. Assuming independent bit errors for simplicity, this is just $1 - (1 - P_e)^3 \approx 3P_e$, $P_e \ll 1$. In the coded case, the bit error probability is of the same form as P_e, but with the signal energy reduced by the factor 7/3. (Why is this so?) Thus, using the symbol p to represent this probability, we have, for the non-fading white-noise (or "white-interference") case, $p = \frac{1}{2}\mathrm{erfc}\sqrt{3E_b/7N_0}$. Hence $p > P_e$, as noted above. (It is left for the reader to repeat this discussion for the fading environment, using the appropriate equations from Chapter 6.) The probability of block error in the coded case is the probability that at least two independent errors will occur in a pattern of seven bits. (Why?) For small p this is approximately $21p^2$. Comparing this expression with the uncoded value $3P_e$, it is clear that coding will only be effective for small enough p, i.e., large enough E_b/N_0, so that the quadratic dependence p^2 begins to take over. Details are left to the reader. A similar analysis appears in Schwartz (1990) for the (7, 4) single-error-correcting code. It is shown there that, for the white-noise non-fading case, the (7, 4) code does not provide significant improvement in error correction until the bit error probability P_e is 10^{-5} or less. The corresponding value of E_b/N_0 is about 9 dB or more. In the cellular environment block coding is thus employed most often to detect errors. Convolutional coding, to be discussed later in this chapter, is used for error correction. The error-detection capability of block codes will become apparent after our discussion of cyclic codes that follows.

Cyclic codes: polynomial representation

The Hamming bound discussed above provides a method for determining the code size needed to provide a specified error-correction capability. It does not provide the means to actually construct the code, i.e., to determine the parity-check bits required. Additional structure is needed to actually allow the specific codes to be determined. The class of cyclic codes does provide this necessary structure. In addition, as we shall see, cyclic codes are describable in polynomial form, allowing them to be generated by shift-register techniques. Block codes used for error detection rather than correction in cellular systems are usually specified by their polynomial representation. We shall provide some examples later in this section.

A cyclic code is defined as one for which codewords are lateral shifts of one another, with mod-2 addition bit-by-bit of any two codewords resulting in another codeword. Thus, if an n-bit codeword is represented by the code vector $\boldsymbol{c} = (c_1, c_2, \ldots, c_n)$, the code vector

$(c_2, c_3, \ldots, c_n, c_1)$ is another codeword, as is $(c_3, c_4, \ldots, c_n, c_1, c_2)$, and so forth. Consider the (7, 3) code we have been using as an example throughout this section. Note that all codewords, except for the all-zero one, may be obtained by lateral shifts of one another. Bit-by-bit mod-2 addition of the codewords (0011101) and (0100111) results in codeword (0111010). The (7, 3) code is thus an example of a cyclic code and does have a polynomial representation, as we shall see shortly.

This definition of a cyclic code results in specific properties of the corresponding code-generator matrix **G**. Thus, it is readily shown, using a contradictory argument, that the last element of **G**, in the kth row and nth column, must always be a 1. For let that element be a 0 instead. Consider the information word $\boldsymbol{d}$ given by (000 . . . 01). Since the corresponding codeword $\boldsymbol{c}$, using (7.3), is given by $\boldsymbol{c} = \boldsymbol{d}\mathbf{G}$, we must then have $\boldsymbol{c} =$ (000 . . . 01 . . . 0), i.e., the last bit of $\boldsymbol{c}$ is a 0. Shifting this codeword once to the right, we get a codeword whose information bits are all 0, while the parity-check bits are of the form 1 . . . . But this is not possible the way we have defined our code structure; an all-zero data information sequence must result in an all-zero parity-check sequence. The last row of the code-generator matrix must thus end in a 1. Note that equation (7.7), the generator matrix of the (7, 3) code example we have been using throughout this discussion, does have this property. We shall see, in addition, that this last row determines the code structure, all other rows being found as cyclic shifts and mod-2 additions of previous rows.

We now proceed by describing cyclic codes in their polynomial form. Specifically, let the n-bit codeword $\boldsymbol{c} = (c_1, c_2, \ldots, c_n)$ be described by an $(n - 1)$-degree polynomial in x, in which each power of x represents a 1-bit shift in time. The highest-order term is then written $c_1 x^{n-1}$, the next term is $c_2 x^{n-2}$, and so on, with the last term being just c_n. The full polynomial is thus given by

$$c(x) = c_1 x^{n-1} + c_2 x^{n-2} + \cdots + c_{n-1} x + c_n \tag{7.20}$$

Successive shifts to generate other code words are then repeated by the operation $xc(x) \bmod (x^n+1)$. For example, shifting once, we get

$$c_2 x^{n-1} + c_3 x^{n-2} + \cdots + c_{n-1} x^2 + c_n x + c_1$$

Notice that the coefficients of this polynomial represent the lateral shift noted above, and provide a second codeword. Shifting a second time, we have $x^2 c(x) \bmod (x^n + 1) = c_3 x^{n-1} + c_4 x^{n-2} + \cdots c_n x^2 + c_1 x + c_2$. We shall show below how one generates the specific polynomial $c(x)$ for a given cyclic code (n, k). But before doing this, we show first how the **G** matrix is represented in polynomial form as well. This is simply done, beginning with the first column, by representing each element of value 1 with the appropriate power of x, leaving blank those elements containing a zero. Consider the (7, 3) code example again. From the code-generator matrix **G** of (7.7), we have, as its polynomial representation

$$\mathbf{G} = \begin{bmatrix} x^6 & - & - & x^3 & x^2 & x & - \\ - & x^5 & - & - & x^2 & x & 1 \\ - & - & x^4 & x^3 & x^2 & - & 1 \end{bmatrix} \tag{7.21}$$

Note that the last row of (7.21) is representable by the polynomial $g(x) = x^4 + x^3 + x^2 + 1$. The power of the leading coefficient is precisely $n - k = 4$ in this case. More generally, it is clear from the way the original matrix **G** was constructed that this last row in polynomial form must always be given by $x^{n-k} + \cdots + 1$. (Recall our proof above that the last element of **G** in the case of a cyclic code must always be a 1.) This polynomial $g(x)$ is termed the *code-generator polynomial*. Using the symbol r to represent the number of check bits $n - k$, $g(x)$ may be written generally in the form

$$g(x) = x^r + g_{r-1}x^{r-1} + g_{r-2}x^{r-2} + \cdots + g_2x^2 + g_1x + 1 \qquad (7.22)$$

The coefficients g_i, $i = 1$ to $r - 1$, are either 0 or 1. As its name indicates, once the generator polynomial is specified, the code and all the corresponding codewords may be determined. In fact, the full generator matrix **G** may be written down, given $g(x)$. Consider our (7, 3) example code again. Starting with $g(x)$ in the last row, as shown in (7.21), to obtain the second row from the bottom we first shift laterally by x, writing $xg(x)$, to obtain $x^5 + x^4 + x^3 + x$. But recall that the first $k = 3$ columns in G must be given by the identity matrix $\mathbf{I}_k$ for systematic codes such as this one (see (7.4)). The shifted polynomial $xg(x)$ in this example has the term x^4 which violates the identity matrix condition. Since the mod-2 sum of two codewords is itself a codeword, one of the conditions of a cyclic code, as noted above, we simply get rid of the undesired x^4 term by adding $g(x) = x^4 + x^3 + x^2 + 1$ to the shifted $xg(x)$, to obtain $x^5 + x^2 + x + 1$. (The reader is asked to check this mod-2 summation.) Note that this polynomial is precisely the one appearing in the second row of (7.21). The second row from the bottom, the first row in this case, is again obtained in the same manner: shift the polynomial just obtained laterally by again multiplying by x. Check to see that the condition for the identity matrix $\mathbf{I}_k$, here $\mathbf{I}_3$, is not violated. If it is, again add the last row, $g(x)$, to the shifted polynomial. Note, in this example, that $x(x^5 + x^2 + x + 1) = x^6 + x^3 + x^2 + x$. There is no undesirable x^4 term, so this polynomial is the one used in that row, just as shown in (7.21). We have thus shown, by example, how, given the generator polynomial $g(x)$, the entire generator matrix **G** may be obtained. Cyclic codes are thus derivable from a generator polynomial.

How does one now find the generator polynomial $g(x)$ for an (n, k) cyclic code? We state, without proof, that the answer very simply is that $g(x) = x^r + \cdots + 1$ is a divisor of $x^n + 1$ (Michelson and Levesque, 1985). (Michelson and Levesque provide a proof that $g(x)$ is a divisor of $x^n - 1$, but the two statements are the same in mod-2 arithmetic. We prefer to use the +1 notation here, in keeping with the notation used throughout this chapter.) Consider the class of $(7, k)$ codes, with $r = 7 - k$, as an example. In this case we need the divisors of $x^7 + 1$. It is left to the reader to show that

$$x^7 + 1 = (x + 1)(x^3 + x^2 + 1)(x^3 + x + 1) \qquad (7.23)$$

There are thus two possible codes of the form (7, 4), with $r = 3$, using either of the two polynomials in x^3 to serve as the desired $g(x)$. To find the generating polynomial for the (7, 3) code we have been using as an example throughout this chapter thus far, we multiply the first and last divisors, using mod-2 addition, to obtain $x^4 + x^3 + x^2 + 1$, the polynomial obtained earlier (see the last row of (7.21). Here we simply note that products of divisors are divisors as well.

Given $g(x)$ as the appropriate divisor of $x^n + 1$, we can find the corresponding generator matrix **G** using the recipe indicated above: use $g(x)$ as the last row of **G**. Then find the row

Table 7.2 *Generation of codewords, (7, 3) code, from* $g(x) = x^4 + x^3 + x^2 + 1$

$a(x)$	$a(x)g(x)$	Codeword
0	0	0000000
1	$x^4 + x^3 + x^2 + 1$	0011101
x	$x^5 + x^4 + x^3 + x$	0111010
$x + 1$	$x^5 + x^2 + x + 1$	0100111
x^2	$x^6 + x^5 + x^4 + x^2$	1110100
$x^2 + 1$	$x^6 + x^5 + x^3 + 1$	1101001
$x^2 + x$	$x^6 + x^3 + x^2 + x$	1001110
$x^2 + x + 1$	$x^6 + x^4 + x + 1$	1010011

above by shifting $g(x)$ and adding $g(x)$ if necessary; then repeating the process, row by row. Given **G**, we can then proceed to find all 2^k codewords, as indicated earlier. An alternate procedure allows us to find the 2^k codewords directly from the generator polynomial. Note that, since $g(x)$ is, by definition, a polynomial of the $r = n - k$ order, we may write the $(n - 1)$-order polynomial $c(x)$ in the form

$$c(x) = a(x)g(x) \tag{7.24}$$

with modular arithmetic implicitly assumed, and $a(x)$ a polynomial of order $k - 1$. But there are 2^k such polynomials possible, including the degenerate forms 0 and 1. Each of these polynomials must therefore correspond to a different codeword. By scanning through each of them and multiplying by the given $g(x)$, mod-2, one generates all possible codeword polynomials. The corresponding coefficients, 0 or 1, provide the desired codewords. Table 7.2 shows how this procedure is carried out for our familiar (7, 3) code. Michelson and Levesque (1985) provide a similar example for the 16 possible codewords of a (7, 4) code using $x^3 + x + 1$, one of the other divisors in (7.21), as the generator polynomial. Note that these codewords are precisely those shown in Table 7.1, although not listed in the same order.

It was noted in introducing the polynomial representation of cyclic codes that this representation lends itself directly to shift-register implementation of these codes. This is demonstrated as follows. We first write the k data bits $d_1, d_2, \ldots, d_{k-1}, d_k$ in the polynomial form

$$d(x) = d_1x^{k-1} + d_2x^{k-2} + \cdots + d_{k-1}x + d_k \tag{7.25}$$

a polynomial of degree $k - 1$ or less. (With $d_1 = 0$ and $d_2 = 1$, for example, this expression becomes a polynomial of degree $k - 2$.) The operation $x^{n-k}d(x)$ then results in a polynomial of degree $n - 1$ or less. Dividing this new polynomial by the generator polynomial $g(x)$ of degree $r = n - k$ produces a polynomial $z(x)$ of degree $k-1$ or less, plus

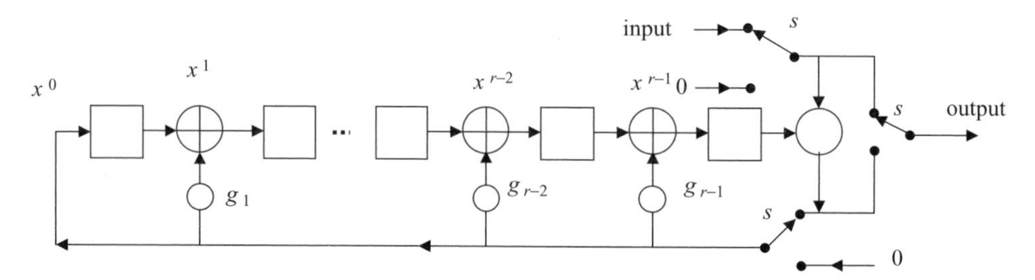

Figure 7.2 Cyclic-code shift-register encoder, *r* check bits. Switches up for information bits, down for last *r* parity-check bits

a remainder polynomial $r(x)$, as indicated in (7.26) following

$$\frac{x^{n-k}d(x)}{g(x)} = z(x) + \frac{r(x)}{g(x)} \tag{7.26}$$

Now we note that, by mod-2 addition, $r(x) + r(x) = 0$. Hence, if we write $x^{n-k}d(x) + r(x)$, the resultant polynomial must be divisible by $g(x)$. From (7.24), this must represent a codeword, and we thus have, again using r in place of $n - k$

$$c(x) = a(x)g(x) = x^r d(x) + r(x) \tag{7.27}$$

Since $x^r d(x)$ represents a left shift by r-bit units of the data bits, the remainder term $r(x)$ must represent the parity-check bits. These bits are thus obtained by carrying out the operation

$$r(x) = \text{rem}\frac{x^r d(x)}{g(x)} \tag{7.28}$$

the symbol "rem" denoting remainder. The (7, 3) code with $g(x) = x^4 + x^3 + x^2 + 1$ again provides an example. Let the data vector be $\boldsymbol{d} = (111)$. The corresponding polynomial is then $d(x) = x^2 + x + 1$. It is then left for the reader to show that dividing $x^r d(x)$ by $g(x)$, with $r = 4$ in this case, results in the remainder polynomial $r(x) = x^2$. This corresponds to the four parity-check bits 0100, so that the full codeword is then 1110100. This is in agreement with both Tables 7.1 and 7.2.

The shift-register encoder representation of the operation of (7.28) appears in Fig. 7.2. The square boxes shown represent 1-bit delays. Modulo-2 sum operations are indicated as well. The weighting factors g_1 through g_{r-1} are the coefficients used earlier in (7.22) in defining the generating polynomial $g(x)$, and are either 0 or 1. The k incoming data bits appear on the line labeled "input," with the switches in the "up" position as indicated. Note that these bits immediately move to the encoder output as desired, thus forming the first k-bits of the n-bit codeword. They also move sequentially through the shift register, the operations indicated being carried out as they do so. As the last of the k data bits leaves the encoder, the switches labeled s are immediately shifted to the "down" positions, and the register bits, representing the parity-check bits, clocked out, a bit at a time, until all r bits have been output. Note that the mod-2 operations continue throughout this period as well. Note also that r-bit 0-sequences are inserted as well at two of the switches to clear the register and await the next data word once the parity-check bits have been output.

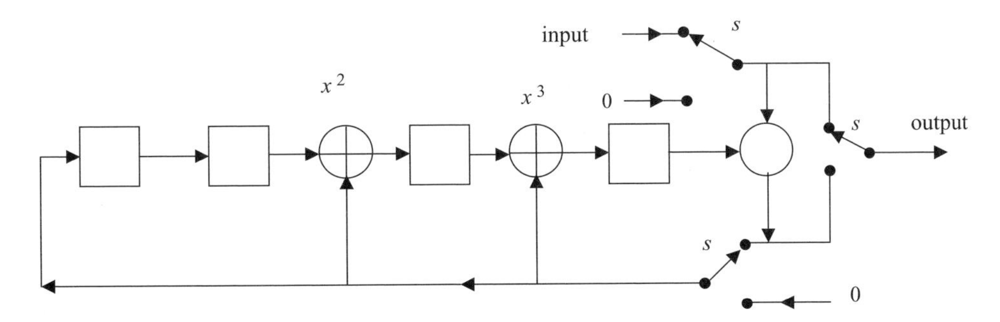

Figure 7.3 Encoder, (7, 3) code, $g(x) = x^4 + x^3 + x^2 + 1$

As an example, the shift-register encoder for the (7, 3) code appears in Fig. 7.3. It is left for the reader to carry out the operations indicated, using several of the eight possible input data sequences to check that the parity bits shifted out agree with those appearing in Tables 7.1 and 7.2. Other examples of polynomial codes designed to correct errors appear in Schwartz (1990) and Michelson and Levesque (1985), as well as in the references cited there.

Error detection with cyclic codes

From the discussion above, it is clear that the polynomial representation of cyclic codes lends itself directly to shift-register implementation of encoders. The receiving system carries out the same shift-register operations to determine the parity-check bits for comparison with the parity bits actually received, to determine whether any bit errors have occurred during the transmission process. The stress above has been on the use of such codes to correct errors. The principal use of block codes in wireless cellular systems, however, has been to *detect*, rather than correct, errors, as noted at the beginning of this chapter. The relatively large information-block size adopted in digital wireless systems, running into hundreds of bits as we shall see, in discussing examples of wireless systems in Chapters 8 and 10, precludes the use of block-based error correction techniques of the type we have been discussing thus far in this chapter. Such large blocks would require inordinately large numbers of parity-check bits to carry out even limited error correction. Parity checking is thus used to detect errors. Packet-switching networks such as the Internet use parity-check calculations for error-detection purposes as well. In the case of wireless systems, it is common to use 8, 12, or 16 parity-check bits appended to the information blocks to carry out error detection. As we shall see in Chapters 8 and 10, convolutional and turbo encoders are used at the transmitter, following the error-detection block parity-check procedure, to provide an error-correction capability in these systems. (At the receiver, the reverse procedure is carried out: the convolutional and turbo decoders first carry out error correction. The block parity-check procedure is then used to detect any errors not corrected.) These codes will be discussed in the next sections. Parity-bit error detection is thus used to detect errors not corrected by the convolutional coding. It is therefore useful, at this point, to focus on error detection using polynomial representations of codes. The block codes are then called outer codes; the convolutional and turbo codes are referred to as inner codes.

Before discussing the use of cyclic codes for detecting errors, it is important to indicate why *detection* of errors, rather than forward-error correction, as discussed thus far in this chapter, can play a significant role in communication systems. We note that error detection is used several ways in telecommunication systems: an information block containing one or more detected errors that have not been corrected may simply be dropped; the receiving system may signal that one or more errors have been detected for further action to be taken; or, as one possible action, the receiving system may signal the transmitter to retransmit the block in question. The first possibility, the simplest, can be used for real-time traffic, such as voice, in which case a receiving system detecting an error in a voice data block may simply discard that block. It is well-known that error rates of up to 5% may be tolerated in real-time voice, i.e., humans participating in a voice call do not find the conversation degraded too much with occasional dropping of voice samples. Second-generation digital wireless systems to be discussed in the next chapter, Chapter 8, are designed principally to handle voice traffic.

Third-generation systems, to be discussed in Chapter 10, are designed to handle packet-switched data traffic as well. Much of this type of traffic does not require real-time delivery. In this case, packet retransmission can be used to correct any errors detected. In particular, a technique labeled generically ARQ, for *automatic repeat request*, is commonly used in packet-switching networks, including the Internet, and has been adopted for packet-switched wireless systems as well. In this technique, a timer is set at the transmitting system on sending a data block. If the receiving system detects no errors, it responds with a positive acknowledgement, to be received by the sending system before the timer times out. On the other hand, if an error is detected, the receiver simply drops the block. The timer at the sending system will then time out, with the data block then automatically resent, and the timer again set. In the case of the Internet, TCP, the transport protocol operating end-to-end, between two end users, carries out an ARQ procedure, the receiving end user checking for errors in data blocks received and sending an acknowledgement back to the sending end user if no errors are detected; otherwise it drops any data block deemed to have been received in error. The sending user then re-transmits that data block after expiration of the timer. It is for this reason that "lower layer" protocols such as the widely adopted Ethernet used in local-area networks need detect errors only. It is left for the "higher-layer" protocols such as TCP to correct errors. (We discuss the protocol-layering concept adopted for packet-switched data transmission briefly in Chapter 10.)

The process of detecting errors in an information block is similar to that carried out for error correction, as implicit from the discussion above: a set of parity-check bits is calculated, using in many cases a generator-polynomial shift register calculation, and the block with parity-check bits added is then transmitted. The receiver carries out the same calculation, and checks to see if the parity-check bits obtained agree with those received. The set of parity-check bits appended to the information block has been variously called the *block check*, *checksum bits*, *error-detection field*, *frame check sequence* (FCS), and *cyclic redundancy checking* (CRC), as well as other designations. We have already seen, at the beginning of this chapter, that a single-parity bit can be used to detect an odd number of bits received in error. More generally, however, as the key to the widespread adoption of parity-check coding for error detection, we emphasize the following statement: *the use of r parity-check bits enables any burst of errors r bits or less in length to be detected.* This is independent of the length of the information block. A constructive proof of this statement appears in Schwartz (1990). Note that an error *burst* of length b consists of a sequence of b bits in which the first and last bits are received in error, with the bits in between received correctly or in error. Thus, the use of eight parity-check bits allows any burst eight bits long or less to be detected; the use of 16 parity bits enables a burst 16 bits long or less to be detected.

Even more interestingly, a cyclic code using r parity bits will detect a high percentage of bursts *longer* than r in length! Specifically, it is readily shown, using the polynomial representation of cyclic codes, that the fraction of bursts of length b greater than r that remain undetected is just 2^{-r} (Schwartz, 1990). This result is independent of the length of the information block. Long blocks may thus be used, and most error patterns detected. If $r = 8$ parity bits are added, for example, the fraction of undetected burst patterns is $2^{-8} = 1/256$; if 12 parity bits are used, this fraction of undetected burst patterns drops to $2^{-12} = 0.00024$; if 16 parity bits are used, the fraction of undetected bit patterns drops to 0.000015! Note, as we shall see in Chapters 8 and 10, that these parity-check bit numbers are those commonly used in transmitting information blocks over the air interface in digital wireless systems.

For example, 16-bit, 12-bit, and 8-bit parity-check sequences have been adopted for use with data and signal information transmission in the third-generation cdma2000 wireless cellular system to be described in Chapter 10. The term Frame Quality Indicator is used in that system to designate the parity-check sequence. The generator polynomials for these three cases appear in equations (7.29) to (7.31) below. The corresponding shift-register implementations, all based on the general implementation of Fig. (7.2), appear in Figs. (7.4) to (7.6), respectively (cdma, 2002).

$$g(x) = x^{16} + x^{15} + x^{14} + x^{11} + x^6 + x^5 + x^2 + x + 1 \tag{7.29}$$

$$g(x) = x^{12} + x^{11} + x^{10} + x^9 + x^8 + x^4 + x + 1 \tag{7.30}$$

$$g(x) = x^8 + x^7 + x^4 + x^3 + x + 1 \tag{7.31}$$

Using the fact that the fraction of undetected bursts is 2^{-r}, as noted above, one can do a "back-of-the envelope" calculation to determine the probability that errors in an information block n bits long will not be detected. Specifically, we assume as the error model that the block error probability is proportional to the block length n. We thus have, as the block error probability np, with p some small parameter. (See Schwartz, 1990 for a discussion of this model. One justification comes from the observation that if bit errors were independent with some small probability p, the probability a block n bits long would be in error would be given by $1-(1-p)^n \approx np$, if $np \ll 1$. In the cellular environment with fading ever-present, bit errors are generally not independent. Fading bursts may, in fact, be much more common. However, as we shall see in discussing second-generation

---

cdma 2002. Physical layer standard for ccdma2000 spread spectrum systems, Release A, 3GPP2 C.S0002-A, Version 6.0, 3rd Generation Partnership Project 2, 3GPP2, February, http://www.3gpp2.org/public_html/specs/

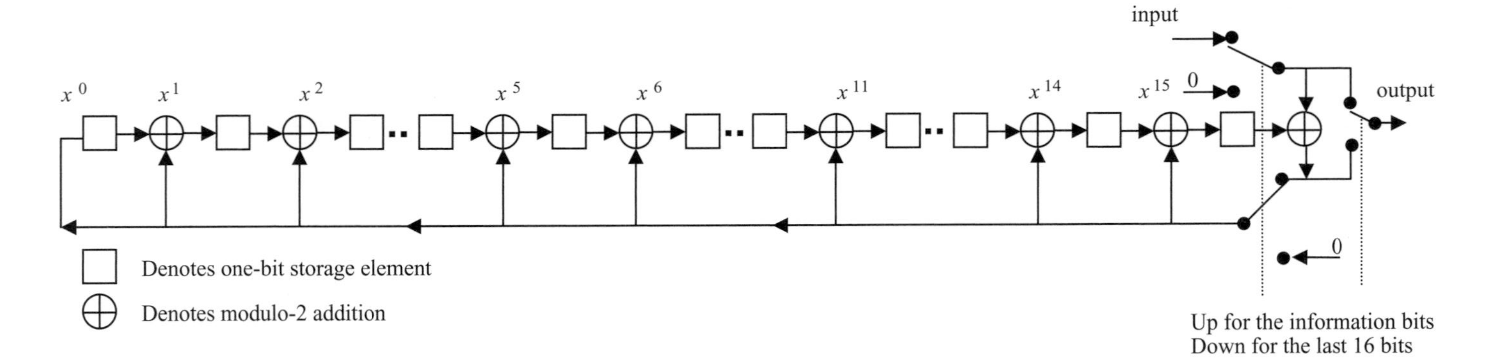

Figure 7.4 Frame quality indicator calculation for the 16-bit frame quality Indicator, cdma2000 (from cdma, 2002: Fig. 2.1.3.1.4.2.1-1)

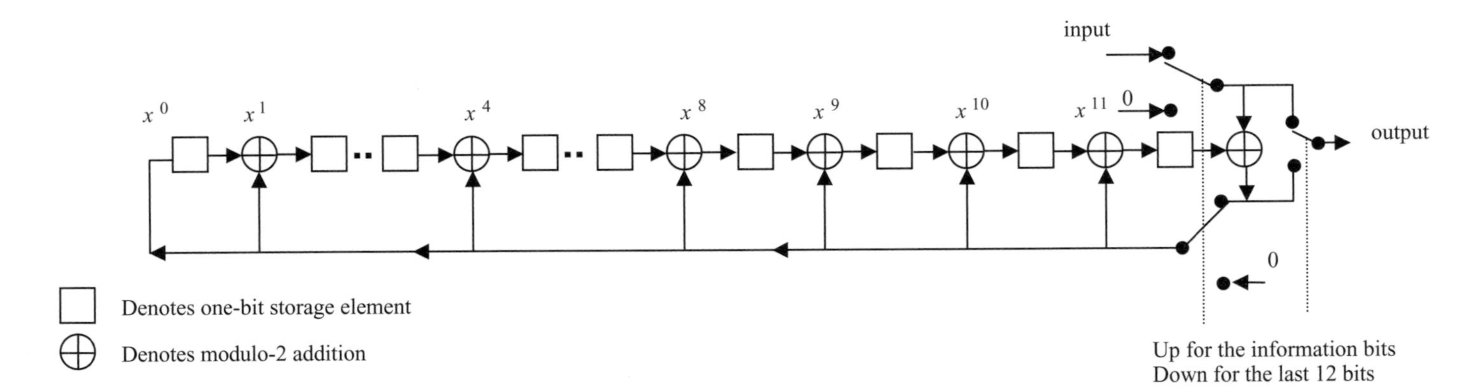

Figure 7.5 Frame quality indicator calculation for the 12-bit frame quality Indicator, cdma2000 (from cdma, 2000: fig. 2.1.3.1.4.2.1-2)

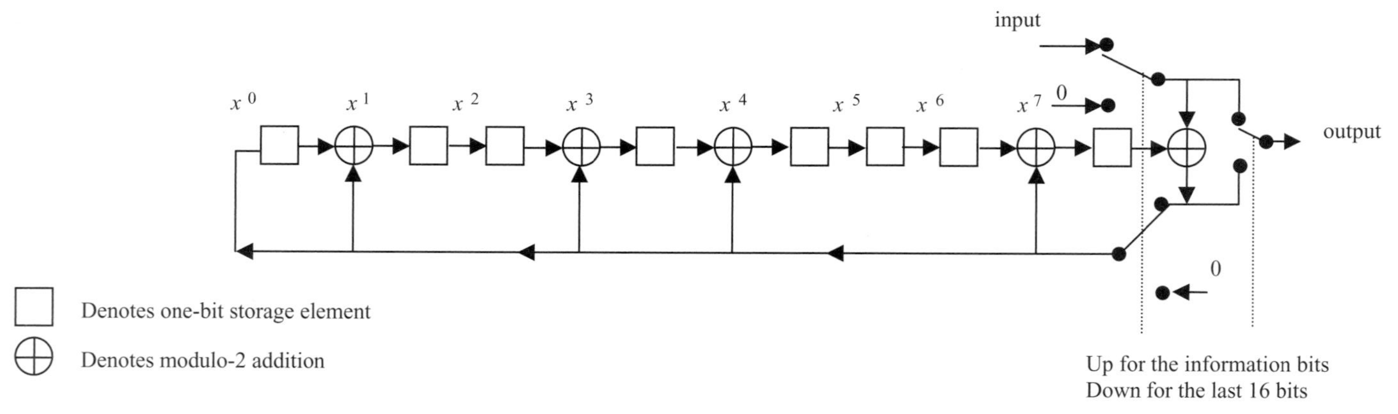

Figure 7.6 Frame quality indicator calculation for the 8-bit frame quality Indicator, cdma2000 (from cdma, 2002: fig. 2.1.3.1.4.2.1-3)

and third-generation systems in Chapters 8 and 10, block interleaving is commonly used to spread each block of data bits out in time, converting possible bursts of errors to quasi-independently generated bit errors. Despite this caveat, the assumption made here that the block error is proportional to the block length seems to be intuitively reasonable, and is based on some experiments on early telephone channels referenced in Schwartz, 1990.) Using this model, the probability P_{bl} that errors in an n-bit block will remain undetected is simply given by

$$P_{bl} \approx np2^{-r} \tag{7.32}$$

Consider some examples: Take $p = 10^{-5}$ in all cases. If $n = 500$ bits, $P_{bl} \approx 2 \times 10^{-5}$ with $r = 8$ parity bits used, and $P_{bl} \approx 10^{-7}$, with 16 parity bits used. If n increases to 1000 bits, these numbers of course double. The error probability is still quite small, however. Note that these numbers do reflect the initially small choice of the parameter p.

This discussion of parity-check error detection concludes our discussion of block coding with particular emphasis on cyclic codes and their polynomial representation. In the next section we move to a discussion of convolutional coding commonly used in wireless systems.

7.2 Convolutional coding

We noted earlier that coding techniques fall into two major categories, block coding and convolutional coding. (Turbo codes, to be discussed briefly in the next section, build on convolutional coding, as we shall see.) Block codes add r parity-check bits to k information bits to form an n-bit codeword. In the case of convolutional codes, a sliding sequence of data bits is operated on, modulo-2, to generate a coded stream of bits. Convolutional coders have the ability to correct errors, and were first introduced for use in satellite and space communication applications. Digital cellular systems now use these types of coders as well, as will be seen in subsequent chapters. The discussion of convolutional coders in this section draws heavily on Schwartz (1990). More extensive tutorial treatments appear in Michelson and Levesque (1985) and Proakis (1995). We shall reference some basic papers in the field as well, as we proceed.

The convolutional coder consists of a K-stage shift register. Input data bits are shifted along the register one bit at a time. Modulo-2 sums of the contents of the shift-register stages are read out at a rate v times as fast by shifting out v bits for every bit in. K is called the constraint length of the coder. Such a coder is classified as a constraint-length K, rate $1/v$ coder. As we shall see later, in discussing the error-correction performance of convolutional coders, performance improvement is obtained by increasing K and the rate v. The price paid is both complexity and increased bit rate of operation. An example of a $K = 3$, rate-$1/2$ encoder appears in Fig. 7.7. The output bit rate is thus twice the input rate. The boxes indicate shift-register stages, just as in Figs. 7.2 to 7.6. Two possible representations are shown. Part (a) of the figure has $K = 3$ boxes; part (b) uses $K - 1 = 2$ boxes. Note that, in both representations, input bits are fed to the two output lines, running

Proakis, J. G. 1995. *Digital Communications*, 3rd edn, New York, McGraw-Hill.

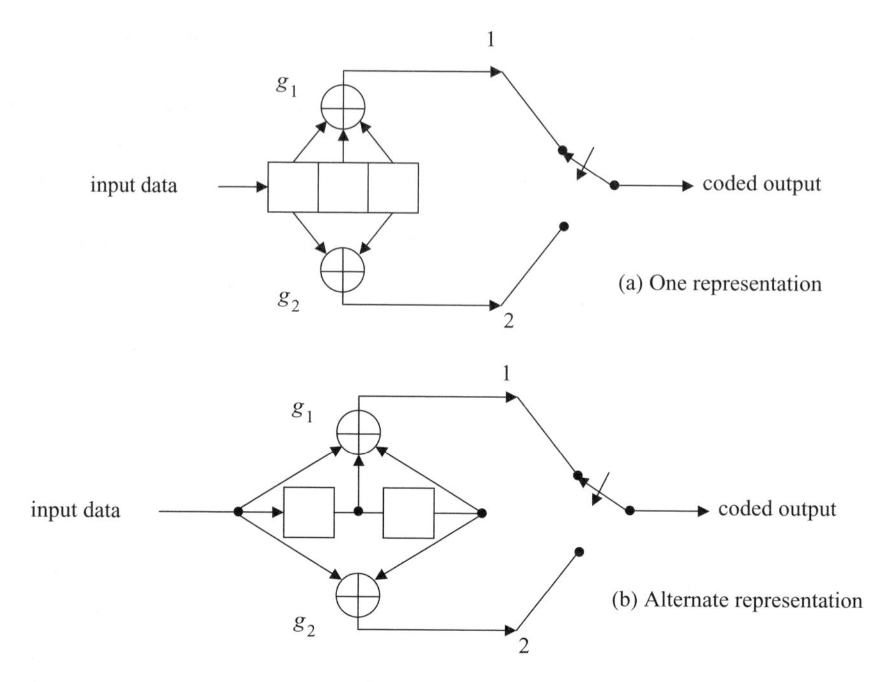

Figure 7.7 Example of a $K = 3$, rate $\frac{1}{2}$ convolutional encoder

at twice the speed of the input line. They are there added, modulo-2, to bits from the other stages, following the connections indicated. The coded output is given first by the bit appearing at output line 1, then by the bit at line 2, then back to line 1 again, and so on, as the switch shown cycles between the two outputs. As bits enter the register, they are shifted through, one bit at a time, again as indicated. The representation of part (b) demonstrates that only $K - 1$ delay stages are needed to represent the encoder. This is due to the fact that the output bits are determined by the connections in the $K - 1 = 2$ rightmost stages of Fig 7.7(b), plus the new bit. Convolutional encoders must be initialized by setting all stages to 0 before operating on an incoming word. This adds K bits to each incoming data word as we shall see in the specific examples shown in Chapters 8 and 10.

Consider now the operation of the encoder of Fig. 7.7 as an example. Say it has been initialized and that the first data bit to appear at the input is a 0. Then clearly two 0s appear at the outputs. Now let a 1 enter at the input. From the connections shown in Fig. 7.7 it is apparent that the two outputs are 1 and 1, respectively. Let another 1 then enter the encoder. It is left for the reader to show that the two outputs are now, in order, 0 1. Letting a 0 now enter, the corresponding outputs are again 0 1. Bits continue to be carried along by the shift register for $K - 1$ stages. The full 16 bits outputted for a particular sequence of eight bits at the input are shown below:

Input bits:	0		1		1		0		1		0		0		1	
Output bits:	0	0	1	1	0	1	0	1	0	0	1	0	1	1	1	1

Note that the particular connections to form the output mod-2 addition of the bits from the various stages may be represented by the two function generators g_1 and g_2 as indicated

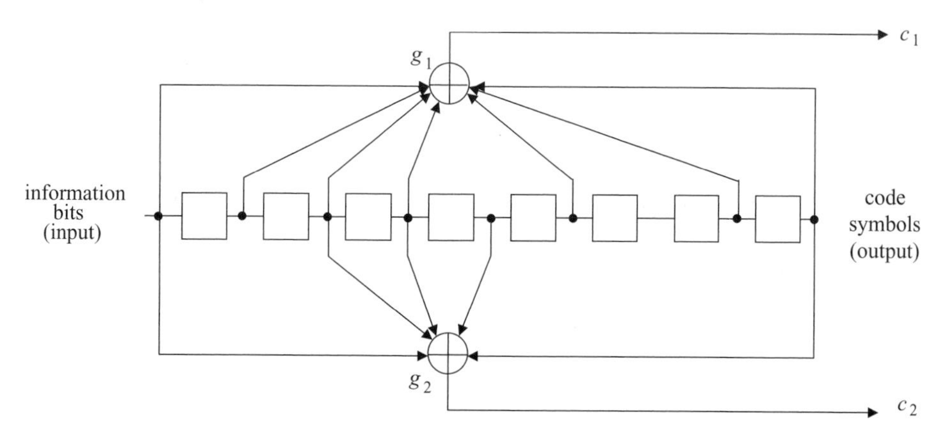

Figure 7.8 $K = 9$, rate ½ convolutional encoder (from cdma, 2002: fig. 2.1.3.1.4.1.3-1)

in Fig. 7.7. Each function generator is represented by a K-bit vector, one bit for each stage of the shift-register, including the input bit. A 1 at a particular stage indicates a connection is made to the output mod-2 adder; a 0 indicates no connection is made. For the example of Fig. 7.7, we would then have $g_1 = [111]$ and $g_2 = [101]$. These function generators serve to represent the convolutional encoder, just as the block-code generator polynomial in its representation of (7.22) and Fig. 7.2 is used to represent a particular block code in shift-register form. Generalizing, a constraint-length K, rate-$1/v$ encoder would have v such function generators, one for each output line, each a K-bit vector indicating how the bits at each stage of the shift-register are to be added to form the output bit.

Most commonly, particularly for longer constraint-length convolutional encoders, the function generators are given in octal form to simplify their representations. Thus, in the example of Fig. 7.7, the two generators are represented, respectively, as $g_1 = 7$ and $g_2 = 5$. The example of a $K = 9$, rate-$1/2$ convolutional encoder used in the third-generation cdma2000 system to be described in Chapter 10 appears in Fig. 7.8 (cdma, 2002).The two sets of coded output bits, c_1 and c_2, are defined as coded output symbols, as shown. The two function generators are given respectively, in octal form, as $g_1 = 753$ and $g_2 = 561$. As a check, note that, for the two sets of connections indicated in Fig. 7.8, the 9-bit connection vectors are, respectively, [111101011] and [101110001]. Grouping three bits at a time to create the octal numbers, we have 753 and 561, as stated. Changing the connections, i.e., changing the function generators, changes the output-bit stream. The particular function generators chosen determine the performance of the convolutional encoder in terms of error probability. We shall summarize the performance characteristics of convolutional encoders shortly.

A particular convolutional coder may be represented three different ways, by a state diagram, by a trellis, or by an ever-expanding tree. These representations provide better understanding of the operation of these encoders and lead directly to their performance characterization. We demonstrate each of these representations, using the example of the simple $K = 3$, rate-$1/2$ coder of Fig. 7.7. We begin with the state diagram representation. Referring back to Fig. 7.7, we note that the output-bit sequence for any input-bit interval is determined by the bit arriving in that interval, plus the bits in the $K - 1 = 2$ rightmost stages

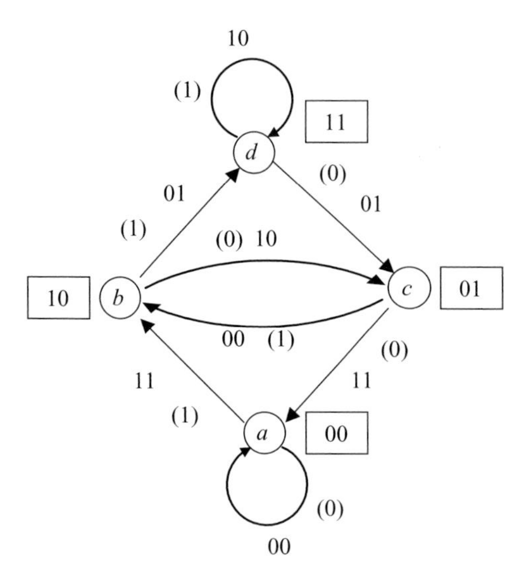

Figure 7.9 State diagram, convolutional encoder of Fig. 7.7

of Fig. 7.7(a) or, equivalently, the $K - 1 = 2$ stages of Fig. 7.7(b). There are $2^{K-1} = 4$ possible arrangements of bits in these $K-1 = 2$ stages. These possible arrangements of bits are termed *states*. Consider, for example, a 0 arriving at the input to the encoder of Fig. 7.7, with the bits in the two stages being 00, as was the case with the example 8-bit input/16-bit output sequence chosen above. The coder is said to be in state 00 when the input 0 arrives, and the output bits are given by 00 as well. Had a 1 arrived at the input, however, the output sequence would have been 11, because of the particular connections chosen. With a 1 appearing at the input, the $K-1 = 2$ bits in the register are now given by 10. The coder is said to have moved to a new state 10. With the 0 arriving, the coder stays in state 00. Figure 7.9 provides the state diagram for the convolutional encoder of Fig. 7.7. The $2^{K-1} = 4$ states in this case are indicated by the $K - 1 = 2$ bits appearing in the rectangular boxes, and are labeled *a,b,c,d*, as shown. Lines with arrows indicate transitions from one state to another, depending on the input bit, shown in parentheses, with the output bits labeled as well, shown following the input bit. Thus, with the coder in state 00 or *a* as shown, a 0 input bit keeps the coder in the same state, with the output bits given by 00 as shown. A 1 at the input, however, drives the coder to state 10 or *b*, as shown, with the bit sequence11 emitted at the output. With the coder in state *b*, a 1 at the input drives the coder to state 11 or *d*, with 01 appearing at the output. If instead, a 0 appears at the input, with the coder in state *b*, the coder moves to state 01 or *c*, with a 10 appearing at the output, as shown. The rest of the possible transitions appearing in Fig. 7.9 are left for the reader to demonstrate. Clearly, with different function generators or coder connections, specific output-bit sequences change, but the state diagram of Fig. 7.9 remains the same. It thus applies to all $K = 3$, rate-1/2 convolutional encoders. A $K = 4$, rate-1/2 coder would have eight states. Note that the state diagram is cyclically traversed, all states being visited: starting at any state at a given time, the finite-state machine representing the convolutional coder eventually returns to that state.

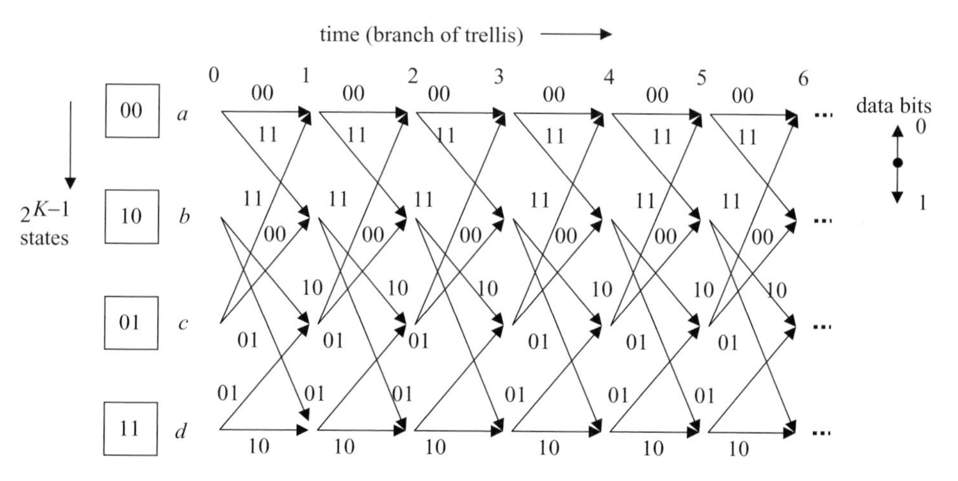

Figure 7.10 Trellis representation, encoder of Fig. 7.7

Consider now the trellis representation of the same $K = 3$, rate-1/2 encoder, as portrayed in Fig. 7.10. The trellis representation of any other encoder can then easily be obtained in the same way. The 2^{K-1} states, the four states of this example, are arrayed along the vertical axis, as shown. Time, in input-bit intervals, is represented by the horizontal axis. Transitions between states are then shown as occurring from one time interval to the next, a transition due to a 0 data input bit always being graphed as the upper line of the two leaving any state. The output bits are shown in order of occurrence alongside each transition, just as in the state diagram of Fig. 7.9. Note how this trellis representation immediately demonstrates the repetitive or cyclical nature of the encoder finite-state machine: starting at any state, one can see at a glance how one moves between states, eventually returning to the state in question. This representation is useful in evaluating the performance of a given convolutional encoder.

The third, tree, representation of convolutional encoders is again best described using the same $K = 3$, rate-1/2 example. The tree representation for this example appears in Fig. 7.11. It is effectively the same as the trellis representation with sequences of transitions now individually separated out and identified as separate paths. As in the trellis representation of Fig. 7.10, just described, upward transitions are shown corresponding to a 0 data input bit; downward transitions correspond to a 1 data input bit. The output bits corresponding to a particular transition along a given path are shown above each transition. This representation too demonstrates the repetitive nature of the convolutional encoder state machine, as is apparent by tracing out the various possible paths, and noting that the $2^{K-1} = 4$ states are entered and reentered as time progresses. The important point to note here is that the number of possible paths, each with a different sequence of output bits, increases exponentially with time. The Hamming distance, the difference in the number of positions between any pair of paths, increases as well. This observation indicates why convolutional codes provide improved performance. As a given path increases in length, it is more readily distinguishable from other paths. One or more errors occurring in transmission are more readily detected and corrected. This simple observation demonstrates

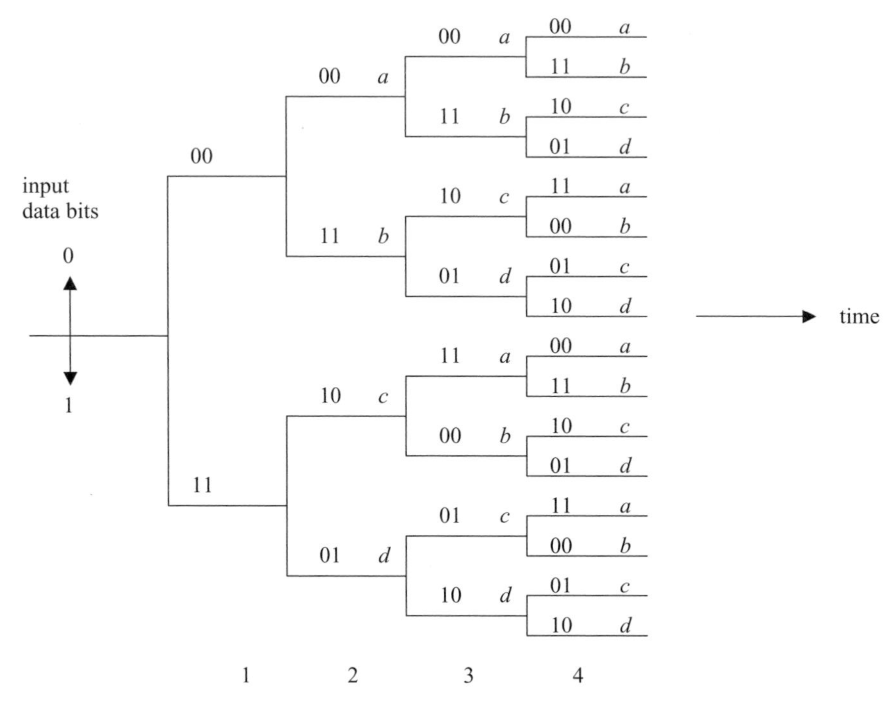

Figure 7.11 Tree representation, coder of Fig. 7.7

as well why increasing the constraint length K and/or the rate ν of the encoder improve the system performance.

Given a particular binary data sequence passed through a convolutional encoder, how does one recover or decode the sequence at the destination, the receiving system? Coding of the information-bearing binary sequence, the subject of this chapter, implies that some transmitted information will have been corrupted during transmission, whether due to interference, fading, or noise. The whole purpose of the convolutional coding we have been discussing is to recover possibly corrupted information with the smallest probability of error. This defines the "best" decoder. With convolutional coding this means, given a received signal sequence, choosing the *most probable path* that would correspond to that sequence. Say, for example, that we select a sequence L input binary intervals long. There are then 2^L possible paths corresponding to the received sequence. The most probable or most likely one of these is the one to be chosen. A decoder implementing such a strategy is called a *maximum-likelihood decoder* (Schwartz, 1990; Proakis, 1995). Consider the following simple version of such a decoder. For each input data binary interval there are ν bits transmitted and ν signals received. Because of fading, noise, and interference, each received signal is essentially analog in form, even though transmitted as a binary 1 or 0. Say, for simplicity, that these ν signals are each individually converted to the best estimate of a 1 or 0. Such a technique is termed hard limiting or *hard-decision decoding*. This procedure simplifies the discussion considerably. (Maximum-likelihood decoding of the actual analog received signals is discussed in Schwartz (1990) and in Proakis

(1995). Studies have shown that one pays a performance penalty of about 2 dB by using hard-decision decoding, rather than retaining the received analog signals and working with these directly (Heller and Jacobs, 1971). This performance penalty corresponds to the added signal power needed to attain the same probability of error. Representation of the received signal by more than one bit and using this representation in making the decision is referred to as soft-decision decoding. The Viterbi algorithm described next for carrying out the decision process lends itself readily to soft-decision decoding (Michelson and Levesque, 1985; Proakis, 1995), thus providing close to optimum performance. We continue with the discussion of hard-decision decoding to simplify the analysis.) The received path now consists of a set of L symbols, each containing ν bits. This received sequence of $L\nu$ bits is now compared with each of the 2^L possible paths and the "closest" one in terms of Hamming distance is now selected as the most likely or most probable one to have been transmitted. The L bits corresponding to this path are then outputted as the most likely data bits to have been transmitted. As an example, consider the tree representation in Fig. 7.11 of the $K = 3$, rate-1/2 encoder of Fig. 7.7 we have been using all along as our example. In the third interval, with $L = 3$, there are eight possible paths to be compared with the received sequence of bits. In the fourth interval, with $L = 4$, there are 16 possible paths to be compared with the received sequence. If we go to the fifth interval before deciding on the path with Hamming distance closest to the received sequence over those five intervals, there are 32 possible paths to be compared with the received sequence.

Increasing L should reduce the probability of error, i.e., the probability of selecting the wrong path. But note that this increase in L results in a major problem. As we increase L, the number of possible paths to be compared increases exponentially as 2^L. If we go to a value of $L = 100$, for example, i.e., compare the Hamming distance over 100 paths, 2^{100} comparisons are needed! What to do? An ingenious maximum-likelihood decoding algorithm due to Andrew J. Viterbi, universally called the *Viterbi algorithm*, provides significant alleviation of this problem, and has found widespread use in satellite and space communications, as well as wireless applications (Viterbi, 1967). The premise of the Viterbi algorithm is a simple one: there are 2^{K-1} states in the binary-input convolutional encoder we have been discussing. Say it is at one of these states during a given binary interval. To get to that state, it could only have been in one of two possible states in the prior binary interval. The $K = 3$, rate-1/2 encoder again provides an example. Referring to the state diagram of Fig. 7.9, say it is currently in state b. Then it could only have been in states c or a in the prior interval. An input binary 1, with 00 appearing at the encoder output, would have driven it from state c to the current state b; an input binary 1, with 11 appearing at the encoder output, would have driven the encoder from state a to the current state b. Say the Hamming distance from the actual received binary sequence is smaller for the path to b coming from the previous state c than for the path coming from state a. Then there is no chance, in future intervals, that the Hamming distance for

---

Heller, J. A. and I. M. Jacobs. 1971. "Viterbi decoding for space and satellite communications," *IEEE Transactions on Communication Technology*, COM-19, 5, part II (October), 835–848.

Viterbi, A. J. 1967. "Error bounds for convolutional codes and an asymptotically optimum decoding algorithm," *IEEE Transactions on Information Theory*, IT-13 (April), 260–269.

the path from state a could ever drop below the one from state c. There is thus no sense keeping the path from state a. We might as well drop it and retain the one from state c. This same comparison can be made at each of the encoder states. This is the essence of the Viterbi algorithm. After each binary interval, the smaller of the two paths into each of the 2^{K-1} states, as compared with the received sequence, is retained as a "survivor." There are thus 2^{K-1} comparisons to be made at each binary interval. Over L binary intervals, the total number of comparisons made using the Viterbi algorithm is therefore $2^{K-1}L$, rather than the 2^L comparisons required using a standard maximum-likelihood procedure.

More generally, if hard-decision decoding is not used, but the analog-received information per binary interval is retained for carrying out the algorithm, an additive Euclidean (geometric)-distance metric is used in place of the Hamming distance to determine the survivors after each interval (Schwartz, 1990; Proakis, 1995). The survivor in this case is the one with the smallest metric, in comparison with the Hamming distance used with hard-decision decoding. Path metrics are augmented each interval. As noted parenthetically above, Viterbi decoders readily accommodate soft-decision decoding at a reasonable cost, and have thus gained wide acceptance in practice.

What value of L should be used? One would obviously like to keep L as small as possible. Studies have shown that a value of L about four or five times K is sufficient to produce performance close to that for very large L, i.e., the best possible (Heller and Jacobs, 1971). These same studies have indicated that a simple decoding procedure is to output the oldest input bit on the most likely of the surviving paths. After each interval, then, after selecting the 2^{K-1} survivors, the oldest input bit of the survivor with the smallest metric is chosen as the one most likely to have appeared at the transmitter coder; the oldest interval of each of the survivor paths is dropped from the path memory. The algorithm is then repeated the next interval. Consider a $K = 9$ encoder such as the one in Fig. 7.8 as an example. Then $L = 36$, if $L = 4K$ is used to choose L. The number of comparisons required to output each bit is then $2^{K-1}L = 256 \times 36 = 9216$ in place of 2^{36}, a considerable reduction!

A very simple example of the operation of the Viterbi algorithm, assuming a hard-decision decoding procedure again for simplicity, is shown in Table 7.3. This example again utilizes the $K = 3$, rate-$1/2$ convolutional encoder of Fig. 7.7. We assume for simplicity that a decision is made every five input-bit intervals. Say the system has been operating for some time. The received sequence of bits (two per input-bit interval) for the past five intervals is shown, and is to be compared with eight possible paths: the four survivors from the previous interval, augmented by the two possible transitions into each of the four states of the current interval. The first entry in the Hamming distance column is the distance from one of the previous surviving paths to the received sequence; the second entry is the distance added due to the current transition. Note that the minimum overall Hamming distance of 2 in this example corresponds to the transition from the previous state b to the current state c. Tracing back the five intervals to the first one, it is readily shown that the input bit at interval 1 must have been a 0. This bit is outputted from the receiver, the four new survivors (starred) are chosen, the first interval is dropped from the path, and the calculation repeated the next interval (not shown).

Table 7.3 *Viterbi algorithm, 5-interval operation, $K = 3$, rate-1/2 coder*

	Possible paths							
interval→	1	2	3	4	5	Previous	Current	Hamming
received bits→	01	01	00	11	10	state	state	distance
	11	01	01	11	00	*a*	*a*	$2 + 1 = 3^*$
	11	01	10	01	11	*c*	*a*	$3 + 1 = 4$
	11	01	01	11	11	*a*	*b*	$2 + 1 = 3^*$
	11	01	10	01	00	*c*	*b*	$3 + 1 = 4$
	00	00	00	11	10	*b*	*c*	$2 + 0 = 2^*$
	11	01	10	10	01	*d*	*c*	$3 + 2 = 5$
	00	00	00	11	01	*b*	*d*	$2 + 2 = 4$
	11	01	10	10	10	*d*	*d*	$3 + 0 = 3^*$

Convolutional encoders were introduced as a particularly efficient and effective way of correcting errors in digital transmission. We described the Viterbi algorithm as an ingenious method of reducing the computational and storage requirements in carrying out maximum-likelihood decoding of convolutionally coded digital signals. How effective *is* convolutional coding? How well do convolutional coders perform in practice? How does their performance depend on the constraint-length K, the rate $1/\nu$, and the function generators used? By performance we mean here the residual probability of bit error after using convolutional coding to both code and decode messages.

To determine the performance of a convolutional coder note that an error occurs when the path selected on reception differs from the correct one, as transmitted. The probability of error is then found by calculating the probability a given (incorrect) path deviates from a specified correct path, and averaging over all such paths. Since no particular path is likely to always be the correct one, we assume all possible paths equally likely to occur – this is, in fact, the basic assumption behind the concept of maximum-likelihood decoding, the decoding procedure implemented by the Viterbi algorithm. In particular, we assume for simplicity and without loss of generality that the correct path is the all-zero one. This implies that the message transmitted is a sequence of zeros. Focusing again on the $K = 3$, rate-1/2 encoder example we have been using throughout this section, note from Fig. 7.10, the trellis representation of this encoder, that, with this "correct" sequence of transmitted bits, the path the encoder follows is the upper horizontal one, staying in state a at all times. An error then occurs if the path chosen at the decoder strays from this horizontal one. The probability of error is then found by determining the possible ways the decoder can deviate from the all-zero path and the probability of doing so. But recall the cyclic nature of entering and re-entering states that we have referred to previously and as readily seen from the trellis representation of Fig. 7.10. We can thus pick any interval to start and determine the various ways of leaving the all-zero path and then returning to it. In particular, we choose to start with interval 0 for simplicity. We now list the various paths that could be taken in leaving the all-zero transmitted path and then returning to it later, in order of increasing Hamming distance with respect to the all-zero path. Note that the only way to leave the all-zero path is to first move to state b with bits 11 received. From state b there are two possible ways to move, the shortest-distance one being to state c. (Note that moving to state d from b also incrementally adds a 1 to the metric, but takes us further away from the all-zero state.) From this state the shortest-distance return to state a is the one moving directly back to a. The Hamming distance of this path is then readily found to be 5. This is then the path with the smallest Hamming distance. This minimum Hamming distance of 5 is termed the *minimum free distance* d_F. It turns out that the performance of convolutional decoders depends critically on the minimum free distance, i.e., the smallest Hamming distance. It is left for the reader to show that the next larger value of Hamming distance for this particular convolutional decoder is 6. There are two possible paths with this distance: a-b-c-b-c-a and a-b-d-c-a. It is left for the reader to show as well that there are four paths of distance 7. Many more paths of higher and higher values of Hamming distance exist as well.

In general, for any convolutional decoder, let the function $a(d)$ represent the number of paths with a Hamming distance of d deviating from, and then returning to, the all-0 test path. The error-probability P_e of deviating from the correct path is then upper-bounded by

$$P_e < \sum_{d=d_F}^{\infty} a(d) P_d \tag{7.33}$$

Here P_d represents the probability that d transmitted bits are received in error. Note again that d_F is, by definition, the minimum free distance. All Hamming distances must thus be equal to or greater than this distance. Why the upper bound in (7.33)? Many of the paths may overlap, as shown in the $K = 3$, rate-1/2 encoder just taken as an example. Note that the branch a–b is common to all paths deviating from the all-zero one. The right-hand side of (7.33), however, assumes the paths are all mutually exclusive, allowing the probabilities to be added together. Equation (7.33) indicates the centrality of the minimum free distance d_F to the determination of the error probability.

The path error probability is not the performance measure we really desire, however. It is the bit error probability P_B that really determines the performance of the encoder. The evaluation of this quantity is somewhat more complex and is found to be upper-bounded as well by a quantity similar to that of (7.33). Details of the analysis to find P_B appear in Schwartz (1990) and Proakis (1995). The critical point to be made is that both probabilities of error, the path error probability and bit error probability, depend on d_F. For a number of channel models, it turns out that the error probability, if small enough, the usual desired condition, reduces exponentially with an increase in the minimum free distance. In that case both error probabilities are about of the same order. Hence the larger the minimum free distance the better the performance. Computer search of convolutional codes for various constraint lengths and rates to find codes with the largest possible minimum free distance was carried out many years ago by J. P. Odenwalder. Many of these codes, as well as others found later by computer search, are tabulated in Michelson and Levesque (1985) and Proakis (1995). In particular, the constraint-length $K = 9$ and rate-1/2 code used in the third-generation cdma2000 cellular system and shown earlier in Fig. 7.8 is precisely the code tabulated in Table 9.1 of Michelson and Levesque (1985). A useful bound on the minimum free distance d_F obtained by Heller and appearing here as (7.34) shows very clearly the minimum free distance increases with both K and ν (Proakis, 1995):

$$d_F \leq \min_{l \geq 1} \left\lfloor \frac{2^{l-1}}{2^l - 1}(K + l - 1)\nu \right\rfloor \tag{7.34}$$

(The notation $\min_{l \geq 1}$ means scan the values of the function following over all values of l greater than, or equal to, 1, and choose the smallest value found. The symbol $\lfloor x \rfloor$ means taking the largest integer in x.) Hence increasing either or both of these parameters improves the performance of the convolutional coder. But recall again that we do not want K too large, since the Viterbi decoding algorithm makes 2^{K-1} comparisons each binary interval and makes a binary decision based on $L = 4K$ prior intervals. Increasing the rate parameter ν requires increasing the transmitted rate as well. Checking (7.34) for $K = 3$, rate-1/2, the example we have been using throughout this section, we find the bound on d_F to be 5, exactly the value we found by tracing the error paths out. This bound is, in fact, quite close to the values of d_F found by computer search (Proakis, 1995).

The actual reduction in bit error probability obtained by using convolutional encoding depends on the type of transmission channel used. Consider, as an example, an additive gaussian noise channel as introduced in our brief discussion in Section 6.4 of Chapter 6. We indicated there that using PSK, the best modulation technique for that channel, a signal-to-noise ratio (SNR) of 9.6 dB is required to ensure a bit error probability of 10^{-5}. The $K = 3$, rate-1/2 encoder introduced in addition to PSK reduces the required SNR to 6 dB. Increasing K to 5 and 7, with the rate kept to 1/2, further reduces the required SNR to 5 and 4.2 dB, respectively. These are sizeable reductions in the signal power required. Table 6.1 of Chapter 6 further indicates that very significant reductions in the required signal-to-interference ratio are obtained over a fading channel when CDMA systems employ dual-diversity RAKE and convolutional coding. As indicated in that table, PSK alone, with no diversity and coding used, would require signal-to-interference ratios of 14 dB and 24 dB for bit error probabilities of 0.01 and 0.001, respectively. Introducing dual-diversity RAKE and convolutional coding reduce these numbers to the more tolerable values of 6 and 7 dB, respectively. These changes represent very sizeable decreases in the signal power required. Convolutional coding can thus make a substantial difference in the performance of cellular systems. In the next section we conclude our discussion of coding techniques by introducing the concept of turbo coding, adopted, in addition to convolutional coding, for some of the third-generation cellular systems.

7.3 Turbo coding

In Section 7.1 we considered error detection and correction using block codes. We focused there on systematic codes in which the data bits are first read out, followed by the parity-check bits. The previous section has summarized material on convolutional codes, which are implicitly non-systematic since output bits are modulo-2 sums of input bits and delayed versions of previous bits. It might have occurred to the reader to ask, why not combine the two types of coders to improve the performance provided by either coding scheme

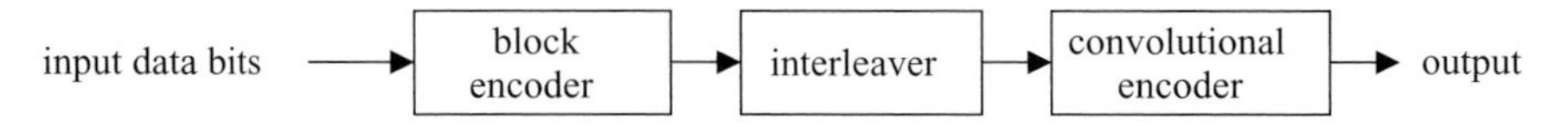

Figure 7.12 Serial concatenation of encoders

alone? Such a technique of combining or *concatenating* two coders was, in fact, first proposed by G. D. Forney many years ago (Forney, 1966), who studied the serial concatenation of two block coders. Later work by many researchers led to a scheme commonly used in deep space communications that concatenates a block code with a convolutional code (Michelson and Levesque, 1985). A block diagram of such a serial-concatenated encoder is portrayed in Fig. 7.12. The block encoder carries out the initial encoding. The resultant encoded blocks are interleaved, or spread out over the space of a number of blocks, to reduce the possible occurrence of bursts of errors, and are then fed into the convolutional encoder. The block coder, usually chosen to be a particular type of block encoder called a Reed–Solomon encoder, is often called the outer coder. The convolutional encoder is, in turn, called the inner coder. On reception, the reverse process takes place: the received bit stream is first decoded by a convolutional decoder, the bits are de-interleaved, and block decoding on the resultant bit stream is carried out. The concatenation of coders provides effective error correction using less-complex and smaller convolutional coders than would otherwise be required (Michelson and Levesque, 1985).

Turbo codes, the subject of this section, involve, as we shall see, the *parallel* concatenation of convolutional coders of the *systematic* type. They provide improved performance over convolutional coding, particularly where higher-speed communication is required, and codes of this type are being adopted for use, in some cases, with third-generation cellular systems, as will be seen in Chapter 10. They represent a relatively recent addition to the error-correction field, having been introduced to the communications community by C. Berrou, A. Glavieux, and P. Thitimajshima (1993; Berrou and Glavieux, 1996).

Their introduction into the coding repertoire has created a tremendous amount of excitement in the coding research community since these codes were shown to provide near-optimum coding performance. Much work on these types of codes has been carried out over the intervening years, with application made to various areas of the telecommunication field, including very specifically to wireless communication. Books are now available providing tutorial introductions. An example is the one by Vucetic and Yuan (Vucetic and Yuan, 2000). To understand why this excitement over the introduction of turbo codes, and their rapid incorporation into wireless systems, among others, we must detour

Forney, G. D. 1966. *Concatenated Codes*, Cambridge, MA, MIT Press.

Berrou, C., A. Glavieux, and P. Thitimajshima. 1993. "Near Shannon-limit error-correcting coding and decoding: turbo codes," IEEE International Conference on Communications, ICC93, Geneva, Switzerland, May, 1064–1070.

Berrou, C. and A. Glavieux. 1996. "Near optimum error correcting coding and decoding: turbo codes," *IEEE Trans. on Communications*, 44, 10 (October), 1261–1271.

Vucetic, B. and J. Yuan. 2000. *Turbo Codes: Principles and Applications*, Boston, MA, Kluwer Academic Publishers.

somewhat to explain the meaning of *optimum* coding performance. The word "optimum" has a very specific meaning here. It is defined in the Shannon sense: Claude Shannon, the eminent engineer and mathematician, and the founder of the field of Information Theory, had shown in his monumental work of 1948, 1949 that communication channels exhibit a *capacity* C, in bits per second, such that, when transmitting at some rate R bps below that capacity, it should be possible with appropriate coding to transmit error free. As an example, consider the most common example of a channel for which Shannon found the capacity explicitly. This is the *additive white gaussian noise* or AWGN channel: the input signals to be transmitted are encoded as continuous, gaussian-appearing waveforms with average power S, and the channel, of bandwidth W Hz, introduces additive gaussian noise with average power N during the signal transmission. The capacity of this channel of bandwidth W Hz is then shown (Michelson and Levesque, 1985) to be given by the familiar equation

$$C = W \log_2 \left(1 + \frac{S}{N}\right) \tag{7.35}$$

Note that Shannon formulated the concept of a channel capacity, of which (7.35) is the most commonly written example, but gave no prescription for the type of coding that leads to the capacity. Many researchers have been engaged over the years in finding coders and codes for them that approach channel capacity as closely as possible. (This capacity is found to be approached by so-called M-ary orthogonal signaling, but only asymptotically for $M \to \infty$ (Schwartz, 1990; Michelson and Levesque, 1985). Coherent M-ary FSK signals provide one example.)

Pursuing this example further, consider the capacity for small values of the signal-to-noise ratio S/N. These small values are obtained by noting that the noise power $N = N_0 W$, N_0 the noise-spectral density, and by letting the bandwidth W increase without limit. (See our discussion in Section 6.4 of these quantities.) The channel is then said to be power limited. For small values of S/N, (7.35) becomes, using the approximation $\log_e(1 + x) \approx x, x \ll 1$

$$C = WS/(N \log_e 2), \quad S/N \ll 1 \tag{7.36}$$

We now write $S = E_b R$, with E_b the signal energy per bit and R the transmission bit rate, as noted above. We also note that $\log_e 2 = 0.69$. It is then left for the reader to show that (7.36) becomes

$$R/C = 0.69/E_b/N_0 \tag{7.37}$$

Since the rate R must be less than the Shannon capacity C to ensure error-free transmission, we have, from (7.37)

$$E_b/N_0 > 0.69 \tag{7.38}$$

In the familiar dB measure, this corresponds to $(E_b/N_0)|_{db} > -1.6$ dB. This limit is referred to as the *Shannon limit*. With E_b/N_0 greater than this limit, the probability of error can theoretically be made to approach zero by proper choice of an error-correcting code. Note that although this limiting expression was obtained by Shannon for an AWGN channel with gaussian-like transmitted signals, it has since been found to be applicable to other types of AWGN channels. The example of M-ary signals was cited above. More to our purposes, a wide class of fading channels exhibits this Shannon limit as well. The one stipulation is that additive white noise be present. A detailed discussion appears in Verdu (2002a) and Verdu (2002b).

Now return to (7.35) and rewrite this equation in terms of E_b/N_0 directly using the relations $S = E_bR$ and $N = N_0W$ noted above. We then get the following equivalent expression for the Shannon capacity

$$\frac{C}{W} = \log_2\left(1 + \frac{R}{W}\frac{E_b}{N_0}\right) \tag{7.39}$$

This expression relates the three parameters C/W (normalized capacity), R/W (normalized transmission rate), and E_b/N_0. In particular, say we operate very close to capacity, so that we again have $R = C$. The Shannon capacity expression (7.39) then gives us the smallest possible value required of the signal bit energy E_b at any capacity level. Call this value $E_{b\min}$. We then rewrite (7.39) in the form

$$\frac{C}{W} = \log_2\left(1 + \frac{C}{W}\frac{E_{b\min}}{N_0}\right) \tag{7.40}$$

This is a parametric expression relating C/W to $E_{b\min}/N_0$. If we plot $E_{b\min}/N_0$ versus C/W, we get the Shannon capacity curve which provides the minimum possible signal energy required to transmit, with zero error probability, at any desired capacity. This minimum energy increases as the capacity, normalized to the bandwidth, increases (Michelson and Levesque, 1985). In the special case where C/W is very small, we again get the limiting $E_{b\min}/N_0 = 0.69$ of (7.38).

How do signals with which we are familiar compare with the limiting curve of (7.40)? We indicated in the previous section that PSK requires a signal-to-noise ratio SNR = 9.6 dB for a bit error probability of 10^{-5}, when transmitted over an additive gaussian channel, just the channel under discussion here. Note, however, that our discussion of Nyquist signal shaping in Chapter 5 indicates that the ratio of PSK bit rate R to the transmission bandwidth W required is close to 1. Setting $R = W$, this says that SNR $= E_b/N_0 = 9.6$ dB for this case. To compare with the Shannon optimum result, we set C/W in (7.40) to 1. (Recall that we took $R = C$ in writing (7.40).) We find the resultant $E_{b\min}/N_0 = 1$ or 0 dB. PSK operating at a bit error probability of 10^{-5} thus requires 9.6 dB more signal power than the absolute minimum required for zero error probability. Note that we also indicated in the previous section that $K = 7$, rate-1/2 convolutional codes require $E_b/N_0 =$ 4.2 dB over the same channel, an improvement of 5.4 dB over uncoded PSK, but still 4.2 dB away from the optimum value of 0 dB. The turbo codes described in Berrou *et al.* (1993) and Berrou and Glavieux (1996) were found, by simulation, to require $E_b/N_0 =$ 0.7 dB at a 10^{-5} bit error probability. This puts these codes within 0.7 db of the Shannon

Verdu, S. 2002a. "Spectral efficiency in the wideband regime," *IEEE Transactions on Information Theory*, 48, 6 (June), 1319–1343.

Verdu, S. 2002b. "Recent results on the capacity of wideband channels in the low-power regime," *IEEE Wireless Communications*, 9, 4 (August), 40–45.

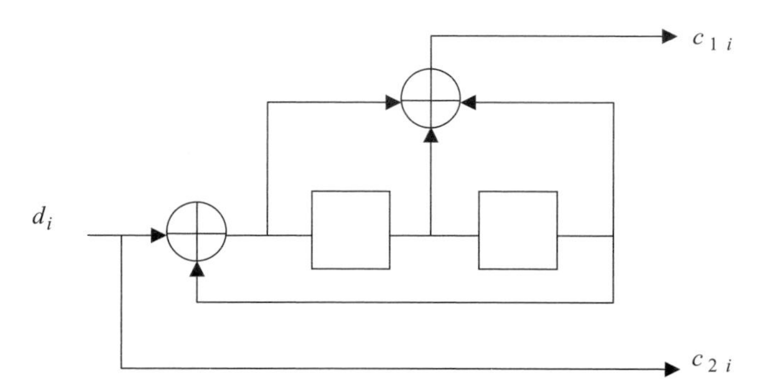

Figure 7.13 Example of a recursive systematic convolutional encoder

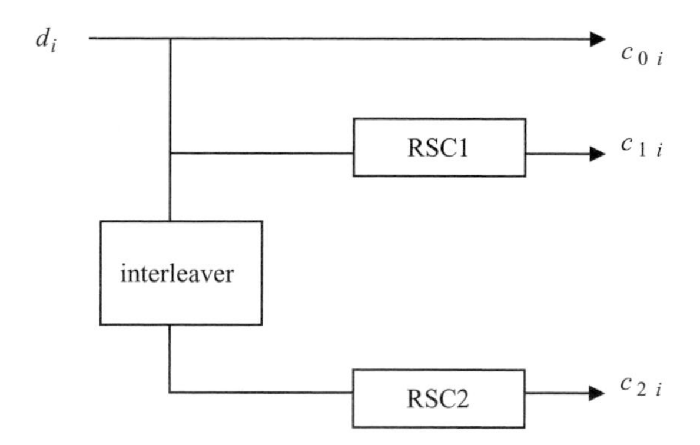

Figure 7.14 Rate-1/3 turbo encoder

optimum. (For the purposes of simulation, 10^{-5} is considered close to 0.) Hence the excitement!

With this digression, we now introduce turbo codes, albeit quite briefly. These codes involve the parallel concatenation of so-called recursive systematic convolutional (RSC) encoders, with an interleaver separating the two. Fig. 7.13 shows the example of a rate-1/2 recursive systematic convolutional encoder (Berrou and Glavieux, 1996). The word systematic means, as already stated, that the input information bits are fed directly to the output. The output line c_2 can be seen to be serving that function in Fig. 7.13, the ith bit on that line being just the ith input bit d_i. The recursive designation means that the output of the convolutional encoder is fed back to the input. This is also shown in the example of Fig. 7.13. A rate-1/3 turbo encoder, as an example, then takes on the form of Fig. 7.14 (Berrou and Glavieux, 1996; Vucetic and Yuan, 2000). The two recursive systematic convolutional encoders RSC1 and RSC2 operating in parallel provide two of the three outputs. Note that the input data bits are fed into the upper convolutional encoder; they are fed into an interleaver before entering the lower encoder. The input bits are also fed directly to the output, as required for a systematic encoder, to provide the third output bit, as shown. The design of the interleaver is critical to the performance of the turbo coder, defined in terms of the bit error probability (Berrou *et al.*, 1993; Berrou and Glavieux, 1996; Vucetic and Yuan, 2000). The purpose of the interleaver is to decorrelate the input sequences of bits

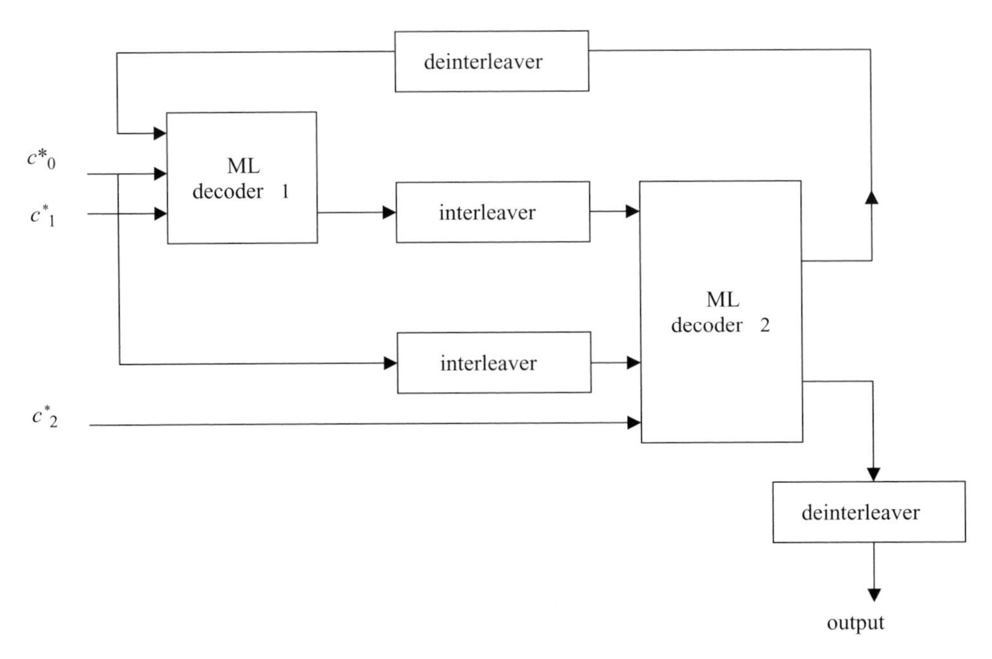

Figure 7.15 Iterative turbo decoder for turbo encoder of Fig. 7.14

fed to both encoders, as well as to the decoders used at the receiver to retrieve the input bit sequence. Pseudo-random scrambling, using a long block size, is found to provide the best interleaver performance: the longer the interleaver size, the better the performance (Vucetic and Yuan, 2000). The performance of turbo codes is again found to depend on a Hamming-type free distance (Berrou and Glavieux, 1996; Vucetic and Yuan, 2000). Good codes can therefore be found through a search procedure involving the Hamming distance (Vucetic and Yuan, 2000).

A maximum-likelihood approach similar to the procedure mentioned in our discussion of convolutional decoding is used to determine the form of the decoder at the receiver. The resultant decoder appears as a serial concatenation of two decoders with the same interleaver as used at the transmitter separating the two (Berrou *et al.*, 1993; Berrou and Glavieux, 1996; Vucetic and Yuan, 2000). Successive iteration is required to attain the best performance. As an example, the decoder for the case of the rate-1/3 turbo encoder of Fig. 7.14 appears in Fig. 7.15 (Vucetic and Yuan, 2000). The symbols c_0^*, c_1^*, and c_2^* are used, respectively, to represent the received signals corresponding to the three transmitted signals shown in Fig. 7.14. The transmitted signals are assumed to have passed through a channel of known properties such as one producing random fading, additive gaussian noise, etc. The specific form of the two maximum-likelihood decoders ML1 and ML2 shown connected in series in Fig. 7.15 depends on the channel properties, as well as the form of the encoders RSC1 and RSC2 of Fig. 7.14. Note that received signals c_0^* and c_1^* are fed into the first decoder ML1; the received signal c_2^* is fed into the second decoder ML2. ML2 also receives as inputs the interleaved output of ML1 as well as the interleaved form of received signal c_0^*. The output of ML2 is fed back to ML1 to form the (iterated) third input to that device. It is this feedback loop that gives the name "turbo" coder to this coding system, based on the principle of the turbo engine. Delays introduced by the ML decoders, interleavers, and the deinterleaver must be accounted for, but are not shown in Fig. 7.15. Iteration continues until convergence to the best possible performance is attained. For large size interleavers of 4096 bits or more, 12 to 18 iterations are typically needed to attain convergence (Vucetic and Yuan, 2000).

The effect of using turbo codes on Rayleigh fading channels has been presented in Hall and Wilson (1998) and is summarized as well in Vucetic and Yuan (2000). Simulation results for rate-1/3 turbo codes show that in this environment they can achieve a performance within 0.7 dB of the capacity possible for this channel. This represents a coding gain of over 40 dB, as contrasted to uncoded PSK! (Recall from Chapter 6 that PSK operating over a fading channel requires $E_b/N_0 = 24$ dB to achieve a bit error probability of 0.001. This increases to 44 dB to achieve a bit error probability of 10^{-5}.) Vucetic and Yuan (2000) contains as well information on some of the turbo coders adopted for use with the third-generation cellular systems to be discussed in Chapter 10.

Problems

7.1 **(a)** Show that one parity-check bit, chosen to be the mod-2 sum of a block of k information bits, can be used to detect an even number of errors among the information bits. Check this result with a number of bit patterns of varying size.

(b) Repeat (a) for the case of odd-parity checking.

7.2 Consider the (7, 3) code of equation (7.2).

(a) Show that this code generates the eight codewords of Table 7.1.

(b) Show that this code is represented by the **P**, **G**, and **H** matrices of (7.6), (7.7), and (7.15), respectively.

(c) Write the $\mathbf{H}^T$ matrix of this code.

(d) Find the syndrome $\mathbf{s}$ of each of the codewords of Table 7.1 if the received code vector has an error in its second digit. Compare this vector with the appropriate column of the **H** matrix.

7.3 The systematic (7, 4) block code has the generator matrix **G** written below.

$$\mathbf{G} = \begin{bmatrix} 1 & 0 & 0 & 0 & 1 & 0 & 1 \\ 0 & 1 & 0 & 0 & 1 & 1 & 1 \\ 0 & 0 & 1 & 0 & 1 & 1 & 0 \\ 0 & 0 & 0 & 1 & 0 & 1 & 1 \end{bmatrix}$$

(a) Find the **P** matrix of this code and compare with that of the (7, 3) code.

(b) Find the parity-check matrix **H** and its transpose $\mathbf{H}^{\mathrm{T}}$ and compare with those of the (7, 3) code.

(c) How many codewords does this code have? Tabulate them all. How many errors can this code correct? Does this agree with the Hamming bound?

(d) Why are the (7, 3) and (7, 4) codes considered dual codes?

(e) Why are there eight correctable error patterns associated with this code, including the all-zero vector? Find the syndrome associated with each pattern.

7.4 Use the Hamming bound (7.19) to show that (15, 11) codes should correct one error; (10, 4), (11, 5), and (15, 8) codes should correct two errors; (15, 5), (23, 12), and (24, 12) codes should correct three errors. Calculate the code efficiency of each of these codes and compare.

7.5 (a) Verify that (7.23) does provide the three divisors of $x^7 + 1$.

(b) Find the two possible codes of the form (7, 4), calculate the matrix **G** for each, and show that one of them does agree with the code of problem 7.3 above.

(c) Use (7.24) to verify the generation of the eight codewords of the (7, 3) code, as carried out in Table 7.2.

7.6 (a) The generator polynomial of the (7, 3) code is given in Table 7.2. Find the parity-check bits for each of the non-zero input data sequences for this code by calculating the remainder polynomial $r(x)$, as given by (7.28).

(b) Check that the (7, 3) encoder of Fig. 7.3 does generate the appropriate parity-check bits for that code.

7.7 Using Fig. 7.2, find the encoder for the (7, 4) code of problem 7.3, similar to Fig. (7.3) for the (7, 3) code. Find the parity-check bits for a number of input data sequences and write out the corresponding 7-bit codewords. Check that the code is single-error correcting.

7.8 Consider the three cdma2000 Frame Quality Indicator shift register implementations (encoders) shown in Figs. 7.4 to 7.6. The generator polynomials for each appear in (7.29) to (7.31), respectively.

(a) Calculate the probability for each of these that errors will remain undetected for blocks 250, 500, and 1000 bits long, in the two cases $p = 10^{-5}$ and 10^{-4}. Comment on these results.

(b) Take the case of the 8-bit Frame Quality Indicator of Fig. 7.6 or (7.31). Choose an arbitrary input bit pattern ten bits long and show the resultant 18-bit systematic code pattern as it leaves the encoder.

7.9 Refer to the $K = 3$, rate-$1/2$ convolutional encoder of Fig. 7.7.

(a) Initialize the encoder so that 0s only appear at its output. Trace the output bit sequence for the following input bit sequence after initialization

1 1 0 0 1 0 1 1

(b) Verify that the state diagram of Fig. 7.9 represents this encoder. Trace through the various states through which this encoder moves for the input sequence of (a), starting at the 00 state. Show the output sequence obtained agrees with that found in (a). Show that the trellis and tree representations of this encoder shown in Figs. 7.10 and 7.11, respectively, also represent this encoder.

7.10 **(a)** Trace through the output bit sequence of the $K = 9$, rate-1/2 convolutional encoder of Fig. 7.8 for an arbitrary input bit sequence. Make sure the encoder is first initialized.

(b) Verify that the two connection vectors may be represented in octal form as 753 and 561.

7.11 Consider a $K = 3$, rate-1/3 convolutional encoder with generators $g_1 = [111]$, $g_2 = [101]$, and $g_3 = [011]$.

(a) Sketch the diagram of this encoder.

(b) Provide the state diagram, labeling the various states, and transitions between states, as is done in Fig. 7.9. Find the output bit sequence for the input bit sequence shown in Prob. 7.9(a). Ensure you again initialize the encoder.

7.12 **(a)** Check the entries of Table 7.3 for the Viterbi algorithm. Verify that the eight survivor paths after five intervals are those shown, with the four new survivors those so indicated. Choose an arbitrary sequence of pairs of received bits over a few more intervals. Repeat the algorithm several more times, making a decision each interval as to the estimate of the input bit five intervals before.

(b) Choose an arbitrary input bit sequence of at least 15 bits. These are to be encoded by the same $K = 3$, rate-1/2 encoder as in Fig. 7.7, the example of (a) above. Determine the encoder output sequence. Then carry the Viterbi algorithm out for two cases: the first in which no errors occur during transmission; the other for which a few bits of the transmitted output sequence are received in error. Compare the results of the estimates of the input bits in the two cases. Use a decision interval L of 5 at first, and then try one of 12. Compare the results in both cases.

7.13 Repeat 7.12(b) for the convolutional encoder of problem 7.11. Develop your own table similar to that of Table 7.3. Again show the resultant input bit estimates for the two cases of no errors during transmission and a few errors occurring. Again try two different intervals of operation such as 5 and 12, and compare results.

7.14 **(a)** Refer to the discussion in the text of the Hamming distance of paths in a convolutional encoder. Show for the encoder of Fig. 7.7, with trellis diagram represented by Fig. 7.10, that the smallest Hamming distance, the minimum free distance, is 5. Show there are two paths with Hamming distance of 6, and four paths of distance 7.

(b) Verify that the upper bound formula of (7.34) gives the value of 5 for the minimum free distance for the encoder of (a). What is this bound for the coders of problems 7.10 and 7.11? How does this bound then vary with constraint length and rate?

7.15 **(a)** Show, following the analysis in Section 7.3, that the Shannon limit, with $C \ll W$, is –1.6 dB. Here C is the channel capacity, W the transmission bandwidth. This represents the case of very wide bandwidth transmission.

(b) Now consider the case of bandlimited transmission, as exemplified by mobile radio communication. Here W is not particularly large as in (a). Show, in this case, that the Shannon optimum is 0 dB. *Hint*: As indicated in the text, R/W is close to 1, using appropriate signal shaping in this case, with R the bit transmission rate over this channel. Note also, that Shannon's theory indicates that it is possible to attain R/C close to 1 with zero probability of error with appropriate coding.

Chapter 8

Second-generation, digital, wireless systems

In this chapter we discuss current second-generation digital cellular systems. Some mention has already been made of these systems in Chapter 6, in the context of TDMA and CDMA, as well as in discussing their capacities comparatively. In particular, we discussed the North-American TDMA-based IS-136 or D-AMPS, the digital follow-on to the first-generation AMPS analog cellular system; GSM, the TDMA-based pan-European system, which is also being deployed in the USA and elsewhere in the world; and the CDMA-based IS-95. We now go over these systems in more detail, describing first the signaling required across the air-interface between the mobile station (MS) and the base station (BS) to establish and receive calls. We then discuss the signaling required within the wired network to which the base stations are connected, to control mobile handoffs from one cell to another, as well as the signaling required to locate and page a mobile in order to deliver an incoming call. We conclude the chapter with a brief discussion of voice signal processing and coding in these systems.

A much simplified diagram of a second-generation cellular system appears in Fig. 8.1. A given base station (BS) will typically control multiple mobile stations (MS). (The North American standard calls the mobile a cellular subscriber station, CSS.) Multiple base stations are, in turn, controlled by a mobile switching center (MSC), responsible for handling inter-cellular handoff, as well as mobile location, paging, and other mobile management and control functions. The home location register (HLR) contains reference and profile information for all mobile subscribers registered with this MSC as their "home location." Visitors to a "foreign" location register with the MSC in that area. The visitors' reference and profile information are then stored in the visitors' location register VLR associated with that MSC, after communicating with the mobile's HLR.

Registration and authentication of a mobile turning itself on, preparatory to either sending or receiving calls, is done by sending appropriate control messages across the air interface between the mobile and BS. These messages are then forwarded to the MSC for authentication. If the mobile is in its home location, the MSC queries its HLR to verify and approve registration of the mobile. If the mobile is in a foreign location, the local MSC/VLR combination will forward the requested registration of the mobile to its HLR. Light dashed lines in Fig. 8.1 indicate the paths taken by the various control signals. Bold

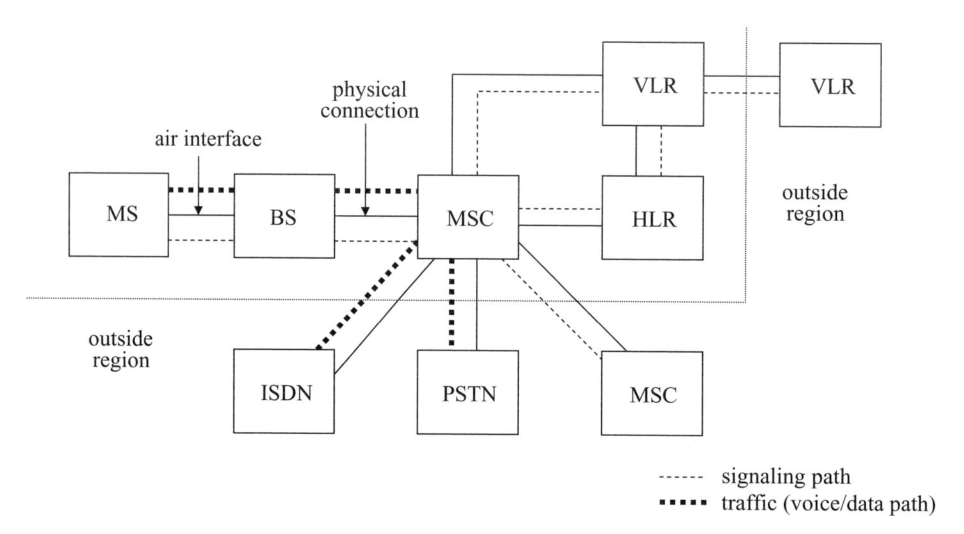

Figure 8.1 Cellular system, simplified

dashed lines represent paths taken by voice and data traffic, once a call is set up and is in progress. Traffic is shown flowing over either the public switched telephone network (PSTN) or over an integrated services digital network (ISDN).

In Section 8.1 following, we discuss, in detail, the various control signals transmitted across the air interface in the GSM system, required for a mobile to register and to set up a call. We follow this description of GSM with a discussion, in Section 8.2, of the corresponding control signals used in IS-36/D-AMPS. Section 8.3 discusses the CDMA-based IS-95 in detail. Section 8.4 describes handoff, location, and paging procedures used in these systems. We conclude this chapter with a section briefly describing voice signal processing and coding. Error-correction coding of the digitized voice traffic, as well as of all control signals, is required to be carried out before transmission across the air interface to ameliorate the effects of fading, noise, and interference expected during transmission.

8.1 GSM

Recall that GSM (Global System for Mobile Telecommunications) operates in the 890–915 MHz band uplink (MS to BS) and the 935–960 MHz band downlink (BS to MS). The 25 MHz of bandwidth in each direction is divided into 200 kHz frequency channels, with guard bands of 200 kHz left unused at the lower end of each band. There are thus 124 200 kHz channels available in each direction of transmission. Each frequency channel uses 8-slot repetitive frames, as noted in Chapter 6 (see Fig. 6.3), to transmit the traffic. A full-rate traffic channel occupies one slot per frame. Mobiles and base stations are capable of hopping from frequency channel to frequency channel each frame, following a specifed hopping pattern, to reduce the effect of fading, if the network operator desires to incorporate this feature. The assumption here is that the fading is frequency-dependent; a mobile experiencing fading on one channel may improve its transmission on hopping

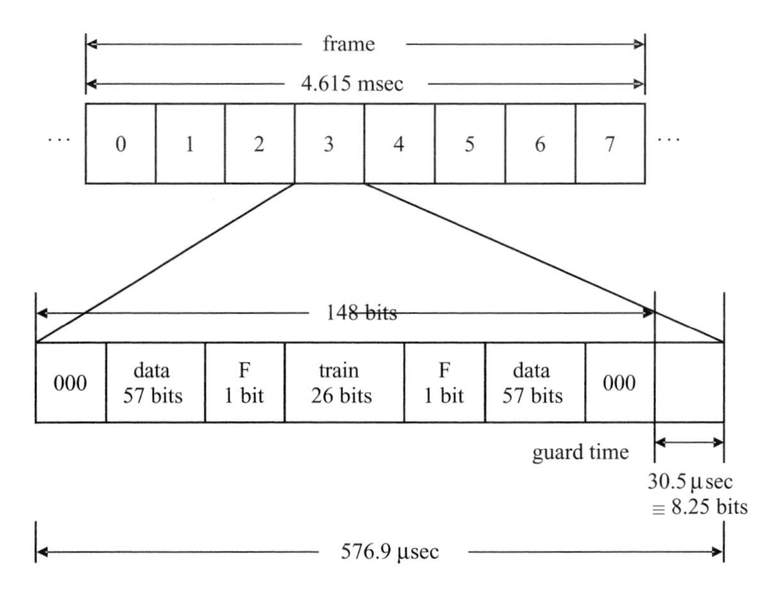

Figure 8.2 GSM frame/slot structure; traffic channel slots

to another channel. This feature clearly requires considerable coordination in each cell as well as in neighboring cells to avoid interference.

The 8-slot frame 4.615 msec long, as well as a blowup of the bit structure of a traffic-carrying slot, repeating Fig. 6.3, is shown in Fig. 8.2. Slots are 148 bits long, as shown, with a guard time of 30.5 μsec equivalent to 8.25 bits separating successive slots. (Control signal slots to be described later have the same slot length in time, but carry different fields with prescribed bit patterns in them.) Note that the slot shown in Fig. 8.2 begins and ends with 3-bit all-zero synchronizing fields. A 26-bit training sequence in the middle of the slot is used to provide an estimate of the radio channel, to be used in training an adaptive equalizer at the receiver to help overcome the multipath fading that may be encountered. The slot carries 114 data bits in two groups of 57 bits each. The two one-bit flag (F) bits indicate whether the data fields carry user or control traffic. (As will be seen later, user data traffic may be interrupted, with special control messages using the traffic slot instead.) Note that the system, in transmitting 156.25 bits over the 0.577 msec slot interval, operates at a 270.833 Mbps bit rate, as already noted earlier in this book. The user data rate is 22.8 kbps: 114 data bits are transmitted once per 4.615 msec long frame, but one frame in 13 is reserved for control signal purposes, as will be seen shortly.

Given this brief summary of GSM user traffic time slots, we now consider the process of setting up a call. A number of control messages, portrayed in Fig. 8.3, must be sent across the air interface for this purpose. These messages will be seen to require special use of the GSM frames and time slots. They are each discussed briefly in the paragraphs following.

A mobile must clearly have power on, before trying to set up a call or to prepare for being ready to receive one. On powering on in any cell, the mobile must first lock on to or acquire the frequencies used in that cell, and then synchronize to the framing-time slot structure. To do this, it first searches for a specific control channel, the *frequency*

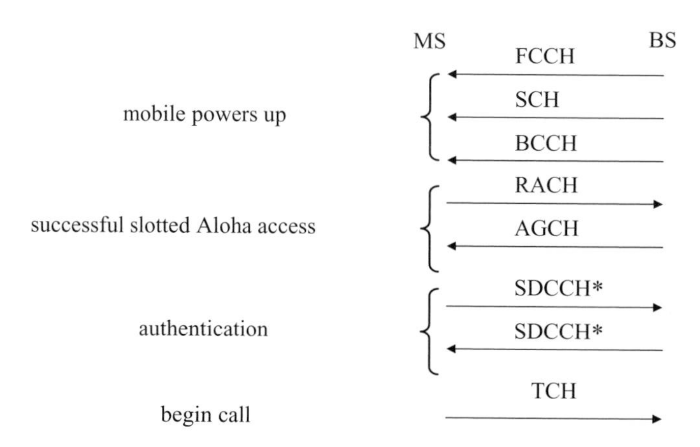

Figure 8.3 Setting up a GSM call
Note: * Can be a series of these messages.

correction channel, FCCH, broadcast by the local BS, which enables it to adjust or synchronize its frequency characteristic to that of the base station. The FCCH message is always followed by the *synchronization channel*, SCH, message, which identifies the BS and provides frame, hence time, synchronization to the mobile. (Note that in this discussion the words *channel* and *message* are used interchangeably. A control channel always carries a prescribed message.) Once frequency and timing information is acquired, the mobile listens to a channel called the *broadcast control channel*, BCCH, which provides information needed to set up the call: the cell configuration, the network to which it belongs, access information, and control channel information, among other items (Steele, 1992: Ch. 8; Goodman, 1997).

With this information available, the mobile terminal is ready to initiate a call. It does this by sending a random access request message over the *random access channel*, RACH, which carries a 5-bit random number plus a 3-bit purpose indicator. (The purpose could be a new call attempt, the example being described here; a response to a paging message; location registration; or a desire to transmit a short message, one of the capabilities of the GSM system.) The slotted Aloha random access protocol, to be described later in this book, is used in sending and receiving acknowledgement of this access attempt. It is called a random access attempt because multiple terminals may randomly attempt to access the base station at the same time, resulting in a "collision." In that case, the base station receives no correct RACH message and does not send an acknowledgement in reply. The slotted Aloha protocol dictates that a transmitting station, the MS in this case, receiving no acknowledgement, retry the access attempt a random time interval later. If the access attempt is successful, the BS acknowledges receipt of the RACH message with an *access grant channel*, AGCH, message, as shown in Fig. 8.3. This message repeats the 8-bit request message, and directs the terminal to a specified *stand-alone channel*, SDCCH, over which the mobile transmits the signaling information required for authentication, as well as to make the desired call connections. If the call setup is approved, the BS

Steele, R. ed. 1992. *Mobile Radio Communications*, London, Pentech Press; New York, IEEE Press.
Goodman, D. J. 1997. *Wireless Personal Communication Systems*, Reading, MA, Addison-Wesley.

Table 8.1 *List of GSM control channels*

Broadcast channels (all downlink, BS→MS direction)
1 Frequency correction channel, FCCH
2 Synchronization channel, SCH
3 Broadcast control channel, BCCH

Common Control channels
1 Paging channel, PCH (downlink, BS→MS)
2 Access grant channel, AGCH (downlink, BS→MS)
3 Random access channel, RACH (uplink, MS→BS)

Dedicated Control channels (bi-directional)
1 Stand-alone dedicated control channel, SDCCH
2 Slow associated control channel, SACCH
3 Fast associated control channel, FACCH

replies to the MS with SDCCH messages directing the mobile to the frequency/time slot traffic channel (TCH) to use for actually beginning the call and sending the desired user information. (This is most commonly digitized voice.)

How are these control channels actually implemented? These channels, plus three other control channels noted below, appear as logical channels within the frame/time slot structure of Fig. 8.2. The control channels are grouped into three categories: broadcast channels, common control channels, and dedicated control channels. These groupings as to category, and the various channels defined in each group, are tabulated in Table 8.1. Note that the three broadcast channels indicated in this table and described above in the context of setting up a call (see Fig. 8.3) are all directed downlink, from BS to MS, as are the paging and access control channels. The paging channel, as the name indicates, is used by a base station to locate a mobile for an incoming call. The random access channel, as already noted, and as shown in Fig. 8.3, is directed uplink, from MS to BS. The dedicated control channels noted in Table 8.1 are bi-directional, allowing mobile signaling, management, and supervisory information to be transmitted in either direction. In particular, the *slow associated control channel*, SACCH, is used by the BS, in the downlink direction, to send transmitter power level and timing advance instructions to the MS. The mobile uses this channel, in the reverse, uplink direction, to send the base station indications of received signal strength, the quality of its traffic control channel, and broadcast channel measurement results from neighboring cells to be used for mobile-assisted handoffs. The *fast associated control channel*, FACCH, is used to send handoff requests and other urgent signaling messages. It is sent on a normal traffic channel, TCH, or an SDCCH, interrupting that channel for this purpose. (Recall the use of the F, flag, bit in Fig. 8.2 for this purpose.)

Consider now the actual implementation of these various *logical* control channels in terms of *physical* time slots and frames in which they appear. Recall that an MS and a BS send user traffic to each other over traffic channels (TCH) consisting of a specified time slot per frame, as indicated in Fig. 8.2, on each of an assigned pair of frequencies.

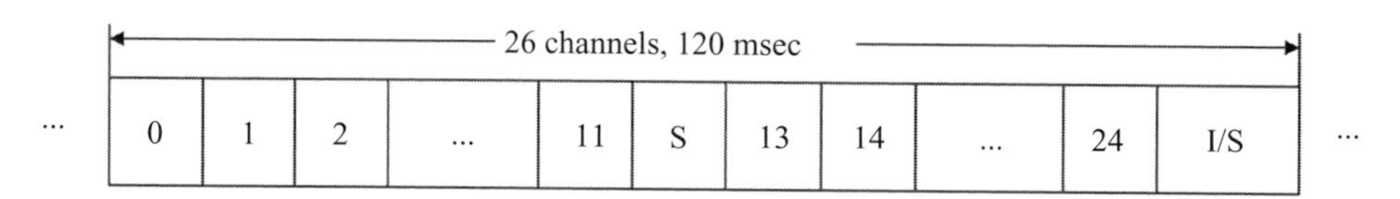

Figure 8.4 Repetitive 26-ch multiframe structure, GSM

This pair could be fixed for the duration of a call, or could be changing every frame interval, following a prescribed hopping pattern. Control channels are implemented the same way, by allocating prescribed time slots using given frequency channels to control channels. The stand-alone dedicated control channel, SDCCH, used in both directions on a temporary basis before assigning dedicated TCHs to a mobile-base station duplex (two-way) connection, as shown in Fig. 8.3, uses dedicated time slots on a prescribed number of frequency channels, allocated for this purpose. GSM uses three different ways of allocating time slots to the other control channels: It sets aside a prescribed number of frequency channels of the 124 available in each direction and "robs" one time slot per frame (time slot 0) to be used for control purposes; it allocates one time slot per 13, as noted earlier, in a normal sequence of traffic channel time slots, as a control channel; it uses a regular traffic channel time slot, when needed, by setting the TCH flag bit, again as already noted. We discuss these various control channel implementations briefly for the control channels tabulated in Table 8.1. Details appear in Rappaport (1996), Goodman (1997), and in Steele (1992: Ch. 8).

The allocation of one time slot in 13, in a sequence of traffic channels, to a control channel is carried out by defining a repetitive 26-frame *multiframe* structure 120 msec long, as shown in Fig. 8.4. Traffic channels occupy frames 0–11 and 13–24. Frame 12, labeled S in Fig. 8.4, is assigned as the dedicated, slow associated control channel, SACCH, of Table 8.1. Frame 25 may also be so assigned, if desired. If not, it is left as an *idle* frame, labeled I. Each of the eight slots in an S frame is associated with the traffic channel for that slot, providing the necessary signaling information for that channel, as indicated earlier. SACCH slots use the same bit structure as the traffic channels, shown earlier in Fig. 8.2. Since each such slot carries 114 information bits, the SACCH bit rate, with one slot per 26 used, is 114 bits/120 msec, or 950 bps. If two slots per multiframe are used, the bit rate doubles to 1900 bps. This is the reason why this channel is designated a *slow* associated control channel. (Note that the user traffic rates are 24 times this rate, 22.8 kbps, as already noted. Half-rate traffic channels may also be defined, being allocated 12 slots per 26-slot multiframe, rather than the full 24 slots. The data rate of these channels is then 11.4 kbps. One SACCH slot per multiframe would then be allocated to each of these half-rate TCH channels.) We shall see shortly that SACCH messages consist of coded signaling messages 456 bits long, sent interspersed (interleaved) over four multiframe time intervals, covering 480 msec. If this time interval is too long for mobility management purposes, as in the case, for example, of a handoff, a TCH channel is instead interrupted by setting the F

Rappaport, T. S. 1996. *Wireless Communications, Principles and Practice*, Upper Saddle River, NJ, Prentice-Hall PTR.

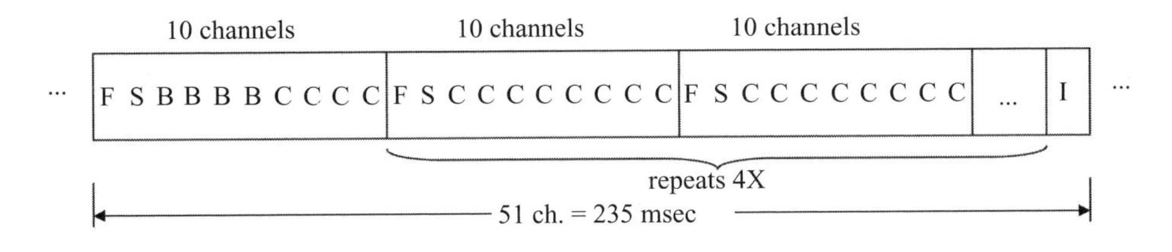

F = FCCH B = BCCH
S = SCH C = PCH or ACH

Figure 8.5 GSM downlink multiframe control structure (slot 0 of selected frequency channels)

bits (Fig. 8.2), and a fast-associated control channel, FACCH, message is sent over that channel.

It was noted above that three methods are used to assign GSM control channels, excluding SDCCH, which uses dedicated time slots on prescribed frequency channels. The two methods just described, allocating one slot in 13 frames to SACCH and setting the F bits in a TCH slot to provide an FACCH message, cover two of the three dedicated control channels of Table 8.1. The third method, "robbing" time slot 0 in a selected number of frequency channels (at most 34) of the 124 available in each direction, is used to implement the remaining non-dedicated control channels of Table 8.1. All time slots 0 in all frames using these selected frequencies in the uplink direction are assigned to random access, RACH, channels. The rest of the broadcast and common control channels shown in Table 8.1, all directed downlink, from BS to MS, are assigned time slot 0 in each frame of a repetitive 51-frame multiframe control structure, using the downlink set of these selected frequencies. (The other time slots are, of course, assigned to TCH or SDCCH channels.) This 51-frame multiframe structure is diagrammed in Fig. 8.5. As indicated in that figure, the 51 frames, repeating every 235 msec, are organized into five 10-frame groups, the first channel in each group, labeled F in Fig. 8.5, corresponding to the frequency correction channel, FCCH. (Recall from Fig. 8.3 that this is the channel a mobile uses to lock on to the base station frequency.) Each FCCH is immediately followed by the synchronization channel, SCH, used by a mobile to establish frame, hence time, synchronization. This channel is labeled *S* in Fig. 8.5. (Note again that the word "immediately" refers to time slot 0 in the next frame. All references here to different frames refer only to time slot 0 in those frames.) The four frames in succession following the SCH frame in the first of the five 10-frame groups carry the broadcast control channel, BCCH, labeled B in Fig. 8.5, used to provide information needed to set up a call. The BCCH thus appears only four times in succession, once per 51 frames. The remaining channels in Fig. 8.5, indicated by the letter C, correspond to either access channels (ACH) or paging channels (PCH). A paging channel, as the name indicates, is used to locate a mobile in a particular cell for an incoming call. Paging will be discussed later, in Section 8.4.

Recall, as sketched in Fig. 8.2, that GSM time slots are all 576.9 μsec. long. The traffic channel slots were shown there to contain 148 bits, followed by a 30.5 μsec guard time.

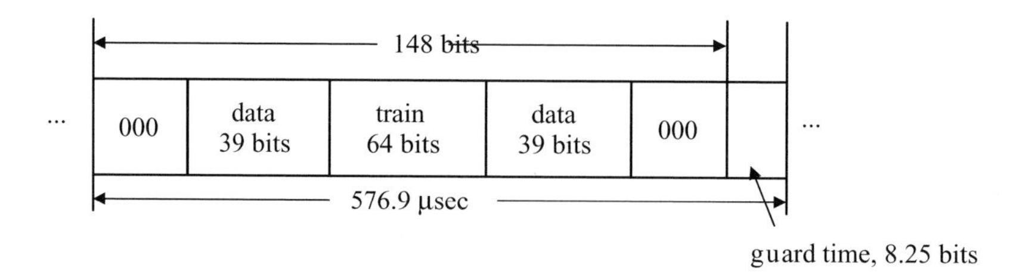

(a) SCH time-slot structure

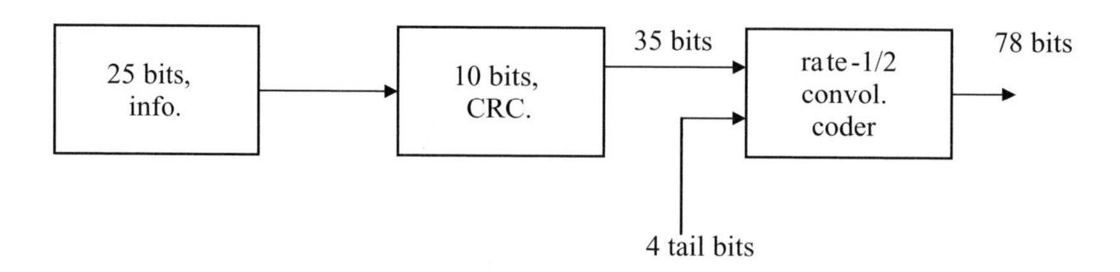

(b) Generation of 78-bit SCH data field

Figure 8.6 SCH implementation, GSM

All control channels, except for the random access, RACH, channel, contain 148 bits per slot as well. All of these, except for FCCH and SCH, use the same slot bit allocation as the TCH allocation of Fig. 8.2. The frequency correction channel, FCCH, transmits a sequence of all 0s over the 148 bits. This results in the GSM base station transmitting an unmodulated sinewave over the entire slot interval that mobiles can use to align their own generated frequency to that of the base station. The SCH, transmitted by the base station, carries a longer, 64-bit, training sequence than the TCH. (Why should this be so?) It thus carries a shorter, 78-bit, data payload, as shown in Fig. 8.6(a). Figure 8.6(b) shows how the 78-bit data field is actually generated from a 25-bit information message. This message is protected against errors, first, by a 10-bit cyclic redundancy code (CRC) which adds ten parity bits, and then by a rate-1/2 convolutional code. Both of these methods of carrying out error control were discussed in Chapter 7. Note that the CRC provides the outer coding mentioned there, while the convolutional coder carries out the inner coding. The convolutional codes discussed in Chapter 7 were assumed to operate on continuously incoming data. The convolutional coders described here, as well as later in the book, are used to protect against errors in finite-length blocks of data. The convolutional encoding must thus be properly terminated at the end of each data block, essentially resetting the convolutional encoder to the same state at the end of each successive data block. A known set of *tail bits* is used for this purpose. In the specific case of Fig. 8.6(b) four tail bits are shown being added at the end of each data block. Recall again, from Chapter 7, that with rate-1/2 convolutional coding two bits are actually sent for each bit of information. The 25-bit information message with ten CRC bits and four tail bits added thus emerges as the 78-bit data field of Fig. 8.6(b).

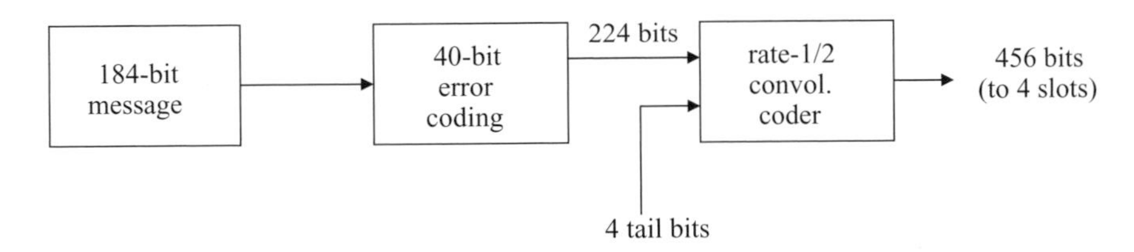

Figure 8.7 Generation of 114 bits/slot data field, BCCH, PCH, AGCH: GSM

The original 25-bit information message contains a 6-bit BS identity code and a 15-bit frame synchronization sequence (Steele, 1992: Ch. 8). MS-BS synchronization in setting up a GSM call (Fig. 8.3) is thus accomplished by having the MS lock on to a series of 15-bit sequences, each coded into 78 bits of data (see Figs. 8.6 and 8.5).

Consider the remaining control channels, other than RACH, appearing in Table 8.1. (These include the broadcast control channel, BCCH, and the paging and access grant channels, PCH and AGCH, respectively.) The 114-bit data field (Fig. 8.2) in each control channel slot is generated by interleaving 456 bits over four successive slots in each multiframe. These 456 bits are themselves generated from a 184-bit information message that is protected by an error-correcting code with 40 parity bits followed by rate-1/2 convolutional coding. Four tail bits are again added to the 224-bit coded sequence entering the convolutional coder. Figure 8.7 portrays the generation of the 456 bits. Note from Fig. 8.5 that the broadcast control channel, represented there by the letter B, uses time slot 0 of the four successive frames shown to transmit the coded 184-bit message, once per 51-channel multiframe. The paging or access grant channels, represented by the letter C in Fig. 8.5, carry the appropriate coded 184-bit message in groups of four successive frames.

The time slot bit allocation structure for the random access channel, RACH, shown in Fig. 8.8, differs from all the other channels, as noted earlier. The actual transmission burst is shorter, covering 88 bits only. The guard interval is thus equivalent to 68.25 bits, or 252 μsec. It is apparent that, since mobile terminals may initially be located anywhere in a cell, this guard interval for the RACH message, the first message sent uplink to the base station by a mobile and requiring a reply, must leave enough time for the maximum possible round-trip propagation delay within a cell. This guard interval value has thus been chosen to allow mobile stations to communicate with the cell's BS at up to a maximum distance of 35 km (Steele, 1992: Ch. 8). (Note that at a propagation velocity of 300 000 km/sec, the round-trip delay for this distance is 233.3 μsec, just under the guard interval length.) Referring back to Fig. 8.8(a), note that eight and three tail bits define the beginning and end of the burst, respectively, while 41 bits are used for synchronization purposes. The remaining data field of 36 bits is a heavily coded version of the original 8-bit random access message, containing a 5-bit random number plus a 3-bit purpose indicator, as noted earlier in discussing Fig. 8.3. Figure 8.8(b) indicates that six parity-check bits are added to the 8-bit message to provide a measure of error correction, the resultant 14-bit coded message plus four trailing bits then being further convolutionally coded to provide the 36 bits of the RACH data field.

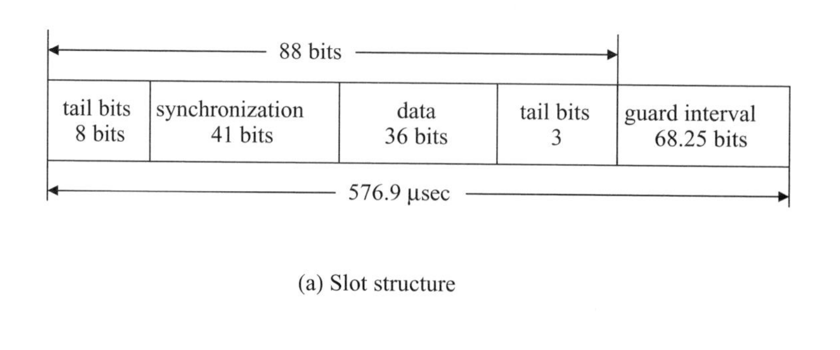

(a) Slot structure

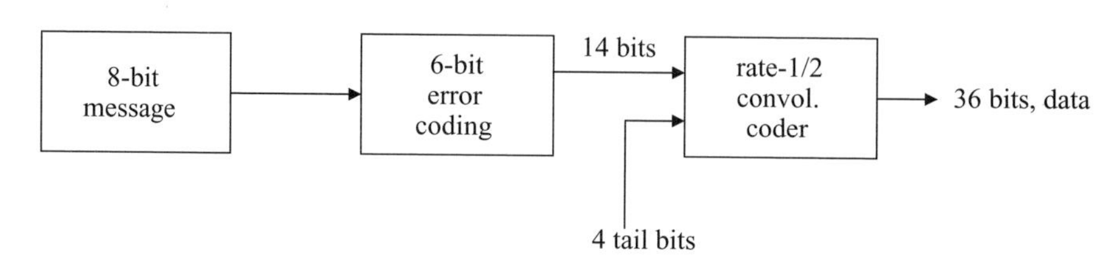

(b) Generation of 36-bit data field

Figure 8.8 Random access channel, RACH: GSM

8.2 IS-136 (D-AMPS)

In this section we discuss the traffic channels, as well as the logical control channels required to set up and maintain a call in the other TDMA-based major second-generation digital cellular system described in Chapter 6, IS-136 or D-AMPS. This second-generation system was developed after GSM, and we shall find, in fact, that the control channels have some similarity to the control channels used in GSM.

Recall, from our previous discussions in earlier chapters, that this system, as deployed in North America, occupies the 25 MHz bands from 824–849 MHz uplink, and 869–894 MHz downlink. Within these bands frequency channels are spaced 30 kHz apart, a frequency channel containing repetitive TDMA frames, carrying six time slots each. An example of the slot structure for the digital traffic channel, comparable with the TCHs in GSM, appears in Fig. 8.9. As discussed previously, two slots per frame are allocated to each full-rate traffic user.

The system transmission rate is 48.6 kbps: 1944 bits/frame (324 bits per slot) are transmitted in 40 msec. Full-rate users get to transmit 520 data bits per frame (260 data bits per slot), so that the data transmission rate is 13 kbps, although sent at the 48.6 kbps rate. These numbers differ considerably, of course, from those for GSM. Early on, it was decided that D-AMPS would use the same frequency channel structure as the original analog AMPS system, so that, in areas of low digital utilization, a dual-mode mobile phone could easily revert to the analog system.

Note from Fig. 8.9 that the slot structure differs in the two directions, uplink and downlink. The 6-bit guard time G in the uplink direction is needed because mobiles in a given cell may be moving with respect to the base station and could be located at varying

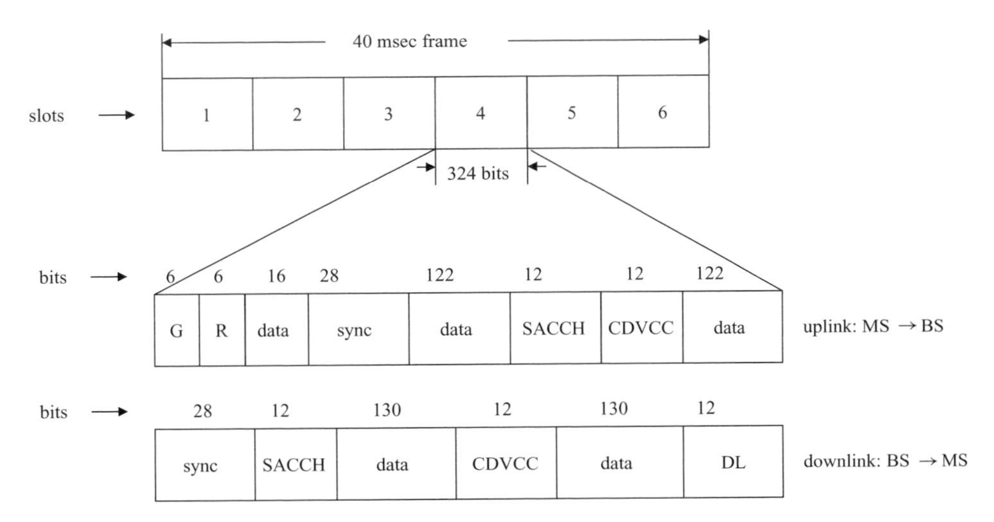

Figure 8.9 IS-136 slot structure, digital traffic channel

distances from that station. It prevents terminals initiating communication at the same time from interfering with one another. The power ramp-up time R in the same, uplink, slot structure is needed to accommodate terminals that may not be on, and that require time for their transmission power to ramp up to the desired value, depending on location within the cell and measured interference power, as dictated by the base station. Since the base station (BS)[1] is always on and always transmitting, these two fields are not needed in the downlink slot structure. Instead the downlink, BS to MS, slot carries a 12-bit DL field. This field carries a 7-bit carrier location message, indicating the carrier frequencies at which digital control channels, to be described shortly, are located. This message is protected with four parity-check bits. One of the 12 bits is left reserved.

The 12-bit CDVCC field, or *coded digital verification color code*, consists of an 8-bit DVCC number plus four parity-check bits to protect it. This field is used as a continuing handshake: the BS transmits the number; the MS replies with the same number. If no reply or an incorrect reply is received, the slot is relinquished. The 28-bit sync field in each direction is used for establishing time synchronization and for training the receiver equalizer. Each of the six-slot sync fields in a frame has a different synchronization sequence. Finally, the 12-bit SACCH fields in each direction carry the *slow associated control channel*. This channel, comparable to the SACCH in the GSM system, is used to send control information between mobile and base station. As an example, the mobile, in response to a measurement order message from the base station, carries out measurements of received signal strength and bit error rate, among other quantities measured, and sends the results of these measurements to the base station. These would be used in determining whether a handoff to another cell is warranted. (This measurement procedure, with the mobile assisting the base station, is then called a *mobile assisted handoff*, or MAHO,

[1] The IS-136 specifications use the terminology *base station, MSC and interworking function*, BMI, in place of base station. We shorten this to BS for simplicity's sake.

Table 8.2 *DCCH logical control channels, IS-136*

Downlink (BS→MS)
1 Broadcast control channels, BCCH: Fast broadcast control channel, F-BCCH Extended broadcast control channel, E-BCCH SMS broadcast BCCH, S-BCCH
2 SMS point-to-point, paging, and access response channel, SPACH: Paging control channel, PCH Access response channel, ARCH SMS channel, SMSCH
3 Shared control feedback channel, SCF
Uplink (MS→BS)
Random access channel, RACH

procedure.) A full-rate traffic channel, covering two slots in a frame, has 24 bits per frame devoted to the SACCH. The SACCH control data transmission rate is thus 600 bps in this case. A *fast associated control channel*, FACCH, is obtained, as in the GSM case, by replacing the 260-bits per slot data fields in the traffic channel with control information. The rate of control data transmission for this channel is then 13 kbps for full-rate channels of two slots per frame.

How is communication between MS and BS now established and maintained in this system? Logical control channels called *DCCH*, for *digital control channel*, are used for this purpose. Full-rate DCCHs correspond to two 48.6 kbps time slots per frame, generally slots 1 and 4, in selected frequency channels. A mobile wishing to establish communication powers on and finds a DCCH. Information on that DCCH enables it to locate its assigned DCCH. It then uses an Aloha-type access procedure, just as in the GSM case, to register, listen for any pages, and originate a call. The logical channels comprising a DCCH for the IS-136 system are tabulated in Table 8.2. It is left to the reader to compare these, after our discussion following, with those appearing in Table 8.1 for the GSM system. Note that three types of downlink logical control channels are defined. The downlink broadcast control channels, BCCH, consist of the *fast broadcast control channel*, F-BCCH, used to carry time-critical information necessary for the MS to establish the structure of the DCCH assigned to it (more about this later), as well as registration and access information; the *extended BCCH*, E-BCCH, carrying less time-critical information; and the *SMS BCCH*, S-BCCH, used to control a *broadcast* short message service defined for IS-136 systems. (GSM has a similar short message service available. These services allow short data messages, in addition to the more common voice messages, to be transmitted over these cellular systems. Data transmission capability will be discussed in detail in later chapters, in connection with our discussion of third-generation, packet-switched systems, as well as data-based wireless LANs. We focus here in our discussion of second-generation systems on voice communications almost exclusively.)

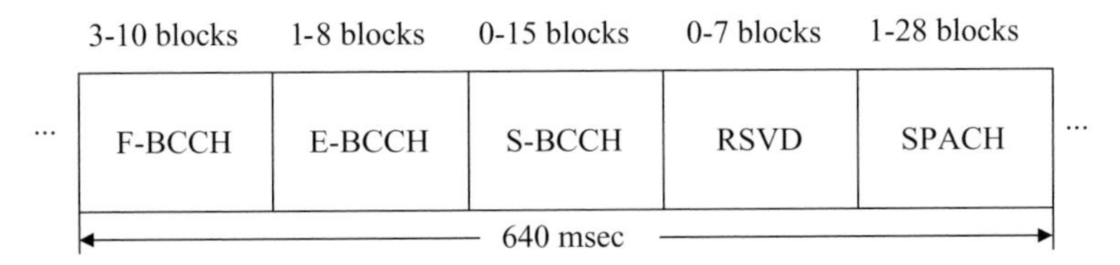

Figure 8.10 Repetitive superframe structure, IS-136

The *SMS point-to-point, paging,* and *access response channel*, SPACH, is a logical channel designed, as the name indicates, to carry paging and access response control information, as well as *point-to-point* messages concerning the SMS service. Finally, the *shared control feedback*, SCF, channel is used to carry downlink information, from BS to MS, as part of the random access procedure. It is used to both indicate to the mobile a random access time slot in which to request access, as well as to provide acknowledgement of the random access request once made. Random access messages, used for call setup or origination, as well as mobile registration and authentication, are carried over the one uplink channel, the random access channel, RACH. Note the similarity with GSM.

How are these various logical control channels implemented? Focus first on the downlink broadcast channels. As noted above, these channels, in the full-rate version, are transmitted using two slots per frame, generally slots 1 and 4, on selected frequency channels. (We focus here only on the full-rate category.) The broadcast channels are transmitted sequentially using a repetitive superframe structure. The superframe format is diagrammed in Fig. 8.10. Each superframe corresponds to 16 consecutive TDMA frames, for a total time interval of 640 msec. The 6-slot TDMA frames are each grouped into two 3-slot blocks and one slot per block (nominally slots 1 and 4, as noted above) is assigned to the DCCH. Each superframe is thus 32 blocks long. As indicated in Fig. 8.10, the first set of three to ten blocks is assigned to the F-BCCH logical channel. This is followed, in order, by the E-BCCH, covering from one to eight blocks, the S-BCCH, assigned from 0–15 blocks, a group of 0–7 blocks left reserved for other functions if needed, and, finally, the SPACH, assigned the remainder of the 32-block superframe. SPACH must, however, be assigned at least one block. Its maximum length is, clearly, 28 blocks. (Why?)

The specific assignment of blocks to each of these logical channels is specified in the F-BCCH message carried by that channel. Recall that a mobile, on powering up, finds a DCCH, which directs it, in turn, to its assigned DCCH. The F-BCCH within that DCCH carries a message containing the structure of the DCCH: the number of F-BCCH, E-BCCH, S-BCCH, reserved, and SPACH blocks within the superframe; access information such as mobile access power, access burst size, parameters for the random access algorithm (maximum number of retries, a random number for authentication, SMS message length, etc.); and registration as well as system id parameters (network type, country code, etc.)

All the downlink broadcast logical channels use 324-bit slot formats similar to those of the downlink digital traffic channel shown in Fig. 8.9. The slot format of these control channels appears in Fig. 8.11. Note that each slot carries the same 28-bit synchronization field and a total of 260 bits of data as does the digital traffic channel. The control slot format differs, however, in having a CSFP field in place of the CDVCC, and 22 bits of SCF in place of the SACCH and DL fields. The SCF bits are, of course, used to carry

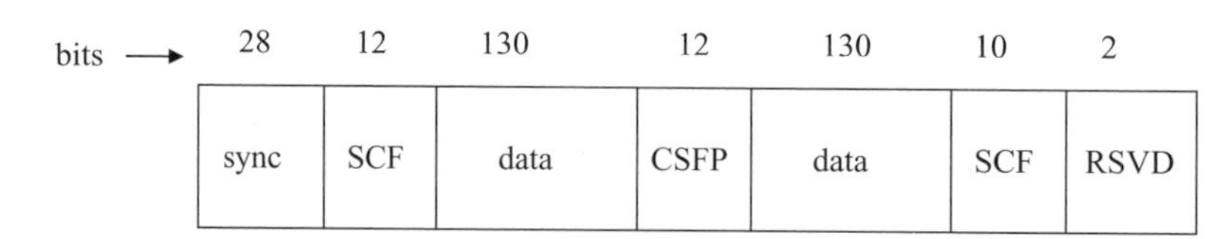

Figure 8.11 IS-136 slot structure, downlink broadcast control channels

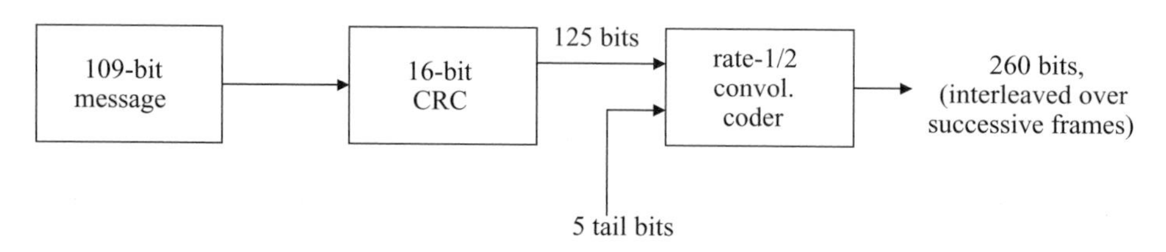

Figure 8.12 Generation of IS-136 control data fields

the SCF logical channel appearing in Table 8.2. SCF messages are thus transmitted as part of a control channel slot, rather than in distinct control slots, as is the case of the other downlink broadcast channel messages. (A discerning reader may have noted, and wondered, on seeing the superframe format of Fig. 8.10, as to how the SCF channel is implemented.) The SCF channel is used, as noted above, to respond to the mobile's random access attempt. We shall have more to say later about the messages carried by this channel. Continuing our discussion of the control channel slot format, we note that the 12-bit CSFP field (for *coded superframe phase*) is used to indicate the location of a TDMA block within a superframe. This field carries an 8-bit SFP (superframe phase) number and four parity-check bits to protect this number against errors. Five of the eight bits are used to keep a modulo-32 count of the blocks within the 32-block superframe. The other three bits are reserved and set to 0. (Note again that we are only considering full-rate DCCHs here. For a half-rate DCCH, only the even SFP numbers are used in the counting.) The number starts with zero in the first F-BCCH slot and is incremented by one every TDMA block within the superframe.

Now consider the 260 data bits in this downlink DCCH slot format. They are generated by beginning with 125 bits, adding five tail (zero) bits, and feeding the 130 bits into a rate-1/2 convolutional encoder, a process similar to that shown for the GSM system in Figs. 8.6(b) and 8.7. The resultant output of 260 encoded bits is then interleaved over successive TDMA frames to further protect against fading. The 125 bits input to the convolutional coder are themselves generated from 109 bits containing both header and information bits, with 16 CRC (cyclic redundancy check) bits added to protect against errors. This error-control encoding process is diagrammed in Fig. 8.12, which is to be compared with Figs. 8.6(b) and 8.7, showing the process used in GSM. Figure 8.13 indicates how the channel control messages appear at the input to the rate-1/2 convolutional coder. (For those readers familiar with layered architectures for networks, it is to be noted that the IS-136 specifications indicate that the information field of Fig. 8.13 represents layer-3 data, while the full 130 bits of data in the data field of the TDMA control slots of Fig. 8.11 represent layer-2 data.)

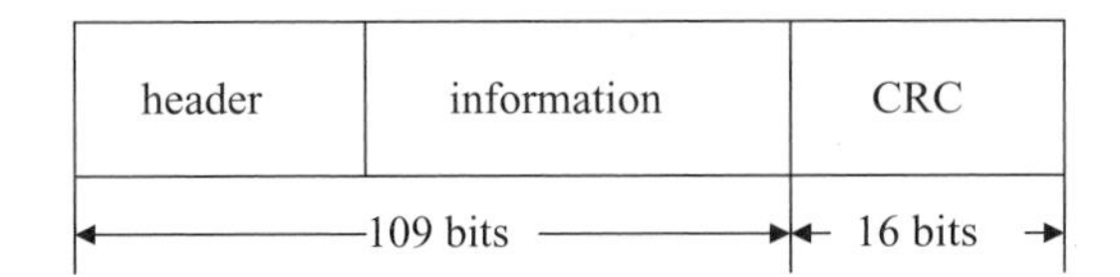

Figure 8.13 Generic form, IS-136 DCCH downlink messages

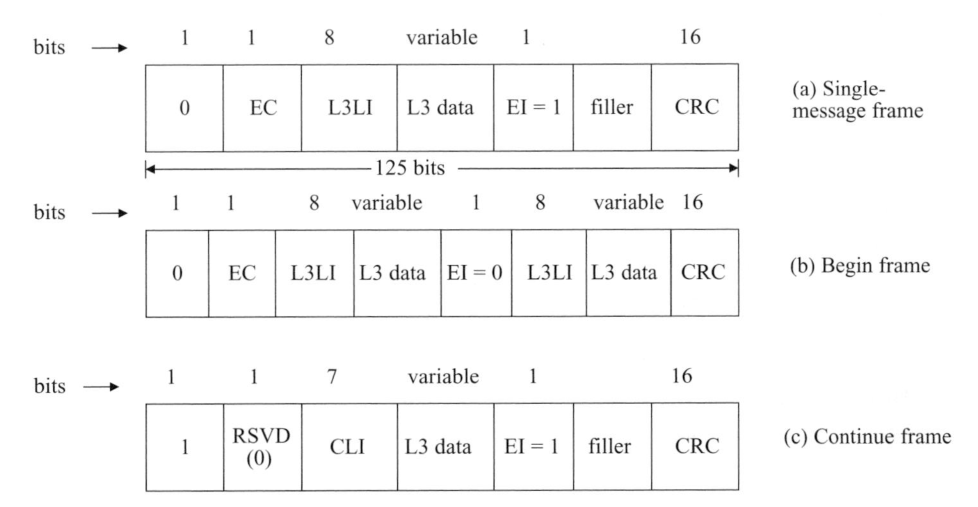

Figure 8.14 Examples of F-BCCH frames

We focus on F-BCCH messages as a specific example. Recall that this control channel is used to transmit DCCH structure, access parameter, registration, and system id information. A number of different types of messages may be sent, encompassing this information. These messages may require more than 109 bits to be transmitted, hence provision must be made for continuing messages over into another frame. Examples of three different types of F-BCCH frames appear in Fig. 8.14. Note that all contain the 16-bit CRC field of Figs. 8.12 and 8.13, overhead information, and varying-size data fields denoted as L3 data. As noted above, this is the layer-3 data field carrying one of the various F-BCCH messages. The 8-bit parameter L3LI is the length indicator, designating the length, in octets, of the L3 data field immediately following. Figure 8.14(a) represents the frame format used when an F-BCCH message can be transmitted completely within one frame. If a message requires more than one frame for completion, the Begin and Continue formats of Figs. 8.14(b) and (c) are used. The one-bit EI flag set to 1 indicates that filler (all 0s) has been used to pad out the frame; EI = 0 says that a new message follows. EC is used to designate a change in the E-BCCH. Frames begin with a one-bit designator 0 or 1, as shown. The 7-bit CLI, or Continuation Length Indicator, in the Continue frame of Fig. 8.14(c), is used to indicate the number of bits belonging to a Continue message.

As examples of F-BCCH messages, the "DCCH structure" message can contain from 43 to 201 bits. The "Access parameter" message is 65 bits long, the "Registration" message varies in length from 16 to 77 bits, and the "System id" message from 27 to (191 + multiples of 17) bits long. Other F-BCCH messages, some mandatory, others optional, are defined as well.

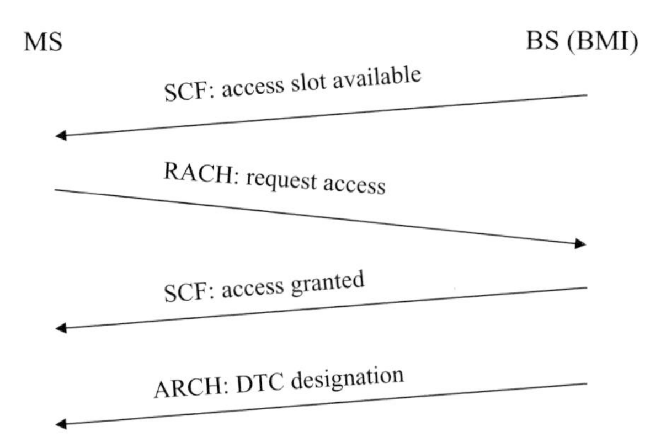

Figure 8.15 Access procedure, IS-136

We now move to considering the one uplink IS-136 control channel, RACH, shown in Table 8.2. This channel is used by the mobile, as already noted earlier, for originating calls, for registration, and for replying to page messages. Its functions are obviously similar to those described previously in discussing GSM. As is the case with the other (downlink) DCCHs discussed above, this channel is carried in time slots 1 and 4 of frames in selected frequency channels. The IS-136 access procedure using the RACH, in conjunction with the downlink channels, is diagrammed in Fig. 8.15. This procedure uses the slotted-Aloha access protocol mentioned in discussing the GSM access procedure (see Fig. 8.3). The mobile station listens to a downlink SCF channel (see Fig. 8.11) to determine a specific future time slot to use to send its RACH message. It then sends the RACH message, as shown in Fig. 8.15. A later SCF message, carried in a specified time slot, will indicate whether the RACH message has been correctly received and access granted. If access is granted, an ARCH message carried on the SPACH channel (Table 8.2 and Fig. 8.10) will follow, indicating the specific digital traffic channel the mobile is to use for communication. If the access is not successful (other mobiles might be attempting access at the same time) the access attempt will be retried a random time later.

Two slot formats are available for sending RACH messages: a normal slot format, 324 bits long, as is the case for the IS-136 traffic and control channels discussed above (Figs. 8.9 and 8.11), and an abbreviated slot format. The two formats appear in Figs. 8.16(a) and (b), respectively. Note that they differ in the length of the data fields. The normal format carries a total of 244 data bits; the abbreviated format carries 200 bits. The latter carries instead an added 38-bit equivalent guard time (AG) plus ramp-up time R of six bits. The overhead fields G, R, and sync are the same guard, ramp-up, and synchronization fields appearing in the uplink traffic channel (Fig. 8.9). PREAM is a preamble field consisting of successive $-\pi/4$ phase changes in the DPSK modulation scheme, used to help with symbol synchronization. The added synchronization field, sync+, is needed because the RACH is not used continuously, as is the case with the digital traffic channel, hence added synchronization time is needed.

The RACH data fields are generated in a manner similar to those discussed for the other DCCH protocols. The normal RACH messages, for example, are 101 bits long to start. 16 CRC bits are added to protect these bits, and the resultant 117 bits, plus five tail bits, are then further protected by passing them through the same rate-1/2 convolutional

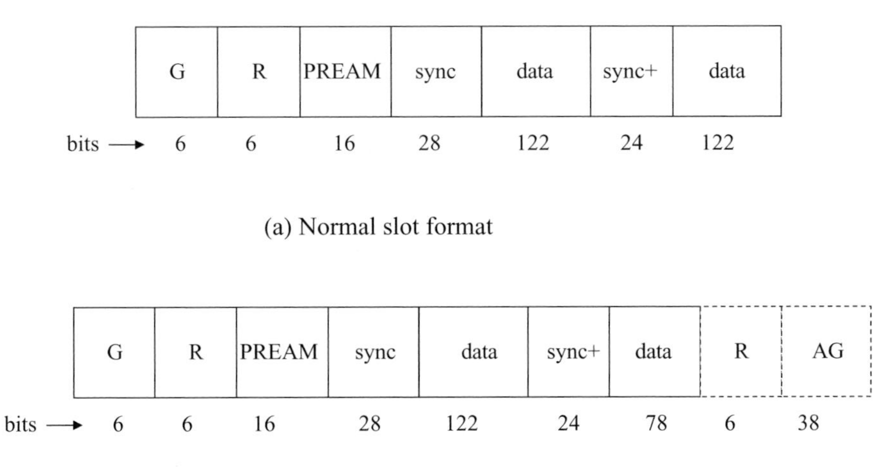

Figure 8.16 Slot formats, IS-136 RACH

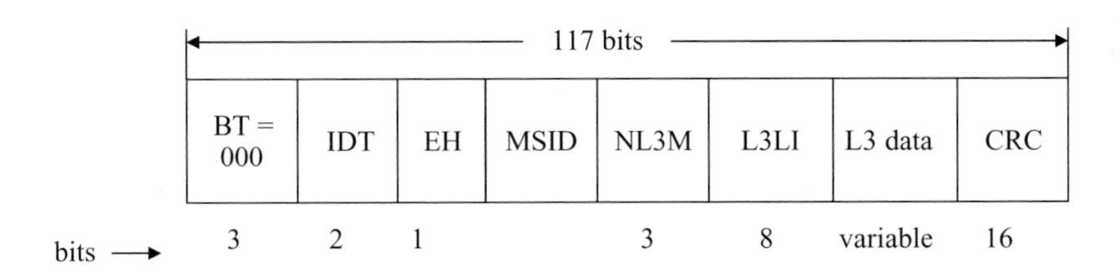

Figure 8.17 BEGIN frame format, RACH message, IS-136

coder used for the other channels (see Fig. 8.12). The initial abbreviated RACH messages are 79 bits long and are then brought up to the 200-bit data field size using the same procedure. There are seven different types of normal RACH frames 117 bits long (including the 16-bit CRC field) that may be transmitted. These include BEGIN, CONTINUE, END, and single BEGIN/END frames, among others. (Compare with the different F-BCCH frames discussed earlier – see Fig. 8.14, for example.) The BEGIN frame format appears, as an example, in Fig. 8.17. The 3-bit BT field is used to distinguish among the seven different types of RACH frames: 000 indicates a BEGIN frame, for example. The mobile station id, MSID, has three possible values: a 20-bit temporary MS id, a 34-bit mobile id corresponding to the normal 10-digit North American telephone number converted to binary format, and a 50-bit international mobile id. The 2-bit IDT field indicates which of these three mobile ids appears in the MSID field. The one-bit EH flag allows an extension header to be present: 0 means no extension header; 1 means it is present. The 3-bit NL3M field indicates how many concatenated layer-3 messages appear in this BEGIN frame. The L3LI field serves the same function as it did in the F-BCCH format example of Fig. 8.14. It indicates the length, in bytes, of the actual message field, L3 data, following. Eighteen different RACH messages are defined. The messages we have implicitly mentioned, Authentication, Page Response, Origination (call setup), and Registration, are included in this group. Messages are distinguished by a 6-bit message type field included in the message formats. The Origination message contains the called party number as part of its format. The calling party number may be optionally included.

Now refer back to the downlink shared control feedback, SCF, channel, as described briefly earlier in explaining the access procedure of Fig. 8.15. We expand here on that brief description. Recall, from Fig. 8.11, that the SCF channel is carried as 22 bits within the DCCHs. As indicated by Fig. 8.15, this channel is used to announce to mobiles wishing to begin an access attempt over a given DCCH the location of RACH slots a specified time in the future to be used for their access attempt. (An access attempt could be in response to a page received over the paging sub-channel PCH of the SPACH; it could be for the purpose of registration; it could be an origination or call setup attempt; it could be for the purpose of sending other RACH messages, as noted above.) Once the RACH message is sent using the RACH time slot specified, a later SCF message, as shown in Fig. 8.15, carries the information as to whether the access attempt has been successful or not. Three separate flags carried by the SCF channel are used to, first, indicate whether a particular RACH slot is currently idle and available for an access attempt; second, to indicate whether an access attempt using a previously designated RACH slot has been successful. (Recall that multiple mobiles might attempt access using the same designated time slot.) The first flag, called the BRI flag, covering six of the 22 SCF bits, is used to indicate that a specified RACH is *Busy* (111100), *Reserved* (001111), or *Idle* (000000). An *Idle* BRI flag indicates that the RACH time slot a specified time interval ahead may be used for an access attempt. If a previous access attempt has been successfully recognized as such by the BS, the BRI flag is set to *Busy*. A 5-bits long second flag, the R/N flag, is used to indicate whether an access attempt has been successful or not: 11111 designates attempt *received*; 00000 indicates *not received*. Finally, seven bits, protected by four parity-check bits, for a total of 11 bits, are designated as the CPE (coded partial echo) flag. These seven bits are the seven least-significant bits of the MSID (see Fig. 8.17), as echoed back by the BS, of a mobile making a successful access attempt. Note that all three flags are used to indicate a successful random access attempt.

8.3 IS-95

Recall from our earlier discussions in this book, particularly in Chapter 6, that IS-95 is a CDMA-based system. Its traffic and control channels are defined as specified codes rather than time slots as in the case of GSM and IS-136. The discussion here will thus differ substantially from the description in the prior sections of those two TDMA-based cellular systems.

We begin by discussing the traffic channel system block diagrams, as sketched in Figs. 8.18 and 8.19. These diagrams are considerably expanded versions of the CDMA block diagram of Fig. 6.5(a), discussed earlier in Chapter 6. The objective there was to portray the process of signal spreading, using a pseudo-random spreading code, as simply as possible. Here we attempt to be more precise, specifically since *codes*, rather than time slots, are used as channels to carry traffic and control information. Comparable control channel system diagrams will be described shortly. Note that the basic portions of the CDMA signal processing sketched in Fig. 6.5(a) still appear in Figs. 8.18 and 8.19: a binary information stream is "multiplied" by a pseudo-noise (PN) chip spreading

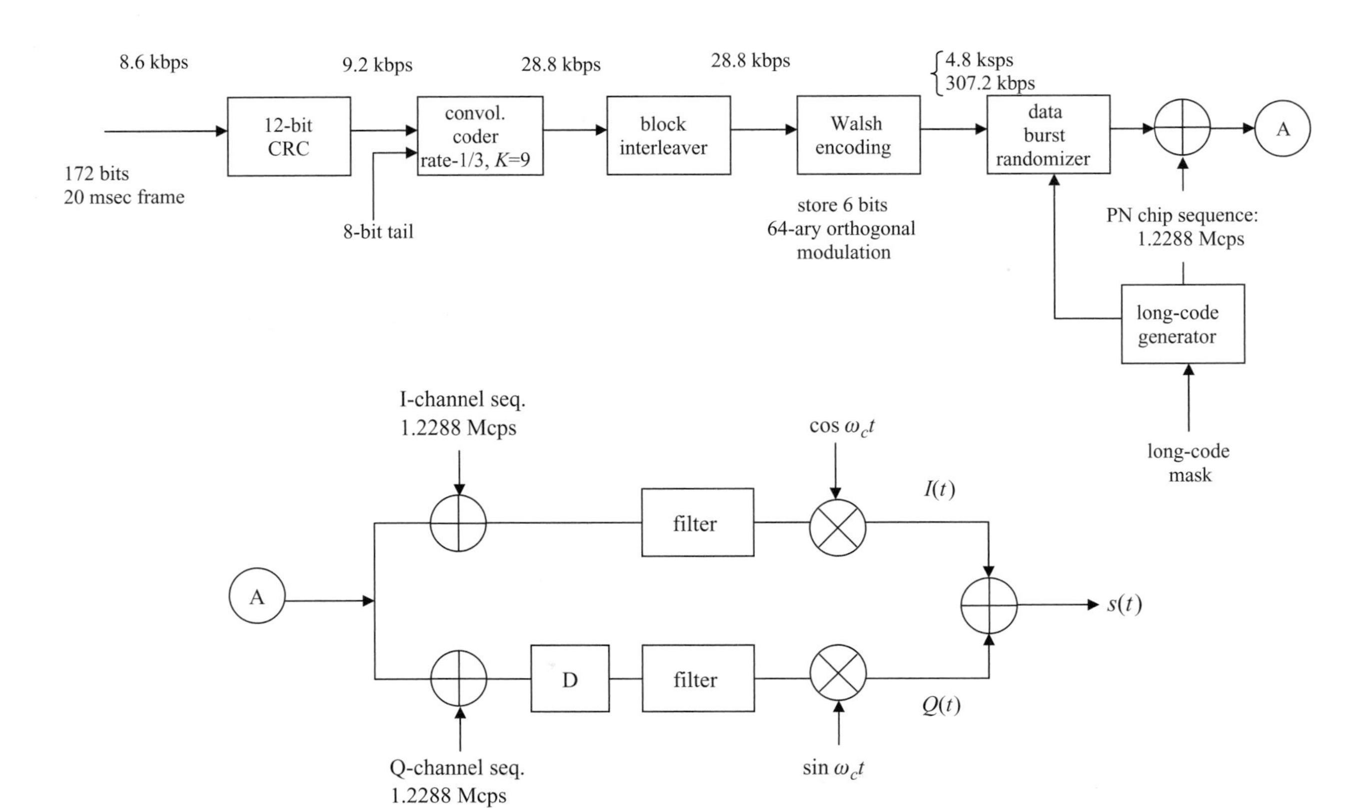

Figure 8.18 IS-95 reverse traffic channel diagram, full-rate case only

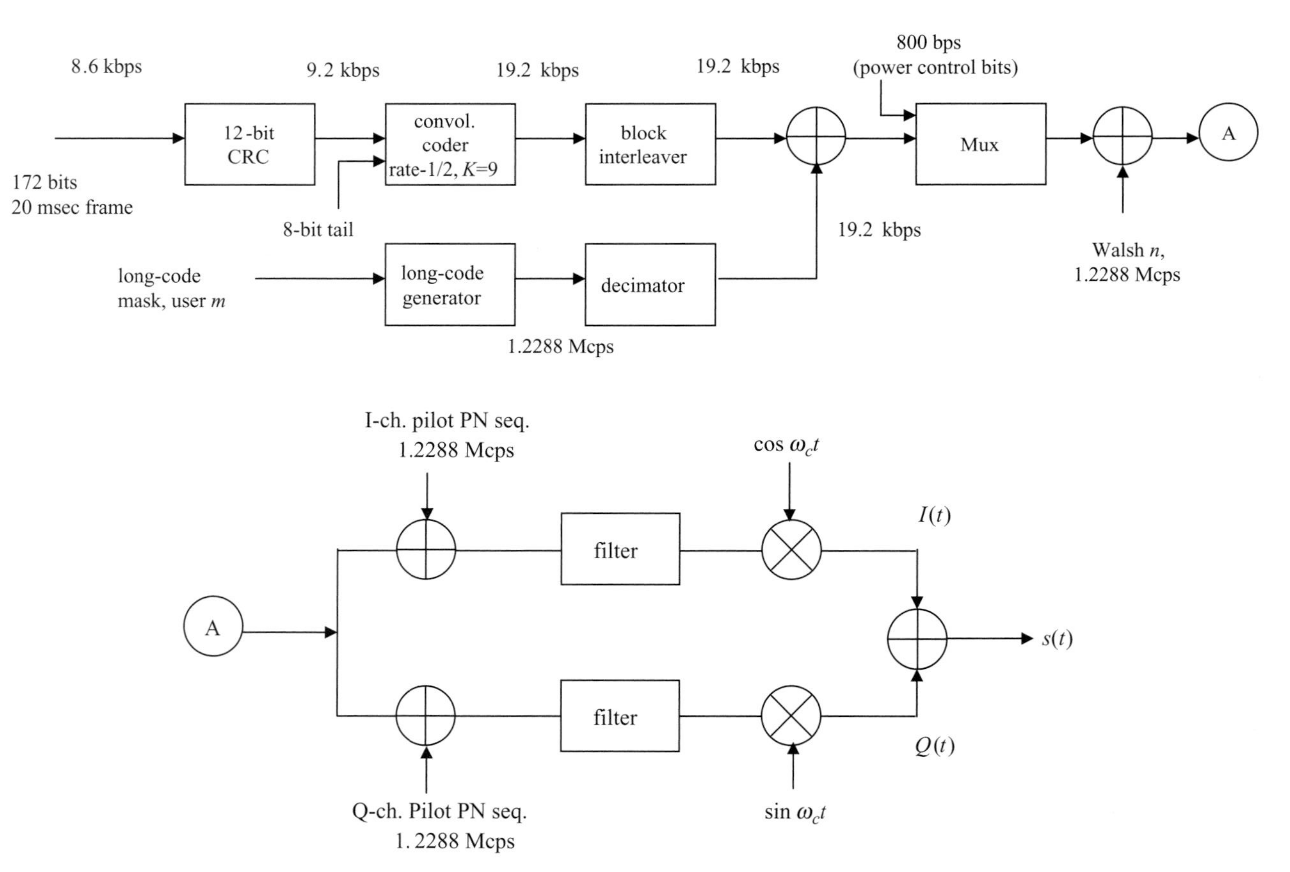

Figure 8.19 IS-95 forward traffic channel, user m, full-rate case

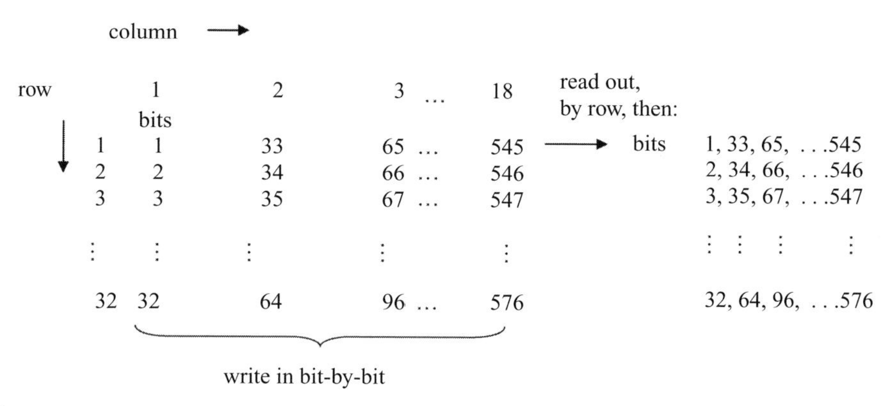

Figure 8.20 Block interleaver operation

sequence, the resulting output shaped by an appropriate low-pass shaping filter, and then fed to a high-frequency transmitter. Modulo-2 (mod-2) addition is used here instead of the multiplication process of Fig. 6.5(a), however. It is left to the reader to show that mod-2 addition of binary 1s and 0s is identical to multiplication of the equivalent bipolar ($\pm$ 1) binary digits. (*Hint*: $+1 \bullet +1 = +1; -1 \bullet -1 = +1; -1 \bullet +1 = -1; +1 \bullet -1 = -1$.) Figure 8.18 shows the reverse or uplink transmitter to be that of the OQPSK type (see Chapter 5), while the forward or downlink transmitter of Fig. 8.19 is of the QPSK type.

We now discuss the two diagrams in more detail, focusing on each of the blocks in succession. We first note that the diagrams of Figs. 8.18 and 8.19 are those for full-rate traffic transmission of 8.6 kbps, as shown at the top left-hand side of each figure. Reduced transmission rates of 4.0, 2.0, and 0.8 kbps are defined for IS-95 as well. As also indicated at the top left-hand side of each figure, for the full-rate case, IS-95 defines consecutive 172-bit traffic frames, 20 msec long. This obviously equates to a traffic input rate of 8.6 kbps. Twelve forward error-correction bits per frame are then added. In the case of the reverse traffic channel of Fig. 8.18 the resultant 9.2 kbps signal is fed into a rate-1/3, constraint length $K = 9$, convolutional encoder, with eight all-zero tail bits appended each frame. (Note that for the forward traffic channel case of Fig. 8.19, the convolutional encoder is of the rate-1/2 type. This encoder provides somewhat reduced error-correction capability than the rate 1/3 type (Viterbi, 1995). Why is this acceptable for the forward or downlink direction?) The 28.8 kbps convolutional encoder output is then fed into a block interleaver to reduce the effect of burst errors, those lasting many bits, as might result from a mobile undergoing a severe fade. This block interleaver operates consecutively on each frame of 576 bits (28.8 kbps times 20 msec). The block interleaver may be visualized as being a 32-row by 18-column array, as shown in Fig. 8.20. The interleaver is filled each frame, one column at a time. It is then read out bit by bit, each row at a time: bits 1, 33, 65, 97, ..., 545, 2, 34, 66, ..., 546, ... in succession. This procedure reduces the possible adverse effect of a burst of errors.

The block interleaver output at the 28.8 kbps rate is now fed into a 64-ary Walsh encoder, as shown in Fig. 8.18. This device requires a side discussion for its explanation.

Viterbi, A. J. 1995. *CDMA, Principles of Spread Spectrum Communication*, Reading, MA, Addison-Wesley.

Six consecutive bits in the 28.8 kbps stream are stored and used to generate one of 64 orthogonal Walsh functions, a set of binary waveforms (1s and 0s), each 64 bits long. The 28.8 kbps input stream is thus converted to a 307.2 kbps output stream. Why use this procedure? The Walsh encoder acts as an orthogonal modulator, analogous to the case of using 64 orthogonal sinewaves, but with binary waveforms used instead. This procedure enables non-coherent demodulation to be used at the receiver, followed by de-interleaving and decoding to recover the original message sent. Straight binary PSK would have provided better performance, but would have required coherent demodulation, with a phase reference, such as a pilot signal, necessary to be sent, accompanying the normal signal. Given the level of coherence of the wireless channels over which the IS-95 is used, this was deemed to be impractical, because of the relatively large energy required to send the pilot signal. Differential PSK, such as is used in the IS-136 system (see Chapter 5), provides performance inferior to the Walsh orthogonal scheme used in this system.

The 64-ary Walsh encoder is generated as shown by the matrix representation following. We start by defining the Walsh matrix $\mathbf{W}_2$ as being given by the two-by-two matrix

$$\mathbf{W}_2 = \begin{pmatrix} 0 & 0 \\ 0 & 1 \end{pmatrix} \tag{8.1}$$

Now let the $L \times L$ Walsh matrix $\mathbf{W}_L$ be defined in terms of the $L/2 \times L/2$ Walsh matrix $\mathbf{W}_{L/2}$ as follows:

$$\mathbf{W}_L = \begin{pmatrix} \mathbf{W}_{L/2} & \mathbf{W}_{L/2} \\ \mathbf{W}_{L/2} & \mathbf{W}'_{L/2} \end{pmatrix} \tag{8.2}$$

Here element w'_{ij} of matrix $\mathbf{W}'_{L/2}$ is the complement of the corresponding element w_{ij} of matrix $\mathbf{W}_{L/2}$, i.e., $w'_{ij} = 1 - w_{ij}$. In particular, Walsh code $\mathbf{W}_4$ is now given, using (8.1) and (8.2), by

$$\mathbf{W}_4 = \begin{pmatrix} 0 & 0 & 0 & 0 \\ 0 & 1 & 0 & 1 \\ 0 & 0 & 1 & 1 \\ 0 & 1 & 1 & 0 \end{pmatrix} \tag{8.3}$$

(One could equally well have started by first defining $\mathbf{W}_1$ as a matrix with the single element 0, and then invoking the recursive relation (8.2) to obtain, first, (8.1) and then (8.3).)

Continuing, Walsh code $\mathbf{W}_8$ is now readily shown to be given by

$$\mathbf{W}_8 = \begin{bmatrix} 0 & 0 & 0 & 0 & 0 & 0 & 0 & 0 \\ 0 & 1 & 0 & 1 & 0 & 1 & 0 & 1 \\ 0 & 0 & 1 & 1 & 0 & 0 & 1 & 1 \\ 0 & 1 & 1 & 0 & 0 & 1 & 1 & 0 \\ 0 & 0 & 0 & 0 & 1 & 1 & 1 & 1 \\ 0 & 1 & 0 & 1 & 1 & 0 & 1 & 0 \\ 0 & 0 & 1 & 1 & 1 & 1 & 0 & 0 \\ 0 & 1 & 1 & 0 & 1 & 0 & 0 & 1 \end{bmatrix} \tag{8.4}$$

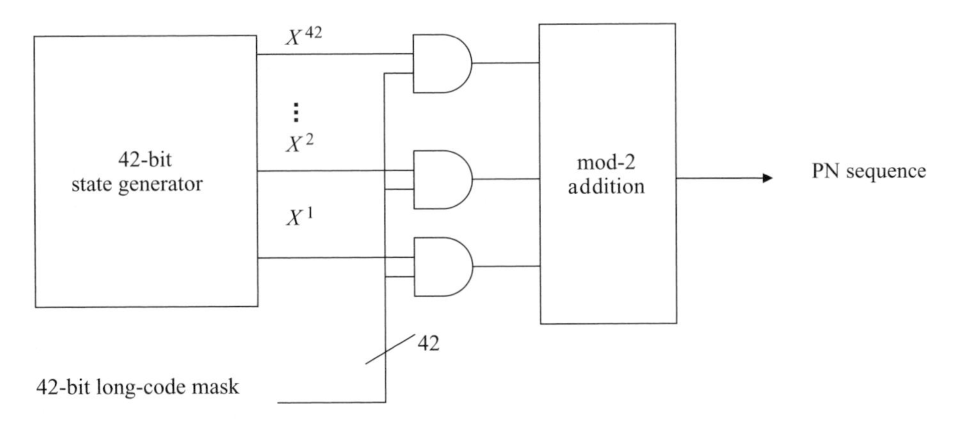

Figure 8.21 Long-code generator

Note how each row differs in $L/2$ (here 4) places. Note also that each row has half $(L/2)$ 0s and half 1s. If each 0 is converted to the equivalent -1, the 1s remaining unchanged, it is clear from (8.4) that multiplying elements of the same column in two different rows together and summing over all columns, one gets 0 as the resultant sum. Specifically

$$\sum_k w_{ik} w_{jk} = 0 \tag{8.5}$$

Here w_{ik} is the matrix element in row i and column k. The sum is taken over all columns. All rows are thus orthogonal to each other. Continuing this process of recursively finding first $\mathbf{W}_{16}$ in terms of $\mathbf{W}_8$, using (8.2), then $\mathbf{W}_{32}$, and, finally, $\mathbf{W}_{64}$, it is readily shown that all of these matrices have the same property noted above, that half the entries are 0, the other half 1, with all rows differing in $L/2$ places. The orthogonality relation of (8.5) thus holds throughout. Details are left to the reader. In particular, for the $\mathbf{W}_{64}$ matrix, the 64 different rows of 64 bits each may be used to provide the 64 Walsh codes required to implement the Walsh encoding box of the reverse-direction traffic channel of Fig. 8.18. Each of the 64 possible 6-bit sequences stored in the encoder thus triggers the output of the corresponding 64-bit long code, given by the appropriate row of the Walsh matrix. The 64 codes thus obtained constitute an orthogonal set, as shown by (8.5).

Continuing with the discussion of the reverse-direction traffic channel of Fig. 8.18, note that the required CDMA spreading of the traffic signal is now carried out by mod-2 addition of the 1.2288 Mcps pseudo-noise (PN) chip sequence and the 307.2 kbps Walsh code sequence. The PN chip sequence is generated by a 42-bit wide long code generator, as indicated by Fig. 8.18. This long code generator consists of a 42-bit wide state generator with 42 taps, the output of each being combined mod-2 with a 42-bit long code mask. The resultant 42 bits are added mod-2 and are fed out as the desired PN sequence of Fig. 8.18. This generator corresponds to the (42-bit long) shift register discussed in Chapter 6. A block diagram of the long code generator is shown in Fig. 8.21. The long code mask is unique to each mobile in a system, so that the corresponding spreading sequence identifies the mobile to the base station. The mask is made up of a 32-bit *electronic serial number* (ESN) assigned permanently to each mobile terminal, plus an added known sequence of ten bits. This ESN must be known to the base station since it must know the PN code sequence in carrying out the receiver operation of decoding the CDMA traffic signal sent uplink by the mobile (see Fig. 6.5). As we shall see shortly, the mobile ESN is, in fact, transmitted uplink, from mobile to BS, in all access channel control messages, as part of the mobile id, during the initial access procedure.

The long code generator PN sequence generated at the 1.2288 Mcps rate provides the signal spreading of 128 discussed in Chapter 6. The IS-95 system baseband frequency response, as will be noted below, results in a nominal transmission bandwidth W of 1.23 MHz, i.e., the system provides 1 bps per Hz of bandwidth. This bandwidth, divided by the 9.6 kbps bit rate at the input to the convolutional encoder of Fig. 8.18, gives the spreading gain of 128. Note also that the PN sequence generated by the long code generator repeats itself every $2^{42} - 1$ chips. At the 1.2288 Mcps chip rate, this results in a PN period of about 1000 hours, or 41.6 days! Additional, quadrature, spreading is carried out in both the inphase (I) and quadrature (Q) channels as shown in Fig. 8.18. The PN sequences used here have periods of 2^{15} chips. At the 1.2288 Mcps rate, this period corresponds to exactly 75 repetitions every 2 seconds.

We conclude the discussion of the reverse traffic channel system diagram of Fig. 8.18 by noting that the baseband lowpass filter shown in both the I- and Q-arms of that figure is defined to have a flat, passband response to within $\pm$ 1 dB up to 590 kHz, and to then drop, by at least 40 dB, to a stopband beginning at 740 kHz. Doubling these values to obtain the transmission response centered at the carrier frequency, this says that the nominal transmission bandwidth of 1.23 MHz falls, as it should, between these values of 1.180 MHz and 1.480 MHz. Finally, the D delay box shown in Fig. 8.18 is precisely the $T/2$ value shown in Fig. 5.15 of Chapter 5, as required to obtain the desired OQPSK transmission. Here the delay is half the chip value at 1.2288 Mcps, or 409.6 nsec.

We now consider the system block diagram for forward, BS to mobile, traffic channels shown in Fig. 8.19. Note that the diagram is similar to the one for the reverse traffic channel of Fig. 8.18 just discussed, with some critical differences. It has already been noted that the convolutional encoder is of the rate 1/2 type, rather than the rate 1/3 type in the reverse direction. The mod-2 addition with the user-defined ESN is done at the 19.2 kbps level to provide data scrambling: the first PN chip of every 64 coming out of the long code generator, provided by the output of the decimator, is added modulo-2 to the output of the block interleaver. (Note that 1.2288 Mcps/64 = 19.2 kbps.) The wideband spreading, on the other hand, is done via mod-2 addition with Walsh code chips at the 1.2288 Mcps rate. A 64-ary Walsh code is used here, as in the reverse direction case, but the 64 possible Walsh code values, each 64 chips long, serve a different purpose: they are used to designate the traffic or control channels over which information is being sent. In the case of Fig. 8.19, as an example, the base station is shown sending user m information over traffic channel n, designated by Walsh code $\mathbf{W}_n$. This technique of using codes to designate channels differs, of course, from the time-slot channel techniques used in the TDMA systems, GSM and IS-136, discussed earlier.

The 64 possible Walsh codes, in defining traffic or control channels in the forward direction, are specified as follows: Walsh code $\mathbf{W}_0$, consisting of the all-0 sequence, corresponds to the pilot channel, which we shall see, shortly, provides to the mobile the necessary carrier phase and timing information on powering up in a given cell. A synchronization (sync) channel may be used to provide to the mobile additional system timing as well as system configuration information. If used, it is represented by $\mathbf{W}_{32}$, the first 32 bits of which are all 0s, the remaining 32 bits all 1s. Up to seven paging channels per forward CDMA channel, to be used as part of the access process, as well as to transmit paging messages to a mobile, may be present, given by $\mathbf{W}_1$ to $\mathbf{W}_7$. The remaining Walsh-defined channels are assigned to forward traffic channels. As an example, one could have three paging channels, $\mathbf{W}_1$ to $\mathbf{W}_3$, and no sync channels. In this case, Walsh channels $\mathbf{W}_4$ to $\mathbf{W}_{63}$ are assigned as traffic channels. If a sync channel is, however, present, it is assigned channel $\mathbf{W}_{32}$, the remaining channels again being assigned as traffic channels. As shown in Fig. 8.19, transmission to a given mobile user is specified by using that user's long code mask, i.e, the long code sequence carrying that user's ESN. User m is so identified, as already noted above, in Fig. 8.19. The particular traffic channel being used is identified by the Walsh code used, in this case, again, as already noted, $\mathbf{W}_n$. The mobile, on its receiving frequency, will carry out the reverse procedure of Fig. 8.19. Details appear later in the discussion of the IS-95 control channels.

Note that Fig. 8.19 shows an 800 bps stream labeled "power control bits" being multiplexed with the 19.2 kbps traffic stream exiting the block interleaver. Recall from Chapter 6 that CDMA systems rely on tight control of mobile power, to ensure that the power received at the BS from all mobiles in a cell is the same. This is accomplished by using a combination of open-loop control provided by each mobile, augmented by a closed-loop correction, involving the base station. It is this power control correction that is provided by the 800 bps power control sub-channel. The BS estimates the received power of each mobile over an interval of 1.25 msec and then transmits a 0 back to a mobile if that mobile's power is too low and is to be increased; a 1 if the power is to be decreased. At the 19.2 kbps traffic bit rate, 24 bits, each lasting 52.083 μsec, are transmitted every 1.25 msec. Two of these bits are replaced every 1.25 msec by the power control bit, 104.166 μsec long, to form the 800 bps power control sub-channel. This is an example of what is termed a *punctured code*. (Not shown in Fig. 8.19 is a further decimation by a factor of 24 of the 19.2 kbps sequence exiting the decimator to provide an 800 bps signal to provide timing control for the traffic channel/power control bit multiplexor.)

Finally, note that the I- and Q-spreading sequences include the label "pilot PN seq." This refers to the fact that base stations each use a specified time offset, of 512 possible, of the pilot PN sequence. The same offset is used on all forward control channels and on all CDMA frequency assignments for a given base station. The time offset thus identifies a given base station. Offsets may be reused, however. Time offsets are specified in units of 64 chips. An offset of 10, for example, refers to $10 \times 64 = 640$ chips from a zero-offset pilot PN sequence. At a chip rate of 1.2288 Mcps, this corresponds to 520.83 μsec of actual time delay.

Consider now the control channels used in IS-95. We have already noted that the forward, downlink, direction has pilot, sync, and paging control channels defined. Traffic channels in this direction carry the power control sub-channel just discussed, as well as downlink signaling messages. The reverse, uplink direction only has the access control channel mentioned earlier defined. This is thus similar to the random access channel, RACH, defined for both IS-136 (Table 8.2) and GSM (Table 8.1). The uplink or reverse traffic channel carries signaling messages as well. Figure 8.22 portrays these various

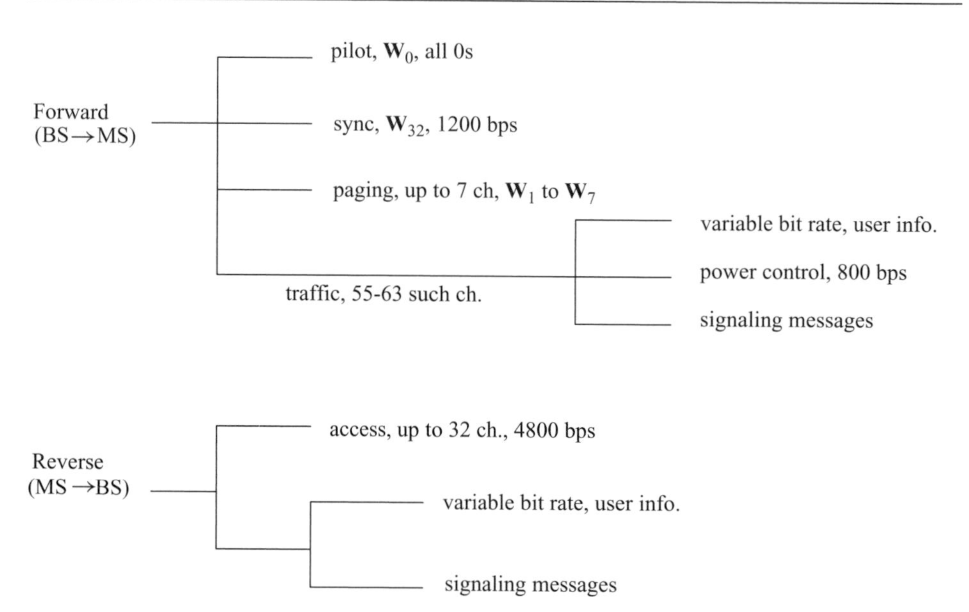

Figure 8.22 Traffic and control channels, IS-95

channels schematically. Distinct codes are used to define all these channels, as indicated above for the forward direction.

We focus first on the uplink access control channel. As just noted, this channel serves the same role in the IS-95 system as the RACH channels in the GSM and IS-136 systems. It is a multi-access channel used by the mobile to initiate communication with the BS, as well as to respond to paging messages carried in the forward direction over the paging channel shown in Fig. 8.22. Up to 32 such access channels are defined for IS-95, each associated with a particular paging channel. At least one access channel must be available for each paging channel. An access channel is generated, using its own unique code to identify it, following the block diagram of Fig. 8.23. We only describe portions of this diagram which differ from the corresponding diagram, Fig. 8.18, for the reverse-direction traffic channel. Note that 88 access channel bits are generated per 20 msec frame. Eight tail bits are added each frame to produce a 4.8 kbps signal entering a rate-1/3, constraint-length $K = 9$ convolutional encoder. The resultant 14.4 kbps encoder output is repeated, bit-by-bit, to obtain a 28.8 kbps bit stream, the same bit rate as in the full-rate traffic channel case of Fig. 8.18. (It is to be noted that symbol repetition is used as well in the less than full-rate traffic channel cases of 4, 2, and 0.8 kbps appearing at the input to the convolutional encoder of Fig. 8.18. In all cases, a 28.8 kbps bit stream is fed into the Walsh encoder shown there, just as in this case of the access channel.) The same 64-ary Walsh orthogonal modulation is used as in the reverse traffic channel case, followed again by PN spreading. The difference here, however, is that the 1.2288 Mbps PN sequence is generated in this case by a 42-bit access channel long code mask made up of a 16-bit base station id; a 9-bit pilot PN-offset number, Pilot_PN, corresponding to the base station's pilot PN sequence offset noted above; a three-bit paging channel number (recall that at most seven paging channels are available per forward CDMA channel); and

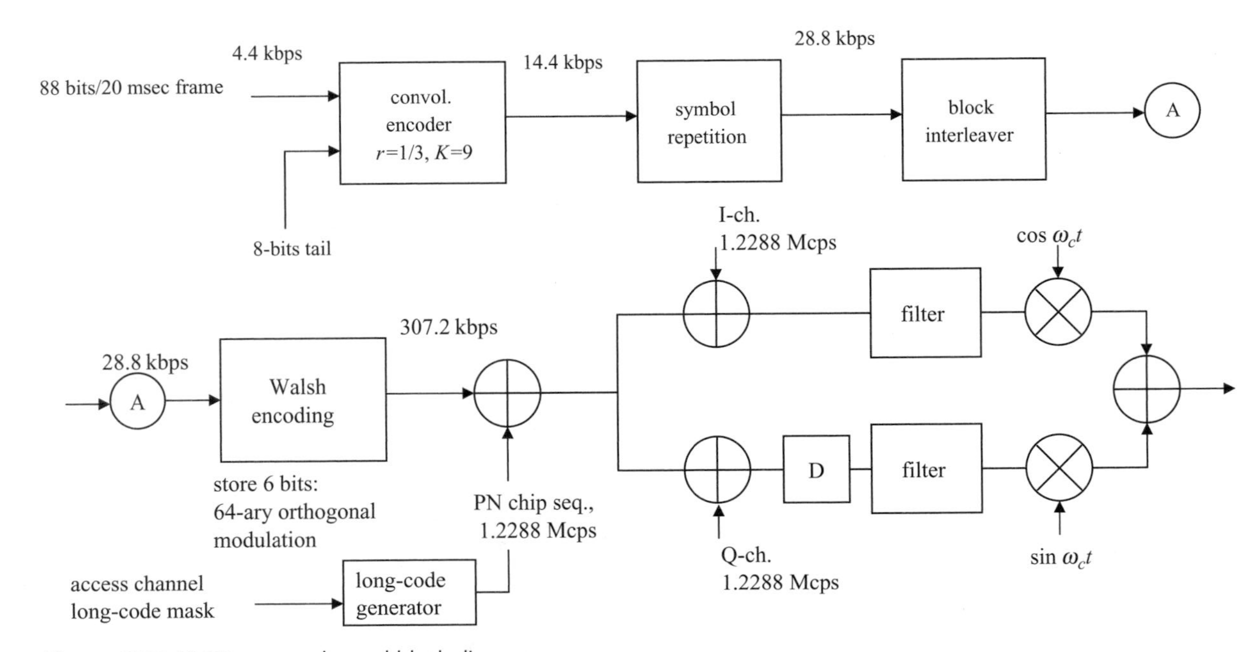

Figure 8.23 IS-95 access channel block diagram

a 5-bit access channel number. These 33 bits are augmented by a prescribed 9-bit field to form the 42-bit mask in this case. These various numbers and ids will have previously been made available to the mobile during initialization, after powering up, through a forward-direction sync channel message and the System Parameters message carried by the paging channel. These messages will be described in our discussion later of system control channels used during call processing.

A modified Aloha random access procedure is used to transmit a message to the BS over an access channel. Just as in the cases discussed earlier of GSM and IS-136, this message could be sent in response to a paging message received over the paging channel, it could be a message from the mobile registering a new location, or it could be a message sent as part of a call origination procedure. Response-attempt and request-attempt procedures differ slightly, but in both cases a sequence of access attempts is made using an access channel each time randomly chosen from the list of available access channels. Each access attempt is actually made up of a number of access probes, each probe increasing in power, and randomly delayed with respect to the previous one. Random backoff delays are also used between each multi-probe access attempt in the sequence. Access is completed at any time during the sequence on receipt of an acknowledgement over the paging channel. Access channel messages all carry a 34-bit *mobile information number* MIN and the 32-bit ESN to which reference was made earlier.

The forward-direction, base station to mobile, control channels specified in Fig. 8.22 are diagrammed in Fig. 8.24. The all-zero pilot channel is spread, as shown, by the all-zero Walsh code $\mathbf{W}_0$, at a spreading rate of 1.2288 Mcps. The sync channel operates at a 1.2 kbps rate, which is converted to 4.8 kbps after rate-1/2 convolutional encoding and symbol repetition. After block interleaving, the sync channel bits are spread by adding them modulo-2 to the Walsh code $\mathbf{W}_{32}$ that identifies this channel, running at a 1.2288 Mcps rate. Finally, the paging channels, up to seven in number, inputting information at initial rates of 9.6 kbps or 4.8 kbps, and being converted up to 19.2 kbps after convolutional encoding and symbol repetition of the 4.8 kbps signal, are spread by a 1.2288 Mcps Walsh code, chosen from $\mathbf{W}_1$ to $\mathbf{W}_7$, depending on the number of paging channels assigned. In all these control channels, as well as the forward-traffic channels of Fig. 8.19, the Walsh codes are 64 chips long, repeating every 64 chips, and running at the 1.2288 Mcps chip rate. Note that the same BS pilot PN sequence offset appears in the I- and Q-spreading signals for all these control channels.

How are these different forward-direction control channels used? The pilot channel is used by mobile terminals within a given cell, as already noted, to acquire carrier phase and timing reference. The sync channel, operating at the rate of 1200 bps, transmits one message only, containing system time, the 9-bit offset number Pilot_PN, system/network ids, and the paging channel information rate being used, whether 4800 or 9600 bps. The paging channel is used to transmit control messages to mobiles that have not been assigned traffic channels. One obvious example of a paging channel message is the Page message to a specified terminal, announcing that another user would like to initiate a call. This procedure will be considered in detail in the next section. The paging channel is used as well, as part of the random access protocol, to acknowledge receipt of a mobile-generated access channel message. Other messages sent over the paging channel include a Channel

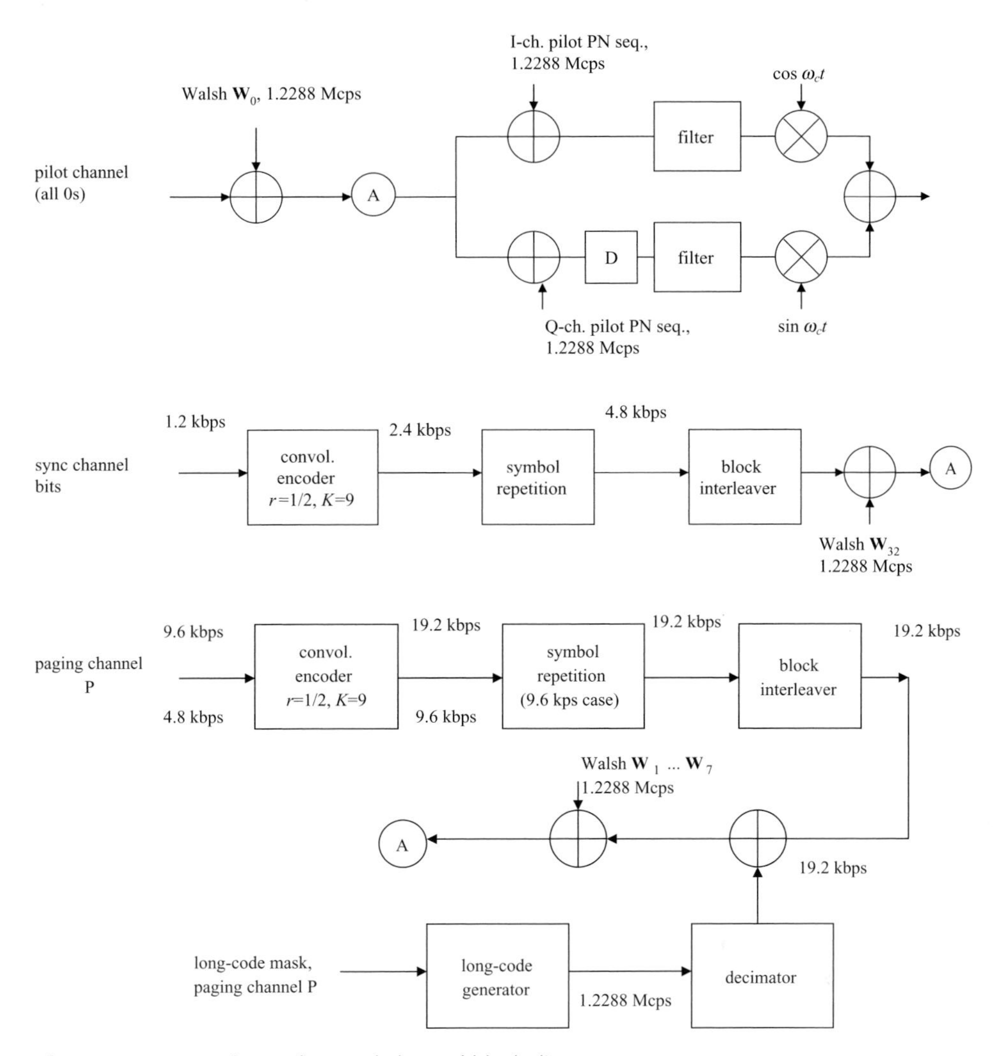

Figure 8.24 IS-95 forward control channel block diagrams

Assignment message, to which reference will be made shortly in discussing the call connection process; a System Parameters message which carries, among other items, the base station id, system/network ids, Pilot_PN, and the number of paging channels available over this CDMA channel; an Access Parameters message, which includes, among other parameters, power information for the mobile, the number of access channels available for this paging channel, Pilot_PN, information about the access procedure (number of access probes and acknowledgement timeout, for example), Access message format, etc.; a Neighbor List message; a CDMA Channel List message; a variety of Order messages; and a number of other message types. Some of these messages are sent periodically, others as needed. We will discuss a number of these further, both in our discussion of call origination that follows, and in providing examples of how these messages are actually transmitted over the paging channel. They may be grouped into such categories as Broadcast messages

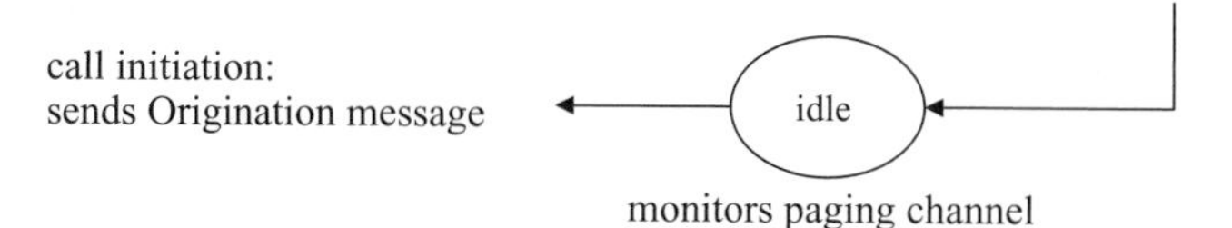

Figure 8.25 IS-95: preparation to set up a call

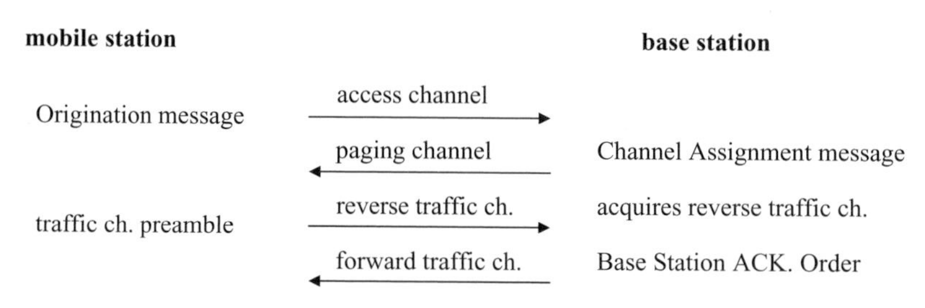

Figure 8.26 IS-95, call origination (simplified)

that provide system and access parameters, Call Management messages, Authentication messages, and Operations and Maintenance messages (Goodman, 1997).

The procedure used to set up a call in IS-95 uses the control channels in both reverse and forward directions just discussed and is similar to that used in the GSM system, as outlined previously in Fig. 8.3, and in the IS-136 system, as specified by Fig. 8.15. A mobile on powering up in a given cell first follows the various processes diagrammed in Fig. 8.25. It sets its receiving code channel for the all-zero pilot channel, searches for, and acquires the pilot channel for that cell. It then sets its receiving code channel for the sync channel, and acquires that channel. The mobile is now in its idle state, in which it monitors the paging channel, to obtain the information such as system and access channel parameters necessary to originate a call. These parameters include the specific paging channel to be used, as well as the corresponding access channels associated with that paging channel. Once these parameters are collected, the mobile is ready to begin the call origination procedure using an access channel.

From the idle state, the mobile can register as well with the base station using the access channel by sending a Registration message. This message, containing the mobile id among other fields, indicates the mobile location to the system. After registration, the mobile then either returns to the idle state, starts the call origination procedure, or responds to a waiting paging message.

We now follow through with the call origination procedure, indicating the messages transmitted and the channels used to carry them. Figure 8.26 diagrams the procedure and is to be compared with Figs. 8.3 and 8.15, showing the comparable call-setup procedures for GSM and IS-136, respectively. It bears repeating that the channels used here are codes rather than the time slots used in the TDMA-based GSM and IS-136. Channels in the forward, downlink direction are specified by the appropriate one of 64 possible Walsh codes (Figs. 8.19, 8.22, 8.24) and by their long-code masks appearing as part of the PN chip sequences in the reverse, uplink direction (Figs. 8.18 and 8.23). As shown in Fig. 8.26, the mobile initiates call setup by sending an Origination message over the access channel. Assuming this message is received correctly, the base station responds by sending a Channel Assignment message over the paging channel. This message indicates to the mobile the forward traffic channel (Walsh code) on which the base station will be transmitting, and to which its receiving code for the traffic channel is to be set. The mobile, on receipt of the Channel Assignment message, begins sending a traffic channel preamble over the reverse traffic channel, consisting of continuous zeros at the 9600 bps rate, which is used by the base station to acquire that channel. The base station now sends a Base Station Acknowledgement Order over the forward traffic channel, and two-way conversation over the traffic channels may now begin.

What is the format of these signaling and control messages? Consider those in the reverse direction. As noted, they are carried either over the access channel or over the reverse traffic channel. We consider first signaling messages sent over the reverse traffic channel, the carrier of data traffic, principally voice in second-generation cellular systems. As we have indicated, the reverse traffic channel can operate at 9600, 4800, and 2400 bps, changing from frame to frame, if necessary. For simplicity, we focus on the full-rate, 9600 bps, case only, as was done in diagramming the reverse traffic channel system block diagram of Fig. 8.18. Successive reverse traffic channel frames 172 bits long and recurring every 20 msec. may carry a combination of data traffic, labeled primary traffic, and signaling traffic. Secondary traffic may optionally be carried as well. Table 8.3 indicates how the 172 bits per frame shown in Fig. 8.18 are allocated between data and signaling or secondary traffic. (Recall from Fig. 8.18 that each 172-bit frame has 12 error-check bits added, the resultant 184-bit frame, with an 8-bit all-zero tail appended, then being fed to the rate-1/3 convolutional encoder.) As shown in Table 8.3, the first bit in the frame, the "mixed mode" bit, if set to 0, indicates that a given frame is carrying primary traffic only. If set to 1, the frame carries a combination of primary traffic and signaling or optional secondary traffic. The following three bits, if the mixed mode bit is set to 1, indicate the division of the rest of the 168 bits between primary traffic and signaling or secondary traffic. An example of a frame carrying 16 bits of primary traffic and 152 bits of signaling traffic appears in Fig. 8.27. The term *dim-and-burst* is used to indicate those frames which carry both primary and signaling traffic, in varying proportions. *Blank-and-burst* indicates a frame carrying signaling traffic exclusively. Section 8.5, which concludes this chapter, discusses how voice calls are handled using both the reverse and forward traffic channels. We therefore focus here on signaling messages only.

Signaling messages sent on the reverse traffic channel vary from 16 to 2016 bits in length. They can therefore occupy multiple frames, depending on the message length and the division of frames between data and signaling traffic. Before being split into multiple frames, however, a message has an 8-bit message-length header added and a 16-bit error check sequence appended. The resultant format appears in Fig. 8.28. The message-length indicator indicates the message length in bytes. An example appears in Fig. 8.29. The signaling message body itself in this example has been arbitrarily chosen to be 37 bytes or 296 bits long. The full message to be carried over the frames, including the 24 bits for header and CRC, is thus 320 bits long. A dim-and-burst division of frames with 80 bits of

Table 8.3 *IS-95 Reverse Traffic Channel, 172-bit Frame (9600 bps case)*

bits per frame:

Format field bits					
Mixed mode	Traffic type	Traffic mode	Primary Traffic	Signaling Traffic	*Secondary Tr. (optional)*
0	–	–	171	0	0
1	0	00	80	88*	0
1	0	01	40	128*	0
1	0	10	16	152*	0
1	0	11	0	168**	0
1	1	00	80	0	88
1	1	01	40	0	128
1	1	10	16	0	152
1	1	11	0	0	168

*Notes: *dim-and-burst*
***blank-and-burst*

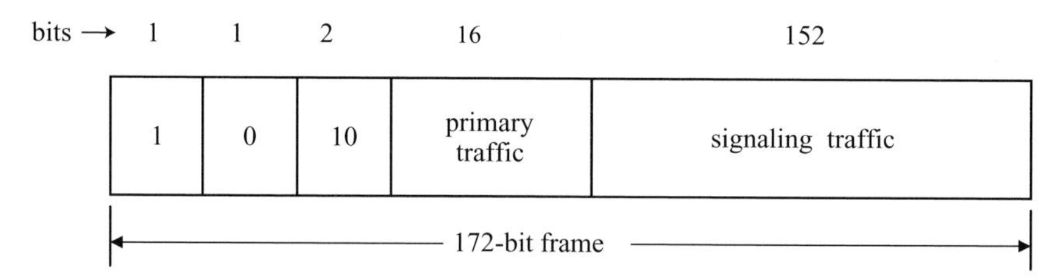

Figure 8.27 Example of IS-95 "dim-and-burst" frame, reverse traffic channel

MSG-LENGTH	message body	CRC
8	16 - 2016	16

bits →

Figure 8.28 IS-95 signaling message format, reverse traffic channel

primary traffic and 88 bits of signaling traffic per frame has been arbitrarily selected. Four frames are thus needed to carry this message. The first bit of the 88 signaling bits in each frame is used to indicate whether the frame in question is the first frame of the message or a continuing frame. Eighty-seven bits per frame thus remain to carry the signaling message. Two hundred and sixty one bits are thus carried in the first three frames; the remaining 59 bits are carried in the fourth and last frame. The 28 bits left over in that frame must then be filled by all-zero padding bits. The message-length indicator is used by the base station, on receiving this message, to delete the padded bits.

There are a variety of reverse traffic signaling messages defined in IS-95, each of varying length, depending on the information to be sent. Examples include a Pilot Strength Measurement message, that provides to the base station a 6-bit indication of the mobile's

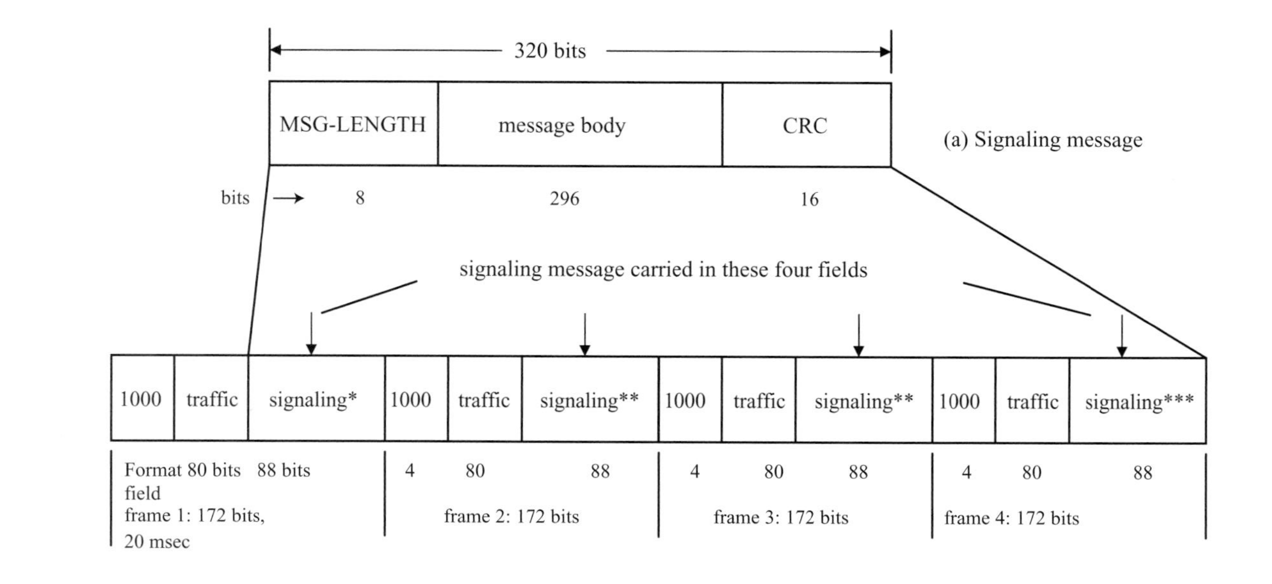

Notes: * bit 1 = 1(first frame) ** bit 1 = 0 (continuing frame)
*** bit 1=0, 28 bits padded (all 0s)

(b) Signaling message carried over four frames

Figure 8.29 IS-95: example of multiframe signaling message, reverse traffic channel

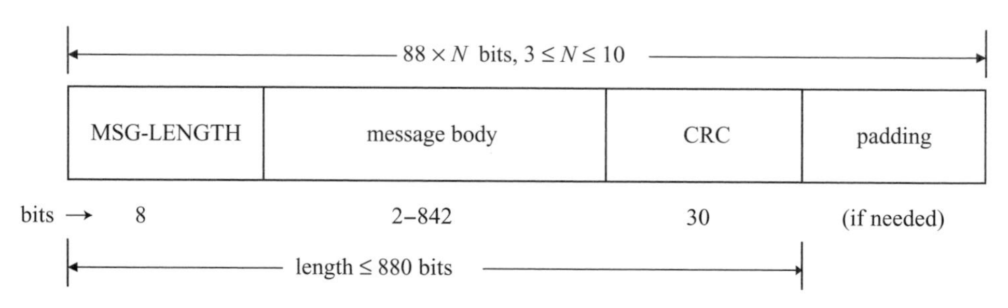

Figure 8.30 IS-95 access channel message capsule

measurement of the signal power of the pilot used by the mobile to derive its time reference; a Power Measurement Report message that provides the base station with the power measured by the mobile for each of the pilots in its active set, as well as the number of frames used in making this measurement; an Origination Continuation message that contains any dialed digits not included in the Origination message sent over the access channel (Fig. 8.26); a Flash with Information message that may include information records such as the calling party and called party's numbers; a Status message that may include such information records as the mobile's identification, its ESN and its station MIN (mobile identification number), terminal information involving the manufacturer's code and model number, and the mobile's security status; and a variety of Order messages, typically simpler messages, such as Release and Connect Orders. Each of these, when sent, would be sent following the format and procedure just described, as summarized in the signaling traffic portions of Table 8.3, and Figs. 8.28 and 8.29.

The discussion above has focused on signaling messages carried by the reverse traffic channel. Consider now signaling messages carried over the access channel. As noted earlier, this reverse control channel is used by the mobile to establish a call (Fig. 8.26), to respond to a page, or to register its location. The three of the various defined access channel messages, including a number of Order messages, that carry out these different tasks are, respectively, the Origination message to which reference has already been made (Figs. 8.25 and 8.26), the Page Response message, and the Registration message. All of these messages carry the mobile id, in addition to other information. Messages may vary in length from 2 to 842 bits. They are embedded in an access channel message capsule defined to be an integral number of 88-bit access frames in length. (Recall from Fig. 8.23 that 20 msec access channel frames carry 88 bits each.) Figure 8.30 displays an access channel message capsule. Note that the message body is preceded by an 8-bit message length field indicating the length of the capsule, including a 30-bit error check field following the message body (but not including the padding shown). Capsules can be no longer than 880 bits or ten frames. The minimum capsule length is three frames. Since capsules must conform with frame boundaries, padding must be used to augment a message length, as shown in Fig. 8.30. The maximum capsule length, in frames, is determined by the base station, and is sent to the mobile over the paging channel in an Access Parameters message when it is in its idle state (see Fig. 8.25). Each 88-bit frame in the capsule is augmented by eight all-zero tail bits to form a 96-bit frame, before being convolutionally encoded, as shown earlier in Fig. 8.23.

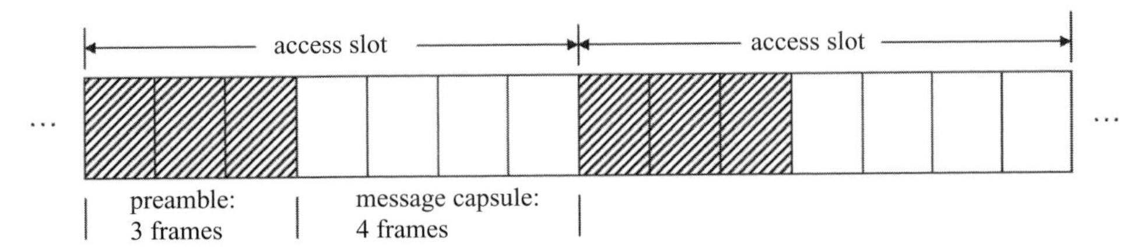

Figure 8.31 IS-95 access channel, transmission slot, 4800 bps

An access channel transmission, called an access channel slot, consists of the augmented message capsule preceded by an all-zero bit preamble, which is a minimum of one frame or 96 bits in length. The preamble is used by the base station to acquire or synchronize to the mobile transmission during the access phase. The preamble length, to a maximum of 16 all-zero 96-bit frames, is also specified by the base station in the Access Parameters message sent over the paging channel. Figure 8.31 portrays a series of access channel slots, each containing, as an example, a three-frame preamble and a four-frame capsule. Transmission is at the rate of 4800 bps entering the convolutional encoder, as indicated in Fig. 8.23.

We have focused thus far on signaling and control messages sent over the reverse channel, from mobile to base station. What is the format of signaling and control messages sent in the forward, downlink direction? Recall from Fig. 8.22 that signaling messages in the downlink direction may be sent over the forward traffic channel as well as over the three forward signaling channels, the sync, pilot, and paging channels, discussed briefly earlier. Signaling over the forward traffic channels is carried out much like signaling over the reverse traffic channels just discussed: the forward traffic channel is also organized into 172-bit frames. These frames carry a combination of primary traffic (principally voice messages), signaling traffic messages, and secondary traffic, and are organized in the same manner as reverse traffic frames, as shown in Table 8.3 and the example of Fig. 8.27. The signaling message format over the forward traffic channel, from base station to mobile, is essentially the same as that portrayed in Fig. 8.28 for the reverse traffic channel case, with the one exception that the message body can be no longer than 1160 bits. Downlink signaling messages requiring multiple frames would thus be sent over a forward traffic channel just as in the reverse traffic channel example of Fig. 8.29. A number of forward traffic channel messages are defined for IS-95. Examples include a variety of Order messages, an Authentication Challenge Response message, a Handoff Direction message, a Neighbor List Update message, and a Power Control Parameters message, among others.

Now consider the forward signaling channels. The pilot channel, as already noted, is used to send an all-zero sequence (Walsh code $\mathbf{W}_0$) by which a mobile acquires carrier phase and timing reference. The one sync channel message, if sent by the base station, is a 162-bit, fixed-length message, sent continuously over the sync channel using Walsh code $\mathbf{W}_{32}$, as noted earlier (Fig. 8.24). As also noted previously, this message contains, among other parameters, a 15-bit system id, a 16-bit network id, the 9-bit Pilot PN sequence offset PILOT_PN unique to each base station, a 36-bit system time parameter, and an indication of the paging channel bit rate. To transmit this message, a format similar to that used in

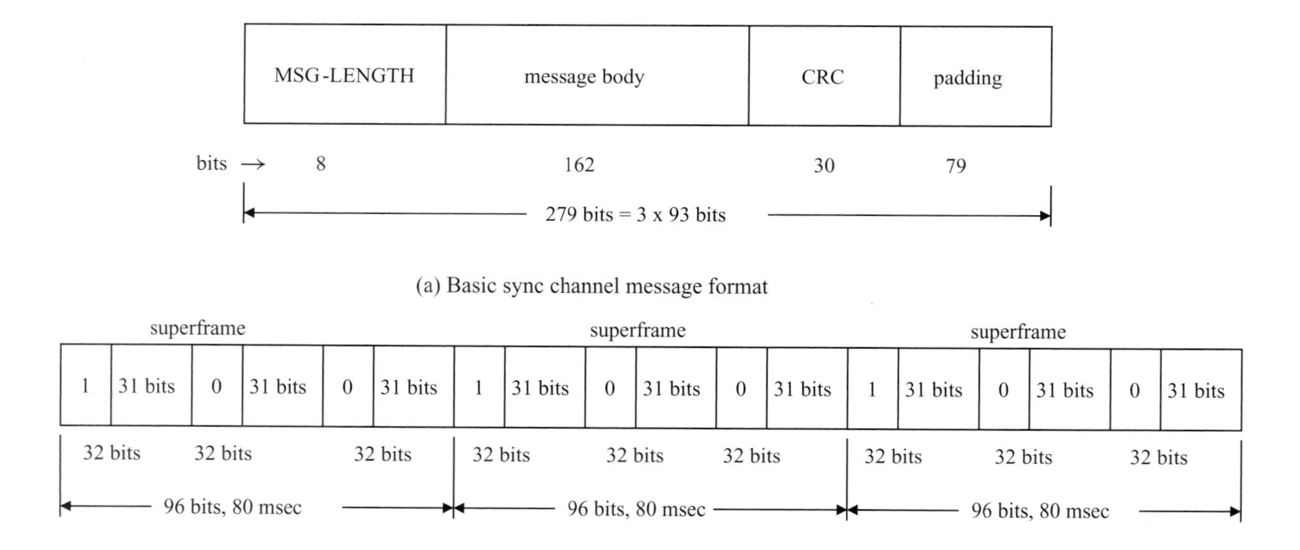

(a) Basic sync channel message format

(b) Transmission of single sync channel message

Figure 8.32 Sync channel message transmission, IS-95

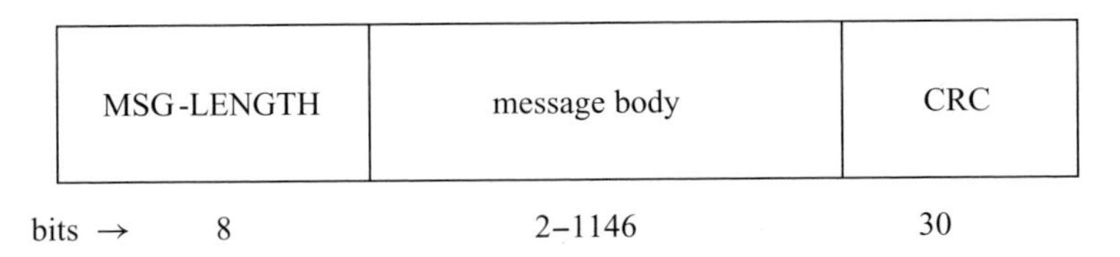

Figure 8.33 Paging channel message format

the traffic channel message case is used (Fig. 8.28): the 162-bit message is embedded in a 279-bit capsule containing, in addition, an 8-bit message-length field, a 30-bit CRC field, and 79 bits of padding (all 0s) for a total of 279 bits. This is shown in Fig. 8.32(a). The capsule is in turn split into three 93-bit segments, with each segment transmitted as three 32-bit consecutive frames forming a superframe, each frame consisting of 31 bits preceded by a 1-bit framing bit: a 1 for the first frame in the superframe group of three, a 0 in the other two frames (Fig. 8.32(b)). One such superframe is transmitted every 80 msec, as shown in Fig. 8.32. Note that this results in the 1200 bps input transmission rate for the sync channel, as indicated in Fig. 8.24. Three superframes are required to transmit each sync channel message, as shown in Fig. 8.32(b).

The paging channel is used to transmit a number of control messages from base station to mobile before assignment of a traffic channel, as noted earlier. We mentioned among others the Page message, the Channel Assignment message used in the call origination procedure of Fig. 8.26, a System Parameters message, and an Access Parameters message. Paging channel messages are sent at a rate of either 4800 or 9600 bps (Fig. 8.24). Transmission is organized into 80 msec slots, each slot carrying 384 bits or 768 bits, depending on the transmission rate used. Slots are grouped into 2048 slot cycles 163.84 sec long, to allow mobile terminals to regularly go into "sleep mode" or idle state and then "wake up" periodically at specified times within a cycle, to determine whether messages are waiting for them. This procedure serves to conserve battery power. Each 80-msec slot is divided into four 20 msec frames, each frame consisting of two 10 msec half-frames. Half-frames, 48 or 96 bits long, have as their first bit an SCI (synchronized capsule indicator) bit. This bit indicates whether message transmission is synchronized or not. Paging channel messages carried by these half-frames each consist of a message body preceded by an 8-bit message length indicator and followed by a 30-bit CRC. The message body, as shown in Fig. 8.33, can vary in length from 2 to 1146 bits. Padding may be added to form a message capsule. If synchronized transmission is used, a synchronized message capsule starts on the second bit, the one following the SCI bit, of a half-frame. Padding is used to ensure this.

8.4 Mobile management: handoff, location, and paging procedures

In the introduction to this chapter we referred to signaling across the air interface between a mobile and a base station, as well as to signaling through the wired network to control mobile communication as mobiles move through a wireless system, or even roam into other systems. This chapter has thus far has focused on signaling across the air interface, with examples provided for the three major standards currently used in second-generation systems: GSM, IS-136 or D-AMPS, and IS-95. We have described in detail in the previous three sections of this chapter how the various signaling messages required to set up a call, as well as carry out other necessary control functions, are transmitted across the air interface between the mobile and base station. In this section we focus on the signaling required as mobiles move between cells, or between larger areas containing multiple cells. The control required to handle the movement of mobiles is referred to as *mobile management*. The movement of mobiles involves essentially three functions: handoff control, location managment, and paging. Handoff control is required as a mobile, involved in an on-going call, moves from one cell to an adjacent one, or from the jurisdiction of one system to another. A new set of transmission channels must be allocated to the handoff call without perceptible notice by the mobile user or interruption to the call. These two types of handoff are called, respectively, inter-cellular and inter-system handoff. (Standards outside North America often use the term handover instead.) Intra-cellular handoff is possible as well, when the mobile/base station combination determines that power considerations or channel fading warrants changing channels within a cell.

Location management is required to handle the registration of a mobile in areas or regions outside its home area, to enable it to be located and paged in the event of an incoming call. Recall from Fig. 8.1 in the introduction to this chapter that a mobile is associated with a home region, with its necessary reference and profile information stored in the data base of the home location register, the HLR of Fig. 8.1, of that area. Each HLR is associated with a mobile switching center (MSC), as shown in Fig. 8.1. The term location or registration area is used to designate a region of multiple cells, each with its base station, controlled by an MSC. On registering in a different region, a visitor's location register, VLR, is used to store the visiting mobile's reference and profile information, after communication takes place with the MSC and HLR of the home area. This registration information is subsequently used to page a mobile when an incoming call is directed to it, after first querying the home MSC and HLR as to the visiting location area. A Page message is sent by the visiting MSC to all base stations under its control, which, in turn, broadcast a Page message over their paging channels.

The signaling messages sent between MSCs to carry out the various mobile management functions alluded to above are carried over the wired network. In North America the various protocols and message descriptions required to handle the inter-system functions are defined by Interim Standard 41, IS-41, of the Telecommunications Industry Association. IS-41 in turn assumes an underlying data network is used to carry the inter-system signaling messages required to carry out mobile management functions. In practice, the data network used conforms to another set of standards called Signaling System Number 7 or SS7, for short. SS7, a layered-protocol architecture involving packet switching at its network layer, is used worldwide to provide call management functions for fixed telephone networks. It was thus natural to adopt this standard and its defined message set for mobile telephone operations as well. A tutorial on SS7 appears in Modaressi and Skoog (1990).

We now study each of the three components of mobile management in more detail. We begin with handoff control, discussing first inter-cellular, and, then, inter-system

---

Modaressi, A. R. and R. S. Skoog. 1990. "Signaling system number 7: a tutorial," *IEEE Communications Magazine*, 28, 7 (July), 19–35.

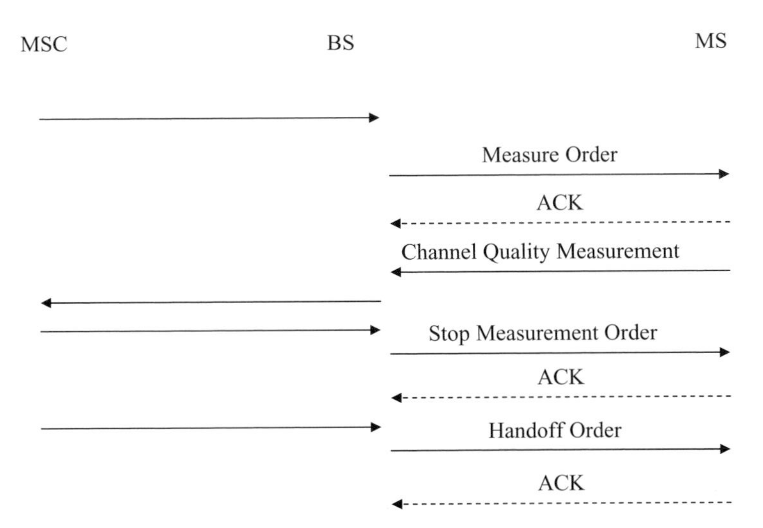

Figure 8.34 Mobile-assisted handoff, IS-136

handoff. We then move on to location management and paging, concluding with a simple consideration of tradeoffs involved in dealing with these two conflicting functions. (Fewer boundary crossings by individual mobile terminals are incurred as the location area size is increased, reducing the number of registrations, but more paging messages must be broadcast to find a given mobile.)

Consider a communicating mobile traveling within a given cell. As it reaches the boundary (generally ill-defined) of that cell, the power received from the cell base station with which the mobile has been in communication across the air interface between them will drop below a pre-defined threshold. In contrast, assume the power received from a neighboring base station exceeds a threshold. A decision to handoff to the neighboring base station and enter the new cell associated with that base station would then be made by the MSC controlling both base stations. Second-generation wireless systems support mobile-assisted handoffs (MAHO) in which power measurements are carried out by a mobile under the command of the MSC and base station, with the results of the measurements transmitted to the base station as uplink signaling messages.

We focus on IS-136 as a specific example. GSM uses a similar handoff strategy. Figure 8.34 diagrams the messages transmitted between the MSC, base station, and mobile as mobile-assisted handoff is carried out. The MSC notifies the base station that channel quality measurements are to be carried out. The base station responds by transmitting to the mobile a Measurement Order message identifying neighboring forward traffic channels, as well as the forward traffic channel over which it is currently receiving messages from the base station, for which channel quality is to be measured. Channel quality measurements consist of received signal strength measurements for the current and neighboring traffic channels, and bit error rate measurements for the current traffic channel. The results of these measurements are reported back to the base station by the mobile, when they are completed, in a Channel Quality Measurement message carried on the SACCH (see Section 8.2). The base station, in turn, forwards the measurement results to the MSC, which then issues a stop measurements command, sent on to the mobile by the base station as a Stop Measurement Order message. If, on analysis of the measurements, handoff is deemed necessary, the MSC so orders, with the base station then signaling the mobile to which new channel to tune. (Figure 8.34 also indicates acknowledgement messages tranmitted uplink by the mobile, acknowledging correct receipt of the corresponding command messages.)

Handoff of a mobile to a new base station thus results in the immediate need to allocate to the mobile a channel within the new cell. Should a channel not be available, the ongoing call would have to be dropped. The dropping probability of a handoff is clearly the same as that of blocking a newly generated call if new calls and handoffs are treated alike in the allocation of channels. The probability of blocking a call was discussed in Chapter 3 (see equation (3.7)). It is generally agreed that mobile system users find the dropping of ongoing calls much more onerous than receiving a busy signal on attempting to initiate a call, i.e., having a new call attempt blocked. After all, it is distinctly unpleasant to be cut off in the middle of a conversation! A number of proposals have thus been made to reduce the probability of dropping a call. These include, among others, giving handoff calls priority over new call attempts or queueing handoff calls for a brief time while waiting for a channel to become available. With a fixed number of channels available, giving priority to handoff calls obviously increases the probability of blocking new calls. There is thus a tradeoff to be considered. We shall return to some of these proposals, comparing them where possible, in the next chapter, in studying call admission control quantitatively.

Mention was made in Chapter 6 that the CDMA-based IS-95 uses *soft handoff* rather than the hard handoff we have been implicitly assuming to this point. It was noted there that soft handoff results in an improvement in signal-to-interference ratio, and hence an increase in system capacity. In the soft handoff procedure, a mobile makes a connection to two or more base stations before choosing the one with which to communicate. It is thus an example of a "make before break" operation. Hard handoff implies the connection with the current base station is broken before connecting to the new one. IS-95 systems, in carrying out handoffs, use mobile-assisted handoffs, as do IS-136 and GSM. But, unlike the handoff procedure adopted for these other systems, in IS-95 it is the mobile that initiates the handoff procedure. It is the mobile that determines handoff may be necessary based on measurements it carries out. It is left to the MSC, however, to actually make the decision to hand off. A handoff is completed when the MSC commands the mobile, by sending messages to the base stations involved, to drop the old base station connection and continue with the new base station connection. But connections to the old and new base stations are maintained until the decision to connect to the new base station only is made. This is the essence of soft handoff.

We now move on to inter-system handoffs. This type of handoff is encountered when a mobile roams, moving from an area controlled by one MSC to that controlled by another. A number of possible handoff scenarios have been defined by IS-41, the North American inter-system mobile communications standard. In Fig. 8.35 we diagram the inter-system procedure in the simplest possible case, that of a mobile m moving from one region controlled by an MSC, labeled MSC-A here, to a region under the control of MSC-B. This procedure is called handoff-forward. As the mobile moves through region A, as shown in Fig. 8.35(a), it eventually comes into an overlap region between A and B

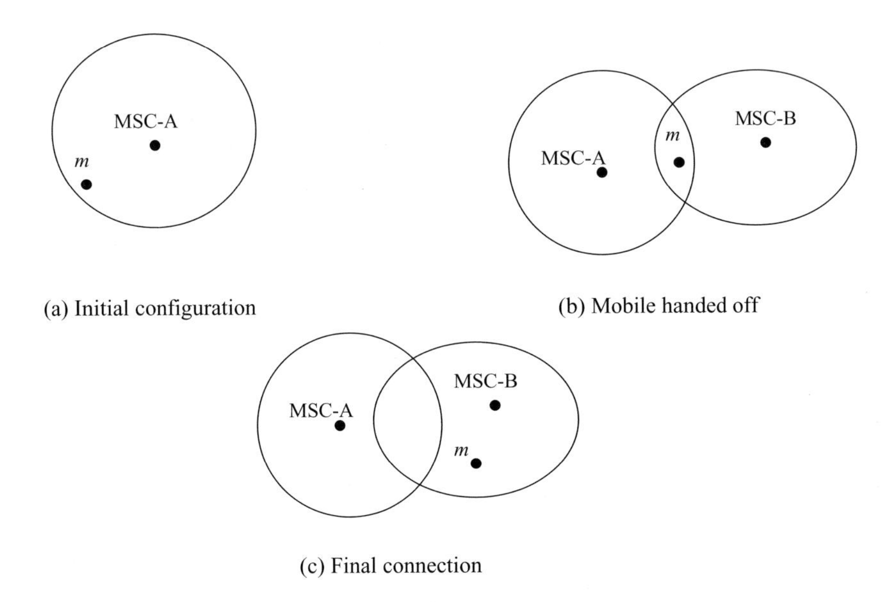

Figure 8.35 Inter-system handoff-forward, IS-41

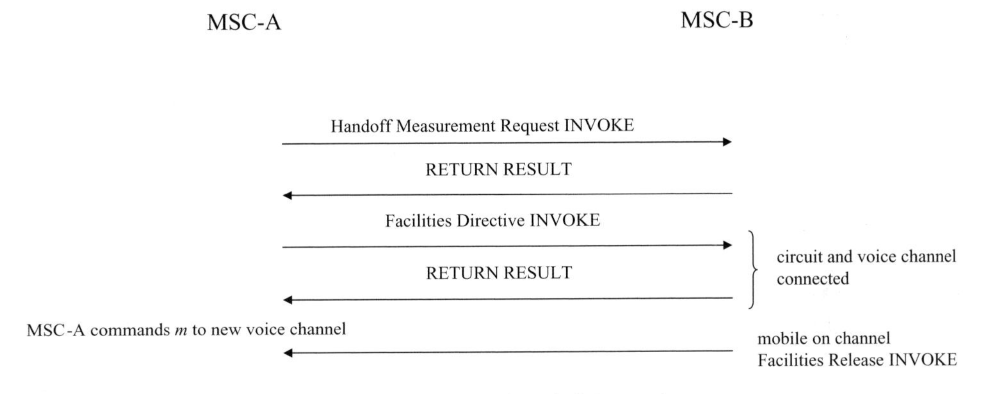

Figure 8.36 Message transmitted, inter-system handoff forward, IS-41

(Fig. 8.35(b)). It is here that MSC-A makes the decision, based on measurements made, to hand the call over to MSC-B. Once the call is handed over, the mobile is under the control of MSC-B, as shown in Fig. 8.35(c). The messages sent and received by the two MSCs while the mobile is in the overlap region of Fig. 8.35(b) are diagrammed in Fig. 8.36. The labels INVOKE and RETURN RESULT represent SS7 message types, used to carry the IS-41 messages. MSC-A, initially serving mobile m's call, is the switch that decides a handoff to the neighboring MSC-B is appropriate. It then sends the HandoffMeasurement Request INVOKE message, carrying the id of the serving cell in MSC-A, to MSC-B. MSC-B replies with a list of one or more of its cells, with the signal quality of each. MSC-A then makes the determination to hand off or not. If it decides affirmatively, it sets up a circuit between the two MSCs and then sends a FacilitiesDirective INVOKE to

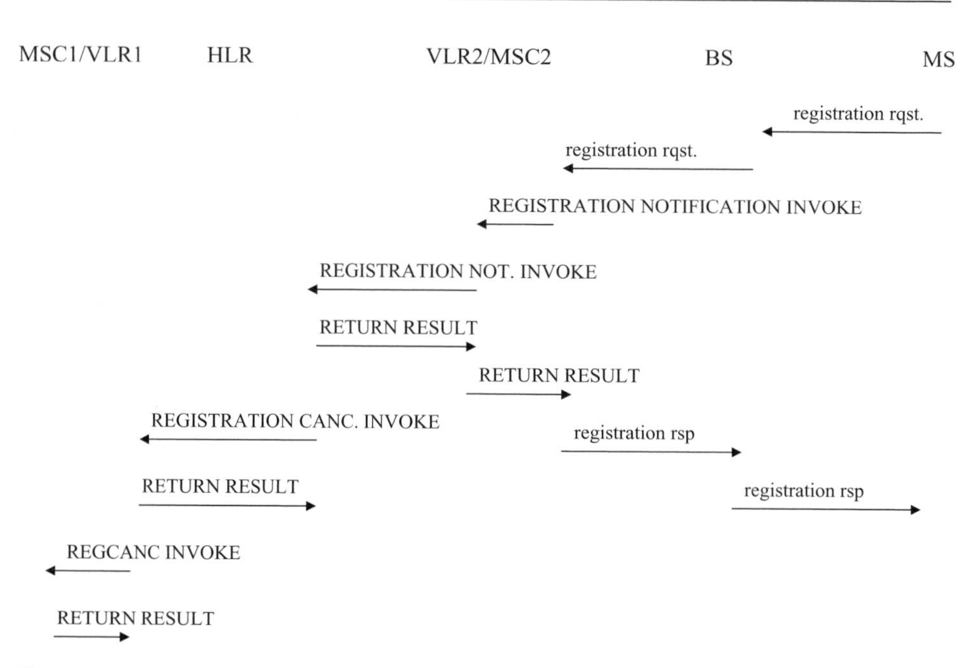

Figure 8.37 Location registration, IS-41

MSC-B, carrying the cell id in MSC-B it has designated as the cell to handle the mobile. If MSC-B finds an available voice channel in the designated cell, it replies with a RETURN RESULT message. MSC-A then commands mobile *m* to switch to that voice channel. With the mobile on that channel, MSC-B sends a FacilitiesRelease INVOKE, releasing the inter-MSC circuit. Handoff is now complete.

Other inter-system procedures defined in IS-41 involve cases in which an initial handoff is returned to the original MSC (handoff-back) or continued on to a third MSC (handoff-to-third).

We now move to a brief description of location management. Recall that location management refers to the requirement that roaming mobiles register in any new area into which they cross. They can then be paged in the event of incoming calls. An area consisting of multiple cells is controlled by an MSC. We again discuss the North American standard as described in IS-41. The procedures prescribed for GSM are quite similar. Associated with each MSC service area, often referred to in the location management literature as a location or registration area, is a Visitor Location Register, VLR, which maintains the data base of foreign mobiles registered in that area. Base stations within a given area periodically broadcast the area id. A roaming mobile, on entering a new area, senses, by listening to the base station broadcast, that it has crossed into a new area and begins the registration process. Figure 8.37 diagrams an abbreviated version of this process. The mobile is assumed in this figure to have earlier been in the area controlled by MSC-1, with its associated VLR1. It has now crossed over to the area controlled by MSC-2 containing VLR2. The mobile's home location register HLR shown, located in its home area, maintains a record of its id, subscriber information, and current location information. (Note that Fig. 8.37 is an expansion of Fig. 8.1.)

The mobile, on crossing into area 2, sends a registration message to the base station whose broadcast it hears, which, in turn, forwards that message to the area MSC. The message contains the mobile identification number MIN. MSC-2 now sends an IS-41-defined REGISTRATION-NOTIFICATION INVOKE message containing the MIN to its VLR. (The VLR could actually be co-located with the MSC.) VLR2, noting this mobile is not currently registered, forwards the message to the mobile's HLR, using the MIN to determine the HLR. The HLR acknowledges receipt of this message with a REGIS.NOT. RETURN RESULT message sent back to VLR2, enters the new location of the mobile in its data base, and then sends a REGISTATIONCANCELATION INVOKE to VLR1, which forwards this message to its MSC. This cancels the mobile's registration in area 1. Other acknowledgement messages, acknowledging receipt of each message sent, appear in Fig. 8.37 as well. Note, in particular, that each registration requires eight SS7 messages, including acknowledgements, to be sent between the various VLRs, MSCs, and the HLR. Four other messages across the air interface and between base station and its MSC are generated as a result of the registration. We ignore here additional messages that would have to be sent to handle any errors or untoward events. The assumption here is that registration proceeds smoothly and correctly. Even without these other messages that might have to be sent, it is clear that registration message activity could end up requiring a sizeable fraction of system capacity, depending on the size of a location area and the density of mobiles within it. Mobile power is also impacted by registration activity. Increasing the size of a location area would obviously reduce a mobile's border crossings requiring registration, hence reduce per-mobile registration activity. But, as we shall see next, in discussing paging of mobiles, an increase in area size results in more paging messages sent to locate a mobile within a location area. There is thus a tradeoff between location management and paging, as noted at the beginning of this section. We shall quantify this tradeoff using a very simple model of mobile movement and determine, quite grossly, a suitable size for the location area.

We now consider call delivery to an idle mobile terminal located outside of the area in which the call originated, roaming beyond its home area. This process, which results in paging of the mobile terminal, is diagrammed in Fig. 8.38. The Call Initiation message from the call-originating terminal (this could be a mobile itself, or could be a stationary phone within the wired network), is directed to the nearest MSC, based on the dialed destination mobile digits carried in the message. The MSC, in turn, forwards the message to the HLR of the destination mobile, using a LocationRequest INVOKE (here abbreviated to LOCREQ). The HLR then sends a RoutingRequest INVOKE (ROUTREQ) to the VLR identified as the last VLR with which the destination mobile registered. The VLR forwards the message to the current serving MSC. The MSC assigns a Temporary Local Directory Number (TLDN) to the intended call and includes this number in a RoutingRequest RESPONSE (RSP) returned to the HLR, via its VLR, as shown. The HLR then sends a LocationRequest Response (RSP) to the originating MSC, including the mobile's MIN. The originating MSC now sets up an end-to-end voice connection to the MSC serving the mobile. That MSC then notifies all base stations in its area via a paging message to page the destination mobile. Each base station, in turn, broadcasts a paging message to all mobile terminals in its cell. The idle roaming mobile being paged, recognizing its id, replies to the base station page using its random access channel, as discussed in earlier

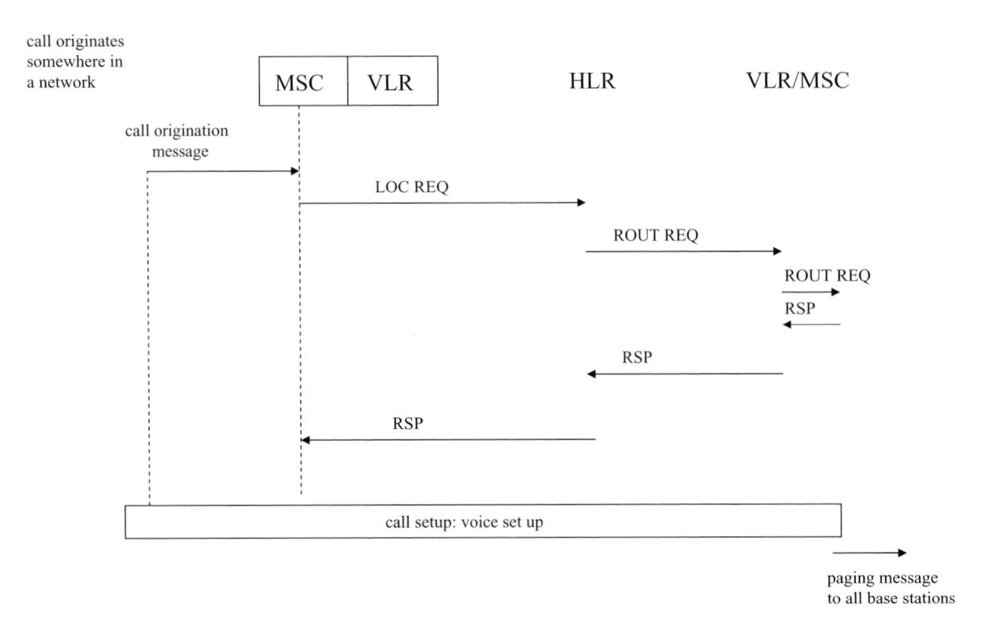

Figure 8.38 Call to roaming mobile, IS-41

sections. Note that this call delivery/paging procedure requires six SS7 messages, a paging message from the serving MSC to each of the base stations in its area, a broadcast paging message by each base station, and a reply to the page from the mobile being paged. For N cells in a location area, there is thus a total of $8 + 2N$ messages transmitted for each page delivered to an idle roaming terminal. (We are again ignoring other message types required to be sent in the event of errors or other abnormal happenings.) As the number of cells in an area increases, effectively increasing the area covered by an MSC, more paging messages are needed. As noted above, this is the opposite effect encountered with location management, where increasing the area covered by an MSC reduced the number of registrations required by a roaming mobile. There thus appears to be an optimum choice of location area size based on minimizing message activity. We provide below one simple approach to determining the "best" choice of location area size. Some references are also provided to papers published on the more general case of reducing location management and paging complexity.

Consider a location area of area A which contains within it N cells, each of area a. Say the area extends, on the average, R m from its center to its border. Call this the "radius" of the area. A circular or hexagonally shaped area would thus have R as the radius. Figure 8.39 shows as an example a square area. Let the cells within it have a corresponding radius of r m. It is then clear that we must have $R = \sqrt{N}r$. Say there are m uniformly distributed mobiles within this area. The average velocity of a mobile is taken to be V m/sec, uniformly distributed over all directions. A little thought will indicate that it takes an average mobile $o(R/V)$ seconds to reach a border ("o" means order of). A typical mobile thus undergoes $o(V/R) = o(V/r\sqrt{N})$ border-crossings/sec. A more detailed analysis, using a fluid-flow model with mobiles represented as infinitesimally small particles of fluid, shows that the average rate of crossing an area of size S is $VL/\pi S$, with L the

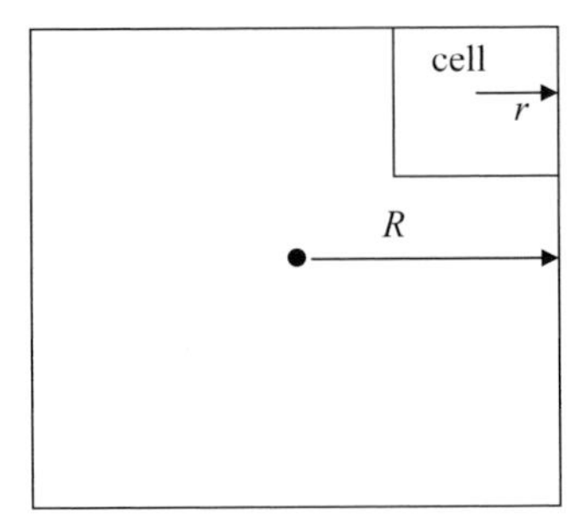

Figure 8.39 Location area

perimeter or boundary length (Jabbari, 1996). For a circle or square (Fig. 8.39) of radius R, this becomes $2V/\pi R$. For a hexagon, this is $2.3V/\pi R$. We shall henceforth use the value $2V/\pi R = 2V/\pi r\sqrt{N}$ for the average rate of location area border crossings per mobile. Note now that if each border crossing results in l location messages being transmitted, the average rate of transmitting location messages is, of course, $2lV/\pi r\sqrt{N}$. (Note that the IS-41 location registration scenario of Fig. 8.37 results in a total of $l = 12$ messages of all types transmitted. If messages transmitted across the air interface only are considered, i.e., those involving the mobile terminal directly, then $l = 2$. The smaller number might be the one used in this analysis if one were concerned solely with the impact of location management on mobile power or processing cost only. We shall, however, focus in our examples below on total messages transmitted.)

Say now that a typical roaming mobile terminal is called, and is hence paged, on the average, λ_p times per second. Let each page require p messages to be transmitted. For the case of Fig. 8.38, $p = 8 + 2N$, as noted above. The average number of paging messages transmitted per unit time is thus $\lambda_p p = \lambda_p(8 + 2N)$. The total number C of messages transmitted per unit time per mobile due to registration and paging is thus given by

$$C = 2lV/\pi r\sqrt{N} + \lambda_p(8 + 2N) \tag{8.6}$$

We now take this number to be the "cost" per mobile of carrying out location management. This number clearly exhibits a minimum value as N, the number of cells per location area, varies. The value of N that results in this minimum cost value is readily found by the usual technique of setting the derivative of C with respect to N equal to zero, if we assume, for simplicity, that N is a continuously varying variable. Carrying out this differentiation, we find the "best" choice of N, the number of cells defining a location or registration area, to be given by

$$N_{opt} = (lV/2\pi r\lambda_p)^{2/3} \tag{8.7}$$

Note, however, that the "optimum" choice of N is to be taken as quite approximate because of the simple model used and approximations made. It thus provides "ballpark" figures only.

Jabbari, B. 1996. "Teletraffic aspects of evolving and next-generation wireless communication networks," *IEEE Personal Communications*, 3, 6 (December), 4–9.

Consider some examples of the use of (8.7). We take three cases: a macrocell example, a medium-sized cell, and a microcell. Let $l = 12$ location messages be required to be transmitted for each border crossing, as suggested by Fig. 8.37. We take the paging rate per mobile to be $\lambda_p = 1$ incoming call per hour. (The effect of doubling this rate will be considered below. Readers can vary this rate and see for themselves its effect on size of location area.) The resultant approximate values of N_{opt} for different mobile speeds and cell sizes are tabulated as follows:

1 *Macrocell*

(1) $r = 10$ km, $V = 60$ km/hr
(2) $r = 15$ km, $V = 100$ km/hr $N_{opt} = 5$ cells per area in both cases

2 *Medium-sized cell* (urban environment), $r = 1$ km

(1) walking case $V = 5$ km/hr $N_{opt} = 4$
(2) vehicle $V = 50$ km/hr $N_{opt} = 20$ or 21

(If the average paging rate is increased to 2 calls/hr per mobile, the numbers of cells per location area decrease to 3 and 13 or 14, respectively.)

3 *Microcell*-$r = 100$ m, $V = 5$ km/hr $N_{opt} = 20$ or 21

These examples demonstrate the expected results, as indicated by intuition and, more specifically, from (8.7): as mobile velocity increases, with cell size fixed, the location area should be increased, to reduce the number of border crossings with attendant registrations required. As cell size is reduced, with mobile speed fixed, the location area size should also increase. As the incoming call rate increases, however, the location area size should be reduced.

These results are, of course, quite approximate, as already noted. They focus on the simplest possible flow model with mobile density assumed uniformly and densely distributed throughout a region. Average velocity and average paging rate only are considered. More detailed studies and analyses have appeared in the literature. Some of these studies have incorporated probabilistic distributions of mobile locations and velocities. A number of different algorithms to reduce location and/or paging cost have been studied and compared as well. Research into location management continues to be an active field. Interested readers are referred to papers incorporating some of these studies and analyses, as well as to the references contained within them. Examples of such papers include Wang (1993), Mohan and Jain (1994), Bar-Noy *et al.* (1995), Akyildiz and Ho (1995), Jain and Lin (1995), Rose and Yates (1995), Lin (1997), Akyildiz *et al.* (1998).

Wang, J. Z. 1993. "A fully distributed location registration strategy for universal personal communications systems," *IEEE Journal on Selected Areas in Communications*, 11, 6 (August), 850–860.

Mohan, S. and R. Jain. 1994. "Two user location strategies for personal communication services," *IEEE Personal Communications*, 1, 1 (1st Qtr.), 42–50.

Bar-Noy, A. *et al.* 1995. "Mobile users: to update or not to update?" *Wireless Networks*, 1, 2 (July), 175–185.

Akyildiz, I. F. and J. S. M. Ho. 1995. "Dynamic mobile user location update for wireless PCS networks," *Wireless Networks*, 1, 2 (July), 187–196.

Jain, R. and Y.-B. Lin. 1995. "An auxiliary location strategy employing forwarding pointers to reduce network impacts of PCS," *Wireless Networks*, 1, 2 (July), 197–210.

Rose, C. and R. Yates. 1995. "Minimizing the average cost of paging under delay constraints," *Wireless Networks*, 1, 2 (July), 211–219.

Lin, Y.-B. 1997. "Reducing location update costs in a PCS network," *IEEE/ACM Transactions on Networking*, 5, 1 (February), 25–33.

Akyildiz, I. F. *et al.* 1998. "Mobility management in current and future communication networks," *IEEE Network*, 12, 4 (July/August), 39–49.

8.5 Voice signal processing and coding

We have focused thus far in this chapter on signaling messages in second-generation systems and how they are transmitted via signaling channels. In this section we finally consider the major application of these systems, that of transmitting voice signals. Voice signals are transmitted at considerably reduced rates compared with the rates used in wired digital telephone systems. This is necessary because of the relatively low bandwidths available in wireless cellular systems. The harsh transmission environment involving fading and interference from mobile terminals requires strong error protection through coding as well. Two steps are therefore necessary in transmitting voice messages over the wireless air interface. The voice signals must first be compressed significantly to reduce the bit rate required for transmission. Coding techniques must then be used to provide the error protection needed. Both steps must clearly result in voice signals that are acceptable to a receiving user.

Voice signal compression has been an ongoing concern for many years. Multiple techniques have been proposed, investigated, standardized, and implemented. Early digital telephony standardized on 64 kbps PCM voice transmission, attained by sampling 3.2 kHz bandlimited voice 8000 times per second and then using eight bits per sample to provide acceptable speech quality for telephone applications. This technique relies on application of the Nyquist sampling theorem, which states that by sampling a bandlimited signal at a rate at least twice the bandwidth the signal can be retrieved with no distortion (Schwartz, 1990). The quantization process, using eight bits or 256 levels of signal amplitude in this case, does introduce some distortion labeled quantization noise (Schwartz, 1990). Differential PCM (DPCM), in which a current voice sample is compared with a weighted (filtered) sum of previous samples and the difference signal then quantized and transmitted, has been used effectively to reduce the telephone voice transmission bit rate to 32 kbps. To go lower in transmission bit rate requires more sophisticated and complex signal processing.

Why is it even possible to compress the transmission bit rate of voice? It is well-known that a great deal of redundancy exists in voice communication. As an example, listeners can often make up for the loss of a spoken word, or of a portion thereof, by noting the context in which these appear. Early experiments with 8000 samples per second PCM showed that native listeners could generally understand voice quantized to only 1 bit per sample. (However, the "noise" level was of course very high!) The objective is then to reduce this redundancy as much as possible, while still maintaining a reasonable subjective quality of the signal resulting. There is an additional tradeoff required in the case of digital voice signals transmitted over harsh media such as those encountered in wireless cellular systems. This has already been noted above: as signal redundancy is reduced, its susceptibility to fading, noise, and interference introduced over the air interface becomes greater. It is thus necessary to code the transmitted digital signal, increasing the transmission bit rate required, and effectively undoing some of the redundancy reduction. The particular signal compression technique used thus becomes critical.

The second-generation cellular systems, choosing to compress voice signals significantly to uncoded data rates ranging from 8 kbps in the case of IS-136 to 13 kbps in

---

Schwartz, M. 1990. *Information Transmission, Modulation, and Noise*, 4th edn, New York, McGraw-Hill.

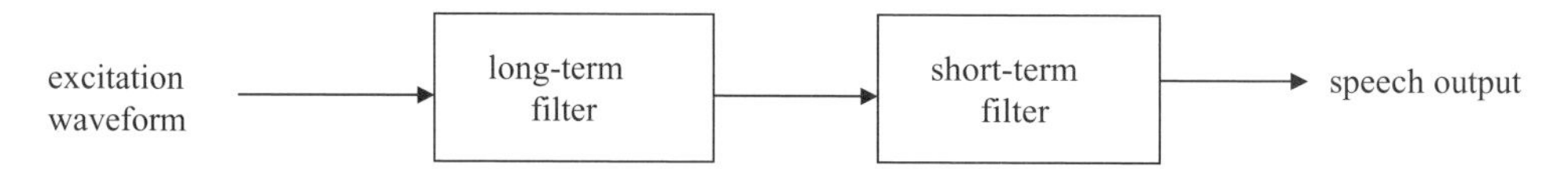

Figure 8.40 LPC speech model

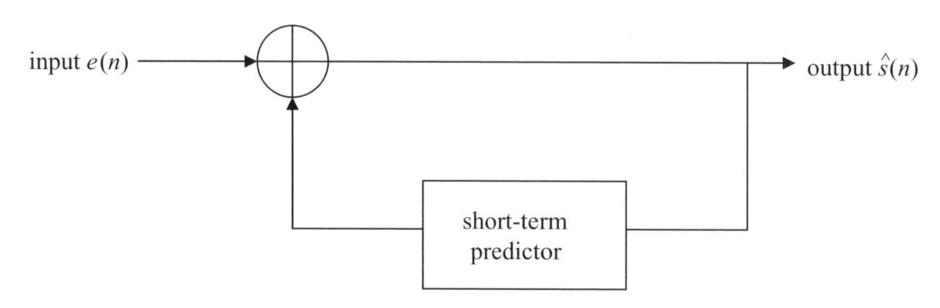

Figure 8.41 Short-term filter: linear prediction

the case of GSM, have all adopted variations of a compression technique called *linear predictive coding* (LPC). LPC basically starts by modeling the human speech generation process as an excitation waveform driving a set of linear filters (Atal and Hanauer, 1971; Jayant, 1992; Rabiner and Hwang, 1993). The LPC speech model is depicted in Fig. 8.40. The excitation waveform and long-term filter model speech sounds generated by the lungs and vocal cords; the short-term filter represents the vocal tract. Two types of sound are generally recognized: voiced and unvoiced. Voiced sounds due to the opening and closing of the vocal cords tend to be quasi-periodic in form; unvoiced sounds are due to turbulent air created by constrictions in the vocal tract. The excitation waveforms appearing in the LPC model of Fig. 8.40 would thus be a combination of periodic pulses and a noise-like (random) signal. The term "long-term" used in describing the first of the two filters in Fig. 8.40 refers roughly to the *pitch period* of the voiced sounds due to the vibration of the vocal cords. The pitch period tends to be in the order of 3–15 msec (Rabiner and Hwang, 1993). "Short-term" implies times in the order of 1 msec.

Why the term "linear predictive coding"? The output of the LPC speech model at a given sampling time is based on the weighted sum of previous samples, plus a new value due to the sampled input. Specifically, focus on the short-term filter of Fig. 8.40. A simple diagram of this filter appears in Fig. 8.41. The output $\hat{s}(n)$ at sampling interval n is given by adding the sampled input $e(n)$ to the output of a linear predictor, as indicated in Fig. 8.41. The linear predictor model may be written quite simply as

$$\hat{s}(n) = \sum_{i=1}^{m} a_i \hat{s}(n-i) + e(n) \tag{8.8}$$

The current value of the output is thus based on m samples in the past, each spaced 125 μsec apart, as we have previously noted. The a_is are weighting factors, found adaptively,

Atal, B. S. and S. L. Hanauer. 1971. "Speech analysis and synthesis by linear prediction of the speech wave," *Journal of the Acoustic Society of America*, 50, 2 (August), 637–655.

Jayant, N. 1992. "Signal compression: technology targets and research directions," *IEEE Journal on Selected Areas in Communications*, 10, 5 (June), 796–818.

Rabiner, L. and B.-H. Hwang. 1993. *Fundamentals of Speech Recognition*, Englewood Cliffs, NJ, Prentice-Hall.

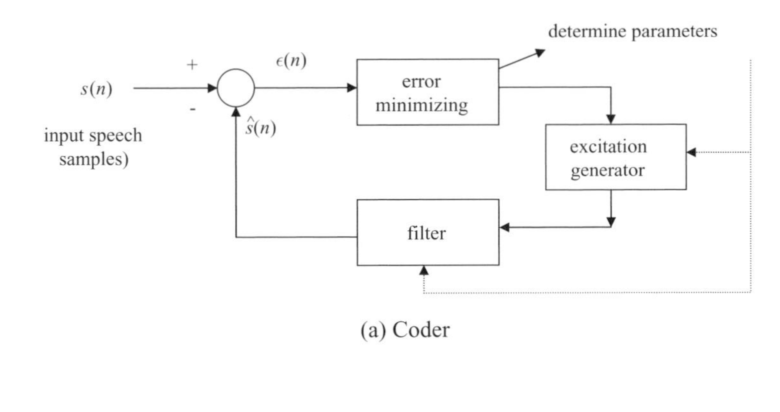

(a) Coder

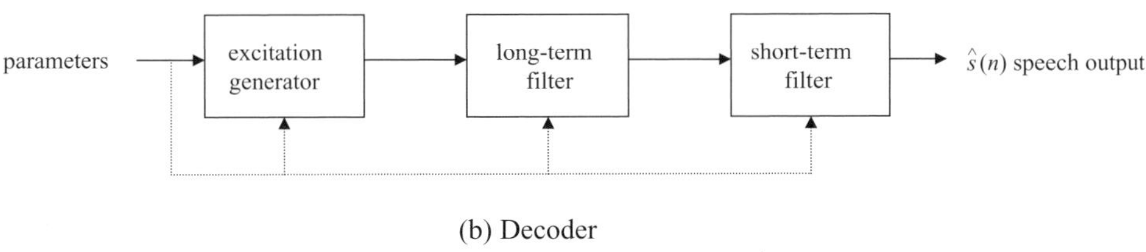

(b) Decoder

Figure 8.42 LPC system

as we shall see. Taking the z-transform of (8.8), it is left for the reader to show that the resultant transfer function, $H(z) = S(z)/E(z)$ is given by

$$H(z) = 1\bigg/\left(1 - \sum_{i=1}^{m} a_i z^{-i}\right) \tag{8.9}$$

This transfer function is exactly that of an all-pole linear filter. The number of samples m is typically ten or so, meaning the filter operates over the relatively "short" time of 1 msec, as already noted. The model for the "long-term" filter is usually taken to be that of an all-pole filter also, but with only one or two poles. Specifically, say it is represented by a one-pole filter. Its transfer function is thus clearly given by

$$H(z) = 1/(1 - bz^{-p}) \tag{8.10}$$

The parameter p is just the pitch period in multiples of sampling time, to be adjusted, by measurement, periodically. It would thus be in the range of 20 to 120 or so, corresponding to 2.5 to 15 msec in actual time, as noted above. The parameter b is a weighting factor.

Using the LPC model of speech generation given by Figs. 8.40 and 8.41, the basic LPC system used to compress speech is given by the block diagrams of Fig. 8.42 (see also Steele, 1992: 227). This system compares the output of the model with the actual speech samples and attempts to minimize the difference (error) signal by adjusting the excitation and filter parameters periodically. It shows both a *coder* at the speech generating side (Fig. 8.42(a)) and a *decoder* at the receiving side (Fig. 8.42(b)). The combined system is normally called a speech *codec*. Consider the coder first. Quantized input speech samples labeled $s(n)$ are generated every 125 μsec. The difference $\varepsilon(n)$ between these and the

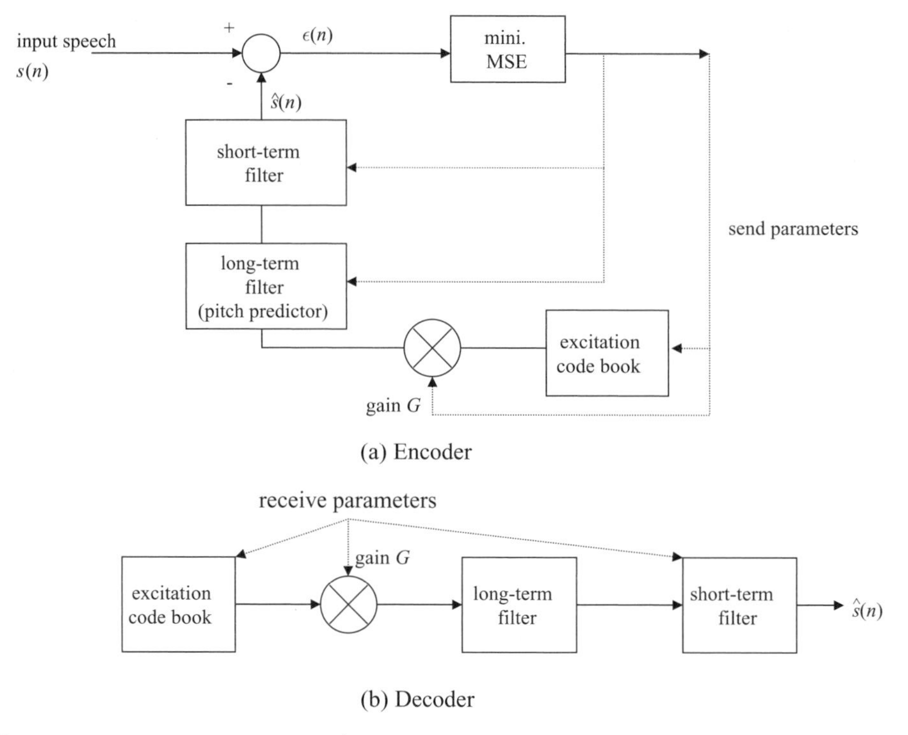

Figure 8.43 CELP system

speech model output $\hat{s}(n)$ is minimized by adjusting the excitation generator and filter parameters. (Normally mean-squared error minimization, over several samples, would be carried out. This corresponds to minimizing $\sum_n \varepsilon^2(n)$.) The resultant *parameters* are then transmitted at sample intervals to the receiving system. The receiving system, the decoder, then carries out the inverse process, using the parameters received to adjust the excitation generator and filters. As we shall see below, sending the parameters periodically, rather than voice samples or their differences, results in a sizeable reduction in the transmission bit rate. The codecs used in second-generation systems use variations of this system, as noted above.

In particular, GSM uses a scheme called *LPC with regular pulse excitation* (LPC-RPE). IS-136 and IS-95 use as their voice codecs variations of the *code-excited linear predictive coding* (CELP) scheme. We now describe these different coding–decoding techniques. Consider the CELP technique first (Schroeder and Atal, 1985; Chen *et al.*, 1992; Steele, 1992).A block diagram of the basic CELP scheme appears in Fig. 8.43. (Not appearing are weighting filters sometimes used to weight input signal samples and/or the error samples.) Note that the system looks very much like the LPC codec representation of Fig. 8.42. The basic difference is that the excitation generator of Fig. 8.43 is given by a selected set

---

Schroeder, M. R. and B. S. Atal. 1985. "Code-excited linear prediction (CELP): High quality speech at very low bit rates," Proc. IEEE Int. Conf. Acoust. Speech, Signal Process., March, 937– 950.

Chen, J.-H. *et al.* 1992. "A low-delay CELP coder for the CCITT 16 kb/s Speech Coding Standard," *IEEE Journal on Selected Areas in Communications*, 10, 5 (June), 830–849.

of gaussian (i.e., random) codewords. The parameter G shown represents a gain factor adjusting the amplitude of the codewords selected. The long-term filter of Fig. 8.43 is commonly designed as the single-pole filter of equation (8.10) with transfer function $1/(1-bz^{-p})$. As previously, the parameter p represents the estimated pitch period of the quasi-periodic voiced sounds. The parameter b is an adjustable weighting factor. The combination of random codewords and this long-term filter serve to model the vocal-cord (glottal) excitation, as noted in discussing LPC. The short-term filter again models the vocal tract.

CELP uses vector quantization (Rabiner and Hwang, 1993): This is a technique in which parameter adjustments are made periodically, after a number of quantized error samples, constituting a vector, have been collected. Vector quantization has been shown to provide compression advantages. As an example, say adjustments are made every 5 msec, with speech sampling carried out every 125 μsec. Vectors of 40 samples each are thus generated every 5 msec. The excitation codebook then consists of a suitable number of random (gaussian) codewords, each a quantized 40-element vector. Mean-square error minimization over the filter parameters, gains, and codewords, as shown in Fig. 8.43, is then used to determine the "best" parameters and codewords each interval. It is these parameters that are sent out over the air interface voice channel and used in the decoder to reconstitute the speech signal, as shown in Fig. 8.43b. As noted above, transmission of these parameters rather than a low-bit version of the actual speech signal results in considerable reduction of the voice channel bit rate. Consider a simple example, taken roughly from the IS-136 codec to be considered below. Say the short-term filter is a 10-pole all-pole filter with transfer function $1/(1-\sum_{i}^{10} a_i z^{-i})$. This filter then has ten coefficients to be determined. Say these coefficients are sent every 20 msec, and 4-bit accuracy only is needed to characterize these ten numbers. The coder thus need only send 40 bits every 20 msec to represent them. Let the other parameters, the pitch p, weight b, gain G, and the codeword selected be represented by seven bits each, to be collected every 5 msec. This corresponds to 28 bits every 5 msec. The total number of bits sent every 20 msec is thus 152 in number, for an aggregate bit rate of 7600 bps. (We shall see below that IS-136 sends 159 bits every 20 msec over its voice channel, for an aggregate bit rate of 7950 bps.) Substantial reduction in bit rate is thus possible, using vector-quantized CELP. With this introduction on generic voice compression techniques, we now focus specifically on the voice-compression/coding techniques adopted by each of the three second-generation cellular systems.

IS-136

The North American second-generation cellular system IS-136 uses for its voice-compression system a variant of CELP called *vector-sum excited linear predictive coding* (VSELP). In this system analog speech is sampled at the 8000 samples/sec rate noted a number of times previously. It is thus sampled at 125 μsec intervals. The samples are, however, quantized to 13 bits/sample, rather than the eight bits/sample used in wired digital telephony. A block diagram of VSELP appears in Fig. 8.44 (TIA, 1992). Note

---

TIA 1992. "EIA/TIA Interim standard, cellular system dual-mode mobile station-base station compatibility standard," 15-54-B, Telecommunication Industry Association.

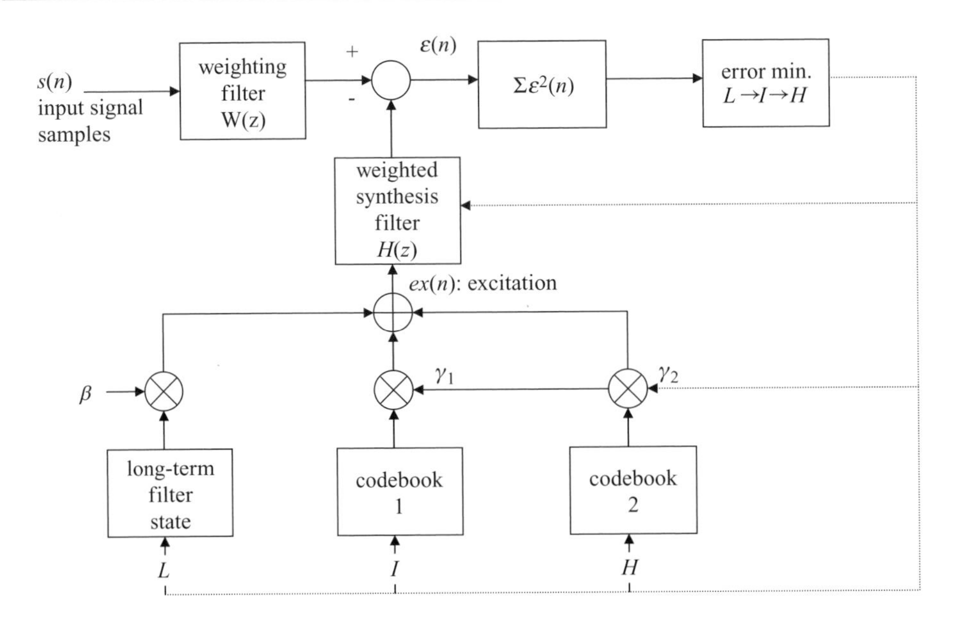

(a) coder

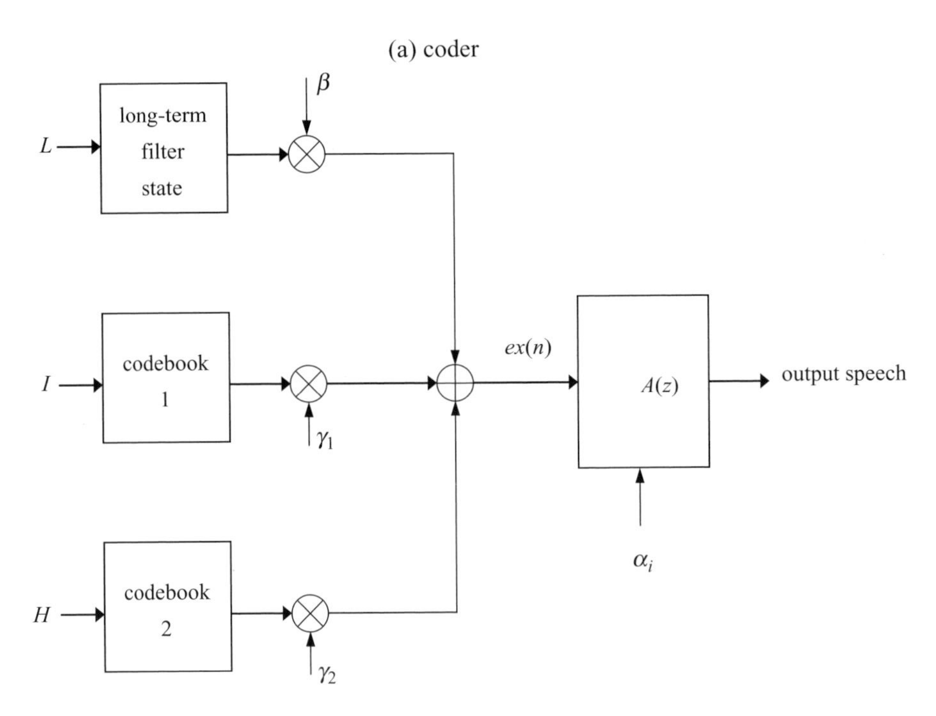

(b) decoder

Figure 8.44 IS-136 codec: VSELP system

from Fig. 8.44 how similar the VSELP codec is to the basic CELP codec of Fig. 8.43. The prime difference is in the use of two stochastic codebooks, rather than one as in CELP. The long-term filter shown is represented by the single-pole filter of (8.10). Its pitch parameter (also called the lag parameter) is given by the parameter L indicated in Fig. 8.44, while the gain parameter β in Fig. 8.44 corresponds to the parameter b in (8.10). L ranges in value from 20 to 146, in units of the sampling interval of 125 μsec, corresponding to 127 possible values. It can thus be represented by seven bits. The weighted synthesis filter $H(z)$ of the encoder and the filter $A(z)$ both correspond to the all-pole transfer function of (8.9), each with $m = 10$ weighting parameters labeled α_i, $i = 1 - 10$, as indicated in Fig. 8.44. Signal samples are weighted as well by the weighting filter $W(z)$ shown. This filter is structured with both zeros and poles, using the same parameters α_i in both the numerator (zeros) and denominator (poles) of $W(z)$. Codewords in each codebook are given by 40-element random vectors, defined in terms of seven 40-element random basis vectors chosen for each codebook. There are thus $2^7 = 128$ possible codewords in each codebook from which to choose an appropriate excitation vector.

Operation of the codec proceeds as follows: the error term $\varepsilon(n)$ is squared and summed over 5 msec intervals, corresponding to 40 signal samples. This is then minimized over the various parameters and codewords, minimizing first with respect to the long-term filter as the only excitation, and then moving sequentially on to the codewords, shown as parameters I and H in Fig. 8.44. The parameters and codewords minimizing the squared error are then transmitted over the air interface between base station and mobile to the receiver decoder shown in Fig. 8.44(b). It is again worth stressing at this point that compressed signal samples or differences in such samples are *not* transmitted over the air interface, as would be the case for PCM or versions of DPCM. It is the adaptively varying CELP *parameters* that are transmitted to the decoder. This decoder of Fig. 8.44(b), just as the one in Fig. 8.43(b) for the CELP system, reproduces the speech model of the encoder of Fig. 8.44(a). The parameters and codewords are actually transmitted every 20 msec or every four 5 msec intervals. Why is a 20 msec interval chosen for sending the VSELP parameters? Recall from our discussion of IS-136 in Section 8.2 that two slots per 40-msec frame, each slot carrying 260 data bits, are used to provide full-rate voice transmission (see also Fig. 8.9). The transmission of VSELP parameters at 20 msec intervals thus provides codec adaptation twice each IS-136 frame. These parameters are sent as 159 bits every 20 msec, as noted above, corresponding to a raw voice bit rate of 7.95 kbps, the IS-136 compressed bit rate mentioned earlier. Coding to be described below results in a coded bit rate for voice of 13 kbps, however. The 159 bits are divided among the various parameters and codewords as follows: the three gains β, γ_1, and γ_2 are allocated a total of eight bits per 5 msec sub-interval, or 32 bits for the 20 msec interval. The two codewords I and H each require seven bits per 5 msec sub-interval, as indicated above, for a total of 56 bits each 20 msec interval. Seven bits per 5 msec sub-interval are needed to transmit the parameter L, as already noted above, for a total of 28 bits over 20 msec. Thirty-eight bits per 20 msec are used to transmit the ten filter coefficients. Finally, the energy per 20 msec frame is transmitted as well, using five bits. The total is thus 159 bits every 20 msec, as stated above.

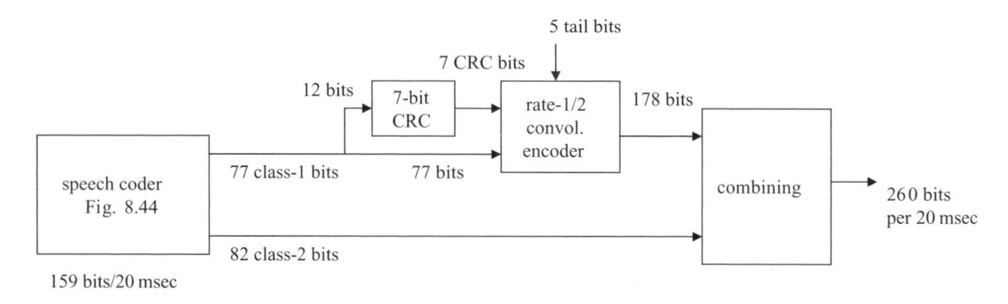

Figure 8.45 IS-136, voice coder encoding

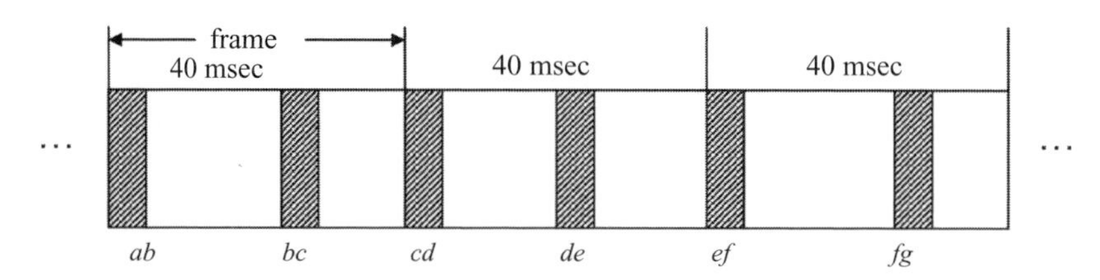

Figure 8.46 Effect of interleaving, IS-136

These 159 bits must be protected before transmission to mitigate against possible channel errors. Three protection mechanisms are used. Forward error control, using cyclic redundancy coding (CRC) and convolutional coding, on some of the bits, added to the remainder of the uncontrolled bits, is used to convert the 159 raw parameter bits to the 260 data field bits to be transmitted per 20 msec interval, the number cited above. (Details of this forward error control procedure appear below.) These 260 bits are then interleaved over two slots of the 40 msec IS-136 frame. Figure 8.45 portrays the full error control procedure. This error control procedure adopted for voice traffic is to be compared with that portrayed in Fig. 8.12 for IS-136 control information. Describing Fig. 8.45 in more detail, note that the 159 bits are first divided into two classes, 77 Class-1 and 82 Class-2 bits, as shown in the figure. Class-1 bits only are protected by the error control mechanisms. As examples, the codeword parameters are all sent as Class-2, unprotected bits. The L parameters are all sent as Class-1 bits. The other parameters are sent as mixed-class bits. Of the 77 Class-1 bits, 12 are selected as the most perceptually significant bits. The 12 bits are chosen from the energy parameter and five of the ten filter coefficients. These 12 bits first have seven CRC bits added to provide some forward-error control, as shown in Fig. 8.45. The resultant CRC-encoded sequence and the remaining Class-1 bits are fed into a rate-1/2 convolutional encoder, to which five tail bits are added, to form 89 bits entering the convolutional coder. The resultant 178 bits exiting the convolutional coder are then added to the 82 unprotected Class-2 bits, as shown in Fig. 8.45, to form the 260-bit output sequence generated every 20 msec.

These 260 bits are now interleaved over two data traffic slots (Fig. 8.9) to further protect the bits, as stated above. The interleaving procedure would appear as shown in Fig. 8.46. Say the successive coded 260-bit sequences are labeled a, b, c, d, e, f, g, ...

Half of each sequence then appears in each of two successive slots assigned to a particular voice channel. (Recall that a 6-slot IS-136 frame lasting 40 msec has two slots per frame assigned to each of three channels – see Fig. 8.9.) Figure 8.46 shows the second half of the 260 *a* bits plus the first half of the 260 *b* bits appearing in one slot; the second half of the *b* bits and the first half of the *c* bits assigned to the next slot of the same channel, etc. The 260 coded bits of each voice user are thus sent in two successive slots, spaced three slots apart.

IS-95

The voice-compression technique used in IS-95 is similar to the one just described for IS-136, except that one codebook only is used (Garg, 2000). It uses CELP-type technology as shown in Fig. 8.43. Speech samples are again taken at 125 μsec intervals, quantized in this case to at least 14 bits/sample, and then compared with the output of LPC-type filters whose input is a random-type vector chosen from an excitation error codebook. Minimum mean-square error determination is carried out every 20 msec, as in the case of IS-136, and the CELP system parameters found to minimize the mean-square are transmitted to the decoder at the receiving system, as indicated in Fig. 8.43, and as described as well in our discussion of the IS-136 VSELP system.

As noted in Section 8.3, devoted to IS-95, a number of voice bit rates are available for IS-95. We focus again on the full bit rate implementation only. Consider Figs. 8.18 and 8.19, which provide the block diagrams of the full-rate IS-95 reverse and forward traffic channels respectively. They indicate that 172 bits every 20 msec, defined as a frame interval, provide the input to these systems. It is these input bits that correspond to the CELP parameters to be transmitted to the decoder over the air interface after further coding and modulation, as shown in these figures. The forward-error correction, convolutional encoding, and block interleaving functions diagrammed in these figures, and discussed in Section 8.3, provide the equivalent of the voice protection functions just described for IS-136, as diagrammed in Fig. 8.45.

More precisely, 171 bits every 20 msec-frame, are outputted by the CELP coder, for an IS-95 speech bit rate of 8.55 kbps. Recall from Table 8.3 of Section 8.3 that the IS-95 full-rate traffic channel may be used to carry both primary and signaling traffic. We consider the case in which the channel carries primary traffic only. From Table 8.3, then, 171 bits per 20 msec frame are actually sent, since one bit is used to designate this case. The 171 bits sent are distributed, as in the case of the IS-136 VSELP coder, among the various CELP system parameters. The IS-95 CELP coder again uses a single-pole long-term filter. Its lag or pitch parameter L is updated, as in the IS-136 case, every 5 msec, or four times a frame, and seven bits per update transmitted for a total of 28 bits. Three bits per 5 msec sub-interval, or a total of 12 bits per frame, are used to transmit the long-term filter gain coefficient. The short-term filter coefficients are transmitted as 40 bits every frame. The excitation codeword and the gain G (Fig. 8.43) are updated eight times per frame, with seven bits and three bits, respectively, used to represent these updated parameters each time. The codebook transmission requirement is thus 80 bits per 20 msec frame. The total

---

Garg, V. K. 2000. *IS-95 CDMA and cdma2000*, Upper Saddle River, NJ, Prentice-Hall PTR.

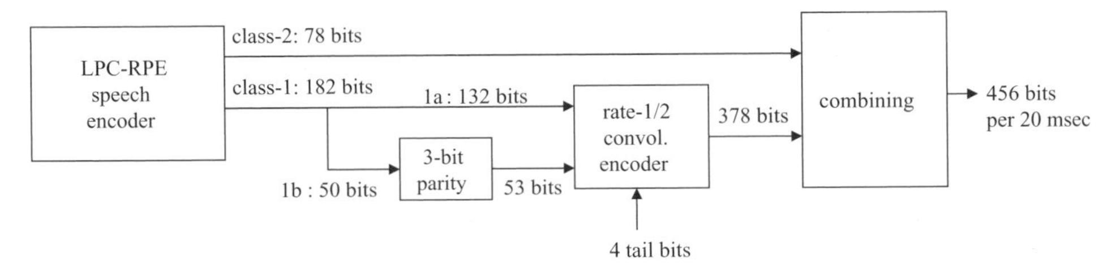

Figure 8.47 GSM, voice signal protection encoding

number of bits required to transmit these CELP parameters is thus 160 bits per frame. However, for further error protection, over and above that provided as shown earlier in Figs. 8.18 and 8.19, 11 parity bits per frame are calculated and appended to the 160 bits, for the total of 171 bits noted above.

GSM

We conclude this section on voice-signal compression and coding adopted for the second-generation digital cellular systems by describing the system used in GSM. As mentioned earlier, the GSM voice-compression scheme is called *linear predictive coding with regular pulse excitation* (LPC-RPE). It is clear from the name that LPC serves as the basis for this scheme, just as it did for the other two compression procedures just described. As such, it follows the basic LPC form of Fig. 8.42. The difference between the GSM technique and the CELP-based techniques of the other two cellular systems lies in the choice of excitation generator. GSM uses regularly spaced pulses of varying amplitude to provide the filter excitation, rather than the CELP random-codeword approach adopted by the other systems. The amplitudes of these pulses are chosen to minimize the mean-squared error, as in Fig. 8.42(a). This GSM approach pre-dates the other two schemes and provides somewhat less compression efficiency: 260 bits per 20 msec interval are generated at the speech coder, for a 13 kbps effective speech transmission rate. More recently, enhanced full rate and half-rate CELP-type coders have also been standardized for GSM. The half-rate system operating at 5.6 kbps is of the VSELP type (Garg and Wilkes, 1999).

Details of the LPC-RPE scheme appear in (Steele, 1992: Ch. 3 and 8). The long-term prediction filter used is again that of the single-pole type. The delay (pitch) and gain parameters are determined, using minimum mean-square error estimation, every 5 msec, and seven and two bits, respectively, are used to represent the delay and gain parameters. This accounts for 36 of the 260 bits assembled every 20 msec. Estimation of the eight coefficients of the short-term filter every 20 msec accounts for another 36 bits. The remaining 188 bits, 47 bits every 5 msec, represent the parameters defining the appropriate excitation signal for that interval (Garg and Wilkes, 1999; Steele, 1992: Ch. 8).

Error protection following the LPC-RPE speech encoding process is portrayed in Fig. 8.47 (Goodman, 1997; Steele, 1992: Ch. 8). The error protection procedure is followed

---

Garg, V. K. and J. E. Wilkes. 1999. *Principles and Applications of GSM*, Upper Saddle River, NJ, Prentice-Hall PTR.

by an interleaving operation. This protection procedure is to be compared with that diagrammed in Figs. 8.6 to 8.8 for the various GSM control channels. As in the IS-136 case described earlier (Fig. 8.45), the 260-bit speech encoder output appearing every 20 msec is broken into two classes of bits. Of the 260 bits, 182 bits are designated as Class-1, significant, bits; 78 bits are designated Class-2, non-essential bits. No error protection is provided to the non-essential bits. The Class-1 bits are further broken into two groups, as in the IS-136 case: 50 of the 182 bits are first protected, weakly in this case, by three parity check bits used to detect, but not correct, errors. The 53 bits are then combined with the remaining 132 bits and convolutionally encoded in a rate-1/2 encoder, with four tail bits added. The resultant block diagram appears in Fig. 8.47. The 378 bits appearing at the convolutional coder output every 20 msec plus the unprotected 78 bits provide 456 bits every 20 msec, for a voice traffic ouput rate of 22.8 kbps. Recall from Section 8.1 that this is precisely the voice traffic bit rate adopted for GSM. (As shown in Fig. 8.2, 114 data bits are transmitted in each traffic slot. A full-rate user occupies one such slot every frame, repeating at 4.615 msec intervals. One frame in 13 is, however, "robbed" for control purposes.) An interleaving operation is carried out following the protection encoding operations just described to provide further signal protection. It is to be noted that 456 is just 8 times 57. In the interleaving operation, 57 bits from each of two 456 bit-long speech blocks, covering 40 msec of time, are combined to fill time slots in eight successive frames (Goodman, 1997; Garg and Wilkes, 1999; Steele, 1992: Ch. 8).

Problems

8.1 Figure 8.3 shows how a GSM call is set up, using the various control channels described in the text and indicated in Table 8.1. The various steps required to set up a call, as shown in Fig. 8.3, are to be mapped, in time, on to timing diagrams such as those shown in Figs. 8.4 and 8.5. For this purpose, prepare two sets of multi-frame timing diagrams, one below the other. One represents transmission from a given mobile attempting to establish a call; the other corresponds to the frame timing for the base station to which the mobile will transmit its messages uplink. Indicate on these two diagrams the various control signals transmitted to and from the mobile, as they appear in each of the frames, as the call is set up.

8.2 Compare the IS-136 control channels shown in Table 8.2 with those of GSM shown in Table 8.1. Consider, in the comparison, the function of a particular control channel, as well as its implementation in the time-slot/frame structure. Which control channels are similar, which different?

8.3 **(a)** Provide a diagram such as that of Fig. 8.3 showing how a call is set up in IS-136. This will require augmenting the access procedure of Fig. 8.15 to include a mobile powering on and obtaining necessary synchronization and access information.

(b) Repeat problem 8.1 above for IS-136, mapping the diagram of (a) to the appropriate multi-frame timing diagrams.

(c) Compare the GSM and IS-136 call set-up procedures.

8.4 Consider the IS-95 traffic channel block diagrams of Figs. 8.18 and 8.19.

(a) Explain why Fig. 8.19 represents a QPSK-modulated system; Fig. 8.18 an OQPSK-modulated system.

(b) Show how the various data bit rates such as 9.2 kbps, 28.8 kbps, 307.2 kbps, and 19.2 kbps, appearing at the tops of both figures, are obtained.

(c) Why does the downlink direction of transmission require less error correction capability than the uplink direction?

(d) Explain why block interleaving reduces the effect of burst errors.

(e) Show how puncturing the 19.2 kbps data sequence of Fig. 8.19 results in an 800 bps power control signal.

8.5 **(a)** Show the Walsh code $\mathbf{W}_8$ is given by (8.4). Find the Walsh code $\mathbf{W}_{16}$.

(b) Demonstrate the validity of the orthogonality relation (8.5) for the two Walsh codes of (a). Show, by inspection, that both codes have half the entries 0, half 1, while all the rows differ in four and eight places, respectively, as indicated in the text.

8.6 **(a)** Explain the details of Fig. 8.23, focusing on the functions of each block, as well as showing how the various bit rates indicated are determined.

(b) Referring to Fig. 8.25, showing the procedure followed by an IS-95 mobile, from powering up to call initiation, explain what information the mobile derives from the pilot, sync, and paging channels.

(c) Compare the IS-95 call initiation and origination procedures of Figs. 8.25 and 8.26 with those of GSM and IS-136 (D-AMPS).

8.7 Figure 8.27 provides an example of the frame structure occurring for one example of a dim-and-burst frame. Repeat this figure for some other examples chosen from Table 8.3.

8.8 Figure 8.29 provides an example of IS-95 multi-frame signaling. Show the sequence of frames required to transmit a signaling message 2016 bits long using blank-and-burst signaling. Indicate the allocation of bits in each frame, as shown in Fig. 8.29(b). Repeat for a choice of dim-and-burst signaling.

8.9 Explain in your own words the distinction between hard and soft handoff. Why does soft handoff provide a performance improvement in CDMA systems? (See Chapter 6.)

8.10 Figure 8.36 diagrams the exchange of messages required to be transmitted when an inter-system handoff-forward procedure is carried out. Reference is made in the text as well to inter-system handoff-back and handoff-to-third procedures. Provide possible message-exchange diagrams for each of these other procedures.

8.11 Let r be the radius in meters of a cell, R the radius in meters of a location area containing N cells. Demonstrate why a typical mobile moving at an average speed of V m/sec undergoes, on the average, $o(V/r\sqrt{N})$ location-border crossings/sec. (The expression o() means "order of".)

8.12 **(a)** Show, for the simple location area tradeoff discussion in the text, that (8.7) provides an estimate of the "optimum" number of cells per location area.

(b) Verify the resultant number of cells per location area is as indicated in the text for the examples chosen there. Repeat the calculations for various other choices of cell size, mobile speed, and paging rate.

8.13 **(a)** Show that the linear-predictor model, Fig. 8.41, for the short-term filter of the LPC speech model of Fig. 8.40, has the form of (8.8).

(b) Show that the z-transform transfer function of the filter of (a) is given by (8.9). Explain the statement, that with $m = 10$ samples used for this filter, the filter operates over a time of approximately 1 msec.

(c) Explain why, in the CELP system of Fig. 8.43, 40-element vectors are generated if adjustments are made every 5 msec.

(d) Carry out the calculation indicated for the example of the CELP codec described in the text and show that a transmission bit rate of 7600 bps results.

8.14 **(a)** Consider the IS-136 VSELP voice codec of Fig. 8.44. Carry out the calculations indicated in the text and show, first, how an uncoded voice bit rate of 7.95 kbps is obtained, followed by the coded bit rate of 13 kbps indicated in Fig. 8.45.

(b) Repeat (a) for the GSM speech-encoder discussed in the text, with protection encoding carried out as shown in Fig. 8.47. Calculate, first, the uncoded voice bit rate and then show the actual transmitted bit rate is 22.8 kbps

Chapter 9

Performance analysis: admission control and handoffs

We have, in earlier chapters in this book, explicitly discussed the concept of the *capacity* of a cellular system. In particular, in Chapter 2 we used the well-known Erlang-B formula to relate the blocking probability of a call attempt to traffic intensity (the call arrival rate times the call holding time) and the capacity, in number of channels, each capable of handling one call. In Chapter 3, we showed how the introduction of the cellular concept increased the overall system capacity, in terms of the number of simultaneous calls that could be handled. We further showed in Chapter 4 that Dynamic Channel Assignment, DCA, could, in principle, be used to increase system capacity for the case of moderate traffic intensity. In Chapter 6 we then compared the capacity of three second-generation digital cellular systems, IS-136, IS-95, and GSM. In that chapter we noted as well that mobile calls had to be *handed off* to adjacent cells as mobiles moved through a specified system. In particular, we noted that the CDMA-based IS-95 offered an improved performance because of the possibility of using *soft* handoff. We followed up on this introduction to handoffs by commenting on the actual process of handoff in Chapter 8.

In this chapter we quantify our discussion of the handoff process by describing cellular *admission control*: new call attempts are accepted in a given cell only if capacity is available to handle the call. Otherwise the call attempt is blocked. Handoffs to a new cell may be handled the same way, in which case their probability of blocking, termed handoff dropping, is the same as that for new calls, or they may be provided some priority, to reduce the probability of dropping. This latter approach is preferred, if possible, since it is universally agreed that customers do not like to have calls interrupted in the middle of a conversation. There is a tradeoff if handoff priority is invoked, however: for a given capacity, reducing the probability of dropping a handoff call must result in an increase in new-call blocking probability. If the resultant new-call blocking probability increases only incrementally, while the handoff dropping probability is reduced significantly, there is clearly an advantage to pursuing such a technique. A number of procedures for handling handoffs have been proposed in the literature and we shall describe, as well as compare, some of them. One scheme is better than another if it results in a lower handoff dropping probability for a given blocking probability. Conversely, with the handoff dropping probability constrained to a specified maximum value, one scheme is better than another if it results in a lower blocking probability.

A common handoff priority scheme we shall study is one in which a specified number of channels is set aside for the exclusive use of handoffs. The number to be set aside can be made adaptable with traffic intensity to satisfy a given handoff dropping/blocking-probability combination. This priority strategy is often termed a *guard-channel* approach, for obvious reasons. Another procedure proposed in the literature is one in which neighboring cells send each other periodically an indication of their channel utilization. By predicting ahead, a given cell can determine the chance of a newly admitted call being denied service in a neighboring cell if it is subsequently handed off. If that probability turns out to be above a given threshold, it is better to deny service to the new call in the first place. Calculations indicate that this strategy provides an improvement over the guard-channel scheme, but it does require periodic communication between cells. Other strategies have been proposed in the literature as well. One simple scheme studied is that of buffering handoff calls up to some maximum time if no channel is initially available. The handoff dropping probability does of course reduce as a result, at the cost of a delay in continuing service. If this delay is not too high, it may be acceptable to the participants in an ongoing call.

Studying and comparing these various schemes requires developing and analyzing performance models for cellular systems. We begin in Section 9.1 by studying simple models for call duration and handoff times, independent of cell geometry. In Section 9.2 we then study performance issues in one-dimensional cell structures. A number of mobility models are introduced and compared in carrying out the analysis in this case. These are then used to compare some of the admission control schemes mentioned above. In Section 9.3 we extend the analysis to the more general and hence more complex case of two-dimensional cells. These increasingly more-complex models complement and extend the limited performance modeling and analysis discussed in earlier chapters that were noted above. Just as was the case in our discussions in previous chapters, we shall be assuming uniform, homogeneous mobile densities and traffic conditions prevail throughout a system. A byproduct of the studies in this chapter should be an improved understanding of performance issues in general in cellular systems.

9.1 Overview of performance concepts

In this section we first focus on simple probabilistic models for call duration and mobile dwell time in a cell. From these models we readily calculate the length of time a channel is held within a cell, as well as the probability of a handoff occurring. The intensity of handoff traffic within a cell is then related to new-call traffic intensity using simple flow analysis. (Clearly, handoff calls must have started as new calls!) Examples for various cell sizes are provided to solidify and explain these concepts. We then introduce call admission control, studying the tradeoff possible between new-call blocking probability and handoff dropping probability through the analysis of the guard-channel control strategy. Note that this introductory analysis is carried out without the need to invoke specific cell geometries. In the sections following, as noted above, we do carry out equivalent performance analyses using, first, one-dimensional and, then, two-dimensional cell models, using different mobile station mobility models.

Consider now a mobile-terminal call in progress in a cellular wireless network. This call could have been initiated in the cell in which the mobile currently finds itself, or could have been handed off from a neighboring cell as the mobile crossed the cell border. We use the parameter $1/\mu$, in units of seconds, to designate the average length of time of a call. This time is often called the average *call holding time*. We assume that the call length or holding time T_n is a random quantity with exponential distribution. This has been a common model for telephone calls for many years. For this exponential case, the call-length probability density function $f_n(t)$ is given by the familiar form $f_n(t) = \mu\exp(-\mu t)$, with the time t given in seconds. The corresponding distribution function, $F_n(t) = \text{Prob}\,(T_n \leq t)$, is then $1 - \exp(-\mu t)$.

The exponential distribution has an interesting property: it is an example of a *memoryless* distribution (Papoulis, 1991). This statement means that the probability of an event taking place is independent of previous events. In the case of the exponentially distributed call length or holding time, for example, the probability of a call ending at any given time is always the same, no matter how long the call has already lasted. Random variables with this memoryless property are termed Markov random variables. The exponentially distributed random variable is a special case of a Markov rv. Probabilistic analysis is simplified considerably, as we shall see, if the memoryless property is found to be reasonably accurate and can thus be invoked.

The mobile carrying this call is implicitly assumed to have been assigned a channel for the call in the current cell, either as it was handed off to this cell, or in response to a new call request within this cell. The mobile is also implicitly assumed to be moving within the current cell (the very definition of a mobile terminal!), and will request a new channel assignment on entering a neighboring cell, if the call has not completed on entering the new cell. This is what we refer to as the process of handoff. The time T_h spent in a given cell, before handing off, is called the cell *dwell time*. In this section we assume this (random) time is also exponentially distributed, with parameter η, to simplify the analysis. The average dwell time is then $1/\eta$. In the sections following, we determine the dwell-time distribution using simple models of mobility and cell geometries. Knowing the call-length distribution and the dwell-time distribution, we can readily find the distribution of the length of time T_c a channel is held or occupied in a given cell. This is an important quantity needed to calculate such performance parameters as call blocking probability and handoff dropping probability.

To determine the channel holding-time distribution, note that the channel holding time is just the smaller of either the dwell time or the call length; i.e., it is the dwell time if the call does not complete while the mobile resides in a given cell; it is the call length if the call is completed while the mobile is still within the cell. (Note that this statement requires the memoryless assumption noted above that call lengths are exponentially distributed.) We thus have the relation

$$T_c = \min(T_h, T_n) \tag{9.1}$$

Papoulis, A. 1991. *Probability, Random Variables, and Stochastic Processes*, 3rd edn, New York, McGraw-Hill.

Now define the various distribution functions of these three independent random variables, respectively, as $F_c(t) = \text{Prob}\,(T_c \leq t)$, $F_h(t) = \text{Prob}\,(T_h \leq t)$, and $F_n(t) = \text{Prob}\,(T_n \leq t)$. It may then be shown from elementary probability theory (Papoulis, 1991) that the three distribution functions are related as follows

$$1 - F_c(t) = [1 - F_h(t)] \bullet [1 - F_n(t)] \tag{9.2}$$

This equality relates the *complementary* distribution function to the product of the two other complementary distribution functions.

Consider now the special case of exponentially distributed random variables, with both the call holding time and dwell time exponential, as assumed above. It is then left to the reader to show, using (9.2), that the channel holding time is exponential as well, with its parameter, defined as μ_c, given by the sum of the other two parameters. This is one of the reasons the dwell time, or time between handoffs in a cell, is assumed here, in this introductory discussion, to be an exponentially distributed variable. We thus have, as the average channel holding time, $1/\mu_c$

$$1/\mu_c = 1/(\mu + \eta) \tag{9.3}$$

(Recall again that $1/\mu$ and $1/\eta$ are the average call length and average dwell time, respectively.)

We can determine the probability P_h of a handoff as well. This turns out to be given by $\eta/(\mu + \eta)$ for the case of exponentially distributed rvs. To demonstrate this simple result, note that this is just the probability that the call duration (holding time) is greater than the dwell time; otherwise there is no handoff. This result is then readily shown by first calculating the probability that the call duration is greater than a specified value of the dwell time, conditioning on that value, and then summing (integrating) over all values of the dwell time. The specific calculation appears as shown in (9.4) following

$$\begin{aligned} P_h = \text{Prob}(T_n > T_h) &= \int_0^\infty f_h(T_h)\left[\int_{T_h}^\infty f_n(t|T_h)dt\right]dT_h \\ &= \int_0^\infty \eta e^{-\eta T_h}\left[\int_{T_h}^\infty \mu e^{-\mu t}dt\right]dT_h \\ &= \eta/(\eta + \mu) \end{aligned} \tag{9.4}$$

The result in the third line of the equation is obtained after carrying out the integration indicated in the second line. (The expression $f_n(t|T_h)$ is just the density function for the call length *conditioned* on a given value of the dwell time T_h.)

We now provide some examples of the use of these results. But note first that the exponential dwell time parameter η, with $1/\eta$ the average dwell time, may also be interpreted as the average rate of crossing cells. This rate is just the quantity mentioned in Section 8.4 of Chapter 8, in discussing the concept of location areas: it is the average rate at which a mobile moving at a speed V crosses a series of cells of area S and perimeter (boundary) length L. It was noted there that this quantity we now call η is given by Jabbari (1996)

$$\eta = VL/\pi S \tag{9.5}$$

In particular, for a circular cell of radius r, we have

$$\eta = 2V/\pi r \tag{9.5a}$$

Say the average call holding time is $1/\mu = 200$ sec. Take three cases corresponding to three different cell sizes: a macrocell of radius 10 km, a smaller cell of radius 1 km, and a microcell of radius 100 m. In the first case we assume vehicles travel at the average speed of 60 km/hr (36 miles/hr); in the two other cases, we assume mobile cell terminals are carried by pedestrians walking at an average speed of 5 km/hr (3 miles/hr). The reader is encouraged to consider other cases as well. The results obtained, by simply "plugging in" to the equations written above, appear as follows:

1 Macrocell, radius $r = 10$ km, speed $V = 60$ km/hr

Average time between handoffs, or average dwell time

$$1/\eta = 940 \text{ sec}$$

Probability of a handoff

$$\eta/(\eta + \mu) = 0.17$$

Average channel holding or occupancy time

$$1/\mu_c = 170 \text{ sec}$$

2 Smaller cell, radius $r = 1$ km, speed $V = 5$ km/hr

Average time between handoffs

$$1/\eta = 1100 \text{ sec}$$

Probability of a handoff

$$\eta/(\eta + \mu) = 0.15$$

Average channel holding or occupancy time

$$1/\mu_c = 170 \text{ sec}$$

3 Microcell, radius $r = 100$ m, speed $V = 5$ km/hr

Average time between handoffs

$$1/\eta = 110 \text{ sec}$$

Probability of a handoff

$$\eta/(\eta + \mu) = 0.64$$

Jabbari, B. 1996. "Teletraffic aspects of evolving and next-generation wireless communication networks," *IEEE Personal Communications*, 3, 6 (December), 4–9.

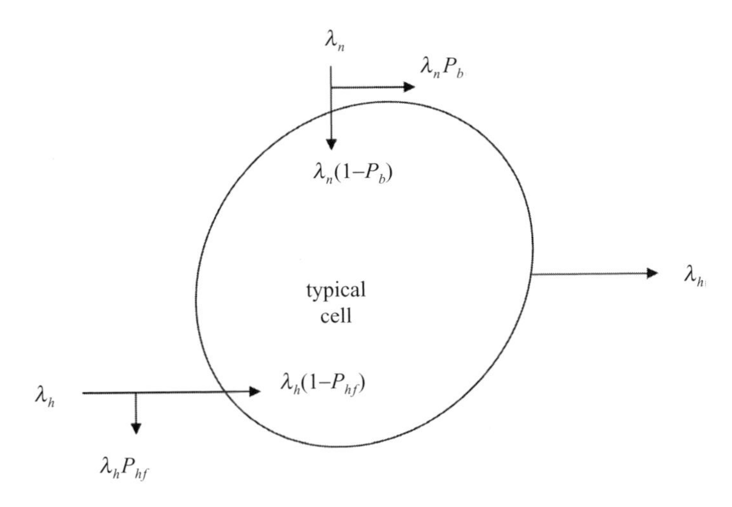

Figure 9.1 Traffic flows in and out of a cell

Average channel holding or occupancy time

$$1/\mu_c = 72 \text{ sec}$$

These results agree, of course, with what is expected. Cases 1 and 2 are quite similar: The radius of the cell in 2 has been reduced to 0.1 of that in 1, but the speed of the mobile has been reduced by somewhat more than that factor as well. The numbers are therefore about the same. There is a relatively long time required to traverse a cell in either case; hence the corresponding handoff time is relatively long as well. Since this time is substantially longer than the average call length, the probability of a handoff is relatively small. (The chances are that the call will terminate before the mobile leaves a cell; hence no handoff would be required.) The channel occupancy time is therefore determined principally by the average call length. Compare the third case now with the second. Here the cell size has been reduced by a factor of 10, but the mobile terminal speed remains the same. The time required to cross the cell, and, hence, for a handoff to take place, is therefore reduced by the same factor. The probability of a handoff taking place increases accordingly, the call holding time remaining fixed. The channel occupancy time within a cell now reduces substantially, since the handoff time drops below the call holding time.

We now discuss the handoff traffic rate and its connection to the new-traffic arrival rate. We do this by invoking continuity of flow considerations (Jabbari, 1996). Figure 9.1 portrays an arbitrary cell within a mobile system. As noted in our introduction to this chapter, we assume stationary, homogeneous conditions prevail. This means that every cell is, on the average, identical, with the same average traffic intensity and density of mobiles. (We have actually invoked this assumption implicitly in writing equation (9.5). Uniform mobile density as well as identical cell geometries and mobile speeds are assumed in deriving that equation.) Let λ_n be the average new-traffic intensity, in calls/sec, generated in each cell. Let P_b be the probability of blocking a new-call attempt. The net new-traffic intensity into a cell is then just $\lambda_n(1 - P_b)$. Now let λ_h be the average handoff traffic intensity, also in calls/sec, due to handoffs attempting to enter a cell from all neighboring cells. On the average, because of the homogeneity assumption, the rate of handoff calls leaving a cell, summed over all neighboring cells, must be the same value λ_h. Let the probability a handoff attempt is dropped for lack of channels in the new cell be P_{hf}. The net rate of handoff traffic entering the cell is then $\lambda_h(1 - P_{hf})$. These quantities are all indicated in Fig. 9.1. From simple flow considerations, the average rate of calls λ_h leaving the cell must equal the average rate of calls requiring handoff, due to both new-call attempts and handoffs entering the cell. The latter is simply the rate of calls entering the cell and acquiring channels, times the probability P_h they will handoff. This statement is an example of flow balance. We shall provide another example later in discussing call admission. The probability P_h of a handoff has been evaluated above in (9.4) for the special case of exponential statistics. Equating the net rate of calls entering a cell and requiring handoff to those leaving the cell, the following flow balance equation may be written

$$\lambda_h = P_h[\lambda_n(1 - P_b) + \lambda_h(1 - P_{hf})] \tag{9.6}$$

Solving this flow equation for the handoff rate λ_h in terms of the new-call arrival rate λ_n, one obtains

$$\lambda_h = \frac{P_h(1 - P_b)}{[1 - P_h(1 - P_{hf})]}\lambda_n \tag{9.7}$$

If the blocking and handoff dropping probabilities are suitably small numbers (P_b, $P_{hf} \ll 1$), one obtains the following approximation for the relation between the two traffic intensities

$$\lambda_h \approx \frac{P_h}{(1 - P_h)}\lambda_n, \quad P_b, P_{hf} \ll 1 \tag{9.7a}$$

In particular, for the special case of exponential statistics, we find

$$\lambda_h \approx \frac{\eta}{\mu}\lambda_n, \quad P_b, P_{hf} \ll 1 \tag{9.7b}$$

Consider two of the examples described above, the macrocellular and microcellular systems, respectively. Applying (9.7b) to these two cases, one finds $\lambda_h = 0.21\ \lambda_n$ in the macrocellular example, and $\lambda_h = 1.8\ \lambda_n$ in the microcellular example. It is clear that, because the probability of a handoff is quite high in the microcellular example for the numbers chosen, new calls will generate substantial numbers of handoffs. The average rate of handoffs is thus almost twice the rate of new-call attempts in that case.

We have defined the quantity P_{hf} as the probability a handoff attempt will be dropped when a mobile carrying an ongoing call enters a new cell. In the case of small cells, with multiple handoff attempts during a call, a more significant measure to users is the probability P_d a handoff call will eventually be dropped as a mobile moves from cell to cell. It has also been called the *forced-termination probability* (Hong and Rappaport, 1999).

Hong, D. and S. S. Rappaport. 1999. "Traffic model and performance analysis for cellular mobile radio telephone systems with prioritized and non-prioritized handoff procedures," *IEEE Transactions on Vehicular Technology*, VT-35, 3 (August), 77–92. See also CEAS Technical Report No. 773, 1 June 1999, College of Engineering and Applied Sciences, State University of New York, Stony Brook, NY 11794.

This parameter may be readily found by noting that it is the probability a handoff call is dropped on the first handoff or the second or the third, and so on. The first probability is just $P_h P_{hf}$, the probability a handoff will occur times the probability it is then dropped for lack of capacity. The probability a handoff will be dropped after the second handoff is given by $P_h^2(1-P_{hf})P_{hf}$, etc. Summing these probabilities, one then obtains, as the forced termination probability

$$\begin{aligned} P_d &= P_h P_{hf} + P_h^2(1 - P_{hf})P_{hf} + P_h^3(1 - P_{hf})^2 P_{hf} + \cdots \\ &= \frac{P_h P_{hf}}{[1 - P_h(1 - P_{hf})]} \end{aligned} \tag{9.8}$$

For small handoff dropping probability P_{hf}, the forced-termination probability is approximated by

$$P_d \approx \frac{P_h}{1 - P_h} P_{hf}, \qquad P_{hf} \ll 1 \tag{9.8a}$$

The forced-termination probability is thus directly proportional to the handoff dropping probability, as expected.

We now focus on the determination of the call blocking probability P_b and the handoff dropping probability P_{hf} used in the analyses above. We have already noted that, if handoffs and new-call attempts are treated alike, the two probabilities are the same, and are given by the Erlang-B blocking-probability formula introduced in Chapters 3 and 4. (See (3.7) and (4.6).) In Chapter 3 we used N to represent the number of channels in a cell; in Chapter 4 we used the letter m. We shall use the letter m in this chapter. There is one basic difference in our discussion here, however. We ignored handoffs in the previous chapters, implicitly focusing on new-call attempts in calculating the traffic intensity, and, from this, the Erlang parameter A used to calculate the blocking probability. The results obtained there for blocking probability are still valid, however, if we add the handoff traffic intensity λ_h to the new-call traffic intensity λ_n in calculating the Erlang parameter A. Note that this is critical in the case of small cells, with multiple handoffs common. We have already seen above that the handoff traffic rate exceeds the new-call attempt rate in that case. Adding these two quantities does raise a question, however. Since the handoff attempt rate is directly proportional to the new-call attempt rate, as seen from (9.7) (after all, a handoff call must have started as a new call!), is it valid to assume the handoffs are independent of the new calls? The answer is yes, since, in calculating the *average* traffic rates, we are implicitly averaging over many independently generated calls. A mobile with a call in progress arriving at a new cell, and requesting a (handoff) channel in that cell, is clearly not connected to mobiles with calls in progress in that cell. New calls in a given cell are independent of handoffs arriving at that cell. On the average, the rate of arrival of handoffs is proportional to the new-call arrival rate, but the two rates may be added together in determining the total cell traffic used to calculate the Erlang-B blocking probability.

If we distinguish between the two types of traffic, as we shall now do, to provide priority to the handoff calls, we treat the handoff calls as if independent of the new traffic, and consider the two average traffic rates separately in calculating the blocking probability and handoff dropping probability. The average handoff rate is, however, still

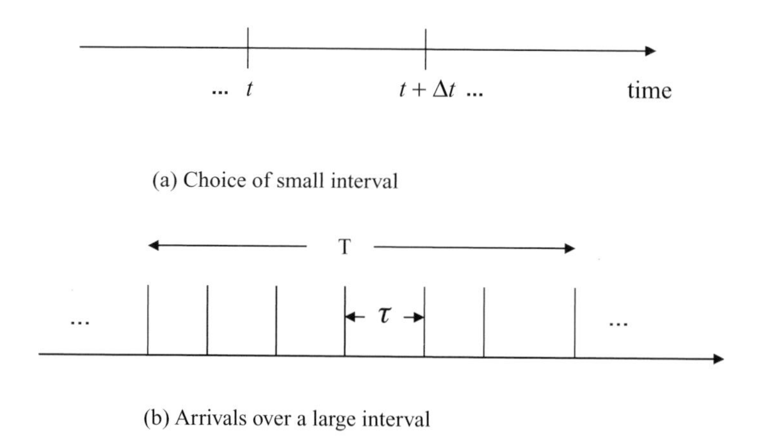

Figure 9.2 Poisson statistics

directly proportional to the new-call traffic rate, and is calculated using equation (9.7). We discuss, in this section, only the guard-channel method of providing priority to handoff calls (Hong and Rappaport, 1999). In the next section we introduce a second method for comparison. The guard-channel approach is to set aside a specific number of channels of the total available in a cell for the use of handoffs only. Let there be m channels in a given cell. Of these, say g are to be used exclusively by handoff calls. This then says that all m channels are available for handoff calls; $m - g$ only are available for new calls. We assume both the new-call and handoff-call arrival statistics are Poisson. Poisson statistics and the exponential distribution are closely related to one another: if arrivals are Poisson, the (random) time between arrivals obeys the exponential distribution (Papoulis, 1991; Schwartz, 1987) . Poisson statistics then provide another example of memoryless, Markov statistics.

The Poisson call-arrival assumption has been used for studying telephone traffic for many years and is well-validated in that environment. Poisson statistics are based on the following postulates:

Consider a very small interval of time Δt spanning the time t to $t + \Delta t$ (see Fig. 9.2(a)).

1 The probability of an arrival (event happening) is $\lambda\ \Delta t \ll 1$; i.e., it is very small and proportional to the time interval.
2 The probability of *no* arrival in the same interval is $1 - \lambda\ \Delta t$; i.e., the probability of more than one arrival in that interval is negligible.
3 Arrivals are *memoryless*: an arrival before time t has no effect on the arrival probability in the interval $t, t + \Delta t$; what happens in that interval has no effect on arrivals after that time.

It may then be shown that the time τ between arrivals (Fig. 9.2(b)) is a random variable with the exponential probability density $\lambda e^{-\lambda\tau}$, with average time between arrivals $1/\lambda$

Schwartz, M. 1987. *Telecommunication Networks: Protocols, Modeling, and Analysis*, Reading, MA, Addison-Wesley.

(Papoulis, 1991; Schwartz, 1987). It may also be shown that the number of arrivals in a larger interval T (Fig. 9.2(b)) obeys the Poisson distribution

$$\text{Prob}(n \text{ arrivals in } T \text{ sec}) = (\lambda T)^n \, e^{-\lambda T}/n!$$

(The probability of no arrivals in T sec is $e^{-\lambda T}$; one arrival $\lambda T e^{-\lambda T}$; etc.) It is then readily shown that the average number of arrivals in the interval T is λT, with the average arrival rate, in arrivals per unit time, just the proportionality parameter λ. So the proportionality factor λ used in postulates 1 and 2 in defining Poisson statistics turns out to be the average rate of (Poisson) arrivals; its reciprocal $1/\lambda$ repesents the average time between arrivals.

We now use the Poisson assumption for both new-call and handoff arrival statistics in calculating both the handoff dropping probability and the blocking probability using the guard-channel procedure. This procedure defines a new-call admission procedure: new calls are admitted only if a channel is available to serve them. By definition, this must be one of the $m - g$ channels serving both new-call attempts and handoffs. The g guard channels are available to serve handoff attempts only, as noted above. This is called an admission control procedure because, by controlling the admission of new calls, blocking them when channels are not available, one can better serve existing calls and handoff-call attempts. Clearly, as the number g increases, the handoff dropping probability will decrease; the new-call blocking probability will, of course, increase at the same time. We shall see that, for given traffic intensity, the relative decrease in handoff dropping probability will generally be more than the corresponding increase in new-call blocking probability, a desirable characteristic.

The analysis of the guard-channel admission control procedure proceeds by use of a local flow-balance procedure commonly used to analyze Markov (memoryless) systems. Note that, with m channels available in a cell, the system can, at any time, be in any one of $m + 1$ states, each corresponding to a particular channel occupancy, including the zero-occupancy case. Because of the memoryless Poisson/exponential assumptions made for both call arrivals and call holding times, we can calculate the steady-state probability of state occupancy of each of the $m + 1$ states. From this calculation, we readily find the probability of occupancy of blocking and handoff dropping states. A similar approach is used in deriving the Erlang-B blocking-probability formula (3.7) of Chapter 3, as well as the blocking state probabilities used in studying the performance of the DCA scheme analyzed in Chapter 4. A discerning reader will note that we explicitly invoked the Poisson statistics assumption in discussing the material in those chapters. We explicitly referred to the "Markov chain" concept there as well. The result of our analysis here will, in fact, be shown to reduce to the Erlang-B blocking formula, stated without proof in those earlier chapters, for the special case of $g = 0$, i.e., no guard channels. We postponed detailed analysis in Chapter 3 to simplify the discussion there. The analysis in this chapter thus provides the missing material in Chapter 3 for those not familiar with telephone traffic theory. (It is of interest to note as well that both the guard-channel concept and its analyis to be presented here are identical to the earlier

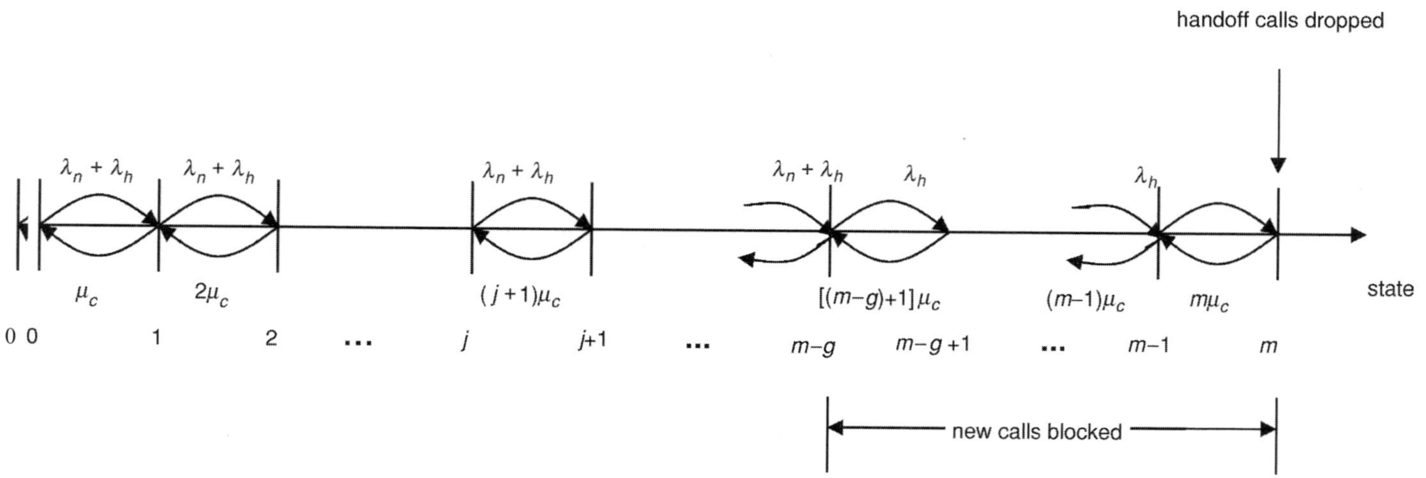

handoff calls dropped

$\lambda_n + \lambda_h$ $\lambda_n + \lambda_h$ $\lambda_n + \lambda_h$ $\lambda_n + \lambda_h$ λ_h λ_h

μ_c $2\mu_c$ $(j+1)\mu_c$ $[(m-g)+1]\mu_c$ $(m-1)\mu_c$ $m\mu_c$ state

0 0 1 2 ... j j+1 ... $m-g$ $m-g+1$ ... $m-1$ m

new calls blocked

Figure 9.3 State diagram, guard channel procedure

trunk reservation strategy developed to handle alternately routed traffic in wired telephone networks (Schwartz, 1987). A *trunk* in telephone system jargon is just a channel for handling a telephone call over a single link of a network. In modern digital telephone systems a trunk would correspond to a time slot of a repetitive frame structure, exactly the same as in the TDM wireless systems such as IS-136 and GSM described earlier in this book.)

Figure 9.3 portrays the state diagram of the guard-channel procedure. New calls are blocked if there are at least $(m - g)$ calls present. Hence the new-call blocking states are those numbered $(m - g)$, $(m - g + 1), \ldots, m$. Handoff calls are dropped if all m channels are occupied. Now define p_j, $j = 0, 1, 2, \ldots, m$, to be the (steady-state) probability of occupancy of state j. It is these $m + 1$ state probabilities that are to be calculated. Once we know these, we determine the new-call blocking probability by summing over the new-call blocking states. The handoff dropping probability is just p_m. The diagram indicates, as well, the various average transition rates between states. We shall have more to say about these transition rates as we pursue the analysis, setting up flow balance equations, one for each state. But consider these transition rates indicated in Fig. 9.3 in both the "up" (right) and "down" (left) directions. Note that the rate of moving "up" one state is $(\lambda_n + \lambda_h)$ in the state region 0 to $(m - g - 1)$, in which either new calls or handoff calls can access available channels. This transition rate arises from the possibility that a call of either type may arrive, moving the system up one state. (Two call arrivals of either type within the same small time interval are precluded because of the Poisson postulate. Transitions between adjacent states only are therefore possible in the model we have adopted.) Once the state $(m - g)$ is reached, however, handoff calls only may access the remaining channels. The transition rate "up" one state is therefore λ_h only in the range $(m - g)$ to $(m - 1)$. Considering the "down" direction now, note that the system will move down, one state at a time, as a call occupying a channel vacates that channel, either by completing or by being handed off to another cell. The rate of doing this is just the parameter μ_c, the reciprocal of the average channel holding time discussed earlier. But, as the number of calls in progress increases, i.e. as we move up in state number, the probability of *one* call vacating a channel goes up correspondingly: if j calls are in progress, the probability of any one of these calls vacating a channel in a small time interval is j times as likely as in the case of only one call in progress. One call only can complete or handoff in any small time interval, however, because of the Poisson assumption. The average rate of vacating a channel, given the system is in state j, is $j\mu_c$ and is so indicated on the diagram.

We use this state-transition diagram to determine the $(m + 1)$ probabilities of state p_j, $j = 0$ to m. These probabilities are readily calculated by invoking the local balance equations, one for each state, mentioned before. Consider state 0, the one with no channels occupied, for example. The only way for the system to leave this state is if either a new call or a handoff call arrives, the system thus moving to state 1 with one channel occupied. Note again, as stated above, that, by the Poisson assumption we have invoked here, at most one call of either type can arrive in any one small time interval. The *average* rate of leaving this state is then just $(\lambda_n + \lambda_h)$, the summed rate of call arrivals, times the probability P_0 of having been in this state. Now consider entry into this state. The only possibility is that the system had one call in progress, and that the call completed or was handed off. The average rate of entering state 0 is then the average rate μ_c of calls completing or handing off times the probability p_1 that the system was in state 1. A system in statistical equilibrium will have the average rates of leaving and entering a state equal. Equating these rates, we have

$$(\lambda_n + \lambda_h)p_0 = \mu_c p_1 \tag{9.9}$$

This equation is the first flow balance equation. Solving for p_1, we have

$$p_1 = \frac{(\lambda_n + \lambda_h)}{\mu_c} p_0 \tag{9.10}$$

Now consider state 1. This state may be entered independently two ways: from state 2 above, by one of the two calls in progress vacating its channel, or from state 0 below by a call of either type arriving. Referring to Fig. 9.3 and summing the average rates of entering state 1 due to these two possibilities, we get, as the average rate of entering state 1, $2\mu_c\, p_2 + (\lambda_n + \lambda_h)p_0$. Proceeding in the same manner, the average rate of leaving state 1, moving independently to either state 0 through a call completing or handing off, or to state 2 through a call of either type arriving, is clearly found to be given by $(\lambda_n + \lambda_h + \mu_c)p_1$. Equating these two transition rates to provide a balance of arrivals and departures from state 1, we get

$$(\lambda_n + \lambda_h + \mu_c)p_1 = (\lambda_n + \lambda_h)p_0 + 2\mu_c p_2 \tag{9.11}$$

This balance equation is readily simplified by using (9.9) to eliminate a term on each side of the equation. We then solve the remaining equation for p_2 in terms of p_1. But we can eliminate p_1 and relate p_2 directly to p_0 by using (9.10). This gives us the equation

$$p_2 = \frac{(\lambda_n + \lambda_h)^2}{2\mu_c \cdot \mu_c} p_0 \tag{9.12}$$

We now simplify the notation by defining the traffic intensity factor $\rho \equiv (\lambda_n + \lambda_h)/\mu_c$. This factor is similar to the Erlang traffic intensity factor $A = \lambda/\mu$, the product of the average traffic arrival rate times the average call holding time, defined in Chapter 3. We then get, in place of (9.12)

$$p_2 = \frac{\rho^2}{2} p_0 \tag{9.12a}$$

Continuing in this manner by setting up a balance equation at each state $j \leq (m - g - 1)$, equating average arrivals and departures at a state to maintain statistical equlibrium, it is left to the reader to show that we get, for the state probability at state j, $1 \leq j \leq (m - g)$

$$\begin{aligned} p_j &= \frac{\rho^j}{j!} p_0, \quad 1 \leq j \leq (m - g) \\ \rho &\equiv \frac{(\lambda_n + \lambda_h)}{\mu_c} \end{aligned} \tag{9.13}$$

Now consider the states $j = (m - g)$ to m. We again write the balance equation for each of the states, noting, however, that, because we are now in the guard-channel region, handoff calls only may access the channels. The traffic arrival rate at each state is therefore λ_h, as shown in Fig. 9.3. This is the only change in the state diagram, as contrasted with the previous non-guard-channel states. We again set up the balance equations, one for each state, beginning with state $(m - g)$. It is left to the reader to show that these equations may then be written, defining the traffic parameter $\rho_h \equiv \lambda_h/\mu_c$, in the form

$$p_{m-g+k} = \frac{\rho_h^k}{(m-g+k)(m-g+k-1)\dots(m-g+1)} p_{m-g},$$

$$\rho_h \equiv \frac{\lambda_h}{\mu_c}, \quad 1 \leq k \leq g \tag{9.14}$$

The state probability p_{m-g} appearing in (9.14) may be written in terms of p_0 using (9.13). Replacing $m - g + k$ as well by j in (9.14), that equation simplifies to

$$p_j = \frac{\rho_h^{j-(m-g)} \rho^{m-g}}{j!} p_0, \quad m - g + 1 \leq j \leq m \tag{9.14a}$$

This completes the evaluation of the m state probabilities p_j, $1 \leq j \leq m$, using the balance equations. All appear written in terms of p_0, the probability the guard-channel system has all m channels available. It now remains to determine this probability. This calculation is readily carried out by invoking the rule that the sum of the probabilities over all $m + 1$ states must equal 1. Doing this by summing (9.13) and (9.14a) over their respective ranges, we get, as the value of p_0 to be used in (9.13) and (9.14a)

$$\frac{1}{p_0} = \sum_{j=0}^{m-g} \frac{\rho^j}{j!} + \rho^{m-g} \sum_{j=m-g+1}^{m} \frac{\rho_h^{j-(m-g)}}{j!} \tag{9.15}$$

The new-call blocking probability and handoff dropping probabilities are now easily calculated, as indicated earlier, in terms of the appropriate state probabilities. In particular, we immediately have the handoff dropping probability P_{hf} given by

$$P_{hf} = p_m = \rho^{m-g} \frac{\rho_h^g}{m!} p_0 \tag{9.16}$$

with p_0 given by (9.15).

The new-call blocking probability P_b is given by

$$P_b = \sum_{j=m-g}^{m} p_j = \rho^{m-g} \sum_{j=m-g}^{m} \frac{\rho_h^{j-(m-g)}}{j!} p_0 \tag{9.17}$$

We do have a problem in using these two equations to carry out numerical calculations. The traffic intensity parameters ρ and ρ_h are both defined in terms of the handoff-call arrival rate λ_h. From (9.7) this rate depends on both P_{hf} and P_b! It is only for small values of these two probabilities that the rate becomes nearly independent of them, as indicated by (9.7a). The answer is to carry out an iterative analysis, first setting both P_{hf} and $P_b = 0$ in the expression for λ_h appearing in the two equations (9.16) and (9.17),

Table 9.1 *Guard channel admission control, macrocell and microcell examples*

Macrocell $P_h = 0.17$, $m = 10$

Blocking probability P_b:

g \ ρ_n	5	6	7	8
0	0.042299	0.083095	0.13135	0.18168
1	0.078954	0.13644	0.1973	0.25609
2	0.12698	0.19545	0.26169	0.32216
3	0.19001	0.26519	0.33291	0.39207

Handoff dropping probability P_{hf}:

g \ ρ_n	5	6	7	8
0	0.042299	0.083095	0.13135	0.18168
1	0.0067967	0.013059	0.020288	0.027685
2	0.0010172	0.0018929	0.0028799	0.0038775
3	0.0001352	0.000241	0.00035682	0.00047215

Microcell $P_h = 0.64$, $m = 10$

Blocking probability P_b:

g \ ρ_n	2	3	4	5
0	0.024982	0.087396	0.15463	0.21613
1	0.060032	0.16876	0.26745	0.34808
2	0.10729	0.24597	0.35706	0.4417
3	0.16754	0.32159	0.43425	0.51629
4	0.24072	0.39807	0.50559	0.58149
5	0.32647	0.47741	0.57533	0.64289

Handoff dropping probability P_{hf}:

g \ ρ_n	2	3	4	5
0	0.024982	0.087396	0.15463	0.21613
1	0.014748	0.048893	0.083461	0.11326
2	0.0080158	0.025415	0.042815	0.057639
3	0.0038816	0.011836	0.019912	0.026888
4	0.0016047	0.0047166	0.0079562	0.010827
5	0.00053252	0.0015088	0.0025497	0.0034975

finding the resultant values of the probabilities, recalculating λ_h, using this new value in (9.16) and (9.17), and repeating until the equations converge. We shall indicate the results of such an analysis shortly, but first we return to the determination of the Erlang-B equation. We set the guard-channel g equal to zero. The blocking and handoff dropping probabilities converge to the same value $(\rho^m/m!)p_0$, as is apparent from both (9.16) and (9.17). The expression (9.15) for p_0 simplifies as well. It is then left for the reader to show that the blocking probability is, in fact, given by the Erlang-B distribution

$$P_b\Big|_{g=0} = \frac{\rho^m/m!}{\sum\limits_{j=0}^{m} \rho^j/j!} \tag{9.18}$$

Note that this is precisely the form of equation (3.7) in Chapter 3, as stated previously. We still have to iterate this equation in carrying out the calculations here involving handoffs, since the traffic intensity does depend on λ_h, which depends in turn on P_b.

Consider the application of the guard-channel admission control scheme to the macrocell and microcell examples described earlier. Recall that the probability P_h of a handoff was 0.17 and 0.64, respectively, in the two cases. Table 9.1 provides the results of calculating the new-call blocking probability P_b and the handoff dropping probability for these two cases, using (9.17) and (9.16), respectively. Equation (9.7) has been used to determine the handoff intensity factor ρ_h in terms of the new-call arrival intensity $\rho_n \equiv \lambda_n/\eta_c$. A value of $m = 10$ channels per cell has been arbitrarily chosen for this example. The blocking probabilities and dropping probabilities are tabulated for a number of values of the new-call arrival intensity and for various values of g, the number of channels allocated exclusively to handoff calls. Results are as expected: for a given new-call intensity ρ_n, the dropping probability does decrease as g increases, with the blocking probability increasing at the same time. Table 9.1 also shows the expected increase of both the blocking probability and the dropping probability with increasing traffic intensity.

Consider the macrocellular example first. It is apparent that the relative improvement in dropping probability in this case is substantially greater than the relative increase in blocking probability. Note that with $\rho_n = 5$, the two probabilities start off at the same value of 0.042 without the use of guard channels. Assigning one of the ten channels as a guard channel, one finds the handoff dropping probability decreasing to 0.0068, a reduction by more than a factor of 6. The blocking probability increases by less than a factor of two, to 0.079. With two guard channels assigned, the handoff dropping probability decreases further to 0.001, a factor of 7 decrease, while the blocking probability increases to 0.13, a much smaller increase. Further increases in the number of guard channels are possible, of course, but the resultant blocking probability becomes quite high. Note, in particular, that with $g = 3$, the handoff dropping probability drops to extremely small values, ranging from 10^{-4} to 5×10^{-4} over the range of traffic intensity from 5 to 8. The corresponding new-call blocking probability ranges from 0.19 to 0.39, however. Increasing the number of channels would, of course, reduce both probabilities. The introduction of guard channels therefore does have the desired effect: on-going calls are much less likely to be dropped during handoff. The resultant increase in blocking probability, if not too high, can be considered a tolerable price to pay for this improvement in service during an on-going call.

What would a good design choice be for this macrocellular example with $m = 10$ channels assigned per cell? It would appear that operating at a new-call traffic intensity of $\rho_n = 5$ would provide tolerable performance, i.e., a relatively good grade of service. A single channel allocated to handoffs only would provide a handoff dropping probability of about 0.7%, while the corresponding new-call blocking probability would be about 8%. Note that the traffic intensity figure here is not quite the traffic intensity in units of Erlangs, as defined in Section 3.3 of Chapter 3. The Erlang definition used there was $A \equiv \lambda_n/\mu$, with $1/\mu$ the average call holding time. Here we have used the parameter $\rho_n \equiv \lambda_n/\mu_c$, with $1/\mu_c$ the average *channel* holding time, equal to $1/(\mu + \eta)$, $1/\eta$ the average cell dwell time. The Erlang load is thus actually higher than the traffic intensity parameter obtained here. In particular, we have $A = \rho_n(1 + \eta/\mu)$. For the macrocell example used here, we have, with $1/\mu = 200$ seconds and $1/\eta = 940$ seconds, $A = 1.21\rho_n$. With $\rho_n = 5$, we therefore have $A = 6$ Erlangs as the new-call traffic allowed per cell. (The three second-generation systems discussed in this book allow much higher traffic intensities, of course, since their capacity figures are substantially higher. See Chapter 6.) With an average call length (holding time) of 200 seconds, the 6-Erlang figure translates into an average call arrival rate of 6 calls/200 sec or 1.8 calls per minute. Increasing the number of guard channels to 2 or 3 results in rather high blocking probability, as noted above. Increasing the new-call traffic traffic intensity to $\rho_n = 6$ and higher also results in relatively high blocking probabilities.

Now consider the second, microcellular example. Table 9.1 indicates that the allowable new-call traffic intensity is considerably smaller, because of the substantial increase in handoffs. In particular, at a new-call traffic intensity of $\rho_n = 2$, the blocking probability and handoff dropping probability are both equal to 2.5% if there are no guard channels. This value provides good blocking performance for new calls, but provides too high a dropping probability for ongoing calls. Assigning one of the ten channels to handoff calls exclusively does reduce the handoff dropping probability to 1.5% and increases the blocking probability to 6%, probably a tolerable value. (Increasing the number of channels would help even more in this case! It is therefore worth repeating that the choice of $m = 10$ channels is arbitrary and has been chosen to provide a simple, easily calculable example, as well as to provide a vehicle for discussion of the perfomance issues involved.) Using two guard channels reduces the dropping probability to 0.8%, but increases the blocking probability to 10%. If the new-call traffic intensity now increases to $\rho_n = 3$, the blocking and handoff dropping probabilities are both 8.7% if no guard channels are assigned. Assigning one guard channel to the handoff traffic reduces the dropping probability to 4.9%, but increases the blocking probability to a rather high value of 17%. So, if only ten channels were available, it would probably be best to stay at a maximum traffic intensity of $\rho_n = 2$ with one channel of the ten set aside exclusively for handoff calls. From our discussion of *Erlang* traffic above, the traffic in Erlangs in this case is $A = 5.6$ Erlangs. (Recall that, for the microcell example, we had $\eta/\mu = 1.8$.) This figure is not much different from the one for the macrocell case. Despite the increase in traffic due to the increased number of handoff calls arriving as the cell size decreases, the channels are held for shorter periods of time as calls hand off. The two effects appear to cancel one another out.

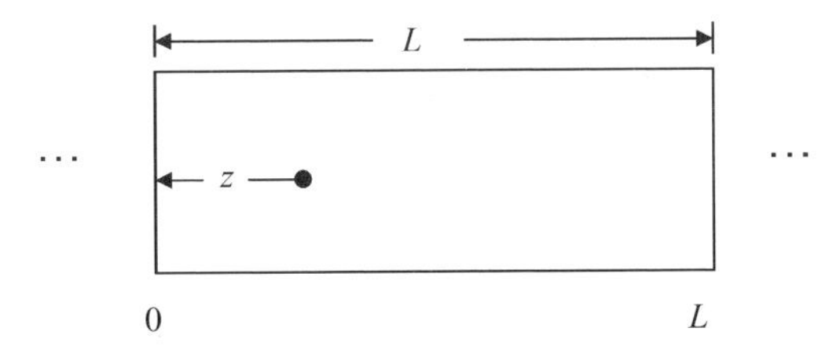

Figure 9.4 Typical cell, one-dimensional system

In the next section and the one following we move on to specific geometries and models of mobility to use in determining system performance. We begin in the next section by discussing the simpler one-dimensional model of a cell. We follow in the section after with a discussion of the two-dimensional case.

9.2 One-dimensional cells

We used the one-dimensional cell model previously in this book, first in carrying out SIR calculations in Section 3.2 of Chapter 3, and then in Section 4.2 of Chapter 4, in providing examples of power control algorithms. The one-dimensional case provides a simple model on which to carry out calculations. It provides a simple model as well for cells located along a highway. The object here, in this section, will be to carry out performance analysis similar to that done in the previous chapter, but with mobility and the one-dimensional geometry specifically included. There are several changes in the analysis that occur as a result of focusing on a specific geometry: by using models of mobile velocity distributions incorporated into the geometry, we can determine the mobile dwell-time distribution from basic principles rather than initially assuming it is exponential, as was done in the previous section; in addition, because the dwell-time distribution is no longer memoryless, the dwell time for newly arrived calls will differ from that for handoff calls. This will be seen to affect the analysis as well. These general considerations are independent of the actual geometry and will therefore carry over to two-dimensional cell geometries studied in the next section as well.

We begin the discussion here by considering the dwell time in the two separate cases of newly generated calls, and those calls carried by mobiles handing off from a previous cell. Note again that the dwell time corresponds to the time a mobile remains in a cell before handing off to another cell. To be specific, we focus on the one-dimensional cell geometry portrayed in Fig. 9.4. The cell length is taken to be L meters long. We again assume a homogeneous situation prevails: All cells are alike and carry the same, uniformly distributed number of mobiles. Mobile velocities in all cells have the same distribution. (We shall focus on two cases, that with all mobiles moving at the same, constant velocity, and that with velocities uniformly distributed up to a given maximum value.) New calls are assumed to be randomly and uniformly generated anywhere within a cell. We use the symbol T_N to represent the resultant, randomly distributed cell dwell time for a mobile generating a new call; the symbol T_H represents the random dwell time for a mobile carrying a call that arrives from, i.e., has been handed off from, another cell. The corresponding probability density and distribution functions to be found are defined, respectively, as $f_N(t)$ and $F_N(t)$; $f_H(t)$ and $F_H(t)$. In the previous section, with an exponential distribution assumed for the dwell time, no distinction was made between these two random variables.

We use these two density functions to calculate the probability of handoff in these two cases of dwell time. Recall, from the previous section, that the probability of handoff is the probability that the call length T_n is greater than the dwell time. We use (9.4) to carry out the calculation in these two cases, but without invoking the assumption that the dwell time is exponential. Thus, once the density functions for the two dwell-time variables are found, the respective probabilities of handoff, $P_N = \text{Prob}\,[T_n > T_N]$ and $P_H = \text{Prob}\,[T_n > T_H]$, are given by the equations following

$$P_N = \int_0^{\infty} e^{-\mu t} f_N(t)dt \tag{9.19}$$

and

$$P_H = \int_0^{\infty} e^{-\mu t} f_H(t)dt \tag{9.20}$$

Note again that P_N is the probability a mobile with a new-call generated handoff; P_H is the probability that a mobile handing off from another cell hands off again. These are then generalizations of the probability P_h of a handoff with the dwell time assumed to be exponential, as given by (9.4). (As an aside, note that (9.19) and (9.20) are effectively the Laplace transforms of the respective functions. They are characteristic functions as well, so that the various moments of the respective random variables, T_N and T_H, may be found by differentiation, if they exist.)

Not only do these two probabilities differ in the case of a specific geometry and mobile velocity distribution, but the relation between handoff call rate λ_h and new-call origination rate λ_n changes as well. This in turn impacts the channel holding-time distribution, as we shall see later in this section. But consider first the relation between the handoff call rate and the new-call origination rate. We extend the flow balance relation of (9.6), based on Fig. 9.1, to obtain the relation here with different handoff probabilities defined, depending on whether we are dealing with new calls or handoff calls. Specifically, by again noting that the average rate of calls handing off from a given cell must equal the total rate of calls handing off, including both new calls generated in this cell as well as handoff calls arriving from neighboring cells, we get the following flow balance equation

$$\lambda_h = P_H \lambda_h (1 - P_{hf}) + P_N \lambda_n (1 - P_b) \tag{9.21}$$

As previously, P_{hf} is the handoff dropping probability and P_b is the new-call blocking probability, the values of which depend on the particular admission control strategy selected. Solving (9.21) for the handoff dropping probability, we find, as a generalization of (9.7)

$$\lambda_h = \frac{P_N(1 - P_b)}{1 - P_H(1 - P_{hf})} \lambda_n \tag{9.22}$$

It behooves us now to find the two dwell-time distributions and the corresponding handoff probabilities P_N and P_H.

Consider first the case of new calls being generated within a typical one-dimensional cell, such as the one shown in Fig. 9.4. We assume mobiles carrying these calls are equally likely to be anywhere in the cell of length L m when the call attempt is generated. We consider two examples of velocity distributions of the mobiles carrying these new calls, as noted above. The first example simply assumes all mobiles are moving with the same constant velocity V_c, equally likely to be moving in either direction. The second example assumes that mobile velocities, although constant throughout the movement within a cell, are randomly chosen from a distribution equally likely to have any value from zero to a maximum value of V_m. On the average, these mobiles are moving at a velocity of $V_m/2$ m/sec. We evaluate the resultant quantities $f_N(t)$, $F_N(t)$, and P_N separately for each example. In particular, for the constant, non-random velocity example, it is left for the reader to show, quite simply, that the cell dwell-time density function is given by

$$f_N(t) = \frac{V_c}{L} \quad 0 \leq t \leq \frac{L}{V_c} \tag{9.23}$$

The corresponding distribution function is then given by

$$\begin{aligned} F_N(t) &= \frac{V_c}{L}t \quad & 0 \leq t \leq \frac{L}{V_c} \\ &= 1 & t \geq \frac{L}{V_c} \end{aligned} \tag{9.24}$$

From (9.19) the corresponding probability P_N of handoff is given by

$$\begin{aligned} P_N &= \int_0^\infty e^{-\mu t} f_N(t) dt = \frac{V_c}{L} \int_0^{L/V_c} e^{-\mu t} dt \\ &= \frac{V_c}{\mu L}(1 - e^{-\mu L/V_c}) \end{aligned} \tag{9.25}$$

This probability thus depends on the dimensionless parameter $\mu L/V_c$. As the cell width L decreases, the velocity and call holding time remaining fixed, the probability of a new call handing off increases. The same observation holds true if the velocity V_c and the average call holding $1/\mu$ increase. In the limit, as $\mu L/V_c$ decreases, the probability of a new-call handing off approaches 1. All of this agrees with intuition. As this parameter increases, the probability of a handoff rapidly approaches $1/(\mu L/V_c)$, and goes to zero as the parameter continues to increase.

Now consider the second example in which mobiles are assumed to be moving with a velocity V chosen randomly from a uniform distribution varying from 0 to V_m. Here we note that the random dwell time T_N is given by z/V, the (random) distance z being the distance a mobile must travel from the point at which the new-call attempt was first made, to the point at the end of the cell at which the mobile enters a new cell and attempts to hand off. The distance z is obviously bounded by $0 \leq z \leq L$. Figure 9.4 shows an example in which a mobile, starting at point z as shown, is moving to the left. Some thought will indicate that it makes no difference, however, in the calculations which direction the mobile is traveling. The random dwell time in this case is given by the ratio of two

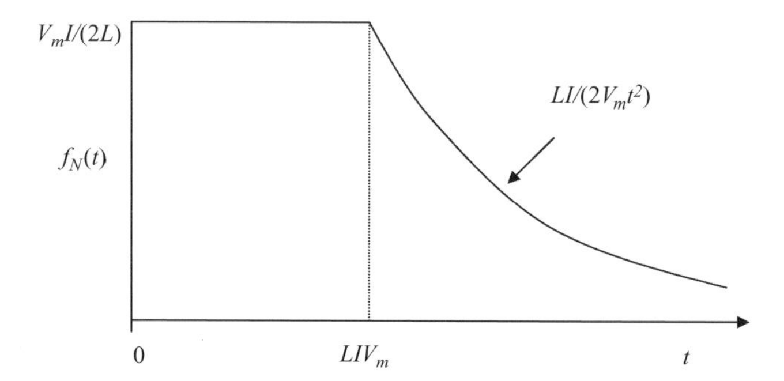

Figure 9.5 Dwell-time density function, one-dimensional system, random velocity

random variables. The probability density function of a random variable defined as the ratio of two random variables is readily found from the basic principles of probability theory (Papoulis, 1991). In particular, the dwell-time density function $f_N(t)$ may be written as

$$f_N(t) = \int V f_z(tV) f_V(V) dV \tag{9.26}$$

In this case both random variables, z and V, are uniformly distributed, $f_z(z) = 1/L$, $0 \leq z \leq L$; $f_V(V) = 1/V_m$, $0 \leq V \leq V_m$. Hence $f_z(Vt) = 1/L$, $0 \leq Vt \leq L$. Inserting these expressions for the two density functions into (9.26), we find $f_N(t)$ takes on two different forms at two different ranges of t

$$\begin{aligned} f_N(t) &= \int_0^{V_m} \frac{V}{LV_m} dV = \frac{V_m}{2L}, \qquad t \leq \frac{L}{V_m} \\ &= \int_0^{L/t} \frac{V}{LV_m} dV = \frac{L}{2V_m t^2}, \quad t \geq \frac{L}{V_m} \end{aligned} \tag{9.27}$$

This dwell-time density function is sketched in Fig. 9.5. The corresponding probability distribution function, found by integrating the density function, is given by

$$\begin{aligned} F_N(t) &= \int_0^t f_N(t) dt = \frac{V_m t}{2L}, \qquad t \leq \frac{L}{V_m} \\ &= 1 - \frac{L}{2V_m t}, \quad t \geq \frac{L}{V_m} \end{aligned} \tag{9.28}$$

Note that there is a probability of 0.5 that the dwell time $T_N \leq L/V_m$. The two distribution functions, for the fixed-velocity case and the random-velocity case, given respectively by (9.24) and (9.28), are sketched in Fig. 9.6. The constant velocity V_c has been chosen equal to the average velocity $V_m/2$ in the random velocity case. Note the similarity of the results, as to be expected.

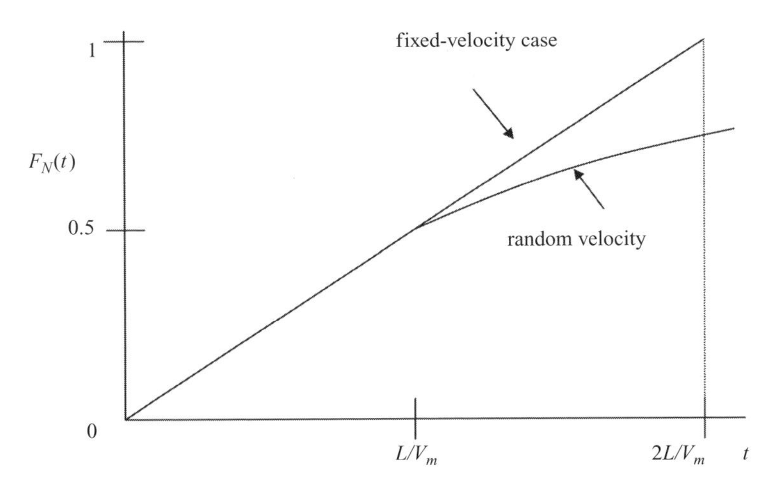

Figure 9.6 Probability distribution functions, dwell time, one-dimenstional system, $V = V_m/2$

Using (9.19), we now calculate the probability P_N a new call will hand off in this case of uniformly-distributed random velocity of mobiles. We have, from (9.27) and (9.19)

$$\begin{aligned} P_N &= \int_0^\infty e^{-\mu t} f_N(t) dt \\ &= \int_0^{L/V_m} \frac{V_m}{2L} e^{-\mu t} dt + \int_{L/V_m}^\infty \frac{L}{2V_m t^2} e^{-\mu t} dt \\ &= \frac{1}{2a}[1 - e^{-a}] + \frac{a}{2} \int_a^\infty \frac{e^{-x}}{x^2} dx \end{aligned} \tag{9.29}$$

Here the parameter $a \equiv \mu L/V_m$. Note that this dimensionless parameter is similar to the parameter $\mu L/V_c$ appearing in (9.25) for the constant mobile velocity case. The discussion there as to the effect of varying the parameter holds true here as well: as the parameter a increases, by increasing the cell length L, decreasing the maximum velocity V_m, or by decreasing the call holding time $1/\mu$, the probability of a handoff decreases. Conversely, as the parameter decreases, the probability of a handoff increases.

Examples

We now provide three examples, using (9.25) and (9.29), of the calculation of the probability P_N of a new-call handoff for various cell sizes and mobile speeds to make these comments more concrete. These examples enable us to compare the fixed (constant) velocity case with that for randomly distributed velocities. We choose $1/\mu = 200$ seconds in all three cases. We also choose $V_c = V_m/2$, i.e., the fixed (constant) velocity is always chosen equal to the average velocity of the random velocity case.

1 Macrocell, $L = 10$ km, constant $V_c = 60$ km/hr, $V_m = 120$ km/hr

$$P_N|_{\text{fixed}} = 0.316 \qquad P_N|_{\text{random}} = 0.30$$

2 Macrocell, $L = 1$ km, constant $V_c = 60$ km/hr, $V_m = 120$ km/hr

$$P_N|_{\text{fixed}} = 0.86 \qquad P_N|_{\text{random}} = 0.78$$

3 Microcell, $L = 100$m, constant $V_c = 5$ km/hr, $V_m = 10$ km/hr

$$P_N|_{\text{fixed}} = 0.84 \qquad P_N|_{\text{random}} = 0.76$$

Note that the probability of new-call handoff for the uniformly distributed velocity case is always less than that of the fixed velocity case. Variation about the average velocity value decreases the probability of a handoff. Cases 2 and 3 are very similar, since both the cell size and velocity have been reduced by almost the same amount. Had the velocity in the microcell case been chosen to be 6 km/hr, the results would have been the same, since the parameters in the two cases would have been the same. But note that handoff in the larger macrocell, that of case 1, is much less likely than that in the smaller cell, that of case 2. This is, of course, as expected.

The dwell-time distributions found above and the calculations of P_N based on them have all been for the case of new calls arriving in a cell. We now repeat the calculations for the case of calls handed off from another cell. Recall that the results would have been the same for these two cases had the dwell-time distribution been exponential or memoryless, as assumed in the previous section. Because of the specific cell geometry used here, the dwell-time distribution is no longer exponential and the two results will differ, as we now show. We refer back to the one-dimensional cell diagram of Fig. 9.4. Consider the fixed-velocity case first. A handoff call will *always* take L/V_c seconds to cross the cell. The dwell-time T_H is thus this constant value, the dwell-time probability density function $f_H(t)$ is the impulse function $\delta(t - L/V_c)$ and the distribution function is $F_H(t) = 0$, $t < L/V_c$; $= 1, t \geq L/V_c$. The probability of handoff P_H is then again the probability that the dwell time T_H is less than the cell holding time T_n, and, from (9.20), is readily found to be given by

$$P_H = e^{-\mu L/V_c} \tag{9.30}$$

This is to be compared with the result (9.25) for the probability of handoff of a new call. Some thought will indicate that P_H is always less than P_N. This is as expected, since a handoff call always has to traverse the full cell length L, while a new call, on the average, has to cover a distance of $L/2$. The chances are thus greater that a handoff call will complete, as compared with a new call, before reaching the other side of the cell and handing off.

Now consider the random velocity case. Recall this was the case in which the velocity of mobiles was randomly chosen from a uniform distribution with maximum value V_m. The average velocity is then $V_m/2$. Since the cell dwell time is always $T_H = L/V$, with V a random variable, we use the common method of determining the density function $f_H(t)$ of T_H by transforming random variables (Papoulis, 1991):

$$f_H(t) = \frac{f_v(V)}{\left|\frac{dt}{dV}\right|} = \frac{L}{V_m t^2}, \quad t \geq \frac{L}{V_m} \tag{9.31}$$

Table 9.2 *Probability of handoff*

P_N		P_H		
Constant V	Random V	Constant V	Random V	$P_H\|_{\text{exponential}}$
1 Macrocell, $L = 10$ km, $V_c = 60$ km/hr, $V_m = 120$ km/hr				
0.316	0.30	0.05	0.074	0.17
2 Macrocell, $L = 1$ km, $V_c = 60$ km/hr, $V_m = 120$ km/hr				
0.86	0.78	0.74	0.64	0.69
3 Microcell, $L = 100$ m, $V_c = 5$ km/hr, $V_m = 10$ km/hr				
0.84	0.76	0.70	0.60	0.64

From (9.20) the probability P_H that a handoff will occur is then given by

$$P_H = \int_{L/V_m}^{\infty} e^{-\mu t} \frac{L}{V_m t^2} dt = a \int_{a}^{\infty} \frac{e^{-x}}{x^2} dx \tag{9.32}$$

The parameter a is again defined to be $\mu L/V_m$. (It is to be noted, by comparing with the equivalent expression (9.29) for P_N, that $P_N = (1 - e^{-a})/2a + P_H/2$. We shall see, from the examples used earlier, that $P_N > P_H$, as was shown above with the fixed-velocity case as well.)

In Table 9.2 we compare the probabilities of handoff P_N and P_H for the three examples considered above, for the two models used, constant and uniformly distributed random velocity, as well as with the results found in the previous section, assuming exponentially distributed dwell time. The average call holding time is 200 seconds in all cases. Note that $P_N > P_H$, as indicated earlier. The probabilities for the two models, with $V_c = V_m/2$, do not differ too much from one another. The probability of handoff assuming exponential dwell time appears to lie between the values of P_N and P_H found using the random velocity model. It is only for example 1, that of the larger macrocell, that significant differences appear between P_N and P_H.

These values for P_N and P_H can now be used in (9.22) to find the average rate of arrival λ_h of handoff calls in terms of new-call arrivals λ_n, particularly if the new-call blocking probability P_b and the handoff dropping probability P_{hf} are small enough to be neglected. In particular, it is of interest to find the ratio λ_h/λ_n, and compare with that obtained using the exponential dwell-time model. The results for the random velocity model compared with the exponential model, with blocking and handoff dropping probabilities neglected, appear as follows for the three cases considered in Table 9.2:

	λ_h/λ_n	
	Random velocity model	Exponential dwell time
1 10 km macrocell	0.32	0.20
2 1 km macrocell	2.2	2.1
3 100 m microcell	1.9	1.8

Note how similar the results are, despite the difference in the models.

As noted, we have neglected the blocking probability P_b and handoff dropping probability P_{hf} in making these comparisons. To find these very significant quantities, by which mobile cell performance is actually measured, we must proceed as in the previous Section 9.1 by defining an admission control strategy such as the guard-channel technique. We shall, in fact, compare the guard-channel admission control strategy with another, predictive, strategy shortly. But to analyze the guard-channel procedure we must first determine the channel occupancy or holding time as was done in the previous section. The difference here is that we have two cases to consider – channel holding time for handoff calls and holding time for new calls. For this purpose, we return to (9.1) and generalize that equation, as well as (9.2) based on it. Let T_{HH} be the channel holding time for handoff calls and T_{HN} the corresponding channel holding time for new calls. Both are random quantities whose probability distributions can be found using (9.2) for each. Specifically, T_{HH} must be the smaller of the dwell time for handoff calls and the call length (call holding time). For a channel in a cell is held until the call hands off to another cell, or the call is completed, whichever comes first. With T_n again representing the random call length with exponential distribution $F_n(t)$ assumed, we have

$$T_{HH} = \min(T_n, T_H) \tag{9.33}$$

Letting $F_{HH}(t)$ be the probability distribution function of T_{HH}, while recalling from our analysis just completed that $F_H(t)$ is the distribution function of the cell dwell time whose determination we just completed, we have, comparing (9.33) with (9.1) and (9.2)

$$1 - F_{HH}(t) = [1 - F_n(t)] \bullet [1 - F_H(t)] \tag{9.34}$$

Note again that this equality from probability theory equates the complementary distribution function to the product of the two other complementary distribution functions.

Similarly, letting $F_{HN}(t)$ be the distribution function of the channel holding time T_{HN} for new calls, we must have

$$T_{HN} = \min(T_n, T_N) \tag{9.35}$$

and therefore

$$1 - F_{HN}(t) = [1 - F_n(t)] \bullet [1 - F_N(t)] \tag{9.36}$$

Let $F_c(t)$ be the overall probability distribution function of the channel holding time under either condition, new calls or handoff calls. This is given by the weighted sum of the two channel holding time distributions $F_{HH}(t)$ and $F_{HN}(t)$, weighting each by the fraction of overall traffic it generates. The average new-call traffic actually carried by a cell is $\lambda_{nc} = \lambda_n(1 - P_b)$; the corresponding handoff traffic carried by a cell is $\lambda_{hc} = \lambda_h(1 - P_{hf})$, with λ_h found in terms of λ_n from (9.22). The overall traffic is just $\lambda_{nc} + \lambda_{hc}$, so that the channel holding time distribution $F_c(t)$ is given by the weighted sum

$$F_c(t) = \frac{\lambda_{nc}}{\lambda_{nc} + \lambda_{hc}} F_{HN}(t) + \frac{\lambda_{hc}}{\lambda_{nc} + \lambda_{hc}} F_{HH}(t) \tag{9.37}$$

Using (9.34) and (9.36), and the assumption that the call-length distribution $F_n(t)$ is exponential with average value $1/\mu$, i.e., $F_n(t) = 1 - \exp(-\mu t)$, the assumption made throughout this chapter, it is left for the reader to show that (9.37) may be written in the following form involving the two cell dwell-time distributions $F_N(t)$ and $F_H(t)$ directly

$$F_c(t) = 1 - e^{-\mu t} + \frac{e^{-\mu t}}{1+\gamma_c}[F_N(t) + \gamma_c F_H(t)] \tag{9.37a}$$

The parameter γ_c appearing in (9.37a) is defined to be the ratio $\lambda_{hc}/\lambda_{nc}$ of handoff to new-call carried traffic.

Note again that the quantities P_b and P_{hf} appearing implicitly in (9.37) and (9.37a) through λ_{hc}, λ_{nc}, and γ_c depend on the admission control procedure specified. As was the case in the previous section, an iterative procedure has to be used to find these quantities once that procedure is specified. There is only one problem with the use of these equations in applying the guard-channel admission procedure described in the previous section. The assumption made there was that the channel holding time was exponentially distributed in order to obtain the state diagram of Fig. 9.3 from which the local balance equations such as (9.11) are obtained. This is clearly no longer the case, as is apparent from (9.37a). We thus follow the approach in Hong and Rappaport (1999) which approximates the channel holding time distribution by an equivalent exponential distribution. The exponential distribution chosen is one that retains the actual average channel holding time $1/\mu_c$. The average channel holding time is clearly given by $\int t f_c(t)\mathrm{d}t$ with $f_c(t)$ the channel holding time probability density function. For well-behaved functions, this may also be written as $\int [1 - F_c(t)]dt = \int F_c^C(t)dt$, $F_c^C(t)$ the complementary distribution function. The equivalent exponential distribution is then $1-\exp(-\mu_c t)$, with average value $1/\mu_c$. It is thus apparent that using the same average value for both distributions is equivalent to writing

$$\int_0^\infty \left[F_c^C(t) - e^{-\mu_c t}\right]dt = 0 \tag{9.38}$$

An alternative interpretation of the approximating exponential is then one with the integral of the difference between the complementary channel holding time distribution and the exponential set equal to zero (Hong and Rappaport, 1999). From (9.38) and (9.37a) we readily find

$$\frac{1}{\mu_c} = \frac{1}{\mu} - \frac{1}{1+\gamma_c}\int_0^\infty e^{-\mu t}[F_N(t) + \gamma_c F_H(t)]dt \tag{9.39}$$

We note again that $\gamma_c \equiv \lambda_{hc}/\lambda_{nc}$, $\lambda_{hc} = \lambda_h(1 - P_{hf})$, and $\lambda_{nc} = \lambda_n(1 - P_b)$ (Hong and Rappaport, 1986). The value of μ_c given by (9.39) is to be used in any calculation of the performance of the guard-channel admission control procedure.

Example

An example of the calculation of $1/\mu_c$ using (9.39) is useful at this point. We take the fixed velocity model for simplicity. We also assume P_b and P_{hf} are both zero. From (9.24) we have $F_N(t) = V_c t/L, 0 \le t \le L/V_c; = 1, t \ge L/V$. We also have $F_H(t) = 0, t < L/V_c; = 1, t \ge L/V_c$. For the 10 km macrocell with a mobile velocity of 60 km/hr, and $1/\mu = 200$ sec, we find from (9.39) that $1/\mu_c = 0.75/\mu = 150$ sec. The comparable result obtained in the previous section, with the exponential handoff model assumed, was 170 sec. For the 100 m microcell with a mobile velocity of 5 km/hr, we find $1/\mu_c = 0.26/\mu = 53$ sec. The exponential handoff model gave a value of 72 sec. As in the exponential handoff model case, the difference in channel occupancy or holding time between the two cell size examples is as expected. The much smaller microcell is traversed much more rapidly by the mobile user, even when traveling at the walking gait of 6 km/hr. A channel is, therefore, held for much less time, on the average.

We now return to admission control mechanisms. Recall that some type of admission control procedure is necessary to ensure cell performance is maintained. A common procedure in telephone systems is to block newly arriving calls if the traffic experienced becomes too great. The traffic in this case is measured in Erlangs, defined as the product of the call arrival rate and the average call holding time. We introduced blocking probability as a performance measure in earlier chapters. Mobile systems introduce handoff dropping probability as well. In the previous section we noted that it is desirable to keep handoff dropping probability quite low, since, once a call is established, a mobile user finds it objectionable to have the call dropped as the user moves into a new cell. For a given cell channel capacity, this implies giving priority to handoff calls. The guard-channel procedure described in the previous section does precisely this. Another handoff priority procedure that has been proposed and studied by a number of investigators queues handoff calls until a channel becomes available (Hong and Rappaport, 1999). A number of other admission control strategies have been proposed and compared as well. We outline one such procedure here, focusing only on the one-dimensional cellular system of this section, and compare its performance in that case with that of the guard-channel scheme (Naghshineh and Schwartz, 1996). The procedure is a distributed one, involving information transfer between adjacent cells, and sets thresholds on the number of users that can be admitted to a cell so that current calls handing off in the near future will be ensured of having channels available to them.

Consider three adjacent cells in this one-dimensional system. Let cell C_n be the cell where a call admission request is made. Let cells C_l and C_r be its neighboring cells, located, respectively, to the left and right of C_n. Let P_i be the probability of having i calls in a cell. Say a cell can N calls. We define the overload probability as the probability that the number of calls in a cell exceeds N. This probability is to be specified at some tolerable value q. In admitting a new call to cell C_n, this objective is to be maintained. To achieve this objective, the cell admission controller uses the knowledge of the current number of calls in cells C_n, C_l, and C_r, as well as an estimate of the number of calls in those cells a time T sec in the future, roughly equal to the time required to handoff. The admission control strategy is thus essentially a predictive one, estimating whether channels will be available to new calls in the event of handoff later on. Specifically, three admission conditions are to be satisfied:

1 T sec in the future, the overload probability of cell C_n affected by handoffs from C_r or C_l to C_n, as well as handoffs from C_n to neighboring cells, must be smaller than q.

Naghshineh, M. and M. Schwartz. 1996. "Distributed call admission control in mobile/wireless networks," *IEEE Journal on Selected Areas in Communications*, 14, 4 (May), 711–717.

2 T sec in the future, the overload probability of cell C_r affected by handoffs from cell C_n or the cell to the right of C_r to cell C_r, and including from cell C_r to any other cell during the interval T, must be smaller than q.
3 Same as 2, focusing, however, on cell C_l instead of cell C_r.

These admission conditions can be quantified in the case of homogeneous conditions as follows:

Let p_m be the probability of handoff from a cell in T sec. If we assume exponential handoff statistics with average rate η, as was done in the previous section, p_m is just $1 - e^{-\eta T}$. With homogeneous conditions, the probability of handing off to the left or right is just $p_m/2$. The probability of *not* handing off in T sec is $1 - p_m = e^{-\eta T}$, defined as p_s. Consider a given cell now. Say there are k calls in the cell at present. The probability that T sec later i of these calls will not have handed off is the binomial distribution

$$B(i, k, p_s) = \frac{k!}{i!(k-i)!} p_s^i (1 - p_s)^{k-i} \tag{9.40}$$

This distribution has the mean value kp_s and variance $kp_s(1 - p_s)$. For large k this may be approximated by a gaussian distribution with the same mean and variance. (Recall that the binomial distribution is the sum of binary random variables. The binary condition here is whether or not a call hands off. By the Central Limit Theorem of probability the distribution of a sum of random variables approaches the gaussian or normal distribution.)

Now consider cell C_n, the one for which the admission control, on the arrival of a new call, is to be invoked. Say it has n calls in progress at the time of arrival of a new call, including that call. Say cells C_r and C_l have, respectively, r and l calls in progress at the arrival of this new call in cell C_n. It is left to the reader to show, based on the discussion above, that the number of calls in cell C_n T sec later can be approximated by a gaussian distribution with mean value $E(m) = np_s + (r + l)\, p_m/2$ and variance $\sigma^2 = np_s(1 - p_s) + (r + l)\, p_m/2\, (1 - p_m/2)$. The overload probability is then approximated by the probability this gaussian distribution exceeds the capacity N. If we set the overload probability to the maximum value q, we can solve for the corresponding maximum value of the number of calls n the cell will support. This provides the first of the three admission conditions outlined above. Specifically, using the gaussian approximation, we find

$$\begin{aligned} q &= \int_N^\infty \frac{e^{-[x-E(m)]^2/2\sigma^2}}{\sqrt{2\pi\sigma^2}} dx \\ &= \frac{1}{2}\text{erfc}(a) \end{aligned} \tag{9.41}$$

where erfc(a) is the complementary error function defined as

$$\text{erfc}(a) \equiv \frac{2}{\sqrt{\pi}} \int_a^\infty e^{-x^2} dx \tag{9.42}$$

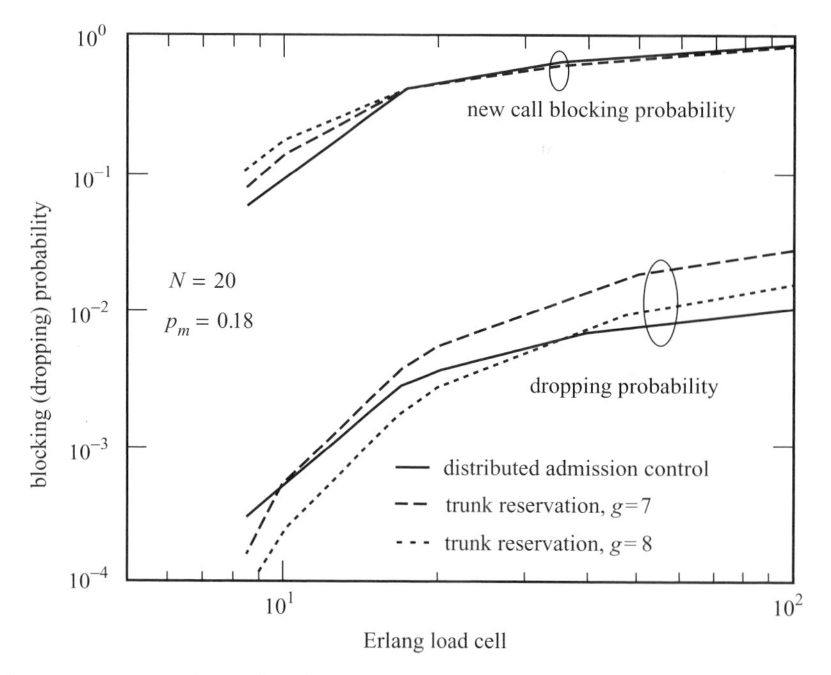

Figure 9.7 Comparison of trunk reservation to state estimation for $N = 20$ (from Naghshineh and Schwartz, 1996, Fig. 6, by permission)

and the parameter a is given by

$$a = \frac{N - E(m)}{\sqrt{2\sigma^2}} = \frac{N - (np_s + (l + r)p_m/2)}{\sqrt{2\left[np_s(1 - p_s) + (l + r)\frac{p_m}{2}(1 - p_m/2)\right]}} \tag{9.43}$$

Given the maximum overload probability q, we use (9.41) and (9.42) to find the corresponding value of a. We then solve for the appropriate value of n, using (9.43). We call this value n_1, since it represents the first admission condition. We repeat the same process for admission conditions 2 and 3, focusing, respectively, on cells C_r and C_l, obtaining values n_2 and n_3 for the allowable values of n in these two cases. The final admission decision comes by selecting the smallest of the three values of n obtained as the maximum number of calls cell C_n will support

$$n = \min(n_1, n_2, n_3) \tag{9.44}$$

Figures 9.7 and 9.8 show the results of simulations made comparing this distributed state estimation admission control scheme with the guard-channel approach (Naghshineh and Schwartz, 1996). Simulations were carried out using a one-dimensional system consisting of ten cells arranged on a circle. Values of the guard-channel reservation parameter g were selected to provide roughly comparable performance between the two

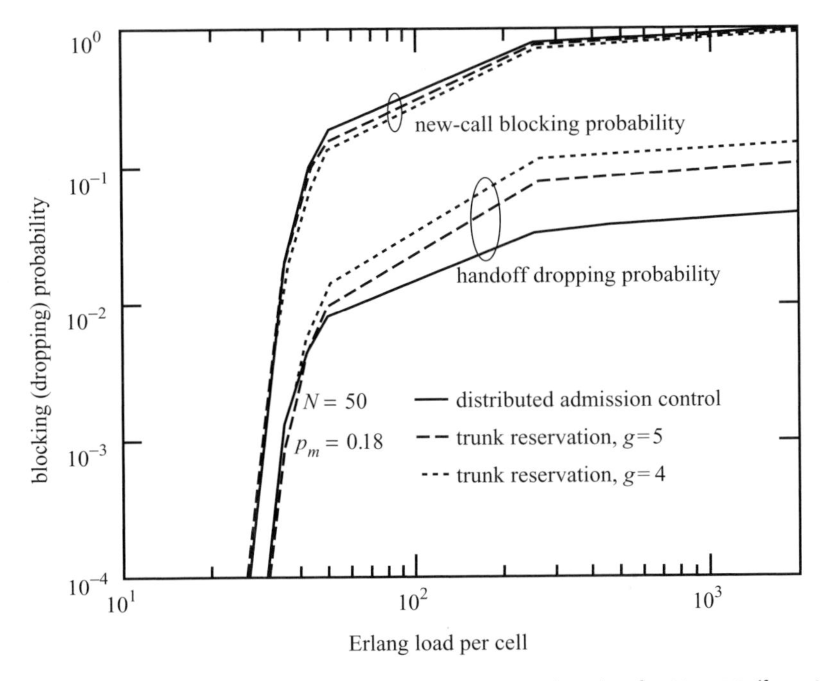

Figure 9.8 Comparison of trunk reservation to state estimation for $N = 50$ (from Naghshineh and Schwartz, 1996, Fig. 7, by permission)

schemes at a nominal Erlang load per cell of $0.85N$. The average call duration was set to 500 seconds. The average handoff time was taken as $1/\eta = 100$ sec, and the estimation time T selected as 20 sec. The resultant probability of handing off in T sec is then $p_m = 1 - e^{-\eta T} = 0.18$. Figure 9.7 compares the two admission control schemes as the Erlang load varies, for the case of $N = 20$ channels per cell; Fig. 9.8 provides the same comparison at $N = 50$ cells. The distributed scheme tends to provide lower dropping probability performance at higher loads. By thus taking the state of neighboring cells into account, system resources may be utilized more efficiently at higher loads than by using the static guard-channel approach. Even better performance may be obtained by modifying the distributed admission control scheme (Epstein and Schwartz, 1998). On a new call arriving, the predicted dropping probability is calculated by each cell, and the call is admitted if the largest of the three probabilities is less than or equal to the parameter q. It is then found that the handoff dropping probability remains almost constant with load. This algorithm has been extended to the case of multiclass traffic, with the different classes requiring different channel bandwidths and having different holding time characteristics (Epstein and Schwartz, 1998). Two other distributed multiclass admission control algorithms have been proposed that use prediction to obtain a load-independent handoff dropping probability, while providing a specified blocking-probability profile

---

Epstein, B. and M. Schwartz. 1998. QoS-Based Predictive Admission Control for Multi-Media Traffic, in *Broadband Wireless Communications*, ed. M. Luise and S. Pupolin, Berlin, Springer-Verlag, pp. 213–224.

(Epstein and Schwartz, 2000). Other admission control strategies have been proposed in Yu and Leung (1997), Choi and Shin (1998), Chao and Chen (1997), Levine *et al.* (1997), among others. Yu and Leung (1997) describe a dynamic version of the guard channel technique. Sutivong and Peha (1997) provide a comparison of a number of call admission control strategies.

In the next section we extend the work of this section to that of modeling two-dimensional cellular networks. The analytic approaches are similar, but somewhat more complex because of the geometry. Some of the equations obtained here for probability distributions are independent of the specific geometry and hence can be used directly.

9.3 Two-dimensional cells

We now extend the one-dimensional analysis of the previous section to the case of two-dimensional cells. In doing this we shall find that some of the analysis and basic equations used in the previous section survive intact, being independent of dimensionality. We follow here the work of Hong and Rappaport (1999), deviating only in carrying out the analysis for the case of constant (fixed) velocity. This does simplify the analysis. Hong and Rappaport provide solutions for the more general case of mobile velocity drawn from a uniformly distributed random velocity distribution. Recall, however, from our analysis in the previous section that results for the fixed velocity case do agree fairly well with those for the uniformly distributed velocity case. We shall also refer to work presented in Guerin (1987) which uses a somewhat different model than that of Hong and Rappaport in studying the performance of two-dimensional cellular systems.

The objective here is to again determine the channel holding time distribution $F_c(t)$, as given by (9.37a). Approximating this distribution by an equivalent exponential distribution with average channel holding time $1/\mu_c$, as was done in (9.38) and (9.39) of the previous section, we can evaluate the blocking probability and handoff dropping probability performance of the guard-channel admission control strategy, as well as other handoff priority admission control strategies proposed in the literature. One such predictive procedure was described in the previous section. Others appearing in the literature were referenced in the previous section. Hong and Rappaport evaluate the performance as well, for the two-dimensional case, of a scheme in which a handoff call is buffered if a

---

Epstein, B. and M. Schwartz. 2000. "Predictive QoS-based admission control for multiclass traffic in cellular wireless networks," *IEEE Journal on Selected Areas in Communications*, 18, 3 (March), 523–534.

Yu, O. and V. Leung. 1997. "Adaptive resource allocation for prioritized call admission in ATM-based wireless PCN," *IEEE Journal on Selected Areas in Communications*, 15, 9 (September), 1208–1225.

Choi, S. and K. G. Shin. 1998. "Predictive and adaptive bandwidth reservation for handoffs in QoS-sensitive cellular networks," Proc. SIGCOMM98, 155–166.

Chao, C. and W. Chen. 1997. "Connection admission control for mobile multiple-class personal communication networks," *IEEE Journal on Selected Areas in Comunications*, 15, 10 (October), 1618–1626.

Levine, D. *et al.* 1997. "A resource estimation and admission control algorithm for wireless multimeida networks using the shadow cluster concept," *IEEE/ACM Transactions on Networking*, 5, 1 (February), 1–12.

Sutivong, A. and J. Peha. 1997. "Call admission control algorithms: proposal and comparison," Proc. IEEE Globecom.

Guerin, R. A. 1987. "Channel occupancy time distribution in cellular radio systems," *IEEE Transactions on Vehicular Technology*, VT-35, 3 (August), 89–99.

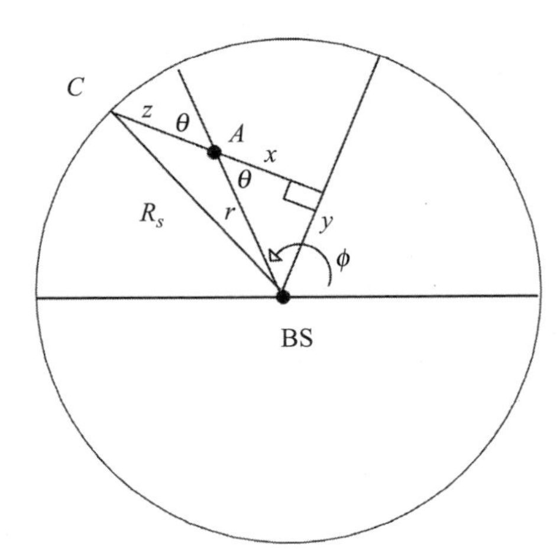

Figure 9.9 New-cell dwell-time calculation, circular-cell model

channel is not immediately available on handoff to a new cell. This procedure obviously trades off handoff dropping probability for a delay in receiving a channel. If the delay is measured in fractions of a second, a mobile user will clearly not notice the effect.

Recall, from (9.37a), that the channel holding time distribution $F_c(t)$ depends on the cell dwell-time distributions $F_N(t)$ and $F_H(t)$, for new and handoff calls, respectively. Calculations of these two distributions were carried out in the previous section for a one-dimensional geometry for both constant (fixed) and uniformly distributed random velocities. We now repeat the calculation for a two-dimensional geometry, following the approach of Hong and Rappaport, but for the constant-velocity case only. Once the dwell-time distributions have been determined, the channel holding time distribution follows directly from (9.37a). As noted earlier, Hong and Rappaport carry the calculations out for the random velocity case. A circular cell geometry is assumed, as indicated in Fig. 9.9, to simplify the calculations. Guerin (1987) uses hexagonal cells, resulting in somewhat more complex calculations. To relate results obtained here to the hexagonal case we equate cellular areas in the two cases, just as was done in Chapter 6, in comparing CSMA interference calculations using these two cellular geometries (see Figs. 6.6 and 6.7). Letting R be the hexagonal cell radius, and R_S be the equivalent circular cell radius, we have $\pi R_s^2 = 3\sqrt{3}R^2/2$, or $R_S = 0.91R$.

Consider new-call arrivals first. Mobiles are assumed to be uniformly distributed anywhere in the cell. Say a new call begins with the mobile at point A in Fig. 9.9. This point is located at an angle ϕ with respect to the horizontal axis, and at a distance r from the base station, taken to be at the center of the cell as shown. The mobile at point A is assumed to be moving with a fixed velocity V_c. The angle ϕ is a random value, uniformly distributed over the range 0 to 2π, because of the assumption that mobiles are uniformly distributed over the cell. The mobile, moving continuously in the same direction, reaches the boundary of the circular cell at point C, a distance z from A. The cell dwell time is then the time for the mobile to cover the distance z. This time, $T_N = z/V_c$, is the random variable whose distribution $F_N(t)$ we would like to calculate. For fixed V_c, the distribution is obviously proportional to $F_z(z)$, the distribution of z. It is this distribution we first have to calculate.

We do this, following Hong and Rappaport, by using the random angle θ shown in Fig. 9.9, in place of ϕ, determining the joint distribution of θ and r, replacing this by the equivalent joint distribution of x and y, the two (random) variables also defined in Fig. 9.9, and then relating z to these quantities. Here $x = r\cos\theta$, while $y = r\sin\theta$. Specifically, note that, because of the assumption of uniform distribution of mobiles throughout the circular cell, θ must be uniformly distributed from 0 to π, while the density function $f_r(r)$ of r is $2r/R_s^2$. (It is clear that these two random variables are independent of one another.) Since the elementary probability of an event in a differential area is independent of the choice of coordinates, whether x, y or r, θ, we must have

$$f_{x,y}(x, y)dxdy = f_r(r)f_\theta(\theta)drd\theta = \frac{2r}{\pi R_s^2}drd\theta \tag{9.45}$$

But $dxdy = rdrd\theta$. We thus have

$$f_{x,y}(x, y) = 2/\pi R_s^2 \tag{9.46}$$

But note from Fig. 9.9 that $(z + x)^2 + y^2 = R_s^2$. Using this equality, one can transform $f_{x,y}(x, y)$ to an equivalent joint density function $f_{z,w}(z, w)$, w a dummy variable, from which one readily finds the desired density function $f_z(z)$ of z. Specifically, let $w = x$. Then we have $y^2 = R_s^2 - (z + w)^2$. We now use the following standard transformation (Papoulis, 1991) of two jointly distributed random variables to another set of two rvs to find $f_{z,w}(z, w)$

$$f_{z,w}(z, w) = \frac{f_{x,y}(x, y)}{|J(x, y)|} \tag{9.47}$$

with $J(x, y)$ the Jacobian of the transformation defined as the determinant

$$J(x, y) = \begin{vmatrix} \frac{\partial z}{\partial x} & \frac{\partial z}{\partial y} \\ \frac{\partial w}{\partial x} & \frac{\partial w}{\partial y} \end{vmatrix} = \begin{vmatrix} \frac{\partial x}{\partial z} & \frac{\partial x}{\partial w} \\ \frac{\partial y}{\partial z} & \frac{\partial y}{\partial w} \end{vmatrix}^{-1} \tag{9.48}$$

This transformation is the extension to the two-variable case of the single-variable transformation used in (9.31) of the previous section. (This transformation could have been used in finding $f_{x,y}(x, y)$ from $f_r(r)f_\theta(\theta)$ above. But the procedure used was much more direct and simpler in that case.)

It is left for the reader to show, by calculating the partial derivatives indicated in the second equation of (9.48) and then evaluating the resultant determinant, that the desired probability density function $f_{z,w}(z, w)$ in the case here is given by

$$f_{z,w}(z, w) = \frac{|z + w|}{\sqrt{R_s^2 - (z + w)^2}}, \frac{2}{\pi R_s^2} \tag{9.49}$$

To find $f_z(z)$, we now integrate over the variable w

$$f_z(z) = \int_w f_{z,w}(z, w)dw \tag{9.50}$$

But recall that $(w + z) = \sqrt{(R_s^2 - y^2)}$. It is clear that we must have $0 \le y \le R_s$ (see Fig. 9.9 and note that $y = r \sin\theta$, with $r \le R_S$). Hence $0 \le w + z \le R_s$. From Fig. 9.9, however, we note that $0 \le z \le 2R_s$, while, with $w = x$, $-R_s \le w \le R_s$. Combining these various inequalities, we must have $-z/2 \le w \le R_s - z$. We thus have, finally, from (9.49) and (9.50)

$$f_z(z) = \int_{-z/2}^{R_s - z} \frac{2}{\pi R_s^2} \frac{(z+w)}{\sqrt{R_s^2 - (z+w)^2}} dw$$
$$= \frac{2}{\pi R_s^2} \sqrt{R_s^2 - (z/2)^2}, \quad 0 \le z \le 2R_s \tag{9.51}$$

The desired distribution of the new-call cell dwell time T_N is now found, as noted earlier, by writing $T_N = z/V_c$, with V_c, the velocity, a given, constant value. In the general case where both z and V are random variables, one of the cases worked out in the previous section, we could again use (9.26) to determine $f_N(t)$ from (9.51) and an assumed velocity distribution. This is what is done in Hong and Rappaport (1999) for the two-dimensional circular cell under consideration here. The calculation, although fairly straightforward, is rather tedious. We thus choose to carry out the much simpler calculation for the constant, fixed velocity case. As noted earlier, based on our discussion in the previous section of the one-dimensional case, one would not expect the results to differ substantially from the constant-velocity case. We do retain some randomness in mobile movement here because mobiles are assumed to be moving in random directions, albeit all with the same velocity. The interested reader can refer to Hong and Rappaport for the calculation of the more general case of uniformly distributed random mobile velocities.

With the velocity of all mobiles chosen to be a constant value V_c, the probability density function of the new-call cell dwell time is found simply from (9.51) to be given by

$$f_N(t) = V_c f_z(V_c t) = \frac{2V_c}{\pi R_s^2} \sqrt{R_s^2 - \left(\frac{V_c t}{2}\right)^2},$$
$$0 \le t \le 2R_s/V_c \tag{9.52}$$

This expression may be written in a somewhat simpler, normalized form by defining a parameter $K \equiv V_c/R_S$. We then get

$$f_N(t) = \frac{2K}{\pi} \sqrt{1 - \left(\frac{Kt}{2}\right)^2}, \qquad 0 \le \frac{Kt}{2} \le 1 \tag{9.52a}$$

The corresponding probability distribution, $F_N(t) = \text{Prob}(T_N \le t)$ is therefore given by

$$F_N(t) = \int_0^t f_N(t)dt = \frac{2}{\pi}\left[\frac{Kt}{2}\sqrt{1 - \left(\frac{Kt}{2}\right)^2} + \sin^{-1}\left(\frac{Kt}{2}\right)\right] \tag{9.53}$$

after carrying out the integration indicated.

A little thought will indicate that $F_N(t)$ ranges monotonically from 0 to 1, as expected. In particular, for $Kt/2 = 1$, or $t = 2/K = 2R_S/V_c$, $F_N(t) = 1$. For $Kt/2 = 1/2$, or

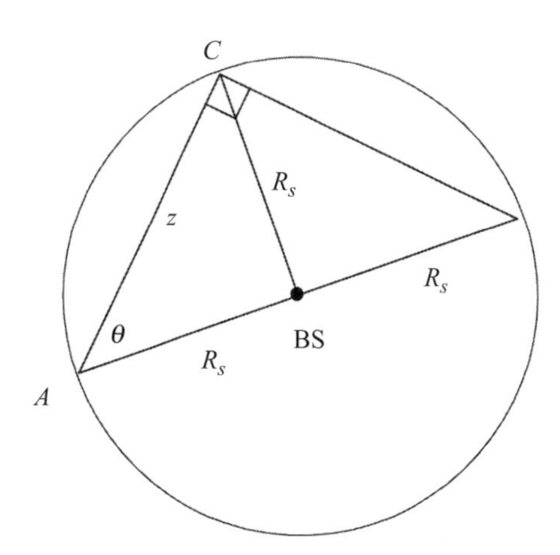

Figure 9.10 Handoff-call dwell-time calculation

$t = R_S/V_c$, $F_N(t) = 0.609$. Thus there is a 61% probability that the mobile, carrying a new call attempt, will leave the cell within the time R_S/V_c. For $Kt/2 = 0.707$, or $t = 1.414R_s/V_c$, $F_N(t) = 0.818$.

Now consider determining the dwell-time distribution $F_H(t)$ of mobiles with handoff calls. Figure 9.10 portrays the geometry involved. A mobile enters a new cell at point A on the boundary, moves a distance z through the cell at the velocity V_c and randomly chosen angle θ, as shown, and exits at a point C, also on the boundary. The dwell time is then just z/V_c, with the velocity again taken to be the constant value V_c. The distance z covered from initial handoff to this cell, at point A, to the next handoff to a neighboring cell at point C may, from the geometry indicated in Fig. 9.10, be written as $z = 2R_S\cos\theta$. The angle θ is uniformly distributed over the range $-\pi/2$ to $\pi/2$ because of our assumption of uniform distribution of mobiles throughout a cell. We first find the distribution $F_z(z)$, and from this the desired distribution $F_H(t)$. The distribution $F_z(z)$ may be determined directly from the uniformly distributed distribution of θ by writing

$$\begin{aligned} F_z(z) &= \text{Prob}(Z \leq z) = \text{Prob}\left(|\theta| \geq \cos^{-1}\frac{z}{2R_s}\right) \\ &= 1 - \frac{2}{\pi}\cos^{-1}(z/2R_s) \qquad 0 \leq z \leq R_s \end{aligned} \tag{9.54}$$

This distribution may also be written in the following equivalent form

$$F_z(z) = \frac{2}{\pi}\sin^{-1}(z/2R_s) \tag{9.54a}$$

The probability density function $f_z(z)$ is just the derivative of $F_z(z)$ and is thus given by

$$f_z(z) = \frac{1}{\pi R_s}\frac{1}{\sqrt{1-(z/2R_s)^2}} \tag{9.55}$$

Since we assume here that the velocity V is a constant value V_c, we immediately have, with $t = z/V_c$, $f_H(t) = V_c f_z(V_c t)$, or

$$f_H(t) = \frac{V_c}{\pi R_s} \frac{1}{\sqrt{1 - (V_c t/2R_s)^2}} \tag{9.56}$$

This handover dwell-time density function $f_H(t)$ can also be written in normalized form by letting V_C/R_S again be defined as the parameter K. We then have

$$f_H(t) = \frac{K}{\pi} \frac{1}{\sqrt{1 - (Kt/2)^2}} \qquad K \equiv V_c/R_s \tag{9.56a}$$

The distribution function $F_H(t)$ is now given by

$$F_H(t) = \int_0^t f_H(t)dt = \frac{2}{\pi} \sin^{-1}\left(\frac{Kt}{2}\right) \tag{9.57}$$

(This result is to be compared with that of (9.54a) for $F_z(z)$, and could, in fact, have been written down directly from (9.54a) in this case of constant velocity, since $t = z/V_c$.) Note, by comparing (9.57) and (9.53), that we can also write

$$F_N(t) = F_H(t) + \frac{2}{\pi} \frac{Kt}{2} \sqrt{1 - \left(\frac{Kt}{2}\right)^2} \tag{9.53a}$$

With the two cell dwell-time distributions, $F_N(t)$ and $F_H(t)$, known, we are in a position, using (9.37a), to calculate the channel holding time distribution $F_c(t)$, just as we did in the previous one-dimensional case. This we shall shortly do, comparing these results with the equivalent exponential distribution, to see how good the exponential approximation is for channel holding time. Recall that this was the assumption made in carrying out the analysis of the guard-channel admission control scheme. But first we calculate the two probabilities of handoff, P_N and P_H, as we did in the previous section, to compare the two-dimensional results with those of the one-dimensional case. We again use (9.19) and (9.20) for this purpose. Consider the probability of handoff P_N for the new-call case first. We then have, using (9.19) and (9.52)

$$P_N = \int_0^\infty e^{-\mu t} f_N(t)dt = \frac{4}{\pi} \int_0^1 \sqrt{1 - x^2} e^{-bx} dx$$
$$b \equiv 2\mu R_s/V_c \tag{9.58}$$

The dimensionless parameter b is comparable with the dimensionless parameter $a \equiv \mu L/V_c$ appearing in the one-dimensional cellular model of the previous section. We shall, in fact, compare one- and two-dimensional examples with the two-dimensional diameter $2R_S$ set equal to the one-dimensional cell length L.

Table 9.3 *Probabilities of handoff*

One-dimensional case			Two-dimensional case		
P_N	P_H	γ_c	P_N	P_H	γ_c
1 Macrocell, $2R_S = L = 10$ km, $V_c = 60$ km/hr, $b = 3$					
0.32	0.05	0.34	0.37	0.23	0.48
2 Macrocell, $2R_S = L = 1$ km, $V_c = 60$ km/hr, $b = 0.3$					
0.86	0.74	3.31	0.88	0.83	5.18
3 Microcell, $2R_S = L = 100$ m, $V_c = 5$ km/hr, $b = 0.36$					
0.84	0.70	2.8	0.86	0.80	4.3

The probability P_H of handoff, given a mobile arriving at a cell with a call already in progress, is similarly found using (9.20) and (9.56)

$$P_H = \int_0^\infty e^{-\mu t} f_H(t)dt = \frac{2}{\pi}\int_0^1 \frac{e^{-bx}}{\sqrt{1-x^2}}dx \tag{9.59}$$

The dimensionless parameter $b \equiv 2\mu R_S/V_c$, just as it was defined above in writing the expression for P_N. In Table 9.3 we provide examples of the calculation of the probabilities of handoff for the same three cases tabulated in Table 9.2 in the previous section. We also include the probabilities of handoff for the one-dimensional case of Table 9.2 to provide a comparison between the one- and two-dimensional cases. Also tabulated for comparison are the values of γ_c defined to be the ratio of handoff traffic λ_h to new-call arrival traffic λ_n, assuming blocking probability and handoff dropping probability are both very small. From (9.22) this is just $P_N/(1 - P_H)$. This gives us a single parameter with which to compare results. These values of γ_c will also be needed in the calculation of the channel holding-time distribution $F_c(t)$, as is apparent from our prior result (9.37a). As noted above, we set the one-dimensional cell length L equal to the circular cell diameter $2R_S$ in carrying out the comparison. The parameter b introduced here is thus made equal to the parameter $a \equiv \mu L/V_c$ introduced in the one-dimensional case. (We choose the constant-velocity model only in making these comparisons.) The average call holding time $1/\mu$ is again taken to be 200 seconds in all three examples.

The results for the one- and two-dimensional examples are comparable, although, in all three examples, $\gamma_c|_{\text{one-dimen}} < \gamma_c|_{\text{two-dimen}}$. Probabilities of handoff are much higher for the smaller macrocell than for the larger macrocell, given the same mobile velocity in each case. These results are as expected, the same comment we made in previous sections. It is also readily shown that if we select the radius R_S of the two-dimensional circular cell to be equal to, rather than one-half, the length L of the comparable one-dimensional cell, the γ_cs obtained are uniformly less than those for the two-dimensional cases. Results for the one-dimensional case are thus bounded by the two-dimensional results. We shall return to these calculations of γ_c later in comparing results for this two-dimensional circular cell model with those appearing in Guerin (1987) using hexagonal cell structures and a somewhat different modeling approach.

We now return to the calculation of the channel holding time distribution $F_c(t)$ and its approximation by an equivalent exponential distribution having the same average value. Recall from the previous section, as reiterated above, that the exponential approximation is necessary to calculate the performance of the guard-channel admission control scheme. Its use, if shown accurate enough, simplifies other performance calculations as well. The expression for $F_c(t)$ for this two-dimensional circular cell model, assuming constant mobile velocity, is written directly from (9.37a), using (9.53) and (9.57)

$$F_c(t) = 1 - e^{-\mu t}\left[1 - \frac{2}{\pi}\sin^{-1}(Kt/2) - \frac{Kt}{\pi(1+\gamma_c)}\sqrt{1-\left(\frac{Kt}{2}\right)^2}\right] \tag{9.60}$$

Recall again that $K \equiv V_c/R_S$, V_c the velocity, assumed constant or fixed, of all mobiles, and R_S the radius of the cell. The parameter γ_c is, as noted above, the ratio of handoff to new-call traffic, and is just $P_N/(1 - P_H)$ for small blocking and handoff dropping probabilities. Note, as required, that (9.60) has as the limits of time t, $0 \leq t \leq 2/K = 2R_S/V_c$. Equation (9.60) may be written in a normalized form better suited for plotting by defining a variable $x \equiv Kt/2 = V_c t/2R_S$. We then get for $F_c(x)$

$$F_c(x) = 1 - e^{-bx}\left[1 - \frac{2}{\pi}\left(\sin^{-1}x + \frac{x\sqrt{1-x^2}}{1+\gamma_c}\right)\right]$$
$$0 \leq x \leq 1 \tag{9.60a}$$

The parameter b is just the quantity $2\mu R_S/V_c$ defined earlier.

We now use (9.39) to calculate the average channel holding time $1/\mu_c$, to be used in the approximating exponential channel holding time distribution, as was done in the previous section. We find, using the normalized form of $F_c(t)$, as given by (9.60a)

$$\frac{1}{\mu_c} = \frac{1}{\mu}\left[1 - e^{-b} - \frac{2b}{\pi}\int_0^1 e^{-bx}\left(\sin^{-1}x + \frac{x\sqrt{1-x^2}}{1+\gamma_0}\right)dx\right] \tag{9.61}$$

As examples of the calculation of $1/\mu_c$, using (9.61), consider the three two-dimensional cases of Table 9.3. For case 1, that of a 5-km radius macrocell with mobile velocity $V_c = 60$ km/hr, $b = 3$, and $1/\mu_c = 0.675/\mu = 135$ sec. For the 0.5 km radius macrocell of case 2, with the same mobile velocity of 60 km/hr, $b = 0.3$, and $1/\mu_c = 0.16/\mu = 32$ sec. These results are, as to be expected, agreeing with those found in the one-dimensional examples of the previous section. As the cell radius decreases, with mobile velocity fixed, the probability of a handoff increases considerably (see Table 9.3), and the average channel connection time decreases rapidly.

How close is the exponential connection-time approximation to the actual probability distribution $F_c(t)$ for the two-dimensional circular cell model, as given by (9.60) or (9.60a), both having the same average connection time $1/\mu_c$, found from (9.61)? We compare the two distributions for examples 1 and 2 of Table 9.3 in Figs. 9.11 and 9.12, respectively. We use the normalized version of the distributions, comparing $F_c(x)$ with $1 - \exp(b\,x\,\mu_c/\mu)$. Note that, for $b > 1$, as is the case for the first example, the exponential approximation is quite accurate. The result for the second example, that of a smaller cell, with $b = 0.3$, is rather poor. The reason is apparent from a comparison of the equations.

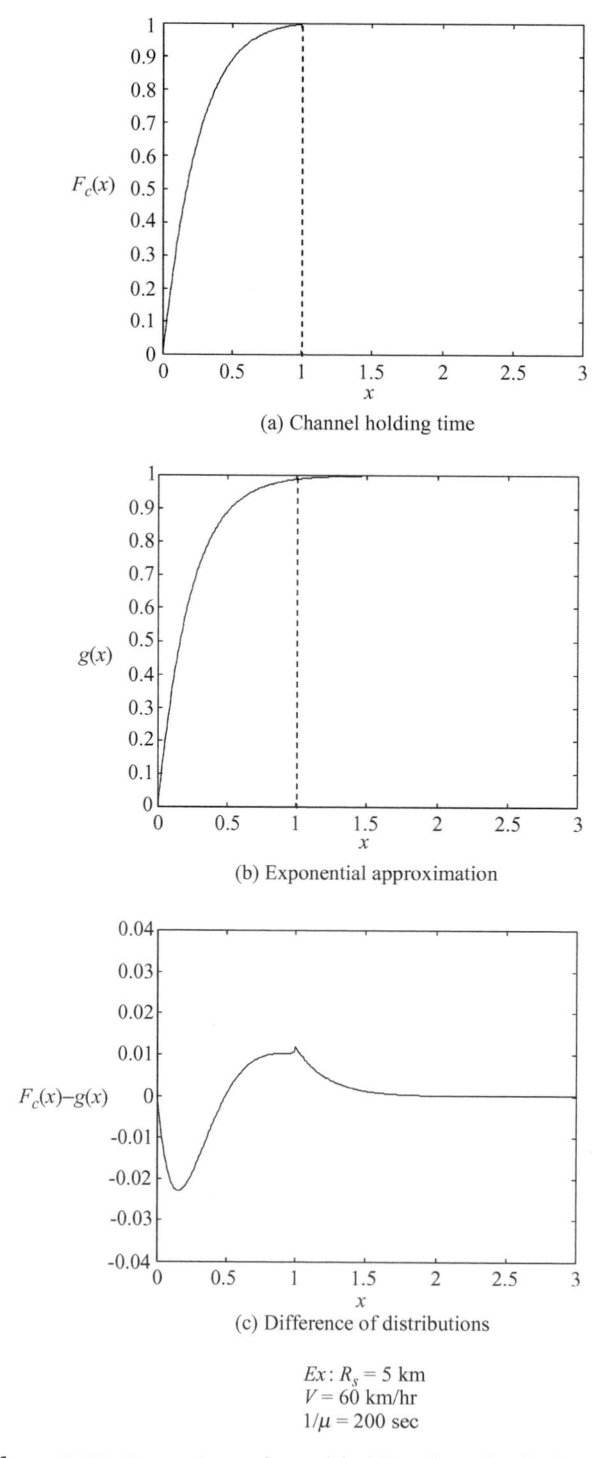

Figure 9.11 Comparison, channel holding time distribution with exponential approximation, case1, Table 9.3, $b = 3$

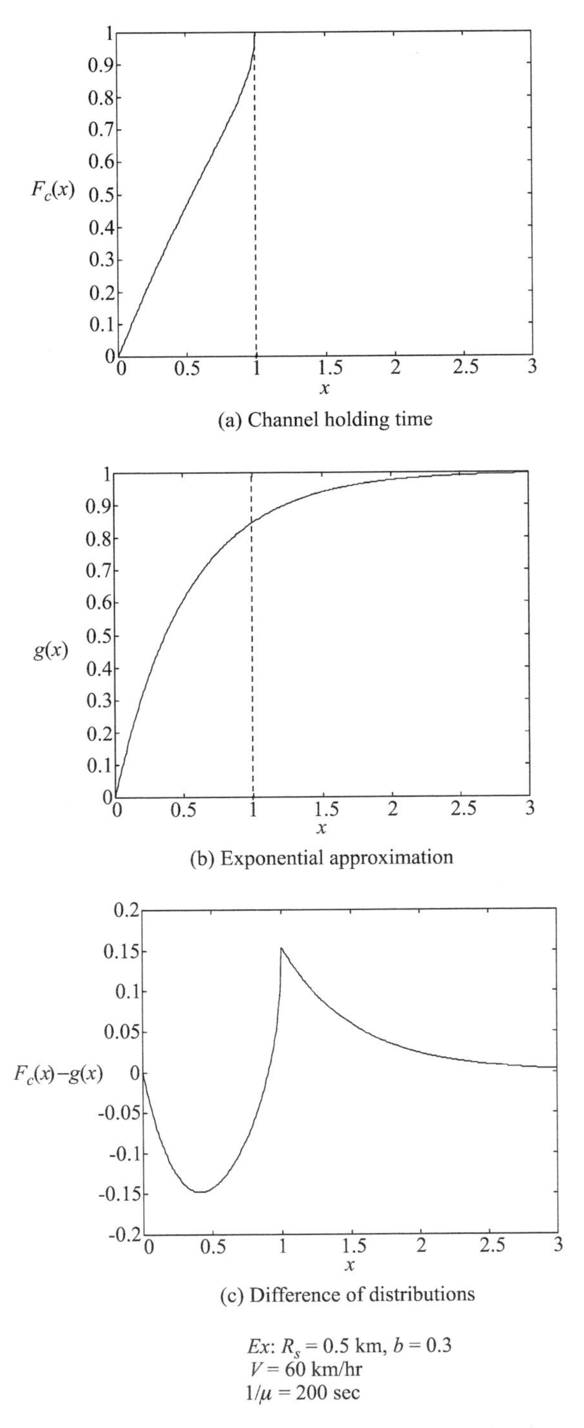

Figure 9.12 Comparison, channel holding time distribution with exponential approximation, case 2. Table 9.3, $b = 0.3$

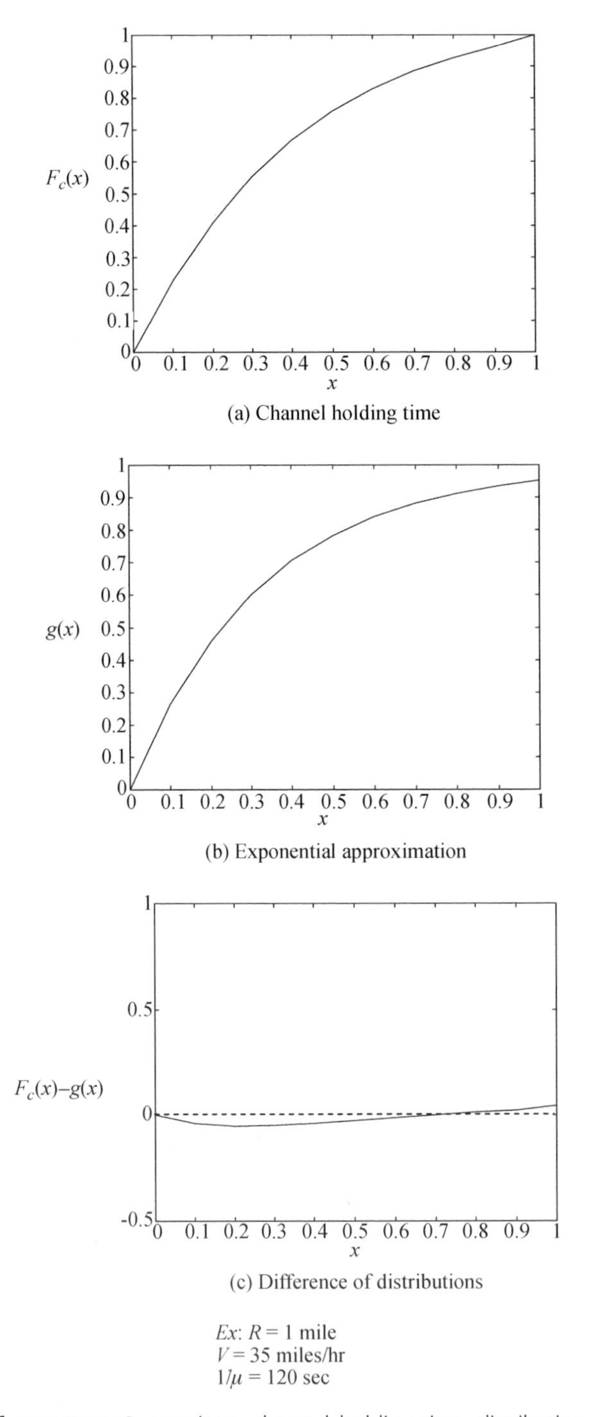

Figure 9.13 Comparison, channel holding time distribution with exponential approximation, $b = 1.56$

The model we have used, that of a constant velocity, results in a channel holding time bound of $2R_S/V_c$, or, in normalized form, $x = 1$. The exponential approximation has no such bound. For relatively large cells, with the velocity not too high, both the exponential approximation and the distribution found using the model approach the limiting probability of 1 rapidly enough so that the fact that the distribution is bounded in time while the exponential approximation is not, does not affect the comparison very much. The result for example 3, with $b = 0.36$, is much like that of the second example. The comparison is quite poor.

Consider now two other examples with the parameter $b > 1$. These are examples considered in Guerin (1987), assuming hexagonal cells, the results for which we shall shortly compare with the analytical results obtained here. For the first example, we let the circular cell approximate a hexagonal cell of $R = 1$ mile in radius; the mobile velocity is chosen to be 35 miles/hr, while the average call holding time is taken as 2 minutes. The equivalent circular cell then has a radius R_S of 0.91 mile, and the parameter $b = 1.56$. The comparison of the resultant two channel holding time distributions, one using (9.60a), the other the exponential approximation with the same average value $1/\mu_c$, appears in Fig. 9.13. Note the exponential approximation is quite good, as expected. The approximation is even better in the case of Fig. 9.14, with the cell radius ten times larger, the velocity and call holding time remaining the same. The exponential approximation for the case of $b > 1$ is thus quite good. This, as we shall now see, agrees with results obtained by Guerin (1987) for the channel holding time distribution using a hexagonal model directly, rather than the circular cell model discussed thus far.

The modeling approach adopted in Guerin may be summarized as follows. Mobiles are all assumed to have the same constant velocity V_c, just the assumption made in this section, but are equally likely to start moving, at call initiation, in any one of four orthogonal directions. They keep their same direction throughout the call, once chosen at the beginning of the call. Detailed analysis of the channel holding time distribution for a two-dimensional hexagonal system is then carried out by direct calculation. Analysis and simulation confirm in this study that the exponential distribution is an excellent approximation to the channel holding time distribution, particularly for larger cell sizes, just the result found above, using the circular cell model. The exponential approximation used in Guerin is one with the average number of handoffs per call $E(h)$ as that calculated using the hexagonal model. Interestingly, as shown below, equating average handoffs per call in the two distributions turns out to be essentially equivalent to equating the average connection holding time $1/\mu_c$ for the two distributions, the approach we have adopted in our analyses in this section and in the previous one! We demonstrate this equivalence in the two approaches shortly.

The detailed analysis for the hexagonal cellular system under the mobile mobility assumptions noted above shows the average number of handoffs per call to be given by the following simple expression

$$E(h) = (3 + 2\sqrt{3})/9\alpha = 0.72/\alpha, \alpha \equiv R\mu/V_c, R \text{ the hexagonal radius}$$

But note that this normalized parameter α appearing throughout the analysis in Guerin is directly proportional to the parameter $b \equiv 2R_S\mu/V_c$ we have been using in the circular cell model adopted for the analysis of this section. In fact, since the circular cell equivalent

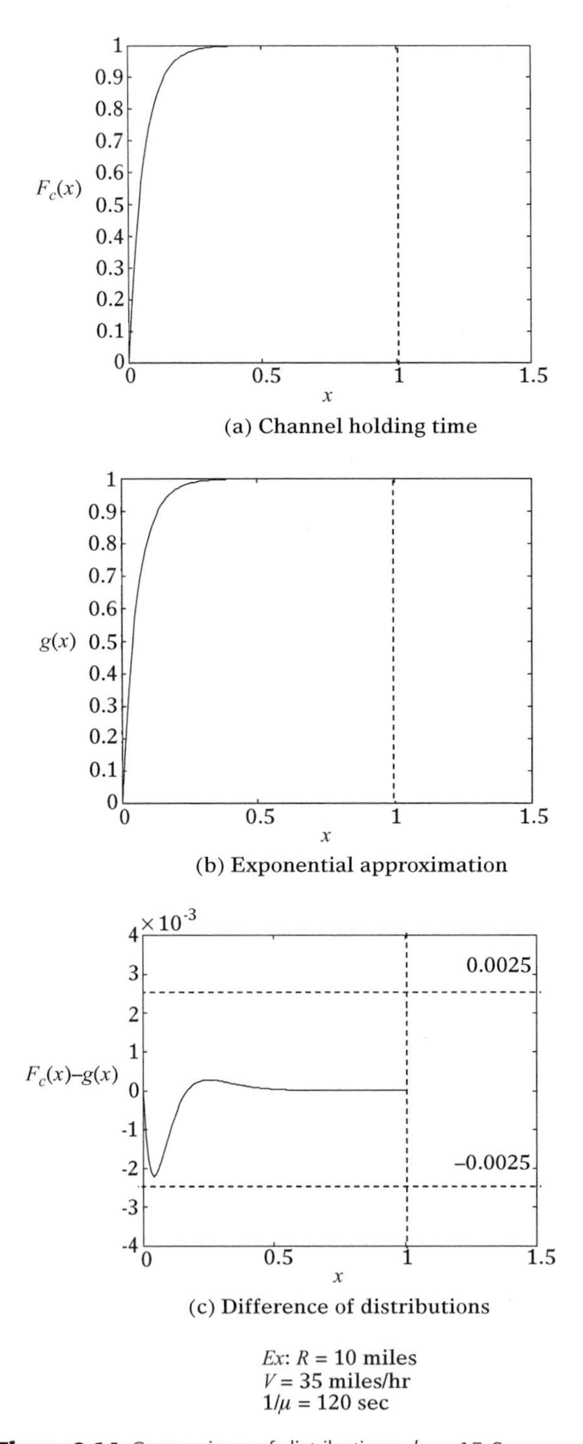

Figure 9.14 Comparison of distributions, $b = 15.6$

Table 9.4 *Comparisons, exponential approx. and analysis*

b	γ_c	$1.5/b$	$1/(1+\gamma_c)$	$1/(1+(1.5/b))$	$1/\mu_c/1/\mu$ (from (9.61))
3	0.48	0.5	0.68	0.67	0.675
0.3	5.18	5	0.16	0.17	0.16
0.36	4.3	4.17	0.19	0.19	0.188

to the hexagonal one has $R_S = 0.91R$, as noted at the beginning of this section, we have $b = 1.82\alpha$, or $E(h) = 1.3/b$, using our notation. But note that we showed, by example, that, for the circular cell model, the exponential approximation was quite good if $b > 1$. This implies that the average number of handoffs per call is the order of 1 or less. Guerin comes to the same conclusion. Such a system requires a large-enough cell, a velocity that is not too high, or a small-enough call holding time, so that not too many handoffs occur. All of these are tied together by the critical parameter $E(h)$.

Since Guerin finds, by direct calculation, that the channel holding time depends on the simple parameter α, we ask whether a similar result might not hold for our case of a circular cell. The answer turns out to be yes! Consider the calculation of the average number of handoffs per call $E(h)$ in the case of a circular cell. Whatever the geometry, this quantity must be given by the ratio of handoff to new-call arrival traffic. This is verified by (9.7b) in which we have neglected blocking and handoff dropping probabilities. The ratio η/μ appearing there is just the average number of handoffs per call or $E(h)$. But recall also, from (9.22), that under the same assumption of small blocking and handoff dropping probabilities, $\lambda_h/\lambda_n = P_N/(1 - P_H) \equiv \gamma_c$. Hence $E(h) = \gamma_c$, the parameter we have found useful in comparing system performance. This was specifically pointed out in discussing the three examples of Table 9.3. Note, however, from Table 9.3, that $b\gamma_c \approx 1.5$ for the two-dimensional, circular-cell examples. Hence $E(h) \approx 1.5/b$ in the circular-cell case. This compares with $1.3/b$ for hexagonal cells, as obtained by Guerin. This is just a 15% difference, and is quite remarkable, considering that the calculation of $E(h)$ is computationally quite complex and tedious in the hexagonal-geometry case, whereas $E(h)$ is quite readily found in the circular-cell model, using the closed-form expressions (9.58) and (9.59) for P_N and P_M, respectively. In fact, one can readily show that, in the two limiting cases, $b \ll 1$ and $b \gg 1$, γ_c approaches $\pi/2b = 1.57/b$ for $b \ll 1$, and $4/\pi b = 1.27/b$ for $b \gg 1$. Hence the two calculations for γ_c, one based on the circular-cell model, the other using a hexagonal-cell model, produce even closer results for $b > 1$. Comparisons between the value of γ_c, as given directly by $P_N/(1 - P_H)$, and as approximated by $1.5/b$, appear in columns two and three of Table 9.4 for the same three examples of Table 9.3. Note how good the approximation is.

We now turn to the exponential approximation to the channel holding time distribution adopted by Guerin. It was noted above that, in that paper, the exponential approximation is defined to be one having the same average number of handoffs $E(h)$ as the distribution calculated from the hexagonal geometry and the mobile mobility model chosen. We now show that this approach is essentially equivalent to one in which average channel holding times are required to be the same, the approach we adopted in this section and the previous one. We recall from (9.3) in Section 9.1 that the average channel holding or occupancy time $1/\mu_c$ in the case of exponential cell dwell-time or handoff statistics is given by $1/\mu_c = (1/\mu)/(1+\eta/\mu)$, with $1/\eta$ the average dwell or handoff time, and $1/\mu$ the average call length or call holding time. As noted above, the ratio η/μ is just the average number of handoffs per call $E(h)$. Hence we have $1/\mu_c = (1/\mu)/[1+E(h)]$. Guerin uses this approximation with $E(h) = 0.72/\alpha$, the value found through the hexagonal cellular analysis. He finds a good fit between this approximation and the calculated channel holding time distribution for $E(h)$ the order of 1 or less, as already noted.

We can also use this simple result for $1/\mu_c$ to compare with our analysis in this section. Since we know that $E(h) = \gamma_c$, we can write $1/\mu_c = (1/\mu)/(1+\gamma_c)$. This gives us another way of expressing the average channel holding time. How does this expression for $1/\mu_c$ compare with that found using (9.61)? It turns out that (9.61) is approximated quite accurately by this simple expression. This is shown in Table 9.4 for the same three examples of Table 9.3. In addition, we also show in Table 9.4 the result of calculating the approximate value of $1/\mu_c$, with γ_c replaced by $1.5/b$. All three values of $1/\mu_c$, normalized to the average call holding time $1/\mu$, are close together! It thus appears that, not only is the exponential distribution a good approximation to the channel holding time distribution, particularly for the case of $b > 1$, as verified by both the circular cell and hexagonal cell models, but that the average channel holding time $1/\mu_c$ is well-approximated by the simple expression $1/(1+1.5/b)$! It thus appears that defining the equivalent exponential channel holding time distribution to be the one with the same average value of handoffs per call as calculated from the model chosen, the approach adopted by Guerin, is equivalent to equating the average channel holding times of the two distributions, the approach adopted here.

Problems

Note: Some of the problems following require, for their solution, the use of a mathematical software package. Any such package will generally suffice.

9.1 Explain, in your own words, and distinguish between, the terms cell dwell time, call holding time, and channel holding time.

9.2 **(a)** The probability distributions of the channel holding time, the cell dwell time, and the call-holding time (call length) are related by (9.2). Show for the case of exponentially distributed dwell time and call holding time that the channel holding time is exponential as well, with the average channel holding time related to the averages of the other two times by (9.3).

(b) Verify that the probability P_h of a handoff in the case of exponentially distributed random variables as in (a) is given by (9.4).

(c) Verify and discuss the results for the three examples of different cell sizes and mobile speeds given following (9.5a). Choose some other examples and calculate the same three parameters, average time between handoff, probability of a handoff, and average channel holding time, for each example.

9.3 Using Fig. 9.1, show how the flow-balance equation (9.6) is derived. Where is the assumption of exponentially distributed random variables used in deriving this expression? (See problem 9.6(a) below.)

9.4 Consider the discussion in the text of the guard-channel admission-control procedure.

(a) Using the concept of local-balance as applied to Fig. 9.3, verify the expressions for the probabilities of state p_1 and p_2 given by (9.10) and (9.12) respectively. Show the general expressions for the probability of state in the two regions indicated in Fig. 9.3 are given by (9.13) and (9.14).

(b) Show the handoff dropping and new-call blocking probabilities are given, respectively, by (9.16) and (9.17).

(c) Table 9.1 shows the results of iterating (9.16) and (9.17), as explained in the text, to find the handoff dropping and new-call blocking probabilities for the two examples of a macrocellular system and a microcellular system. A macrocell in these examples has a radius of 10 km, a microcell has a radius of 100 m. Mobile speeds are taken to be 60 km/sec in the macrocellular case, 5 km/sec in the microcellular case. The number of channels available in either case is $m = 10$. Carry out the iteration indicated for each system, starting with the two probabilities set to 0, for the case of, first, $g =$ one guard channel, and then two guard channels. Do this for a number of values of new-call arrival rate, as indicated in Table 9.1. Verify and discuss the appropriate entries of Table 9.1. Show, in particular, how the introduction of the guard channels reduces the handoff dropping probability at the expense of some increase in blocking probability.

(d) Repeat the calculations of (c) for the other examples used in 9.2(c).

9.5 **(a)** Show the Erlang-B formula for the probability of blocking is given by (9.18).

(b) Plot the Erlang-B blocking probability versus ρ (or A, in Erlangs, as defined in Chapter 3) for $m = 1$, 5, 10, and 20 channels.

(c) A useful recursive relation for calculating the Erlang-B blocking probability of (9.18) is given as follows

$$\frac{1}{P_B(m)} = 1 + \frac{m}{\rho P_B(m-1)}, \qquad P_B(0) = 1$$

Derive this relation and use it to calculate and plot $P_B(m)$ for $1 \leq m \leq 20$ and $1 \leq \rho \leq 20$ Erlangs.

9.6 This problem relates to the handoff of calls for which the cell dwell time is non-exponential.

(a) Why should there, in general, be different cell dwell times, one for newly generated calls, the other for calls handed off from another cell? (*Hint*: Where, in a cell, can a new call be generated? Where is a handoff call, newly arriving at a cell, generated?) Why does the memoryless property of the exponential distribution result in a single dwell-time distribution?

(b) Use Fig. 9.1 and the definition of the two types of handoff probabilities, P_N for newly arriving calls and P_H for calls handed over from another cell, to derive the flow-balance equation (9.21). Compare this equation with that found assuming exponential-distributed cell dwell times. (See problem 9.3 and eq. (9.6).)

(c) Explain why these two handoff probabilities are given by P_N = Prob $[T_n > T_N]$ and P_H = Prob $[T_n > T_H]$.

(d) Show, starting with (9.4) and the definitions of P_N and P_H in (c) above, that these two probabilities are, in general, given by (9.19) and (9.20), respectively.

9.7 Consider the case of the one-dimensional cell of Fig. 9.4, with all mobiles moving at the same velocity V_c, equally likely in either direction.

(a) Show that the new-call cell dwell-time density function, distribution function, and probability of handoff, are given, respectively, by (9.23), (9.24), and (9.25).

(b) Take the three examples considered in the text: a macrocell whose length $L = 10$ km with mobiles moving at 60 km/hr, a macrocell of length $L = 1$ km and the same mobile velocity, a microcell of length $L = 100$ km with 5km/hr mobile velocities. Calls are all exponentially distributed, with a 200 second average value. Show the new-call handoff probabilities are given, respectively, by 0.316, 0.86, and 0.84 (see Table 9.2). Do these numbers agree with intuition? Do they agree with the finding that the probability of new-call handoff in this case of constant mobile velocity depends on the dimensionless parameter $\mu L/V_c$, $1/\mu$ the average call length (call holding time)? Repeat the calculations for some other examples, with one or more of the three parameters changed.

9.8 Consider a one-dimensional cell of length L, as in Fig. 9.4. The mobiles move at velocities randomly chosen from a uniform distribution with maximum velocity V_m. Focus on the new-call dwell times and probability of new-call handoff.

(a) Fill in the details of the derivations in the text of the dwell-time density function, probability distribution, and handoff probability P_N for this case, and show they are given, respectively, by (9.27), (9.28), and (9.29).

(b) Repeat the calculations of P_N for the three cases of 9.7(b), with $V_m = 2V_c$ in all cases. The average call length is 200 seconds in all cases. Show the results obtained agree with those shown in Table 9.2: $P_N = 0.30, 0.78, 0.76$. Show how these results agree with those expected from the dependence on the dimensionless parameter $a \equiv \mu\, L/V_m$. Select some other examples and repeat the calculations, comparing the results obtained with those expected, based on variations in this parameter.

9.9 This problem treats the one-dimensional cell example of problems 9.7 and 9.8, but focuses on the determination of the dwell-time distribution and probability of handoff P_H for calls handed off from another cell.

(a) Show that, for mobiles all moving at the same velocity V_c, P_H is given by (9.30).

(b) Show that, if mobiles move at speeds randomly chosen from a uniform distribution of maximum value V_m, the density function of the dwell-time distribution is given by (9.31), while the probability of handoff is given by (9.32).

(c) Use the three examples of problem 9.7(b) to find the probability of handoff in each case. The average call length is 200 seconds. Show the results given in Table 9.2 are obtained. Choose some other examples, as in problems 9.7(b) and 9.8(b), and show how the results obtained verify the dependence on the parameter $a \equiv \mu L/V_m$ in the random velocity case.

9.10 Consider the three one-dimensional cell examples of problem 9.7(b). (These appear in the text as well.) Find the ratio of handoff call arrival rate to new-call generation rate in each case. Neglect the blocking and handoff dropping probabilities in doing these calculations. Compare results with those provided in the text.

9.11 The text describes a procedure for determining the probability distribution of the channel holding time distribution in the case where the mobile dwell times within a cell are not exponentially distributed. This is described through the use of equations (9.33)–(9.37a).

(a) Derive (9.37a) following the procedure outlined in the text. Note that call lengths must be assumed to be exponentially distributed to obtain this result.

(b) Focus on the one-dimensional cell model discussed in the text. The channel holding time is to be approximated by an exponential distribution with the same average value. This leads to (9.39) as the equation for the approximating average value. Evaluate this expression for the three examples of problem 9.7(b). Mobiles are all assumed to be moving at the same fixed speed, the speeds indicated in problem 9.7(b). Compare the approximating average channel holding times found with those found by assuming an exponentially distributed channel holding time directly.

(c) Repeat the guard-channel calculations of problem 9.4(c), assuming one-dimensional cells, and using the approximating average channel holding times found in (b) above. Compare the results with those found in problem 9.4(c). (See the appropriate entries of Table 9.1 as well.)

9.12 This problem focuses on the two-dimensional cell geometries considered in Section 9.3.

(a) Consider the circular cell geometry of Fig. 9.9. Filling in the details of the calculations outlined in the text, show that the probability density function of the new-call cell dwell time is given by (9.52) or (9.52a), while the distribution function is given by (9.53). Note that the analysis assumes all mobiles travel at the same constant velocity V_c.

(b) Using the geometry indicated in Fig. 9.10, show, following the analysis indicated in the text, that the probability density function of the handoff call dwell-time distribution is given by (9.56a), with the distribution function given by (9.57). Mobiles are again assumed to be traveling at a constant velocity.

(c) Calculate the probability of new-call handoff P_N and the probability P_H of a handoff call again handing off for the three examples of Table 9.3 and compare with the results indicated there.

9.13 (a) Use (9.61) to calculate the approximating average channel holding time for the three two-dimensional examples of Table 9.3.

(b) Plot the channel holding time distribution (9.60) or (9.60a) for the three examples of Table 9.3. Superimpose, if possible, the approximating exponential distribution using the results of (a) and compare. Repeat for some other examples with the normalized parameter $b > 1$. Does the approximation improve with increasing b, as indicated in the text? Can you explain, for your examples, why the approximation might be expected to improve?

Chapter 10

2.5G/3G Mobile wireless systems: packet-switched data

10.1 Introduction

Previous chapters of this book have focused on second-generation (2G) wireless systems designed principally for wireless telephony, i.e., to carry voice calls, interfacing with wired telephone networks. We discuss in this chapter worldwide efforts to develop and deploy more advanced cellular networks, designed to provide higher bit rate wireless data services to interface with the Internet and other data networks. The objective is to provide wireless networks capable of carrying multimedia traffic such as voice, video, images, and data files interfacing with wired networks to present the user with seamless communication, where possible, end-to-end. These cellular networks extend the 2G systems into what is generally characterized as the third-generation (3G) or, for some cases, 2.5G systems. Much higher bit rate wireless local-area networks (W-LANs) have been designed as well to provide some of the same services, and are already beginning to pervade the business and academic sectors. These are discussed in Chapter 12. The new generation of cellular networks treats voice communication essentially the way the second generation does – as circuit-switched telephone traffic. Data, however, are to be carried in packet-switched format. The data bit rates used are higher than those currently available in the 2G systems. The wireless LANs (WLANs) discussed in Chapter 12 use packet switching exclusively. Studies are going on concurrently on fourth-generation cellular systems as well. Those systems would be expected to be all-packet-switched, interfacing seamlessly with packet-switched wired networks such as the Internet. Although voice transmission will continue to play a critical role in 3G systems, the stress in this chapter will be on the use of packet-switching technology to transmit data in these systems.

We have used the two sets of words *circuit switching* and *packet switching*. Circuit-switched communication is the type of communication used worldwide in telephone networks, whether wired or wireless. The appellation *circuit* implies that a dedicated connection has been made end-to-end for a given call, and that the connection is held for the length of the call. In the wired network portion of a circuit, links along the path comprising the circuit carry a repetitive TDM frame structure, just as in the case of GSM and IS-136, with frames consisting of a fixed number of time slots. Each time slot, called a *trunk*, is assigned to a given call for the length of that call. A circuit thus consists of a sequence of dedicated trunks along the path of the circuit. In the case of a CDMA system such as IS-95, a call receives a dedicated code to use on the radio interface portion of its path; the rest of the circuit, carried on the wired portion of its path, is carried on dedicated TDM trunks, just as in the case of the TDMA systems, IS-136 and GSM.

Packet switching, the other type of communication technique, and the one that is deployed in the Internet as well as other data networks, bundles data into blocks of bits called packets. These packets may be of fixed or variable length and are *individually* transmitted across a network, from source to destination, along some path. Packets from different data users *share* the links comprising the end-to-end path. A given time slot on a TDM frame on a particular link is thus not dedicated to a particular user, as in the case of circuit switching. In the case of packet-switched CDMA, a code is only assigned to a given packet for the time required to transmit that packet.

We noted above that the newer 2.5G/3G cellular systems we will be discussing in this chapter have been designed to provide higher bit rate, packet-based data transmission. Data transmission capability, as contrasted with voice communication, has been available for the three 2G systems we have discussed in earlier chapters. GSM, for example, does provide for *short message services* (SMS), with data transmitted, as is voice, in circuit-switched form. This service has had good acceptance in Europe, but is limited in the data bit rates it can provide. CDPD, as another example, is an overlay mobile networking data service developed by IBM for use with the IS-136 networks in the USA. It is designed, where available, to transmit packet-switched data over unused time slots in an IS-136 frame, returning these slots to voice users on demand. It has, however, had limited acceptance. Note that this service, in common with SMS, provides a relatively low bit rate transmission capability service. Both services are thus not completely appropriate for the demands for multimedia broadband (high bit rate) services brought about by the extraordinary increase in use of Internet Web browsing, so common now throughout the world.

As noted above, multimedia Internet services, particularly those related to Web browsing, involve transmitting wideband (high bit rate) files, video, and images in packet form from source to destination, very often from a server to a requesting client. In the case of mobile communications, the client would normally be a mobile data terminal of some type, with the wideband data sent in reply to a request transmitted downlink from an access point on the wired side of the radio link (the base station in the case of cellular systems) to the requesting mobile. Note that Internet data transmission differs in two ways from the type of transmission currently available over second-generation systems we have been discussing up to this point: the data appear in the form of packets and are generally of much higher bit rate or bandwidth. It is this difference in requirements that has led to the deployment of wireless LANs as well as to the international standards activity devoted to developing and deploying third-generation (3G) cellular systems to be discussed in this chapter.

The requirement for packet-based, higher bit rate data communications in 3G cellular systems as well as W-LANs comes about for several reasons. First, as just noted, there is a desire to extend wireless services, most commonly geared to voice traffic, to the newer world of multimedia traffic which is inherently packet-based in form. Connecting cellular systems to the packet-based Internet, or extending the Internet to seamlessly incorporate mobile terminals with data capability, as an example, requires introducing packet-based communications into the mobile wireless world. Second, packet communications has long been recognized as being a much more efficient way of handling the disparate transmission requirements introduced by this newer world of multimedia traffic. Data traffic can be bursty, meaning that the time between successive occurrences of data packets may be very long compared with the time to transmit the packets; there may be frequent transmission required of small volumes of traffic; finally, there may be occasional transmission of long files. (Think of the obvious Internet applications such as email transmission, "surfing" the Web, downloading of files from a Web site, etc.) These variations in transmission requirements are vastly different than those seen in voice-based, circuit-switched traffic. Packet transmission is more efficient in handling bursty traffic than is circuit-switched transmission since multiple packets from various users may share transmission facilities as already noted, unlike circuit switching with its use of dedicated facilities. But to ensure efficient use of the wireless resources such as frequency bandwidth, time slots, or codes in CDMA systems, transmission setup must be very fast and access times made very short. Connect times must be much less than those in the voice-based circuit-switched 2G cellular systems described in previous chapters.

Packet-switched data communication introduces significant differences in performance requirements for the user as well. We focused in Chapter 9 on such performance objectives as call blocking and handoff dropping probabilities. Packet-switched data transmission introduces a completely new set of user performance objectives, in addition to the two just mentioned. These are described under the general rubric of *quality of service* or QoS. These objectives include, among others, packet priority level (different types of traffic may require different handling), probability of packet loss, packet delay transfer characteristics, and data throughput rates. We shall find these packet-switching performance requirements and objectives recurring throughout our discussion in this chapter of 3G systems.

It was thought at first that one worldwide 3G cellular network standard would be developed to which the three current 2G standards we have been discussing would converge. It turned out, as has so often been the case with the development of international standards, that agreement on one standard was not possible. Instead, a number of standards have been developed, with the hope that convergence will still be possible in the future. In Sections 10.2 and 10.3 following, we describe three of the four 3G standards that have been developed, two based on CDMA technology, the other based on TDMA technology. CDMA technology appears to be preferred for a number of reasons. It has the capability, as will be noted later in this chapter, of allocating bandwidth dynamically to different users. This is a desired attribute for packet-switched systems, enabling multimedia traffic, among other traffic types, to be supported. Studies carried out worldwide have indicated as well that this type of technology results in higher-capacity cellular systems with the requisite bandwidths necessary to support higher bit rate multimedia traffic in addition to voice. The two CDMA standards, wideband CDMA or W-CDMA and cdma2000, are described in Section 10.2, after an introductory discussion of increased bit rate CDMA systems in general. cdma2000 has been designed to be backwards compatible with IS-95.

W-CDMA is considered to be an outgrowth of, and possible replacement for, GSM. The various countries and organizations worldwide currently deploying and supporting GSM developed the concept of Universal Mobile Telecommunication Services or UMTS as the objective for the third-generation replacement for GSM. The acronym UMTS is thus often considered synonymous with W-CDMA, and the W-CDMA standard is also referred to as the UMTS/IMT-2000 standard. This standard was developed by the 3rd Generation Partnership Project, 3GPP. The cdma2000 standard was developed by a follow-on project, 3GPP2. These standards projects were created in 1998 and 1999, respectively, as a joint effort of standardization bodies in Europe, Japan, Korea, USA, and China.

The TDMA standard, also dubbed a 2.5/3G standard, is designed to provide a packet-switching capability for GSM as well as enhanced bit rates across the radio interface, while maintaining as much as possible the current characteristics of GSM, and allowing for compatibility in the transmission of voice, as well as other circuit-switched services. It thus represents a less far-reaching change than does W-CDMA. The introduction of packet switching for data transmission over GSM networks does, however, require changes in the GSM wireless core network infrastructure, the resultant packet-switched core network standard being called GPRS (general packet radio service). The enhanced bit rates come from adopting a technique called EDGE (enhanced data rates for global evolution). The EDGE technique is being adopted for enhanced versions of IS-136 as well, with the expectation that there will be a future convergence of the two TDMA systems, GSM and IS-136. Section 10.3 provides an overview of GPRS and EDGE. The concept of a layered architecture plays a significant role in the standardization of GPRS, as it does in all packet-switching networks. We thus review this concept briefly and describe the layering adopted for GPRS in Section 10.3.

Note that we indicated we would only be discussing two of the three CDMA standards in Section 10.2. W-CDMA comes in two flavors, a frequency-division duplex, FDD, version and a time-division duplex, TDD, version, to be used principally in indoor environments. Frequency-division duplexing is the technique used in all second-generation systems, with different frequency bands assigned to uplink and downlink transmission. Time-division duplexing is a technique in which downlink and uplink transmission alternate in time in using the same frequency band. We shall be focusing on the FDD version in our discussion of W-CDMA. The cdma2000 system is designed to be backwards compatible with the circuit-switched IS-95, as noted above, yet is capable of handling much higher bit rate packet-switched data.

In the remaining two chapters we continue our focus on packet switching in wireless systems. In Chapter 11 following, we move away from specific wireless network standards and discuss the problem of combining or multiplexing voice and packet-switched data over a common radio link in more general terms. This problem is generally referred to as the multi-access problem. It is basically a resource allocation problem, that of allocating bandwidth most effectively to the different users, as well as to competing types of traffic such as voice, video, images, and files. It is often referred to as well as the *scheduling* problem. In the case of CDMA systems this corresponds to the appropriate allocation of time and codes; in TDMA systems, time slots in each frame represent the resource to be allocated. Much work has been done in comparing different resource allocation strategies and we describe some of this work in both the CDMA and TDMA domains. An example of a resource allocation strategy that has been studied for cdma2000 appears here at the end of the next section, devoted to the various CDMA schemes proposed for third-generation systems.

In the uplink or reverse direction these various multi-access techniques rely to a great extent on random access procedures. Random access, as we shall see, plays a key role in the 3G systems to be described in the sections following. Recall that initiation of a call in the 2G systems we have already described is done on a random access basis as well. Slotted Aloha is the most common form of random access. In Chapter 11 we therefore provide a brief introduction to the slotted-Aloha concept, as well as to variations on this technique. Finally, in the concluding Chapter 12, as noted above, we discuss packet-switched W-LANs, wireless LANs, with particular emphasis on the IEEE 802.11 wireless LAN standard. This scheme, which has been widely deployed, as noted earlier, in the commercial sector as well as in academic institutions, to provide wireless connectivity to the Internet and to other data networks, uses a modified random access procedure labeled CSMA, for carrier sense-multiple access. We conclude that chapter with a brief introduction to sensor-type wireless networks such as the Bluetooth standard.

10.2 3G CDMA cellular standards

As noted in the previous, introductory, section, we will be describing two of the three 3G CDMA standards in this section. These are wideband CDMA or W-CDMA in its frequency-division duplex, FDD, version, and cdma2000. We begin with some general comments on wider-band (higher bit rate) CDMA, and then focus on each system separately, beginning with W-CDMA.

Wider-band (higher bit rate) CDMA systems

A significant question with 3G CDMA systems is that of handling the much higher bit rates required for the transmission of packet-switched multimedia data traffic. In general, there are three basic ways of increasing the bit rate in CDMA: one could opt for adding frequency bands. This technique of using multiple narrower frequency bands is sometimes called multicarrier CDMA (Milstein, 2000) and is the technique adopted for cdma2000. (As will be noted in discussing cdma2000 later, however, initial versions of cdma2000 use a single band only.) A second technique is to increase the chip rate sufficiently so that some higher bit rate data can be transmitted with a relatively high spreading gain. (Recall from Chapter 6 that the performance and capacity of CDMA systems depend on establishing a high spreading gain, the ratio of spread bandwidth W to the bit rate R. An alternate definition of spreading gain, sometimes used, is the ratio of chip rate to bit rate, since the spread bandwidth is directly proportional to the chip rate.) To handle still higher bit rates, one simply increases the bit rate with the (relatively high) chip rate maintained constant. Higher bit rate signals are thus transmitted with lower spreading gains. Such a scheme is called *variable spreading gain CDMA* (Chih-Lin I and Sabnani, 1995a and b). Since the signal-to-interference ratio, SIR, and hence the performance of a CDMA system,

---

Milstein, L. 2000. "Wideband code division multiple access," *IEEE Journal on Selected Areas in Communications*, 18, 8 (August), 1344–1354.
Chih-Lin I and K. Sabnani. 1995a. "Variable speading gain CDMA with adaptive control for true packet switching wireless network," IEEE International Conference on Communications, ICC95, Seattle, WA, June, 725–730.
Chih-Lin I and K. Sabnani. 1995b. "Variable speading gain CDMA with adaptive control for integrated traffic in wireless networks," IEEE 45th Vehicular Technology Conference, VTC95, Chicago, IL, July, 794–798.

is directly proportional to the spreading factor, it is clear that the performance of such a scheme must deteriorate unless the signal power is increased to compensate for decreasing SIR as the bit rate increases. This is not always possible in the CDMA power-controlled environment and there must come a point where a maximum possible bit rate is reached. This point depends on the type of data traffic to be transmitted, since different traffic types will have different bit error rate requirements, with corresponding requirements on the SIR. W-CDMA uses variable-gain spreading to attain some of its increased bit rate performance. As we shall see, in discussing W-CDMA in the next paragraph, the system design of W-CDMA increases the chip rate to 3.84 Mcps from the value of 1.2288 Mcps used in IS-95 systems, requiring a concomitant increase in transmission bandwidth to 5 MHz.

An immediate advantage of using a higher chip rate is that multipath diversity using a RAKE receiver is improved. Recall the brief discussion of the RAKE receiver at the end of Chapter 2, as well as in Chapter 6, in which the improvements in CDMA system performance through the use of a RAKE receiver are documented. Summarizing the discussion in Chapter 2, we note that the RAKE scheme provides time diversity, where applicable, resolving individual paths in a multipath environment, the number resolved corresponding to the "fingers" of the "RAKE." Signals corresponding to each path can then be combined to provide enhanced signal estimation and corresponding improvement in the signal-to-interference ratio SIR. As noted in Chapter 2, for a RAKE receiver to be effective, the spread signal bandwidth must be greater than the coherence bandwidth. Frequency-selective fading is then incurred, leading to the resolution of multipath echoes. Increasing the chip rate provides this increase in signal bandwidth. As an example, consider the multipath resolution capability of W-CDMA (WCDM, 2000). At a chip rate of 3.84 Mcps the chip duration is 0.26 μsec. Paths differing by this value in time, or the corresponding path difference of 78 meters, using 300,000 km/sec as the speed of the radio waves, can then be resolved. This corresponds to about 240 feet, or to a typical block length in an urban environment. The RAKE receiver diversity improvement can thus be obtained in urban microcells of at least a block in length. We noted in Chapter 6 that IS-95 did have the capability to resolve multipath echoes and use the RAKE finger concept to thereby improve signal detectability. But it turns out that, because the chip rate is considerably less than that used in W-CDMA, multipath resolution is essentially limited to macrocell environments. As we shall see later, cdma2000 retains the IS-95 chip rate to provide backward compatibility with IS-95. Specifically, IS-95 and the cdma2000 3G system use a chip rate of 1.2288 Mcps, for a chip duration of 0.81 μsec. This length in time corresponds to 244 meters in length. The minimum path difference in this case is then the order of 240 meters or so, which implies RAKE receiver diversity improvement for that distance or more. (It is to be noted, however, that as the spread bandwidth increases, increasing the number of paths that can be resolved, the signal energy per path received will decrease, reducing to some extent the expected diversity gain (Milstein, 2000).)

The third technique used to attain higher bit rates in CDMA is to use multiple codes in parallel. This technique is labeled simply *multicode CDMA* (Chih-Lin and Gitlin,

---

WCDM. 2000. *WCDMA for UMTS*, ed. Harri Holma and Antti Toskala, John Wiley & Sons.

1995). In this wideband CDMA procedure each of a group of successive signal bits is chip-encoded in parallel and a different pseudo-random code used for spreading each bit in the group. The encoded spread sequences are then added together and transmitted using one carrier. Conversion from serial to parallel transmission effectively reduces the higher bit transmission by the size of the group. There is thus a "base bit rate" used for transmission and a constant spreading gain results. Two successive bits grouped together and chip encoded in parallel, each with its own chip sequence, then provide a means for transmitting at double the bit rate, but with the same spreading gain. Three successive bits grouped together would provide transmission at three times the basic bit rate, and so on. This technique thus avoids the problem of decreasing spreading gain as the bit rate increases. As we shall see shortly, cdma2000 uses this technique. The W-CDMA standard uses this technique as well to attain its highest data bit rates. Clearly, the codes used in parallel must be orthogonal to one another for the individual bits to be correctly detected, and in order to prevent "self-interference" at the receiver (Chih-Lin I and Gitlin, 1995; WCDM, 2000).

What are the tradeoffs between using multicode CDMA and variable spreading gain CDMA? The basic problem with multicode CDMA, in addition to the requirement that the pseudo-random chip codes used be orthogonal to one another, is that this technique results in wider envelope variations during transmission, resulting in more stringent requirements on the linearity of the transmitter power amplifier (Dahlman and Jamal, 1996). The use of multiple codes requires multiple RAKE receivers as well, one for each code. The SIR performance of the two schemes, variable spreading gain CDMA and multicode CDMA, appears to be about the same, however (Dahlman and Jamal, 1996; Ramakrishna and Holtzman, 1998).

Milstein and co-workers have carried out a number of studies comparing wideband CDMA with multicarrier CDMA, the overall spectrum or frequency band covered being kept fixed. In essence, one is comparing a single wideband CDMA signal with a number of parallel narrowband CDMA signals covering the same band. The results of these studies are summarized in Milstein (2000). As noted earlier, wideband CDMA, defined as a system for which the spread bandwidth exceeds the signal coherence bandwidth (see Chapter 2), results in resolvable multipath components that can, if coherently combined, improve the system performance using a RAKE receiver. This is the point made earlier in this section. (The received energy per path is necessarily reduced, as already noted.) The operative word, however, is *coherent* combining. If non-coherent combining must be used, a non-coherent combining loss is experienced (Milstein, 2000). (As we shall see, W-CDMA uses coherent combining.) In addition, these studies assume perfect estimation of channel parameters. Imperfect estimation of the channel parameters deteriorates the wideband system performance further. Multicarrier systems can be designed using a RAKE receiver at each frequency,

Chih-Lin I and R. D. Gitlin. 1995. "Multi-code CDMA wireless personal communication networks," IEEE International Conference on Communications, ICC95, Seattle, WA, June, 1060–1064.

Dahlman, E. and K. Jamal. 1996. "Wideband services in a DS-CDMA based FPLMTS system," IEEE Vehicular Technology Conference, VTC96, 1656–1660.

Ramakrishna, S. and J. M. Holtzman. 1998. "A comparison between single code and multiple code transmission schemes in a CDMA system," VTC98, May, 791–795.

the result being improved system performance comparable with that obtained by using the single carrier wideband system. The receiver structure is more complex, but lower-speed parallel-type processing may be used while the wideband system requires higher-speed serial processing (Milstein, 2000). Both schemes, the wideband one and the multicarrier one, provide comparable performance for an SIR greater than 0 dB. As the SIR deteriorates, however, the multicarrier system is found to outperform the wideband one (Milstein, 2000).

W-CDMA

We now discuss each of the two CDMA system standards in more detail, beginning with the frequency-division duplexed, FDD, version of wideband CDMA, W-CDMA. This system has been designed to be a direct-sequence CDMA system like the second-generation (2G) CDMA system, IS-95, discussed in previous chapters, but, as already noted, with a much higher chip rate of 3.84 Mcps, instead of the rate of 1.2288 Mcps used in IS-95. This high chip rate allows higher bit-rate data to be transmitted, with relatively high spreading gains maintained. This higher chip rate in turn results in a much higher transmission bandwidth as well, chosen to be 5 MHz. The frequency bands allocated to W-CDMA must thus be moved to higher values to accommodate the higher required transmission bandwidths. In Europe and Asia (Japan and Korea) the 2 GHz band has been selected for this purpose. (This band is not available for 3G systems in the US, however.) Specifically, the choice has been made to use the 60 MHz-wide band 1920 MHz–1980 MHz for uplink transmission, mobile to base station, and the 60 MHz-wide band 2110 MHz–2170 MHz for downlink transmission, base station to mobile. The intent of the W-CDMA system design is to have this system conform to existing GSM networks as much as possible, despite the choice of the CDMA technology and higher frequency bands of operation. It is thus designed to handle handoffs from W-CDMA networks to GSM networks and vice-versa. Within the W-CDMA environment, however, soft handoffs are supported, just as in the case of IS-95. (Note again that the word "handover" is used in place of "handoff" outside North America. Most international documents thus refer to *handovers* rather than handoffs, the term we will be using.)

W-CDMA is designed to provide up to 384 kbps data transmission rate in a wide-area environment and up to 2 Mbps in a local-area environment. Note how much higher these numbers are than the (circuit-switched voice) rates provided in the 2G systems we have discussed in earlier chapters. It is important to note, however, that actual data throughput rates may be less than data transmission rates because of overhead, packet queueing delay, variable-length packet sizes resulting in inefficient use of frame and time slot structures, etc. We shall quantify this comment briefly below, as well as later in discussing IS-95B, the first member of the cdma2000 family, in discussing GPRS, and in discussing the EDGE technique at the end of this chapter.

How does W-CDMA attain these higher data transmission rates? As noted above, W-CDMA uses a combination of the variable spreading gain technique with a fixed chip rate of 3.84 Mcps and the multiple code technique using the same chip rate. The variable spreading gain technique is used to transmit data over the air interface at variable rates ranging, in multiples of 15 kbps, from 15 kbps to 960 kbps uplink and a maximum of double that rate, 1.92 Mbps, downlink. The higher rate downlink, from base station to mobile, is due to the use of QPSK as the modulation technique in that direction; PSK is used in the uplink direction. All data transmission uses rate-1/2 convolutional encoding to protect user data prior to transmission. The actual rates of transmission of user data are then less than half the values just quoted, i.e., they range from less than 7.5 kbps to less than 480 kbps downlink, once tail bits and added forward error correction bits are included as well. Note that the spreading gains, defined in this case as the ratio of chip rate to transmission bit rate (WCDM, 2000), then vary from a high value of 256 (3.84 Mcps/15 kbps) to a low value of 4 (3.84 Mcps/960 kbps). Up to six orthogonal codes can be used in parallel, if needed, to attain still higher bit rates. (This procedure is just the multicode procedure described in the previous section.) The maximum possible bit rate over the uplink radio channel is then $6 \times 960 = 5.76$ Mps, with a spreading gain of 4. The user information rate is somewhat less than one-half of this value, or the order of 2 Mbps, the figure quoted above. Because of the low spreading gain, this highest rate of transmission is limited to local-area environments. The maximum rate of transmission of user data in wide-area environments is limited to the 384 kbps figure quoted above.

It was noted in the previous, introductory, section that the 3G systems are being designed with quality-of-service, QoS, performance objectives in mind. We also noted there that W-CDMA is often considered synonymous with Universal Mobile Telecommunication Services or UMTS; these services, and hence W-CDMA, project handling four QoS traffic classes, two real-time and two non-real-time (WCDM, 2000). The real-time traffic category covers conversational (i.e., two-way interactive) traffic such as voice and video, and streaming multimedia traffic. The conversational class has, as its QoS objectives, low end-to-end delay of traffic delivery and limited time variation between successive information entities comprising the traffic stream (WCDM, 2000). (It is to be noted that it has long been recognized that real-time voice cannot be delayed more than 100 msec in delivery end-to-end. Otherwise the human recipient at the receiving end finds the transmission difficult to follow.) The streaming class requires preservation of the timing between successive information entities (WCDM, 2000).

Non-real-time traffic includes interactive traffic, such as web browsing, and "background" traffic, such as e-mail (WCDM, 2000). Both classes require preservation of data integrity, meaning that data should arrive at the recipient with very low probability of error. (Voice traffic, on the other hand, is still recognizable even when as many as 5% of voice samples sent are dropped or deemed to be in error.) Background traffic, as distinguished from interactive traffic, is that class of data traffic, such as e-mail, which is relatively insensitive to delays in delivery time. In the Internet world this type of traffic has been labeled "best-effort" traffic. Non-real-time interactive traffic is often characterized as being "bursty" in nature, since relatively long periods of time may elapse between successive transmissions of short bursts of data by either side of a two-way connection. As an example, a mobile client connected to an Internet server such as a web site may be sending brief request messages to the server, requiring some time to process the replies from the server, and then replying, in turn, with a new request. We shall see that provision is made in the W-CDMA system for sending small amounts of packet data on an infrequent basis.

As noted above, W-CDMA employs PSK modulation uplink, from mobile to base station, and QPSK modulation downlink. The use of QPSK means that data transmission rates downlink, from base station to mobile, can be double those of rates uplink. This is a desirable feature when receiving from an Internet server or Web site, for example. Coherent detection is employed, using pilot information to provide estimates of the channel. Such information is required in setting up the RAKE receiver.

A repetitive frame structure is employed to transmit data over the radio channel. Data frames are 10 msec long and the specific bit rate used may be changed each frame. The voice codec (coder/decoder), however, uses 20 msec frames, as in the 2G systems discussed in Chapter 8. Voice is transmitted in circuit-switched form, as noted earlier. One of eight different bit rates may be selected to transmit voice signals, ranging from 12.2 kbps down to 4.75 kbps. The rate used may be changed every 20 msec frame. The value of 12.2 kbps has been chosen for compatibility with GSM; a value of 7.4 kbps provides voice bit rate compatibility with IS-136; 6.7 kbps is the same as the bit rate of the Japanese PDC voice codec (WCDM, 2000). The W-CDMA voice coder uses a technique called ACELP, algebraic code excited linear prediction, and processes 8000 voice samples/sec or 160 samples per 20 msec frame in a manner similar to the schemes described in Chapter 8 (WCDM, 2000).

We focus now on variable-rate, packet-switched data traffic. How is this traffic transported over the radio channel? Two types of logically defined dedicated channels (DCH) are used for this purpose. They are, in turn, supported by, or mapped on to, dedicated physical channels. One DCH type is used to transport user information (user data) as well as higher-layer control traffic distinct from control traffic involved with physical operations over the radio channel. (The concept of layering in packet-based communication architectures is reviewed in the next Section, 10.3, in connection with a discussion of GPRS.) The physical channel to which this logical channel is mapped, the dedicated physical data channel or DPDCH, consists of a specific code at a given frequency, reserved for a single user. It is the bit rate of the DPDCH that may change, using the variable spreading factor technique and the multicode technique (when necessary), from one 10 msec frame to the next 10 msec frame. This channel thus supports the four different traffic classes noted above, providing "bandwidth on demand."

The second DCH is a common logical channel shared by all, or a specific sub-set of, users in a cell. It is designed to provide the necessary signaling and control support for the variable-rate data transport DCH, including information as to the data transmission rate in use by that DCH. This signaling/control DCH maps on to the DPCCH. The DPCCH also consists of a specific code at a given frequency, but is always operated at a fixed bit rate with a spreading gain factor of 256. It is transmitted in parallel with the DPDCH. The DPCCH carries the information about the transmission bit rates of the DPDCH occurring during the same frame time, among other parameters. The DPCCH structures differ in the uplink and downlink directions. Figure 10.1 portrays the uplink DPCCH structure mapped on to the W-CDMA 10 msec repetitive frame (WCDM, 2000). (The DPDCH carrying user data and/or upper-layer control information is also shown being transmitted at the same time, in parallel with the DPCCH.) Note that a frame is divided into 15 slots, each 2/3 msec wide and coded into 2560 chips of the chip code sequence selected.

Each DPCCH slot, in turn, consists of four fields, of which the FBI field is optional. The pilot field is used to provide estimates of the radio channel as required for coherent detection and RAKE reception. The TFCI field, transport format combination indicator, is the field that indicates the data transmission rate used in the accompanying DPDCH

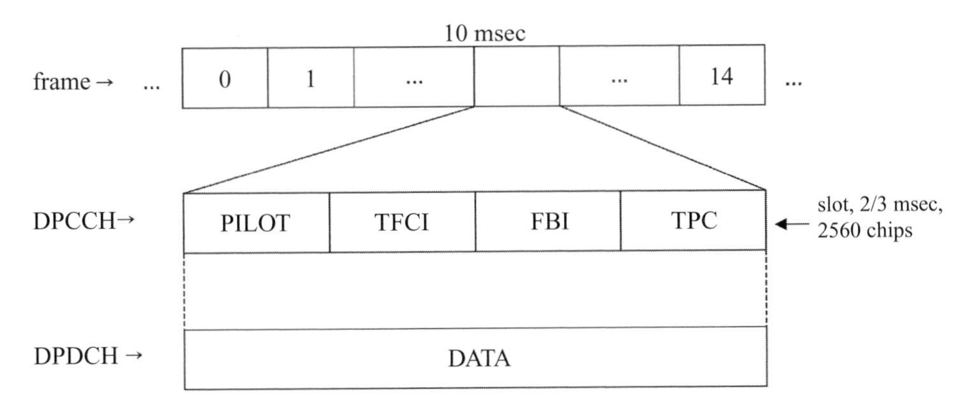

Figure 10.1 W-CDMA uplink dedicated physical channels (from WCDM, 2000: Fig. 6.11)

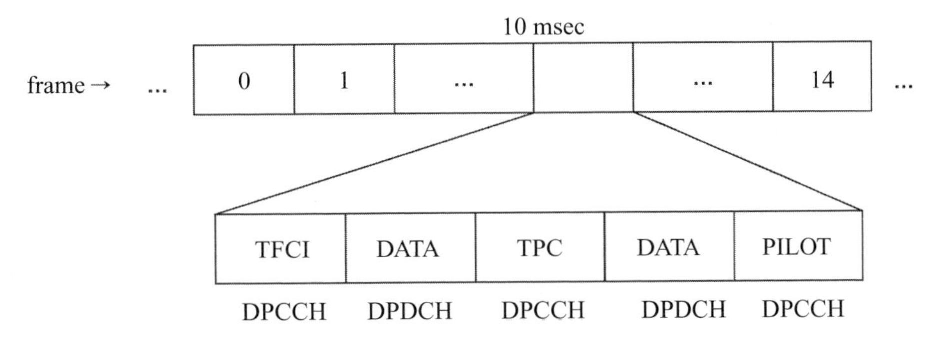

Figure 10.2 W-CDMA downlink dedicated physical channel: control and data channels time multiplexed (from WCDM, 2000: Fig. 6.14)

frame. The TPC field, transmission power control, provides information required for the downlink power control. The optional feedback information, FBI, field is used to control antenna phases and amplitudes if multiple antennas are used to obtain diversity array improvement. Bit interleaving over a 10 msec frame to provide some protection against fading is used in transmitting the DPDCH frame. Additional interleaving of 20, 40, or 80 msec is also possible if the information being transmitted allows the additional delay in receiving it that results. (Conversational traffic, for example, has strict limits on the end-to-end delay between sending and receiving systems. Non-real-time traffic, on the other hand, may have much looser, if any, limits on end-to-end delay.) Since the DPCCH channel carries DPDCH rate information, the DPDCH bits must be buffered at the receiving base station to be processed after the accompanying DPCCH fields are first extracted and interpreted. Details appear in WCDM (2000). The downlink frame structure differs from the uplink one. Both the DPCCH and DPDCH are time multiplexed into one downlink physical channel of 15 slot 10 msec frames. There is no FBI field used. Instead, each slot carries TFCI, TPC, and pilot fields alternating with DPDCH, data, fields. Figure 10.2 indicates this downlink dedicated physical structure. The reader is again referred to WCDM (2000) for details, including information as to the formation of up- and down-link frames as well as to their handling at the respective receivers.

Table 10.1 *W-CDMA common signaling/control channels*

Logical channel	Direction	Physical channel
BCH	D	PCCPCH, primary common control physical ch.
FACH*	D	SCCPCH, secondary common cntrl. phys. ch.
PCH*	D	
RACH	U	PRACH, physical random access ch.
Optional		
DSCH	D	PDSCH, physical downlink shared ch.
CPCH	U	PCPCH, physical common packet ch.
	D	SCH, synch. ch.
	D	CPICH, common pilot ch.
	D	AICH, acquisition indic. ch.
	D	PICH, paging indic. ch.
	D	4 additional ch. associated with CPCH procedure

Note: * both logical channels may share the same physical channel, as shown.

The discussion thus far has focused on W-CDMA packet data transmission using dedicated physical channels. As in the case of the 2G systems, control channels are defined as well to carry physical layer information required for terminals to synchronize to base stations, to obtain necessary code information, to respond to pages, to initiate call setup using a random access procedure, etc. These are all common channels, to be utilized by all terminals, rather than dedicated channels as in the case of the data transmission channels just discussed. Six logical channels mapped on to physical channels are defined for this common transport purpose, two of which are optional, in addition to a number of purely physical channels (WCDM, 2000). (Recall again from our discussion of IS-95 in Chapter 8 that physical channels are always defined as specific codes at given frequencies.) Table 10.1 lists these channels, indicating which are optional as well as the direction of transmission uplink (U) or downlink (D). Note that the only required uplink channel is the random access one, indicating that random access is the technique used by a mobile terminal to establish communication with the base station. We draw on WCDM (2000) for much of the discussion following of these channels.

The downlink broadcast channel, BCH, carries random-access codes and access slots available in a cell, as well as other information needed by mobile terminals. All terminals, no matter what their capability, must be able to receive this channel and decode the information carried. In addition, this channel must be able to reach all mobile users in the cell coverage area. It is thus transmitted using a low, fixed, data rate, and at relatively high power. The forward access channel, FACH, carries signaling and control information to terminals once they have completed the random access procedure described below, and have thus been identified as being actively located in the cell. Some packet data may be transmitted on this channel as well. Bursty data, of which mention was made earlier, might be an example of data transmitted over this channel from base station to mobile terminal.

There may be several such channels in a cell, but at least one must have a low-enough bit rate, as was the case with the BCH, in order to be received by any mobile-terminal type. Additional FACH may be transmitted at higher bit rates. Fast power control is not used on this channel. Note from Table 10.1 that the FACH is shown sharing the same physical channel as the downlink paging channel, PCH. The two channels may be time multiplexed on the same physical channel, the normal mode of operation, or the FACH can have its own stand-alone physical channel.

The PCH has functions similar to the paging channels discussed earlier in our descriptions of the 2G systems. For example, it handles paging messages required to deliver incoming speech calls directed to a terminal expected to be situated in a given location area. The paging channel is used in conjunction with the paging indicator channel, PICH, to enable terminals to operate in a "sleep mode" to conserve power. Terminals "wake up" and listen periodically to the PICH to determine if there are paging messages waiting for them. Details appear in WCDM (2000).

The uplink random access channel, RACH, is used, as in 2G systems, to signal the desire by a mobile to initiate a call, to indicate its presence in a new area, or to register with the network on powering up. The acquisition indicator channel, AICH, is used by the base station to reply to a RACH message. It does this by echoing back a "signature sequence" appearing in the RACH message. (Up to 16 different signature sequences can be acknowledged in the same AICH reply.) The common pilot channel, CPICH, transmitted downlink from the base station provides the phase reference and channel estimation information to be used by the terminal to detect the occurrence of the AICH. The RACH message can be 10 or 20 msec long, and can be used to transmit a brief burst of packet data, containing at most about 600 bits: spreading factors ranging from 256 down to 32 may be used over the random access channel. With a spreading factor of 32, the transmission bit rate is 120 kbps. For a 10 msec message this means 1200 bits are transmitted, corresponding to somewhat less than 600 bits of packet data, taking rate-1/2 convolutional coding into account. If a spreading factor of 128 is used because of power considerations, the amount of packet data that can be transmitted in a RACH message clearly reduces to less than 150 bits.

The synchronization channel, SCH, is used for cell search and to obtain necessary frame and slot synchronization. Two synchronization channels are actually used, a primary one using a 256-chip spreading sequence identical in all cells in a system, and a secondary one involving different code words. Both channels are transmitted in parallel. The SCH is time multiplexed with the primary common control channel, PCCCH, used to transmit the BCH (see Table 10.1). It uses one-tenth of a slot in a 10 msec frame, corresponding to 256 chips of the 2560 chips in a slot.

Note that Table 10.1 lists an optional common packet channel, CPCH, procedure with both an uplink channel and a number of downlink channels defined for it. This procedure, if adopted, provides a modified random access technique with collision detection capability better than that of the usual RACH procedure. As a result, more packet data can be transmitted with an access attempt than with normal random access. We shall quantify this improvement in packet data delivery in discussing slotted-Aloha type random access procedures later in the next chapter. The optional downlink shared channel, DSCH, shown in Table 10.1 provides for increased transmission of dedicated user data downlink from base station to mobile terminal. It is particularly useful for the transmission of high bit rate, highly bursty data. As noted earlier, such data are defined to consist of messages in packet form transmitted only occasionally or a fraction of the time. The example provided earlier was that of server responses to client-based web accesses. File transfers, in general, also tend to be of this type. This optional channel is always associated with a downlink dedicated channel, DCH. The channel can be shared by several users. Like the DCH, it provides for variable bit rates on a frame-by-frame basis and supports fast power control.

The access procedure, using the channels indicated above, may be summarized as follows: a terminal desiring to access the mobile network decodes the BCH and determines which RACH are available, as well as the scrambling codes and set of signature sequences associated with each. It selects both a RACH and a signature pattern randomly. The terminal also measures the downlink power level, and uses this measurement to set the initial RACH transmission power level. It then transmits a 1 msec RACH preamble, with the specific signature pattern chosen modulating the preamble, and waits for an AICH reply from the base station. If the RACH preamble has been received correctly by the base station, the AICH message will carry the same signature pattern. The mobile will then respond by sending a full RACH message, 10 or 20 msec long, as noted above. If, after a few time slots, the mobile receives no AICH message, it will retransmit the 1 msec preamble carrying the same signature sequence, but using a higher transmission power, and repeat the access process. (The successive power increments, as needed, are measured in multiples of 1 dB, as specified by the base station in the original BCH message.)

cdma2000

The cdma2000 system is designed to be backward-compatible with IS-95, the second-generation (2G) CDMA system discussed in earlier chapters. The development and standardization of this newer system may be visualized to have occurred in three chronological phases, leading to successively higher data rates, with a different system designation associated with each, although they are grouped together under the general designation of the cdma2000 family. We shall be discussing each of these systems in turn, focusing on their packet-data capability. These systems all use the CDMA multicode approach described in the previous Section 10.1 to increase the data rate, as well as making some use of variable spreading gain codes and the multicarrier approach.

The first extension of IS-95 to higher bit rate packet-data service is labeled IS-95B. This system provides up to 115.2 kbps packet-data service in the forward or downlink direction, BS to mobile. cdma2000 proper represents a further extension of IS-95 to provide much higher bit rate transmission capability. cdma2000 comes in two air interface modes, cdma2000 1X, which uses the same chip code rate of 1.2288 Mcps and 1.25 MHz bandwidth as IS-95 and IS-95B, and cdma2000 3X. cdma2000 3X is a multicarrier version of the cdma2000 family, using three 1.25 MHz carriers covering a 5 MHz bandwidth in the forward direction to provide the higher data rate service. Each carrier uses the IS-95 chip rate of 1.2288 Mcps. A single carrier is used in the reverse (uplink) direction with a chip rate of either 3.6864 Mcps (3×1.2288 Mcps) or 1.2288 Mcps. Finally, the third member of the cdma2000 family, $1 \times \mathrm{EV} - \mathrm{DV}$, is a high bit rate system designed specifically for packet-data communication.

Just a word on notation before we proceed with our discussion of the cdma2000 family. The cdma2000 terminology uses the words *forward* and *reverse* to refer to directions across the radio or air link, rather than the equivalent words downlink and uplink. We shall adopt the same terminology in this paragraph describing the cdma2000 family. In discussing the cdma2000 traffic and signaling channels shortly, we shall, in keeping with cdma2000 terminology, use the letters F and R to refer to forward-direction channels and reverse-direction channels respectively.

We begin with a discussion of IS-95B (Sarikaya, 2000; Tiedmann, 2001). As noted above, this is a packet-mode version of IS-95 and a precursor of cdma2000. It allows up to 115.2 kbps transmission in both forward and reverse directions, using the IS-95 bandwidth of 1.25 MHz and chip rate of 1.2288 Mcps. This system supports 64 physical channels, Walsh codes, in each of the two radio link directions. These channels come in two types, common channels and dedicated channels, as in the case of W-CDMA. (It is left for the reader to compare the channels, as we briefly describe them, with those described in the IS-95 discussion of Chapter 8.)

Consider the common channels first. These are essentially the same as those used in IS-95, as described in Chapter 8. In the forward direction, these channels include a forward pilot channel, F-PICH, which is always generated, a forward synchronization channel, F-SYNCH, and up to seven forward paging channels, F-PCHs. Reverse common channels consist of random access channels, RACHs, only, one for each paging channel, as was the case with IS-95. The F-PICH is used to provide channel estimation, including power control measurements, as well as other information required for channel acquisition and handoff. The F-SYNCH is a 1.2 kbps channel, providing the system time and paging channel data rate, among other parameters. The RACHs are again used to initiate a call, to respond to paging messages, and to register on a location update. These operate at 4.8 kbps.

Now consider the dedicated traffic channels. These come in two types, fundamental channels, FCHs, and supplemental code channels, SCCHs, up to seven of each associated with each FCH. It is the introduction of the SCCHs, using the multicode technique, that provides the increase in bit rate over that of IS-95. The number of FCHs (and hence SCCHs as well) available for data transport varies, depending on how many of the 64 channels are used for control and signaling purposes. Consider the forward direction first. A F-PICH must always be generated. If a F-SCH and F-PCHs are generated as well, the number of forward fundamental channels, F-FCHs, is reduced by the sum of all these signaling and control channels. The number of FCHs in the reverse direction would be 64 less the number of RACHs, one for each PCH, as noted above.

FCHs in either direction transmit at a number of data rates, with a maximum of 9.6 kbps or 14.4 kbps. (Note that 14.4 kbps is higher than the transmission rate of IS-95.) A SCCH always transmits at the maximum rate of the FCH with which it is associated. In particular, if seven SCCHs are made available for a FCH which transmits at its maximum rate of 14.4 kbps, each SCCH transmits at 14.4 kbps, and the rate of transmission becomes

Sarikaya, B. 2000. "Packet mode in wireless networks: overview of transition to third generation," *IEEE Communications Magazine*, 38, 9 (September), 164–172.

Tiedmann, E. G., Jr. 2001. "cdma20001X: new capabilities for CDMA networks," *IEEE Vehicular Technology Society News*, 48, 4 (November), 4–12.

8 × 14.4 or 115.2 kbps, the figure quoted above. Note how multicode usage increases the bit rate availalable.

Packet data transfer is initiated by a mobile using the reverse forward channel, R-FCH (Sarikaya, 2000). If the mobile requires a higher rate of transmission in the reverse direction, it sends a supplementary channel request message to the base station on the R-FCH channel. The base station replies with a supplementary channel assignment message, assigning a number of reverse supplemental code channels, R-SCCHs, to the mobile. If, on the other hand, the base station determines it needs F-SCCHs to increase the rate of transmission to the mobile, it sends a supplementary assignment message to the mobile, and starts transmitting using the F-FCHs and F-SCCHs assigned.

The effective throughput rate of data packets is, however, less than the value obtained using the bit rate capacities quoted above, first, because of necessary overhead introduced; second, because packets do not necessarily fit into IS-95 slots (Sarikaya, 2000). This reduction in actual data throughput rate is a common characteristic of all packet-switching systems and networks. This point was made earlier in our initial discussion of the W-CDMA bit rates, and will be encountered in the GPRS and EDGE discussion following as well.

Example

Consider a simple example to clarify this point. Frames or slots are 20 msec long, as in IS-95. At a 14.4 kbps data rate this means 288 bits per frame. Say a packet 1200 bytes or 9600 bits long is to be transmitted over the radio link. This packet must be cut into smaller segments to fit into the successive frames. In segmentation, no more than 32 bytes per frame are allowed. This turns into 31 bytes of data plus a 1-byte sequence field needed to number successive segments of a given packet. Say 7 SCCHs plus the FCH are to be used for the maximum transmission rate of 115.2 kbps. Then a little thought will indicate that five 20 msec frames are required to transmit the packet. The time required to transmit the packet is then 0.1 sec, and the effective rate of transmission is 96 kbps. Longer packets would result in higher effective rates of transmission; shorter packets reduce the effective transmission rate. As an example, it is left for the reader to show that the effective rate of transmission of an 800-byte packet, again transmitted at the maximum rate of 115.2 kbps, would be 80 kbps. If the random access procedure in the reverse direction is taken into account, the time to transmit a packet from mobile to base station increases even more, further reducing the effective rate of transmission (Sarikaya, 2000).

We now move to cdma2000 proper, the 3G version of the cdma2000 family (cdma, 2002). As noted above, this system comes in two air interface modes, the 1X mode using the same spreading rate and bandwidth as the earlier IS-95 and IS-95B systems, 1.2288 Mcps and 1.25 MHz respectively, and the 3X, multicarrier, mode mentioned above. We discuss cdma2000 1X only, since work on the high data rate system $1 \times \mathrm{EV} - \mathrm{DV}$ has reduced the need for the 3X mode system (Tiedmann, 2001). This system carries many of the features of IS-95 and IS-95B to maintain backward compatibility. For example, the modulation and type of codes used are similar to those of IS-95. The dedicated and common channels of IS-95B described above are maintained, but with new ones added to provide a greatly increased data rate capability as well as improvements in performance. In particular, a new set of higher bit rate channels, supplemental channels, SCHs, have been introduced, offering bit rates as high as 307.2 kbps per channel, using the variable

Table 10.2 *Reverse channel radio configurations (RC), cdma2000 1X*

RC	Data rates (kbps)	Convolutional/turbo encoder (rate *r*)	Spreading gain
1	1.2, 2.4, 4.8, 9.6	Convol., $r = 1/3$	128
2	1.8, 3.6, 7.2, 14.4	Convol., $r = 1/2$	85.33
3	1.2, 1.35, 1.5, 2.4, 2.7, 4.8, 9.6	Convol., $r = 1/4$	
	$9.6 \times N$, $N = 2, 4, 8, 16$ (9.6, 19.2, 38.4, 76.8, 153.6)	Convol./turbo, $r = 1/4$	$128/N$
	307.2 (9.6×32)	Convol./turbo, $r = 1/2$	4
4	1.8, 3.6, 7.2, 14.4	Convol., $r = 1/4$	
	$14.4 \times N$, $N = 1, 2, 4, 8, 16$ (14.4, 28.8, 57.6, 115.2, 230.4)	Convol./turbo, $r = 1/4$	$85.33/N$

spreading gain approach. At most two SCHs can be associated with a FCH. Performance improvement features introduced include, for the forward link, a fast 800 bps power control capable of reacting to shadow fading and improved transmit diversity capability taking advantage of space–time spreading. Diversity is carried out, where desired, by repeating modulation symbols on two antennas such that they are orthogonal without requiring added Walsh functions. The performance improvement obtained by this procedure is particularly noticeable at lower mobile terminal speeds. Improvements made available in both directions of transmission include the availability of both rate-1/2 and rate-1/4 convolutional coding, and turbo coding to be used on the higher bit rate SCHs with frames carrying more than 360 bits (cdma, 2002; Tiedmann, 2001; Willenegger, 2000). As noted in Chapter 7, the use of turbo coding can improve system performance considerably. Rate-1/4 convolutional coding provides performance improvement as well, again as noted in Chapter 7. The most common frame length used in cdma2000 systems is 20 msec, as was the case with IS-95, as described in Chapter 8. Frame lengths of 5, 40, and 80 msec are available as well.

A variety of radio configurations have been defined for cdma2000 (cdma, 2002). These configurations each specify the range of data rates to be made available over the traffic channels, error correction techniques to be used (type of convolutional or turbo coder), and modulation technique used. We discuss the reverse direction case first. There are six radio configurations specified in that direction, four for the 1X system, two for the 3X system. Of the 1X configurations, either configuration 1, corresponding essentially to a IS-95B system with FCH and associated SCCH channels, or the higher bit rate configuration 3, supporting the SCHs in addition to FCH channels, must be used. A configuration 1 system can optionally support configuration 2 as well; a configuration 3 system can optionally support configuration 4 as well. Table 10.2 provides a summary of the four

cdma 2002. Physical layer standard for ccdma2000 spread spectrum systems, Release A, 3GPP2 C.S0002-A, Version 6.0, 3rd Generation Partnership Project 2, 3GPP2, February, http://www.3gpp2.org/public_html/specs/

Willenegger, S. 2000. "cdma2000 physical layer: an overview," *Journal of Communications and Networks*, 2, 1 (March), 5–17.

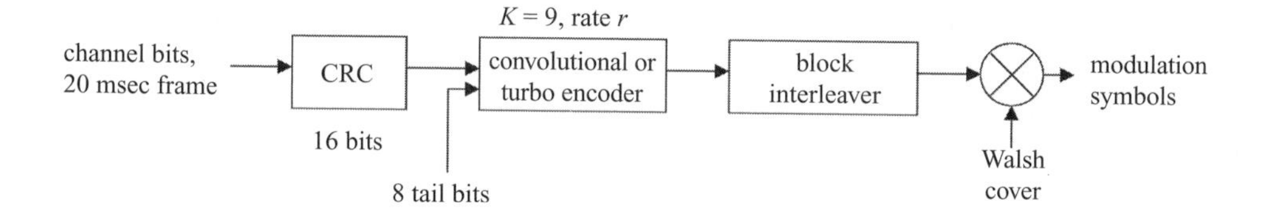

no. of bits in	data rate (kbps)	r	symbol rate (ksps)
360	19.2	1/4	76.8
744	38.4	1/4	153.6
1512	76.8	1/4	307.2
3048	153.6	1/4	614.4
6120	307.2	1/2	614.4

Figure 10.3 Reverse supplemental channel structure, radio configuration 3, cdma 2000 (adapted from cdma, 2002: Fig. 2.1.3.1.1.1.–8)

reverse direction radio configurations specified for cdma2000 1X (cdma, 2002). Note that configuration 1 provides a maximum data rate per FCH of 9600 bps. The associated SCCHs, at most seven per FCH, would then operate at 9600 bps each. Configuration 2 is the one providing a maximum of 14.4 kbps per FCH. There are at most two SCHs per FCH in configuration 3, each supporting data rates up to a maximum of 307.2 kbps. The FCH, if transmitted, operates at a maximum data rate of 9600 bps. There are five possible SCH data rates above 9600 bps, all multiples of 9600 bps, given by $9600 \times N$, $N = 2, 4, 8, 16, 32$: 19.2 kbps, 38.4 kbps, 76.8 kbps, 153.6 kbps, and 307.2 kbps. The variable spreading gain technique is used to generate these higher bit rates, the chip rate remaining fixed at 1.2288 Mcps, as the ratio of chips per bit decreases. In particular, as shown in Table 10.2, the spreading gain decreases as $128/N$ to a minimum value of 4 for $N = 32$. Configuration 4 is similar to configuration 3, also supporting SCH channels, but with the higher data rates derived from multiples of 14.4 kbps, to a maximum of 16×14.4 kbps, or 230.4 kbps. The corresponding spreading gain at this maximum bit rate is $85.33/16 = 5.3$. Turbo coding may be used in place of convolutional coding for the five highest bit rates of configuration 3 and the four highest bit rates of configuration 4.

In Fig. 10.3 we diagram the generation of modulation symbols for the five highest bit rate cases of the SCH for configuration 3 (cdma, 2002). A 20 msec long frame length is assumed. This figure is comparable with the upper portion of Fig. 8.19 in Chapter 8, describing the handling of the much lower bit rate forward traffic channel of IS-95. Note that the number of channel (information) bits input each 20 msec frame interval varies depending on the data bit rate desired. Sixteen cyclic redundancy check (CRC) bits, obtained using a cyclic polynomial code, and referred to as the frame quality indicator, are added to the input channel bits and the resultant data block is fed into the convolutional or turbo coder, whichever is used. Eight tail bits per frame are added as well, as shown. The data rates of Table 10.2 are then defined as the resultant bit rates at the input to the convolutional or turbo coder. Data rates may be changed every frame interval. The encoder used in cdma2000 is always of constraint length $K = 9$. Its rate r varies depending on the radio configuration and the data rate, as shown in Table 10.2. In particular, the SCH for

RC 3 has $r = 1/4$, except for the highest data rate, for which $r = 1/2$ as shown. Data rates may be changed every frame interval.

The encoder output bits are combined into varying-length blocks, depending on the data rate used, and are then block-interleaved as shown in Fig. 10.3. They are then Walsh-encoded, as shown. After Walsh encoding, the SCH output modulation symbols of Fig. 10.3 are added to the FCH modulation symbols if an R-FCH is used, as well as to a reverse pilot channel, R-PICH, signal transmitted at the same time. Orthogonal Walsh encoding is used on the reverse supplemental channel, R-SCH, and R-FCH symbols, to ensure they can be uniquely separated out at the base station receiving system. The summed signals are then spread by the 1.2288 Mcps pseudo-random chip sequence. Each such sequence is generated by a long code generator unique to each mobile, similar to the technique described in the IS-95 discussion of Chapter 8. The R-PICH is a continuously transmitted, constant-value (dc) reference signal enabling coherent demodulation to be used at the receiving base station, thus improving reverse channel detection capability.

Other cdma2000 reverse channels include a reverse dedicated control channel, R-DCCH, used in radio configurations 3 and 4 to transmit bursty data or signaling information at 9600 or 14, 400 bps, a reverse access channel, R-ACH, similar to the one used in IS-95 and IS-95B, a reverse enhanced access channel, R-EACH, and a reverse common control channel, R-CCCH. All these channels, except for R-ACH, are always transmitted together with the R-PICH. The R-EACH supports two modes of access, a basic access mode similar to the ACH, but allowing short messages to be transmitted as part of the random access procedure, and a reservation access mode which speeds up the access procedure and allows longer messages to be transmitted as part of the access procedure (Tiedmann, 2001; Willenegger, 2000). In the reservation access mode, the basic access mode is first used by the mobile to request allocation capacity from the base station. The base station replies with an allocation message on the forward common assignment channel, F-CACH, a channel designed to speed up access acknowledgement and resolve access collisions more quickly. Once the terminal receives this allocation message, it transmits the desired data message using the R-CCCH. The mobile terminal power over that channel is controlled by the base station through a sub-channel transmitted over the forward common power control channel, F-PCCH, which carries multiple such power control sub-channels, each associated uniquely with a R-CCCH (Willenegger, 2000). This use of power control as part of the reservation access mode serves to reduce interference during the access procedure and hence channel reliability. Figure 10.4 summarizes the reverse channels used in cdma2000 1X (cdma, 2002). The block diagram structures of these various channels, similar to the channel structures presented in the IS-95 discussion of Chapter 8, and extending the diagram of the reverse supplemental channel, R-SCH, structure of Fig. 10.3, are all presented in cdma (2002).

Consider now the cdma2000 1X forward channels. The traffic channels used and their characteristics are again defined by radio configurations (cdma, 2002; Willenegger, 2000). There are now five radio configurations defined for cdma2000 1X, instead of the four defined for the reverse direction, as presented in Table 10.2. These forward-direction configurations are presented in Table 10.3, where they are also shown associated with the corresponding RC for the reverse direction. Thus, configurations 1 and 2 are, respectively,

Table 10.3 *Forward radio configurations (RC), cdma2000 1X*

RC	Data rates (kbps)	Reverse ch. RC	Convol./turbo encoder (rate r)
1	1.2, 2.4, 4.8, 9.6	1	Convol., $r = 1/2$
2	1.8, 3.6, 7.2, 14.4	2	Convol., $r = 1/2$
3	1.2, 1.35, 1.5, 2.4	3	Convol., $r = 1/4$
	2.7, 4.8, 9.6		
	19.2, 38.4, 76.8, 153.6		Convol./turbo, $r = 1/4$
4	1.2. 1.35, 1.5, 2.4	3	Convol., $r =1/2$
	2.7, 4.8, 9.6		
	19.2, 38.4, 76.8, 153.6		Convol./turbo, $r = 1/2$
	307.2		
5	1.8, 3.6, 7.2, 14.4	4	Convol., $r = 1/4$
	28.8, 57.6, 115.2, 230.4		Convol./turbo, $r = 1/4$

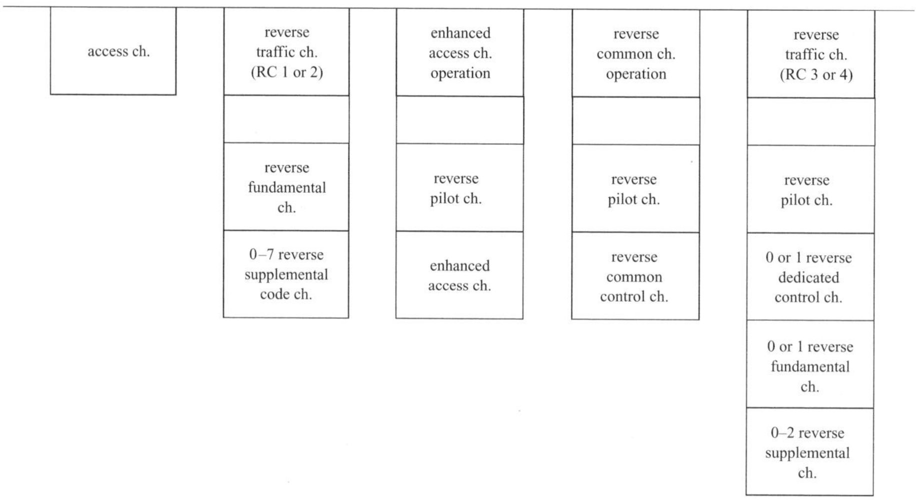

access ch.

reverse traffic ch. (RC 1 or 2)
reverse fundamental ch.
0–7 reverse supplemental code ch.

enhanced access ch. operation
reverse pilot ch.
enhanced access ch.

reverse common ch. operation
reverse pilot ch.
reverse common control ch.

reverse traffic ch. (RC 3 or 4)
reverse pilot ch.
0 or 1 reverse dedicated control ch.
0 or 1 reverse fundamental ch.
0–2 reverse supplemental ch.

Figure 10.4 Reverse direction channels, cdma2000 (from cdma, 2002: Fig. 2.1.3.1.1–1)

each associated with RC 1 and 2 in the reverse direction. These again represent legacy configurations, providing backward compatibility with IS-95 and IS-95B. Forward RC 3 and 4 are both associated with RC 3 in the reverse RC 3; forward RC 5 is associated with reverse RC 4. The traffic channels for RC 1 and 2 are forward fundamental channels, F-FCH, and forward supplemental code channels F-SCCH, up to seven of the latter per forward traffic channel. These channels thus have the same characteristics as in the reverse traffic channel case, and are again backwards-compatible with IS-95 and IS-95B. The traffic channels for the higher bit rate forward radio configurations RC 3 to RC 5 are also similar to those in the reverse direction: each forward traffic channel may carry a lower bit rate F-FCH and up to two higher bit rate forward supplemental channels. The data rates of the SCHs may be changed from frame to frame, just as in the reverse traffic channel case. The variable spreading gain technique is again used to generate the various bit rates shown in Table 10.3, the spreading chip rate always remaining fixed at 1.2288 Mcps. Generation of the forward supplemental channel, F-SCH, is carried out in a manner similar to that of the reverse supplemental channel, R-CH, shown in Fig. 10.3. Block diagrams of the generation of all the forward traffic channels as well as forward signaling and control channels appear in cdma (2002).

The forward control and signaling channels consist of forward control channels familiar to us from our discussion of IS-95 in Chapter 8, plus newly defined channels such as the F-PCCH and F-CACH mentioned above in connection with the reservation access technique. Recall from Chapter 8 that IS-95 has, in addition to forward traffic channels, pilot, sync, and paging channels in the forward direction. These channels appear as well in the cdma2000 architecture. Newly defined forward common channels, in addition to F-PCCH and F-CACH, include a forward common control channel, F-CCCH, and a forward broadcast channel, F-BCCH used to transmit paging messages, channel assignments, short broadcast messages, and overhead messages. Paging is carried out in a two-step process to reduce mobile power usage and increase battery life. Mobiles are normally in sleep mode. A new quick paging channel, F-QPCH, consisting of successive single-bit indicators, is used to signal arrival of a page message. Mobiles wake up to check the indicators. This uses very little energy. The mobile to which a paging message has been sent then follows up by reading the message on the appropriate paging channel, the IS-95-type paging channel, F-PCH, for systems backward-compatible with IS-95, or F-CCCH on newer cdma2000 systems. Other, more specialized, common channels include the forward transmit diversity pilot channel, F-TDPICH, the forward auxiliary pilot channel, F-APICH, and the forward auxiliary transmit diversity pilot channel, F-ATDIPCH, designed for base stations equipped with spot beams and/or multiple antennas allowing space diversity techniques to be implemented (cdma, 2002; Tiedmann, 2001; Willenegger, 2000).

New forward dedicated control channels have been designed for cdma2000 as well. These include a dedicated control channel, F-DCCH, and a forward dedicated auxiliary pilot channel, F-DAPICH. The former would be used for the transmission of bursty data or signaling information. The latter serves a purpose similar to that of the common control channel, F-APICH (Willenegger, 2000). All forward channels are shown listed in Fig. 10.5 (cdma, 2002). Also indicated in that figure is the number of channels per base station, if specified.

Consider now how some of these channels are used as a mobile powers up and acquires system information in beginning the process of registering as part of the call initiation process (Willenegger, 2000). Base stations continually transmit the forward pilot channel, F-PICH, represented by the continuously transmitted constant value 1, as scrambled by a PN sequence. Just as in the case of IS-95, different base stations are identified by a different PN offset. The mobile, on powering up, searches for the strongest pilot channel, and, on acquiring it, next decodes the forward synchronization channel, F-SYNCH. This sync channel is synchronized with the pilot channel and scrambled with the same PN sequence, enabling coherent demodulation to be used. The sync channel identifies the base station and system with which it is associated, as well as other necessary system information such as system timing, broadcast channel, and paging channel information.

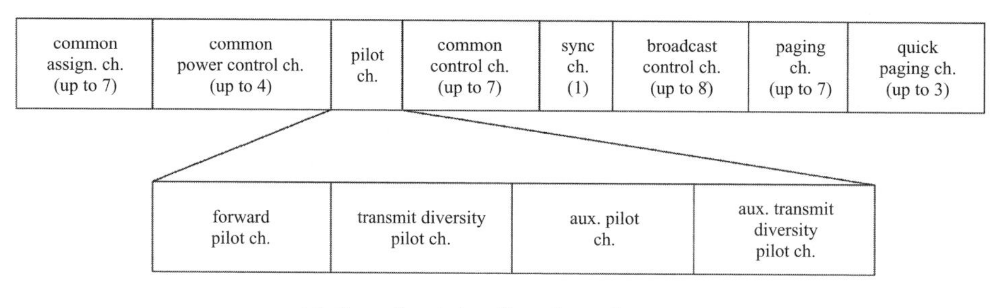

(a) Control and signaling channels

dedicated control ch. (0–1)	fundamental ch. (0–1)	power control subch.	supplemental code ch. (0–7) RC 1,2	supplemental ch. (0–2) RC 3,4,5

(b) Traffic channels

Figure 10.5 Forward channels, cdma2000 (from cdma, 2002: Fig. 3.1.3.1.1–1)

Additional detailed system information is now obtained by listening to either the forward paging channel, F-PCH, in the case of cdma2000 systems backward-compatible with IS-95, or to the broadcast channel, F-BCCH, if compatibility is not required.

With system parameters obtained, the mobile can proceed with registration, using one of the access modes noted earlier. In particular, in the basic access mode a preamble, consisting of a sequence of all-zero frames, and a registration message following, are sent repeatedly, with increasing power each time, until an acknowledgement message on a forward common assignment channel, F-CACH, is received from the base station. A random backoff procedure is used in repeating non-acknowledged access attempts. Finally, the base station notifies the mobile of successful registration by confirming the registration using the forward common control channel, F-CCCH. The mobile can now go into idle mode awaiting a page message, as indicated above, or request the setup of a dedicated channel for information-message transmission, again using one of the access modes described above.

We conclude this section on the cdma2000 family by briefly describing the 1×EV–DV standard, designed specifically to handle high-rate packet data (cdma, 2001; Bender *et al.*, 2000). We focus on transmission in the forward direction, from base station to mobile terminal, since this is the direction in which most high data rate traffic will presumably be flowing, in response to mobile users' requests to Internet web sites. The forward channel carries the forward traffic channel or the control channel, plus the forward pilot channel and the forward medium access control (MAC) channel, all time-multiplexed together. User packets are transmitted over the forward traffic channel at rates ranging from 38.4 kbps

---

cdma 2001. cdma2000 high rate packet data air interface specification, 3GPP2 C.S0024, Version 3.0, 3rd Generation Partnership Project 2, 3GPP2, December, 5, http://www.3gpp2.org/public_html/specs/

Bender, P. *et al.* 2000. "CDMA/HDR: a bandwidth-efficient high-speed service for nomadic users," *IEEE Communications Magazine*, 38, 7 (July), 70–77.

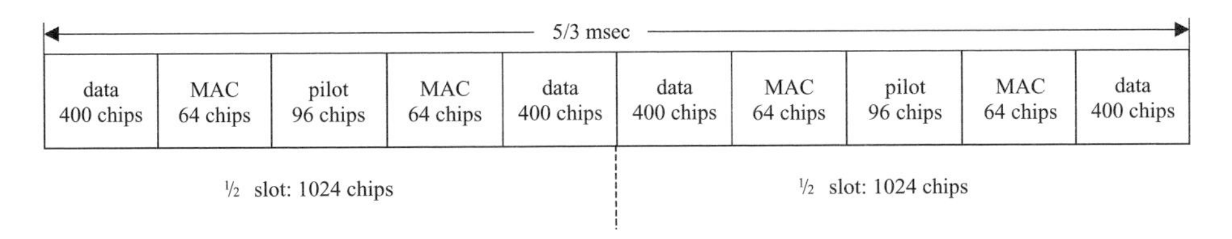

Figure 10.6 Slot configuration, cdma2000 high rate packet data air interface (cdma, 2001: Fig. 9.3.1.3.1–2)

to 2457.6 kbps (2.4576 Mbps). (For comparison, the data rates available in the reverse direction range from 9.6 kbps to 153.6 kbps.) The control channel carries control messages and may carry user traffic as well. The pilot channel, consisting of all zeroes mapped into a sequence of +1s and then coded into the Walsh 0 codeword, as in IS-95 and cdma2000 1X, is used by the mobile for initial acquisition of the base station signal, just as in these other systems. In addition, it is used by the mobiles for phase and timing recovery, as well as to estimate the SIR at that mobile. This estimate is then fed back to the base station to control the data rate to be used in transmitting to the mobile. The MAC channel consists of a set of 64 Walsh channels. Three subchannels are transmitted over this channel, the reverse power control channel and the DRCLock channel which are time-multiplexed over one of the Walsh channels, from 5–63, assigned to a particular mobile terminal, and a reverse activity channel, assigned to Walsh channel number 4.

The forward channel carrying these various time-multiplexed channels is transmitted as a repetitive series of time slots 5/3 msec in length. Each slot is therefore associated with 2048 chips of the 1.2288 Mcps pseudo-noise chip stream. Figure 10.6 portrays the slot configuration of an active channel with a traffic or a control channel present and occupying the four data positions indicated, for a total of 1600 chips. Note how the various channels are time-multiplexed symmetrically about the center of the slot, the slot being considered to be made up, virtually, of two symmetric halves. The pilot channel occupies 192 of the 2048 chips, at the center of each of the two slot halves. It is surrounded, in each half of the slot, by the MAC channel components, as shown in Fig. 10.6. An idle slot, always sent when neither traffic nor control information is available to be transmitted downstream to any of the mobiles, would be empty in the data positions, but would still carry the MAC and pilot channels in the same positions as shown.

Data transmitted on the forward traffic channel and directed to a particular mobile terminal is encoded into fixed-length packets and transmitted at varying rates, the rate chosen for a given packet signaled by the mobile in a previous reverse transmission. That rate, in turn, depends on the SIR measured by the terminal on receipt of pilot channel signals from its base station and interfering base stations, as noted above. An example of the choice of the data transmission rate will be described below, based on simulation results appearing in Bender *et al.* (2000). The various data rates available, the packet lengths prescribed for each, and the type of modulation chosen for each data rate, are presented in Table 10.4 (cdma, 2001). (The control channel operates at the two lowest data rates only.) Some of the data rates, particularly those at lower values, require multi-slot transmission per packet. The number of slots required to transmit a packet is also shown in Table 10.4. Multiple slots, when used, are spaced four slots apart. Turbo encoding is

Table 10.4 *Forward traffic channel parameters, cdma2000 1×EV–DV: high rate packet data system*
(from cdma, 2001; Table 9.3.1.3.1.1-1)

Data rate (kbps)	Packet length (bits)	Slots/packet	Code rate	Modulation type
38.4	1024	16	1/5	QPSK
76.8	1024	8	1/5	QPSK
153.6	1024	4	1/5	QPSK
307.2	1024	2	1/5	QPSK
614.4	1024	1	1/3	QPSK
307.2	2048	4	1/3	QPSK
614.4	2048	2	1/3	QPSK
1228.8	2048	1	1/3	QPSK
921.6	3072	2	1/3	8-PSK
1843.2	3072	1	1/3	8-PSK
1228.8	4096	2	1/3	16-QAM
2457.6	4096	1	1/3	16-QAM

specified for this system, and the turbo encoding rate appears in Table 10.4 as well. The packet lengths shown in Table 10.4 are those appearing at the input to the encoder, just as in the case of the other cdma2000 systems described earlier. (The one difference here is that a varying-length preamble, covering the first portion of the data-chip field, is sent prior to transmitting the packet (cdma, 2001).)

What is the system configuration required to obtain the various data rates shown? Note, in particular, that the highest data rates exceed the pseudo-noise chip rate. Multi-level modulation, as indicated in Table 10.4, as well as parallel transmissions, are used to obtain the high data rates. The detailed block diagram of the forward channel structure appears in Fig. 10.7, as taken from cdma (2001). We summarize portions of this system configuration: data packets are fed into the encoder, the output of which is scrambled and interleaved, as in other cdma2000 systems. After the interleaving process, the resultant bit stream is fed into a modulator which carries out QPSK, 8-PSK, or 16-QAM modulation, as required, as shown in Table 10.4. (See Chapter 5 for a discussion of these different modulation alternatives. 8-PSK is discussed briefly later when describing the EDGE scheme for enhancing GSM performance.) For example, two successive bits are combined to form the QPSK-modulator input, three successive bits are combined to form the 8-PSK input, and four bits are combined to form the 16-QAM input. Modulator output symbols are repeated as necessary, to bring the effective modulator output symbol rate up to the same value for all data rates. (This is thus the case for the lower data rates.) The resultant modulator output sequence symbols are then demultiplexed to form 16 pairs of in-phase (I) and quadrature (Q) parallel streams (cdma, 2001). Each stream is then Walsh-encoded using one of 16 Walsh codes operating at 76.8 ksymbols/sec each, and the 16 separate I and Q streams are individually summed to form a 1.2288 Msymbols/sec pair of

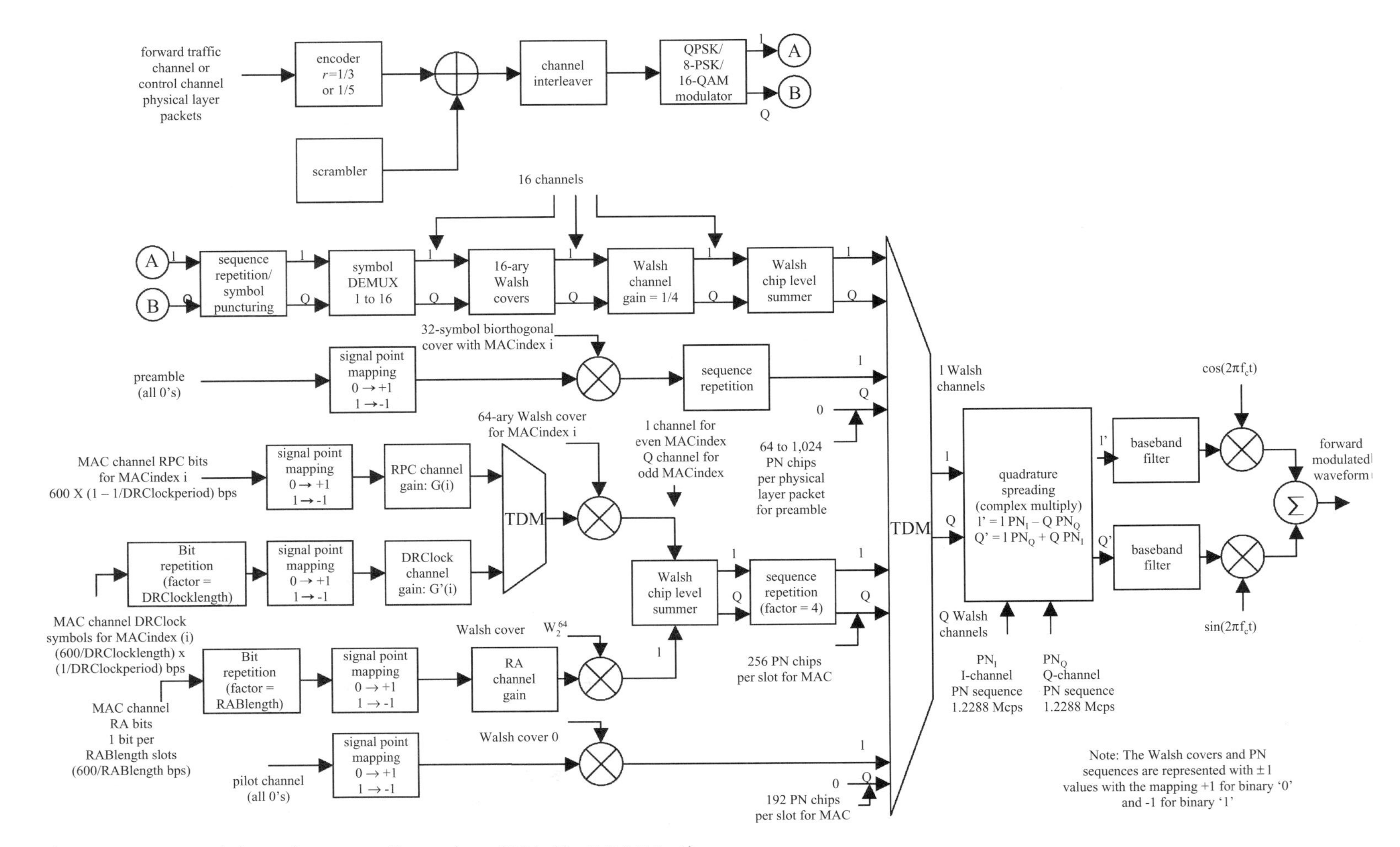

Figure 10.7 Forward channel structure (from cdma, 2001: Fig. 9.3.1.3.1–1)

I and Q streams. Each of these two traffic streams is then time-multiplexed with the pilot and MAC channels, as required, as well as with the data preamble symbol stream. The composite I and Q time-multiplexed sequences are then separately encoded with 1.2288 Mcps pseudo-noise sequences, and multiplied by the desired high-frequency cosine and sinewave carrriers, respectively, for transmisson over the air interface. We now provide a number of examples

Example 1

Consider the highest data rate, 2.4576 Mbps, as an example. This data rate corresponds to a packet of 4096 bits transmitted in a 5/3 msec time slot, as indicated in Table 10.4. The rate at the output of the rate-1/3 encoder used is then three times this rate, or 7.3728 Mbps. Passing this bit stream through the 16-QAM modulator, i.e., collecting four bits at a time to drive the modulator, results in a modulator output symbol rate of 1.8432 Msymbols/sec, or 3072 symbols per 5/3 msec time slot. This symbol stream is now demultiplexed into the 16 parallel pairs of Walsh-encoded streams noted above.

Example 2

The case of the 1.8432 Mbps data rate at the encoder input similarly corresponds to 3072-bit data packets per 5/3 msec time slot at the encoder input. Since an 8-PSK modulator is used in this case, we again have a modulator output rate of 1.8432 Msymbols/sec, or 3072 symbols per time slot demultiplexed into the 16 parallel pairs of Walsh-encoded streams. However, for the same transmitter output power, the 2.4576 Mbps data rate stream requires a higher SIR at the receiver than the 1.8432 Mbps data rate stream because of the use of 16-QAM modulation as compared with 8-PSK.

Example 3

Finally, consider the 1.2288 Mbps data rate case. Note, from Table 10.4, that there are two options. In one case a 2048-bit packet can be transmitted in one time slot, with QPSK used as the modulating scheme. In the other case, a packet of twice the length, 4096 bits, can be transmitted using two time slots, spaced four time slots apart, but using 16-QAM. Added latency is introduced in this case because of the time slot spacing. Moving to the other extreme in Table 10.4, note that the lowest data rate indicated is 38.4 kbps. In this case a 1024-bit data packet would require 16 time slots to be transmitted, with four-slot spacing between each of the packets. The latency, i.e., the time required to transmit and hence receive a packet, not including any processing and packet buffering time, is now quite high. But note that, both because bit widths are larger and QPSK can be used rather than 8-PSK or 16-QAM, the SIR required at the receiver is considerably less than that required for the higher bit rate transmissions.

Users would obviously prefer as high a data rate as possible using this high rate packet data system. What data rates are actually possible, and what are the tradeoffs with respect to the latency? Bender *et al.* (2000) report on a simulation study carried out on a large cdma2000 data network incorporating most of the same features as those defined in cdma (2001). (Encoder rates differ somewhat from those shown in Table 10.4. For example, lower bit rate systems are assumed to use a rate-1/4 rather than a rate-1/5 encoder, as shown in Table 10.4; the higher bit rate systems use rate-1/2 rather than the rate 1/3 encoder of Table 10.4. All data rates except for the two highest, 1843.2 kbps and 2457.6 kbps, use QPSK modulation; the latter two use 8-PSK and 16-QAM, respectively. Note that the 921.6 kbps rate in Table 10.4 is shown using 8-PSK.) The terminals in the simulation study each estimate the received SIR, including, in the interference parameter estimated, both intracellular and intercell interference as well as receiver thermal noise. This estimated value of SIR is then converted by the terminal to the maximum data rate possible for a given packet error probability, and that rate is transmitted to the base station, just as described earlier in our discussion of this system.

Table 2 in Bender *et al.* (2000) lists the resultant SIR values found, in this study, to be required to maintain a 1% packet error rate. Interestingly, these are found to increase almost linearly with data rate from 38.4 kbps until 1228.8 kbps. Thus, a 38.4 kbps data rate requires an SIR of −12.5 dB, with roughly 3 dB increases, or a doubling, of SIR required for each doubling of the data rate until 1228.8 kbps, requiring an SIR of 3 dB, is reached. (Note that the overall dB difference is 15.5 dB while the ratio of data rates is 32, or 15 dB.) Doubling the rate further to the highest rate of 2.4576 Mbps requires, however, a further 6.5 dB increase of SIR to the value of 9.5 dB. Also shown in Bender *et al.* is a plot of the cumulative distribution function (cdf) of the SIR, as measured in the simulation study. This plot indicates, for example, that the SIR is less than 0 dB 50% of the time and less than 5 dB 80% of the time. (Conversely, there is a probability of 0.5 that the SIR will exceed 0 dB and a probability of 0.2 of exceeding 5 dB.) Combining the measured cdf of the SIR and the SIR required for transmission at a given data rate, the probability of being able to transmit at each data rate can be calculated. The resultant probability distribution or histogram of the data rates is also tabulated in Bender *et al.* The most probable data rate is found to be 307.2 kbps, with a probability of 0.24. The probability distribution of the other data rates is roughly symmetrically distributed in probability about the 307.2 kbps probability value. For example, the probability of transmitting at a 153.6 kbps rate is about 0.08, while the probability of transmitting at a 921.6 kbps rate is about 0.09. The average system throughput, given by summing the various transmission rates possible, each weighted by the probability of achieving that rate, turns out to be about 570 kbps.

Is it possible to exceed that average throughput value? Bender *et al.* suggest a technique based on constraining the latency. Note that this simple calculation of average system throughput just described ignores latency completely. Latency will thus vary, depending on the rate of transmission. From Table 10.4, for example, it is clear that the number of slots required to transmit a given-sized packet is inversely proportional to the data rate. Those users forced to transmit at lower rates because of lower SIR will suffer correspondingly longer delays for transmission. Bender *et al.* propose a transmission strategy with the ratio of maximum to minimum latency fixed at some specified value. They then show how the average throughput may be maximized given this specified latency ratio. Increasing the latency ratio will increase the average throughput. If users are all required to have the same latency value, irrespective of data rate, the average throughput turns out to be 360 kbps for the system simulated. If the latency ratio now increases to 8, the average system throughput increases to 750 kbps.

This discussion of the appropriate transmission strategy in the forward direction is an example of the problem raised at the end of the previous, introductory, section of this chapter on how best to allocate system resources. It was noted there that the resource allocation problem is also called the scheduling problem. We address this problem further in Chapter 11, in discussing, by example, the general issue of multi-access: how both mobile users and different types of traffic in both the forward and reverse directions can appropriately be assigned system resources to achieve desired performance objectives. A specific example applicable directly to the cdma2000 high data rate system just described appears in Andrews (2001) in which a throughput-optimal scheduling scheme is discussed, using somewhat different performance objective functions than the one involving the latency ratio noted above, appearing in Bender *et al.* (2000).

10.3 2.5/3G TDMA: GPRS and EDGE

Having discussed the 3G CDMA systems in the previous section, we now move on to the 3G systems using TDMA technology. We focus on systems arising out of, and compatible with, GSM. As noted earlier in the introduction to this chapter, and, as is apparent from the choice of title for this section, these TDMA-based 3G systems are often considered 2.5G systems as well: they are designed to provide GSM with a packet-switching capability for data transmission as quickly as possible, with as few changes to the GSM architecture as possible. There are two basic sets of standards defined to provide these changes, GPRS (general packet radio service) and EDGE (enhanced data rates for global evolution).[1] GPRS consists, as we shall see below, of two basic parts:a core network portion and an air interface standard. EDGE, as indicated by the meaning of the acronym, is designed to provide substantially increased bit rates over the air interface. It was also noted earlier that it is expected that IS-136 systems will move to incorporate these new technologies and architectural features as well (IEEE, 1999). The combination of GPRS and EDGE is sometimes called EGPRS, for enhanced GPRS. We first describe GPRS and then EDGE. The core network portion of GPRS uses a layered architecture with the Internet-based TCP/IP protocols playing a key role. We thus take time out to briefly review the concept of layered architectures in packet-switched systems, using TCP/IP layering as the prime example. (It is to be noted that both W-CDMA and cdma2000 use layering in their packet-switching architectures, but, because the stress in this chapter thus far has been on the air interface, we have not had occasion to discuss this concept. Layering is, in fact, used in describing the various messages transmitted in second-generation cellular systems, but we chose not to discuss this issue in Chapter 8 to avoid unneeded complexity at that time.)

GPRS

As is true for the 3G CDMA systems discussed in the previous section, the GPRS technology and architecture have been specifically designed with packet-based applications in mind. In the introduction to this chapter, we briefly discussed the characteristics of these applications and the requirements they place on the 3G packet-based systems. We noted, for example, that a number of types of data transmission have been identified as resulting from these applications: bursty data traffic, implying the time required to deliver the data

---

[1] Alternate interpretations appearing in the literature for the letter G in this acronym include "GSM" and "GPRS." Recall, from Chapter 8, that the acronym GSM itself stands for Global System for Mobile Communication, although originally coined from the French phrase Groupe Speciale Mobile. Acronyms tend to stay on forever, while their interpretations have a way of changing!

Andrews, M. *et al.* 2001. "Providing quality of service over a shared wireless link," *IEEE Communications Magazine*, 39, 2 (February), 150–153.
IEEE. 1999. "The evolution of TDMA to 3G," *IEEE Personal Communications*, 6, 3 (June).

(the "burst") is much less than the time between successive bursts; frequent transfer of small volumes of data; and occasional transmission of larger files, as might occur when accessing a Web site. These applications give rise to specific attributes desired of GPRS (GPRS, 1998; Kalden *et al.*, 2000). (The other 3G systems discussed previously obviously incorporate these attributes as well.) They include efficient use of the radio resources, one of the reasons for moving to packet-based transmission, as noted previously, with connect times being much less than those encountered in handling circuit-switched voice traffic; efficient transport of packets; the ability to connect seamlessly to other packet networks (PDNs) such as the Internet; and, finally, the provision of desired quality-of-service, QoS, characteristics.

We noted, in discussing W-CDMA/UMTS, that a number of traffic classes were defined for that system, each characterized by such QoS objectives as delay, time variation between information entities, and data integrity (probability of error). QoS in GPRS is defined as one of a set of service characteristics (GPRS, 1998). Users subscribe to a specified subscriber profile for an agreed contractual period, this profile containing, among other information, a QoS profile. QoS attributes included in this profile are priority of service, reliability, delay, and throughput (GPRS, 1998). Three levels of priority are prescribed: high precedence, normal precedence, and low precedence. Low precedence would correspond to the W-CDMA background class or the Internet best-effort class, with no specific constraint on time of arrival of information transmitted. Three reliability classes are specified, reliability being defined by four different parameters relating to delivery of a user information unit: the probability of loss of the unit of data, probability of duplication, probability of mis-sequencing (data units arrive out-of-order), and probability the data unit will be corrupted, or delivered with an undetected error. Class 1 of the three reliability classes requires these parameters to be of the order of 10^{-9}, i.e., extremely low. Applications in this class are sensitive to errors, have no error-correction capability, and have limited error tolerance capability. Reliability class 2 is characterized by moderate values of these parameters, ranging from 10^{-4} for probability of loss to 10^{-6} for corrupt data unit probability. Applications associated with this class are error-sensitive, but are capable of limited error correction, and have good error-tolerance capability (GPRS, 1998). Finally, class 3 is to be used by applications that are not error-sensitive. The probability of loss prescribed for users of this class is the relatively high value of 10^{-2}; the corrupt data unit probability is set at this value as well. Duplicate data probability and out of sequence probability are both specified as 10^{-5} (GPRS, 1998).

There are four delay classes defined in the GPRS specifications. These classes are spelled out in Table 10.5, taken from Table 5 of GPRS (1998). By "delay" is meant the time required to transfer a data unit through the GPRS network(s) only. It does not include time required to transfer data units through external networks, such as the Internet, to which the GPRS network(s) might be connected. This time includes the time required to access the radio channel and then transfer a data unit uplink, the time to traverse the

GPRS. 1998. "Digital cellular communications system (Phase 2+): general packet radio service description," Stage 1, EN 301 113 v6.1.1 (1998–11) (GSM 02.60 version 6.1.1 Release 1997), European Telecommunications Institute (ETSI), Sophia Antipoli, Valbonne, France.

Kalden, R. *et al.* 2000. "Wireless internet access based on GPRS," *IEEE Personal Communications*, 7, 2 (April), 8–18.

Table 10.5 *GPRS delay classes*
(from GPRS, 1998; Table 5)

	Maximum delay (sec)			
	128 octets, data unit size		1024 octets, data unit size	
Delay Class	Mean delay	95 percentile	Mean delay	95 percentile
1 (Predictive)	<0.5	<1.5	<2	<7
2 (Predictive)	<5	<25	<15	<75
3 (Predictive)	<50	<250	<75	<375
4 (Best effort)	(Unspecified)			

GPRS(s) networks, and the time to schedule and then transfer a data unit downlink to a receiving mobile. Note from Table 10.5 that the delays specified are defined in terms of both the mean transfer delay and the 95-percentile delay, and are given for two different data unit sizes, 128 and 1024 octets. Note also how high many of the delay values are. They are clearly not acceptable for real-time interactive voice, which, as we indicated earlier, would be expected to have a maximum end-to-end delay of about 0.1 sec. An average delay of minutes might, however, be acceptable for certain types of data traffic.

The final QoS parameter appearing in a user profile is the user data throughput. This parameter represents the data throughput requested by a user. It is defined by two negotiable parameters, the maximum bit rate and the mean bit rate (GPRS, 1998). The mean rate includes, for bursty transmissions, intervals in which no data are transmitted. A user transmitting data at a negotiated maximum rate of 384 kbps, for example, but which does so only 5% of the time, on the average, would have a mean throughput rate of 19.2 kbps. The maximum bit rate allocated to a user is clearly limited by the maximum bit rate available over a given system. Since GPRS is designed to adapt GSM to packet switching of data, it uses the same 8-slot repetitive frame structure of GSM we described in Chapter 8. Unlike circuit-switched GSM, with users each allocated one time slot per frame, packet-switched GPRS users can be allocated up to all eight slots per frame to increase their throughput capability. (This clearly reduces the number of users capable of accessing a given system, but note that the slot occupancy assigned is for one frame at a time, unlike circuit switching, in which a slot, the trunk, is held for much longer times.) It is this maximum possible allocation of eight slots per frame that provides the absolute throughput limit.

Depending on the application and market requirements, some mobile users might want to avail themselves of the GPRS packet-switched and GSM circuit-switched capability simultaneously; others might want to use GPRS or GSM at different times. For this reason, three different GPRS mobile station classes have been defined (GPRS, 1998). Class A supports the simultaneous use of both GPRS and GSM services. "Simultaneous" implies that both services will be using the same set of time slots to transmit information.

Hence, for this class, a minimum of one time slot per frame must be made available, when required, for each of the two services.[2] Class B does not support simultaneous transmission of information. Class B mobile users can, however, make and/or receive calls on either of the two services sequentially. The appropriate service to be used would be done automatically. Finally, class C supports one or the other service only, but not both. A mobile can, however, be reconfigured to another class.

The discussion to this point has focused on the types of applications supported by GPRS, with QoS attributes associated with each, as well as the different classes of mobile stations supported. We now describe the GPRS architecture required to handle these applications and provide the desired quality-of-service profile. The implementation of GPRS requires changes in the GSM core network architecture, as well as changes in the structure of messages transmitted across the GSM air interface. We discuss the GPRS core network architecture first, then move to the transmission of messages across the air interface. Two new packet-based support nodes have been added to the GSM system architecture in establishing the GPRS core network. These are the *serving GPRS support node*, or SGSN, and the *gateway GPRS support node*, or GGSN. The SGSN is connected to base stations in a given serving area, delivering packets to, and receiving packets from, GPRS mobile stations associated with those base stations. It thus serves a function similar to that of the mobile switching center (MSC) in the circuit-switched GSM world. The SGSN is, in turn, connected to other SGSNs in a given public land mobile network (PLMN) or, through the GGSN, to external packet data networks such as the Internet. The SGSN carries out packet routing and, like its circuit-switched counterpart MSC, is responsible for mobile management, location management, and user authentication, among other functions required in its serving area. Location information for all GPRS users registered in the SGSN service area is stored in the SGSN. This includes the home location register (HLR) and visitor location register (VLR).

Figure 10.8 shows how the SGSN and GGSN might be deployed in a very simple data networking architecture. Two SGSNs, each controlling a number of base stations, are shown attached to the same public land mobile network, PLMN1. As a special case, SGSNs could be directly connected to one another. (For simplicity, one of the SGSNs is shown controlling one base station only.) Connection to the external public data network, PDN, is made through a GGSN, as shown. PLMN2 is shown connected to the public data network as well. Only three data mobile stations M1–M3 are specifically identified, to simplify the figure. Note that data mobiles may be connected via land-based data networks to other data mobiles, to data users, data terminals, and data bases in various packet-switched data networks, as indicated. Other possible configurations, specifically

[2] The deployment of this class poses problems in ensuring interworking between existing, widely deployed GSM systems and the GPRS systems being discussed here. Included are issues involving the use of higher frequency bands for the 3G GPRS systems, with different propagation characteristics which result in poorer signal penetration within buildings. This would thus presumably result in increased handovers between GPRS (or W-CDMA for that matter, if used) and GSM as users penetrated buildings. Another significant issue is the need to keep mobile terminal costs low, and yet allow support of both GSM and GPRS. The need to resolve these problems has led to the proposal to create the class A Dual Transfer Mode (DTM). Details of the issues raised and the solutions proposed by class A DTM are discussed in Pecen and Howell (2001).

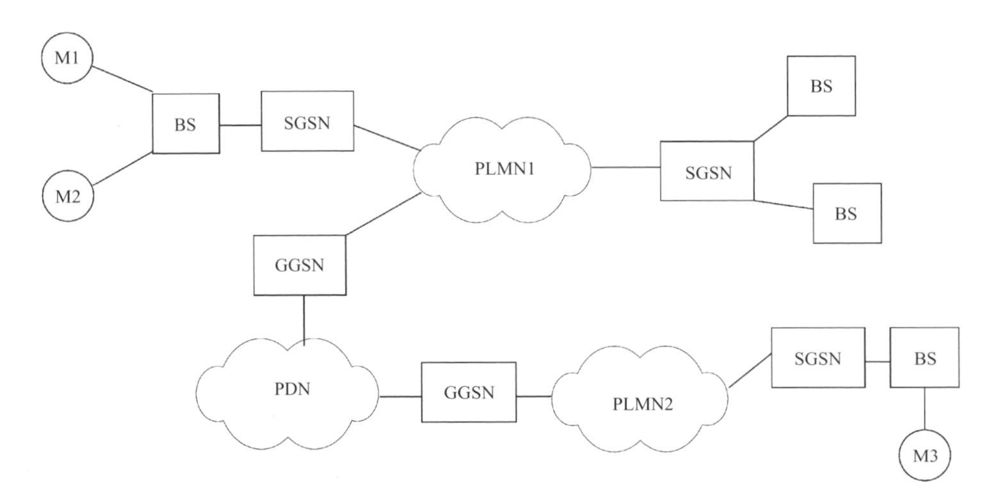

Figure 10.8 Simple GRPS networking configuration

allowing multiple routing paths between mobile and/or fixed data terminals and data bases appear in Bettstetter *et al.* (1999).

Layered architectures reviewed

How are data units transmitted across a packet-switching network such as that portrayed in Fig. 10.8, from a data mobile user to a destination such as a web site associated with the Internet or, perhaps, to another data mobile user? Packet-switching networks use layered architectures to ensure data units arrive in appropriate form at the desired destination (Kurose and Ross, 2002; Schwartz, 1987). Operations at each layer of the architecture are specified by protocols establishing the desired rules of procedure to be followed at that layer. (The word *protocol* obviously comes from the diplomatic world where it denotes a procedure whereby individuals are enabled to communicate with one another.) A protocol at a given layer is used to logically connect devices at each end of that layer. The core network of GPRS uses a layered architecture to ensure correct communication across the network and beyond, as we shall shortly see. But before focusing on the GPRS architecture and protocols we first review briefly the concept of layered architectures. We do this through a simple example involving the Internet layered architecture.

Generally speaking, in this concept a layer below is designed to provide a service for the layer above. A simple example using the Internet architecture appears in Fig. 10.9. Figure 10.9(a) shows a very small portion of a network, with two user hosts, terminals or computers, connected to one another through two routers. Each router has additional output links shown, connected to other routers or to other hosts (not shown). Figure 10.9(b) indicates the architecture layers appearing at each of the devices indicated in part (a) of the figure. The two hosts each have five layers of the architecture, as indicated; the two

---

Bettstetter, C. *et al.* 1999. "GSM Phase 2+ general packet radio service GPRS: architecture, protocols, and air interface," *IEEE Communication Surveys*, 3rd qtr.

Kurose, J. F. and K. W. Ross. 2002. *Computer Networking*, Boston, MA, Addison-Wesley.

Schwartz, M. 1987. *Telecommunication Networks: Protocols, Modeling, and Analysis*, Reading, MA, Addison-Wesley.

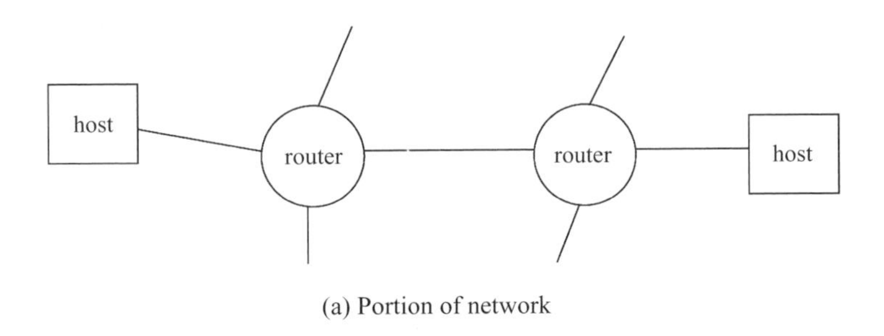

(a) Portion of network

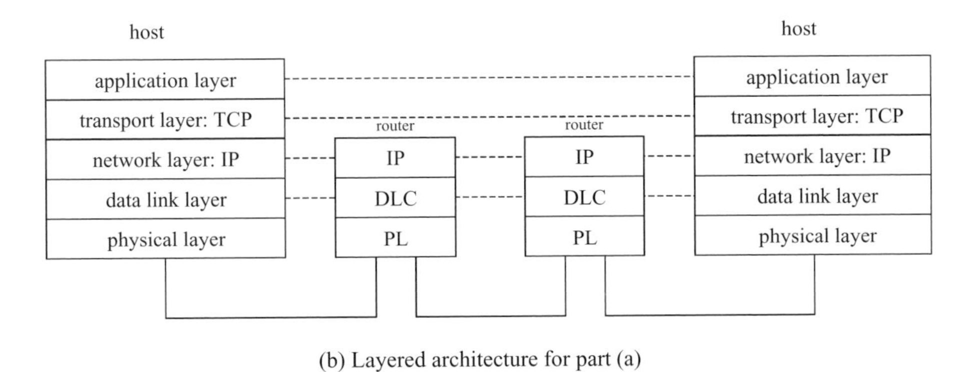

(b) Layered architecture for part (a)

Figure 10.9 Internet TCP/IP layered architecture

routers carry three layers of the architecture only. Protocols corresponding to each layer would normally be implemented in software, or, at the lowest layers only, in hardware. The dashed lines connecting devices or systems at a given layer indicate logical connections. The solid lines at the lowest, physical, layer represent the only links, whether radio, wire, or fiber, across which bits are actually transmitted.

The application layer is the one at which users enter their specific messages to be sent from one host, the user terminal or computer, to another. These messages in the form of information data units, often called the information payload, must be routed across a network or multiple networks to arrive at the desired destination. Layers below the application layer are used to ensure the data units arrive at the destination as desired. Consider first the layer below the application layer. This is called the transport layer. In the Internet suite of protocols this can be either TCP, transmission control protocol, most commonly used for non-real-time applications, or the user datagram protocol, UDP, used for real-time applications such as voice. We refer to TCP only for simplicity of discussion. This protocol provides a service for the application layer above, in carrying out data unit sequencing and error correction, among other operations, to ensure correct, orderly delivery of messages across the network. Error correction is carried out using the automatic-repeat-request, or ARQ, procedure described briefly in Chapter 7. (Recall that, in this procedure, a timer is set at the sending system, the host in Fig. 10.9, as a message is transmitted. If the destination host receives this message correctly, it transmits a positive acknowledgement back to the sender. Otherwise it drops the message. The sending host, on receiving the acknowledgement before a specified timeout interval, assumes the message has been correctly received, and discards the message which it has previously stored. If no acknowledgement is received before the timeout has expired, the sender repeats the message.) Among other functions, TCP also carries out flow control operations to ensure orderly transmission and reception of messages without overflowing receiving buffers.

As noted above, the dashed lines shown connecting the application and TCP layers at the two hosts at each end of the network(s) indicate that a layer at one end is logically connected to its peer at the other end of the network(s). These two layers operate end-to-end, between sending and receiving hosts, with no knowledge of the actual network in between. It is as if a logical pipe connected each of these layers at one end to its peer at the other end. The objective then is to insert messages at one end of the pipe and have them arrive correctly, and in the sequence in which transmitted, at the other end. That, as noted, is the service provided by TCP to the application layer above. But messages do not flow through "pipes." The data units carrying the messages must be routed across a network. This routing function is the service provided to TCP by the network layer just below. This layer, in the case of the Internet, is implemented as IP, the internet protocol. Note from Fig. 10.9(b) that IP is installed at each of the hosts and at each router in the network, as shown in Fig. 10.9(b). Dashed lines are thus shown interconnecting each of the systems at which routing decisions are made. The dashed lines again indicate logical, rather than physical, connections between the devices involved.

To physically get from one device to another, however, the data units must be transmitted across the links interconnecting the devices, indicated by the solid lines in Fig. 10.9(a), as noted above. A data link protocol at the data link layer is used to ensure the block of bits constituting a data unit does arrive recognizably at the receiving end of a link. (This layer is often implemented as two sublayers, as we shall see in discussing the GPRS layered architecture below.) Among the functions of the data link protocol is that of recognizing the beginning and end of a given data unit. The protocol is usually designed as well to determine whether a data unit has been received error free at the end of a link, either correcting messages received in error, or often simply signaling to the layers above that errors have been detected. It is at a sublayer of this layer that access procedures such as the Aloha protocol we have described a number of times are carried out across the radio link in the case of wireless communication. Data link layer connections are logical connections as shown by the dashed lines in Fig. 10.9(b).

As noted above, the actual physical connections between devices and systems in the network(s) are made at the lowest layer of the architecture, the physical layer, as shown by the solid lines connecting devices in Fig. 10.9(b). This is the layer that actually carries out physical transmission of the bits constituting a data unit. Radio transmission across the mobile-base station interface is an example of a physical layer procedure. The various transmission techniques described in this and earlier chapters, such as PSK, QAM, and even CDMA, are each incorporated as part of a physical layer procedure. The objective of the physical layer, in support of the data link layer above, is to ensure that individual bits are transported across a link, whether that link is a radio link, a pair of wires, or an optical fiber link in a transport network.

Note that we have been using the single term "data unit" throughout this section to denote a block of bits at each layer of the architecture. We used that term as well elsewhere in this book to denote a block of bits. In actual practice, even though a single message may be sent by the application layer, the block of bits carrying that message will change as it moves down through the layers. Messages may be too long to be transmitted as one data unit; they may thus be split into multiple units. We saw examples of this in Chapter 8, in describing message transmission in circuit-switched second-generation cellular systems. Alternately, messages may be too short to be economically transmitted across a network. Messages may then be combined to form larger data units. But more significantly, as the message works its way "down" through the various layers, the protocols at each layer add so-called "headers" (and sometimes "trailers" as well) to the message. These headers carry information specific to a given layer and are used by the corresponding protocol at the receiving end of a given layer, whether across a link or across the entire network, to carry out the function of that layer. The header plus the data unit received at a given layer from the layer above are together referred to as a *protocol data unit* or PDU. At the receiving system, protocol data units move "up" the layers, with headers being stripped off as they do so. Protocol data units thus change as they move "up" or "down" the layers, deleting or adding headers as they do so. For this reason protocol data units at a given layer are given names specific to that layer. We thus have the application protocol data unit (A-PDU), the transport protocol data unit (T-PDU), the network protocol data unit (N-PDU), and the data link protocol data unit (DL-PDU). Each of these PDUs may be visualized as flowing across the logical link connecting peer protocols at a given layer. Actual physical transport occurs at the physical layer only. The PDU actually transported across a physical link then carries the original information message from the application layer plus all the headers added in moving "down" through all the layers.

Figure 10.10 shows these various protocol data units defined above moving "down" through the various layers, with headers being added at each layer. Transmission of the composite DL-PDU carrying all the headers, including the one added at the data link layer, takes place at the physical layer. Conceptually then, the sending host thinks it is transmitting information in the form of an application protocol data unit. The receiving host "sees" only this A-PDU as well. The layers below add various fields to the message, in the form of the headers shown in Fig. 10.10, to ensure the original message is appropriately routed across the network and delivered in the form desired. Specific examples follow shortly as we describe the GPRS layered architectures in more detail.

The discussion thus far has focused on PDUs. Where then does the commonly used term "packet" fit in? Strictly speaking, the word *packet* is synonymous with the network protocol data unit, the N-PDU, the PDU moving, and being routed, across the entire network(s). In the Internet architecture, therefore, it is the IP layer that deals with packets, usually referred to as IP packets, and handles their routing through the network(s). Informal usage, however, often has the data units of other layers being called packets as well. (As an example, Ethernet data units are sometimes referred to as Ethernet packets, even though Ethernet operates at the lower portion of the data link layer, called the medium access or MAC layer, as well as at the physical layer.) Where there is no confusion as to which layer is being discussed, this informal usage poses no problem. Where data units of

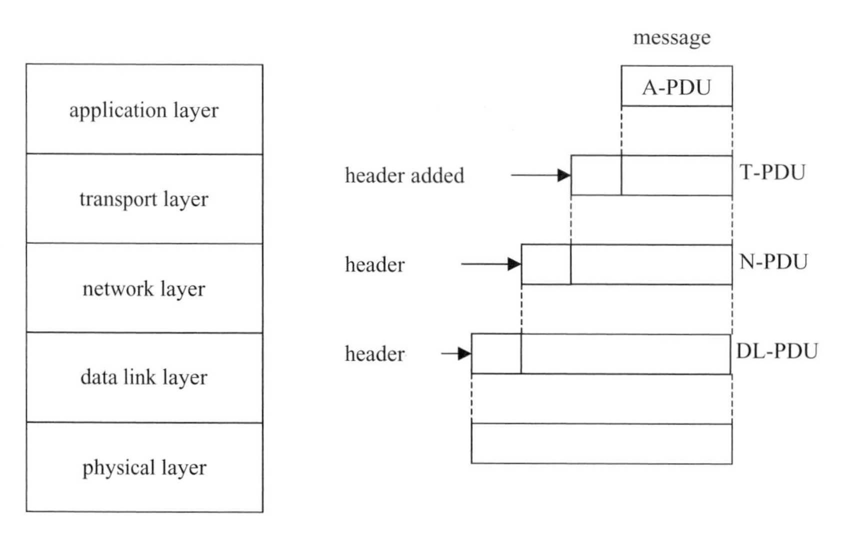

Figure 10.10 Layering and protocol data units

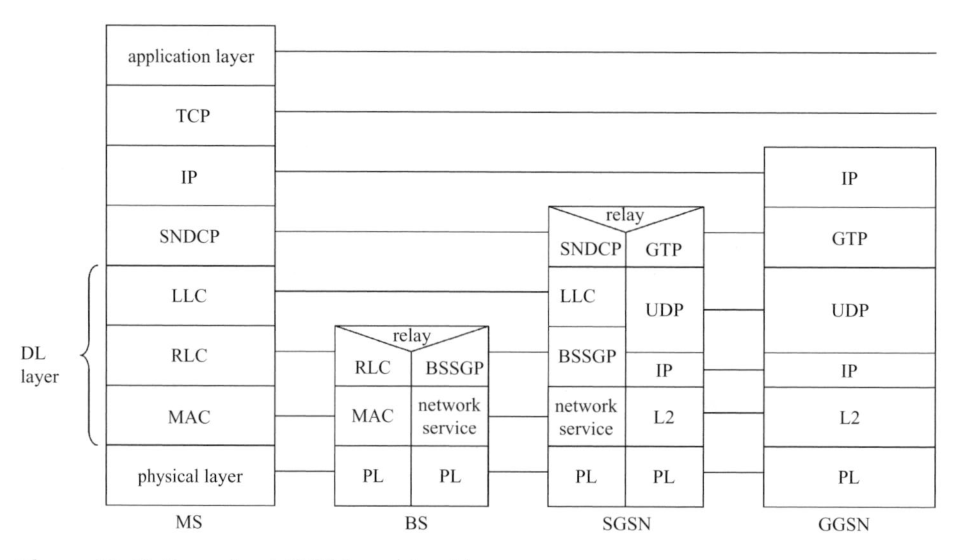

Figure 10.11 Example of GPRS layered architecture

different layers are being discussed at the same time, however, care has to be taken and the appropriate nomenclature used.

GPRS layered architecture

With this review of layered architectures we are in a position to discuss, albeit briefly, the layered architecture adopted for the GPRS core network. We follow this brief introduction to the GPRS architecture with a discussion in more detail of the protocols used to define the interactions across the air interface between mobile and base station.

Figure 10.11 portrays a simplified view of the GPRS layered architecture, focusing on one mobile station and its base station, the SGSN with which the BS is associated, and a single GGSN connecting the SGSN to an external packet data network (Kalden *et al.*, 2000; Cai and Goodman, 1997). We note that connections at each layer are all shown as solid lines to simplify the diagram, although it is understood that the only actual transmission of bits takes place between the various physical layers shown. The other connections shown are all logical ones, just as in Fig. 10.9 discussed above. We focus here on TCP/IP-based communication with the Internet or some other IP-configured packet data network. The GPRS architecture provides for the use as well of other packet-based network services such as X.25 (Kalden *et al.*, 2000; Cai and Goodman, 1997; Bettstetter *et al.*, 1999). (The X.25 protocol is discussed in Schwartz, 1987.) Note first that, because the destination host is not shown in Fig. 10.11, the application and TCP layers are shown as not terminating in this figure. The IP layer supporting these layers terminates at the GGSN. Hence the GPRS network is considered, logically, as one hop in a packet-switched network. IP packets are therefore transmitted logically from the mobile station to the GGSN to which it is connected over the GPRS core network. The subnetwork-dependent convergence protocol layer, SNDCP, shown underlying the IP layer, allows IP packets to be multiplexed on one logical connection, carries out segmentation of these packets into shorter data units when necessary, and provides a compression capability. In the event another network protocol were to be used in place of IP, only SNDCP would have to be changed. The other protocols shown in Fig. 10.11 would not be affected. Note that the SNDCP layer provides a logical connection between the mobile station and the SGSN, bypassing the base station. Although the SNDCP layer terminates at the SGSN, the SGSN serves as a relay station in relaying the IP packets on to the GGSN which connects it to the packet data network(s) outside the GPRS core network.

Underlying and servicing the SNDCP layer is the data link layer. This layer, as shown in Fig. 10.11, is made up of two sublayers, the logical-link control sublayer, LLC, and the combination of the radio link control RLC and the medium access control MAC sublayer. Splitting of the data link layer into LLC and MAC sublayers is common in packet switching networks, with the IEEE 800 series of layer 2, data link control, protocols as the most widely adopted set of examples (Kurose and Ross, 2002; Schwartz, 1987). (Local-area networks or LANs such as the ubiquitous Ethernet all use layer-2 protocols. Ethernet is synonymous with IEEE 802.3. We discuss two wireless LAN standards, 802.11 and 802.15, in Chapter 12.) The combination of RLC and MAC sublayers is unique to GPRS. Note that the LLC sublayer, like the SNDCP layer just above it in the protocol stack, is shown connected logically between the mobile and the SGSN. LLC-PDUs appearing at the base station are relayed directly to the SGSN. The RLC/MAC sublayers are, however, logically connected across the air interface between mobile and base station. The LLC sublayer provides a number of services to the SNDCP layer above. These include, among others, flow and sequence control, as well as error control. The RLC protocol of the RLC/MAC sublayer provides for segmentation of LLC PDUs when necessary, as well as error control of RLC/MAC PDUs. Error control can incorporate error correction using an ARQ technique or error detection only, with upper layers signalled of any errors detected. (These techniques have been mentioned previously in Chapter 7.)

Cai, J. and D. J. Goodman. 1997. "General packet radio service in GSM," *IEEE Communications Mag.*, 35, 10 (October), 122–131.

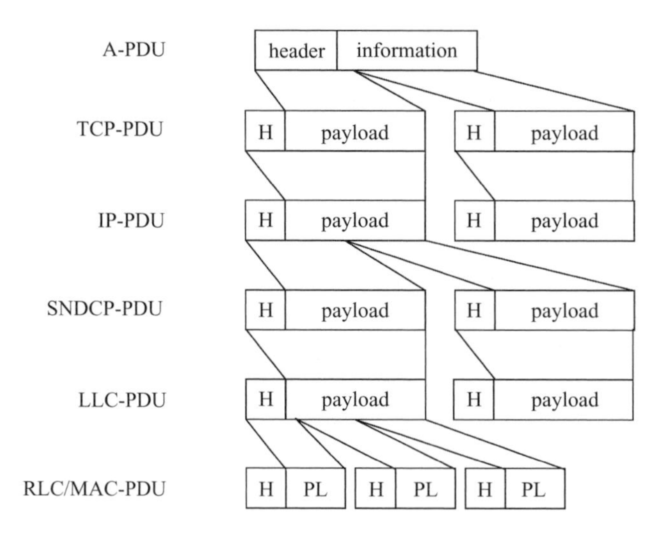

Figure 10.12 Generation of GPRS air interface PDUs (from Kalden *et al.*, 2000: Fig. 3)

The medium access control or MAC protocol is the one that incorporates the access control for multiple mobile stations accessing the base station, using a modified form of the slotted Aloha access technique we have discussed a number of times in this book. It is also responsible for providing multiple time slots (channels) to a mobile when needed and when available. The physical layer at the MS-BS radio interface is actually composed of two sets of sublayers. The upper one of the two, the physical link sublayer, is the one at which forward-error correction is carried out, using convolutional coding. It is also responsible for the monitoring of signal quality across the air interface and carries out power control functions. The lower sublayer, the physical radio frequency layer, is the one that provides the actual frequency allocation and modulation characteristics of the signals transmitted over the air interface. This sublayer is the same as that used in GSM. We return to a more detailed discussion of the lower-layer protocols defined across the air interface after concluding our discussion of the various GPRS layers and protocols appearing in Fig. 10.11.

(As an aside, it is worthwhile noting that much of our earlier discussion of cellular systems in this chapter, as well as in Chapter 8 on second-generation circuit-switched systems, focused on the physical layer across the air interface, with some implicit discussion of the MAC layer as well, in discussing access procedures and logical messages transmitted. We made no comment on the protocol layering used in these systems to avoid confusion at the time. The reader is now encouraged to re-read this earlier material and see where the layering concept is incorporated in the design of these systems.)

Before continuing with the discussion of the other protocol layers shown in Fig. 10.11, it is instructive to see how PDUs are formed at the various layers just discussed which comprise the protocol stack at the mobile station. The PDU at the lowest layer is then the one actually transmitted across the air interface between MS and BS. These PDUs at various layers are displayed in Fig. 10.12 (Kalden, 2000). The segmentation of PDUs indicated at some of the layers has been chosen purely arbitrarily to provide examples of this process. Thus, the application PDU generated at the mobile is shown being split into two, as an arbitrary example, to form the TCP payload. (Application PDUs have no prescribed maximum length. TCP-PDUs are limited to a maximum length of 64,000 octets or bytes.) Each of the two segments so generated then has a standard TCP header indicated by the symbol H appended to form the TCP-PDU. This TCP-PDU now forms the IP payload, with the IP header H added by the IP. The resultant IP-PDU or network packet is then transmitted logically to the GGSN, as shown in Fig. 10.11. The IP packet is, again as an example, shown being split into two segments, with an SNDCP header added to each to form the SNDCP-PDUs. (SNDCP-PDUs are limited to 1560 octets (bytes) in length, while IP packets may be up to 64 koctets in length.) The resultant SNDCP-PDUs are now handed over to the LLC protocol which adds its own header. Finally, the LLC-PDUs are shown being segmented into multiple RLC/MAC-PDUs for final transmission across the air interface, using the services of the physical layer: as we shall see shortly, in the next paragraph discussing the GPRS air interface, RLC/MAC-PDUs are limited to at most 54 octets in length, including the headers added at those layers. It was noted above, in beginning our discussion of Fig. 10.11, that the base station relays LLC-PDUs from a mobile to the SGSN. This means that the RLC/MAC-PDUs into which a given LLC-PDU may have been segmented must be recombined at the base station to form the LLC-PDU, after first stripping off the RLC/MAC headers. The LLC-PDU is then forwarded unchanged to the SGSN. The RLC/MAC header information carried by the RLC/MAC-PDU is used by the base station to carry out the necessary RLC/MAC protocol functions.

Returning now to Fig. 10.11, we describe briefly the other protocol layers shown in that figure. Figure 10.8 showed a very simple example of a GPRS networking configuration. SGSNs, each serving multiple base stations, may be interconnected through public land mobile networks (PLMNs), and, through these networks, to external packet networks by means of GGSNs interfacing the external networks. The GPRS architecture provides that these PLMNs be IP-based. This means that all routing through the PLMNs must be carried out using IP and its associated routing protocols. An information-transmission connection between an SGSN and a GGSN corresponds logically to one hop, however, as shown in Fig. 10.11. This implies that the routing through the PLMN between SGSN and GGSN must be hidden from the information-bearing IP packets being relayed from the mobile station to the GGSN connected to the external packet network. This "hiding" of the underlying routing through the PLMN is referred to as "tunneling": The information-bearing IP packets of Fig. 10.11 are encapsulated at the SGSN within a header carrying the destination GGSN address and are then routed through the PLMN to the GGSN using the IP protocol associated with the PLMN. The encapsulation or tunneling is carried out by the GPRS tunneling protocol or GTP shown in Fig. 10.11. Logically, this protocol sees the SGSN and GGSN directly connected together. Physically, multiple routing hops may be necessary. This direct logical connection is shown being served by the transport-layer protocol UDP, since IP-based packet data networks do not expect reliable network transmission. (The TCP layer shown in Fig. 10.11 protects the mobile-public data network connection. X.25 does expect a reliable data link, and were it to be used to provide the network service between mobile and GGSN in place of IP, TCP would be used to service the GTP layer, and logically connect SGSN with GGSN instead of UDP (Bettstetter *et al.*, 1999).) The PLMN IP layer is then used to route the encapsulated information through the PLMN, as shown in Fig. 10.11. (It is important not to confuse the TCP and IP protocols at the mobile with the internal UDP/IP protocols of the PLMN. We note again that the mobile-generated IP packets destined for an external packet network such as the Internet are tunneled through the PLMN using the internal protocols of the PLMN, which are specified to be UDP and IP.) Finally, we note that the data link and physical layer protocols in a given PLMN are not specified and are therefore shown as L2 (layer 2) and PL, respectively.

The last protocol still to be discussed in Fig. 10.11 is BSSGP. The base station subsystem GPRS protocol BSSGP provides a logical connection between base station and SGSN, and is used to convey necessary routing and QoS information between the two. The connectivity between the two is provided by some (unspecified) network service and physical layer protocol associated with that network service.

The procedures used by a mobile station to connect to the GPRS network and then to set up a session with an external packet data network such as the Internet provide a specific example of how the layered architecture of Fig. 10.11 comes into play. Negotiation of the QoS profile desired for this session takes place during this process as well. To connect to the network, the mobile station must first register with the SGSN. This procedure is called GPRS attach (Bettstetter *et al.*, 1999; Koodli and Punskari, 2001). Once successfully attached, the mobile begins the process of activating a so-called *packet data protocol (PDP) context*. In the IP-based environment we have been assuming throughout this section, as indicated in Fig. 10.11, this PDP context corresponds to an IP-based virtual connection between the mobile and the GGSN which provides the desired entry to the external packet data network (Bettstetter *et al.*, 1999; Koodli and Punskari, 2001). During the GPRS Attach procedure, the mobile user authorization is validated and the user profile information is copied from the HLR to the SGSN. Note from Fig. 10.11 that the messages exchanged between mobile and the SGSN during this procedure involve the lower layers of the architecture only.

The activation of the IP-type PDP context in setting up the desired session with an external packet network involves determination of the mobile IP address if none has previously been assigned to the mobile station and negotiation of the QoS to be supported over the session. (The mobile home network may permanently assign an IP address to the mobile or an address may be assigned dynamically during an IP session setup. In this latter case, the address assignment is made by the GGSN.) The activation process involves the transmission of a series of request-response messages between the mobile station, its SGSN, and the GGSN (Bettstetter *et al.*, 1999). These messages are portrayed in Fig. 10.13 in the case of a successful context-activation process (Bettstetter *et al.*, 1999: Fig. 4). The mobile first sends an Activate PDP context request message to its SGSN, as indicated. This message contains the PDP type, IPv4 for example, the IP (PDP) address if previously assigned, the QoS profile requested, and some other message parameters. (If no IP address has been assigned to the mobile, the address field is left blank.) This message is sent using the facilities of the SNDCP layer of Fig. 10.11 which connects the mobile

Koodli, R. and M. Punskari. 2001. "Supporting packet data QoS in next-generation cellular networks," *IEEE Communications Magazine*, 39, 2 (February), 180–188.

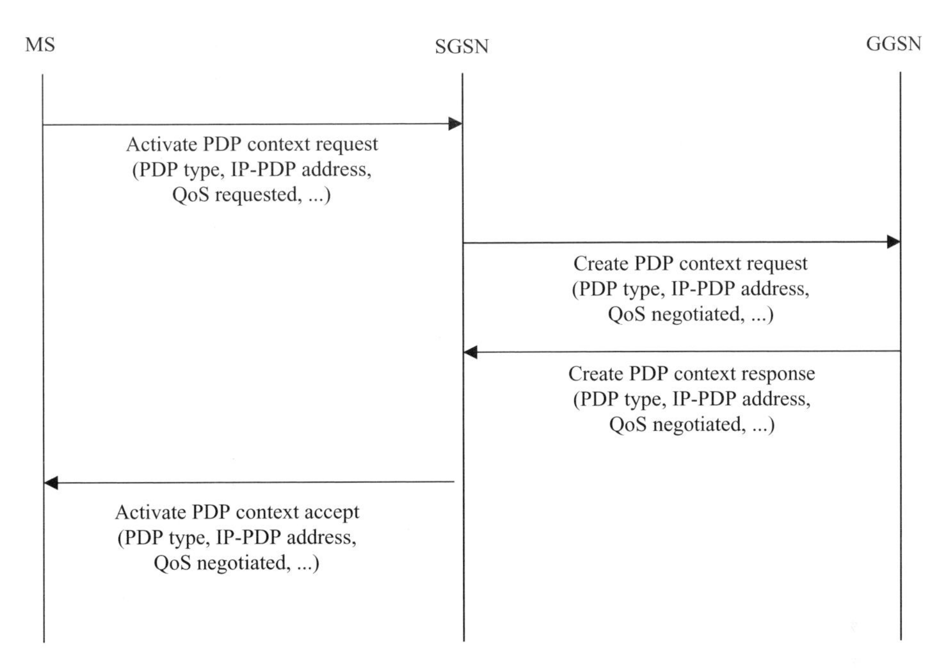

Figure 10.13 PDP context activation (from Bettstetter *et al.*, 1999: Fig. 4)

station and the SGSN logically, as shown in that figure. The *physical* transmission of the message requires relaying it through the base station, creating and attaching the LLC header as shown in Fig. 10.12, and then splitting the resultant LLC-PDU into multiple RLC/MAC-PDUs as required, each with a MAC/RLC header, for transmission across the MS-BS air interface, also as shown in Fig. 10.12. (Details of how these RLC/MAC-PDUs are actually transmitted across the air interface appear in the next section.) User authentication is carried out by the SGSN and, if the user is granted access, a create PDP context request message is sent by the SGSN to the GGSN. This message also carries the mobile IP (PDP) address, if assigned (otherwise the address field is again left blank), a QoS profile field, and other parameters. As shown in Fig. 10.11 this message is tunneled from SGSN to GGSN using GTP and the layers below supporting GTP. Negotiation of the QoS parameters requested may be carried out as part of this context activation procedure. Network resources must be available to provide the QoS attributes finally agreed on. If they are available and the negotiation is successful, a create PDP context response message is returned by the GGSN to the SGSN, as shown in Fig. 10.13. The SGSN, in turn, forwards an activate PDP context accept message to the mobile station. In the case where there is no IP address assigned to the mobile, the GGSN assigns one, so indicating in its create PDP context response to the SGSN. This address then appears as well in the address field of the activate PDP context accept message sent from SGSN to the mobile. The SGSN maps the QoS successfully negotiated into an RLC/MAC radio priority level and indicates in its accept message to the mobile the priority level to be used in its subsequent uplink access request. That request in turn is used by the base station to determine mobile access priority. The QoS profile negotiated is also mapped by both the SGSN and GGSN into IP QoS parameters if the GPRS core network provides that capability (Koodli and Punskari,

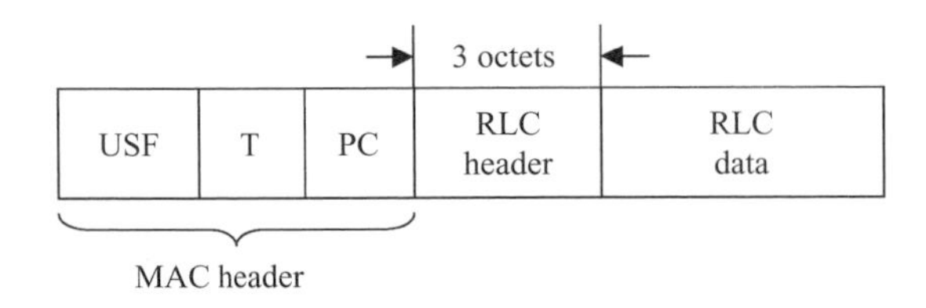

Figure 10.14 RLC/MAC protocol data unit

2001). IP packets bound to, or moving from the mobile in question are then appropriately marked by the GGSN and SGSN, respectively.

GPRS air interface

With the GPRS core network architecture and protocol layers now described, we are in a position to focus in more detail on the GPRS air interface. We begin with the RLC/MAC-PDUs (Fig. 10.12) and show how they are mapped into the GSM time slot-frame structure. We noted at the beginning of this section that GPRS packets share the GSM frame structure with GSM circuit-switched calls. Recall from Chapter 8 that GSM frames are 4.615 msec in length, each frame divided into eight time slots, with 114 data bits being made available per slot. In the GPRS packet structure an RLC/MAC-PDU, after appropriate coding for error detection and correction, is mapped into the data portion of four time slots, one each from four consecutive frames, called a *radio block*. A coded RLC/MAC-PDU will therefore be 456 bits long. This radio block represents the basic transmission unit of a GPRS packet data channel, or PDCH, on to which GPRS logical traffic and control channels, to be described later, are mapped. It is clear from our discussion earlier that PDCHs share time slots, the GSM physical resources, with the GSM circuit-switched traffic channels. Since there are eight time slots per frame, there can be up to eight PDCHs made available to GPRS users, depending on demand and QoS desired. Capacity is thus available "on demand." Unlike GSM traffic channels, which are assigned to a given user for the duration of a (circuit-switched) call, PDCHs are assigned to transmit individual packets only, and are released at the conclusion of transmission. PDCHs are further defined in terms of a multiframe structure consisting of 52 frames, hence 240 msec long. As in GSM, one of every 13 frames is not used for a PDCH. There are thus 12 radio blocks available per 240 msec multiframe. On the average, then, it takes 20 msec to transmit a radio block.

We now return to the transmission of RLC/MAC-PDUs using radio blocks. Figure 10.14 shows the basic structure of an RLC/MAC-PDU, prior to coding, carrying user data. A 3-octet RLC header is added to the RLC data portion, or payload, which, from Fig. 10.12, might consist of a segment of an LLC-PDU. (An RLC/MAC-PDU carrying signaling or control information would carry this information in place of the RLC data and header (Cai and Goodman, 1997).) A MAC header containing a power control field PC, a block-type indicator T, and a 3-bit uplink state flag USF is added as well. The USF field designates the mobile station assigned to a particular radio block (Cai and Goodman, 1997; Kalden *et al.*, 2000). Figure 10.15 shows how this PDU is processed for transmission onto a 456-bit radio block. (Interleaving is also carried out, but not shown

Table 10.6 *Comparison of GPRS coding schemes*

Scheme	PDU size (bits)	BCS (bits)	Add. coded USF bits	Tail bits	Convol. coder (bits) Input	Output	Punctured bits	Eff. rate
CS-1	184	40	0	4	228	456	0	0.5
CS-2	271	16	3	4	294	588	132	0.64
CS-3	315	16	3	4	338	676	220	0.74
CS-4	431	16	9	–	–	456	0	1

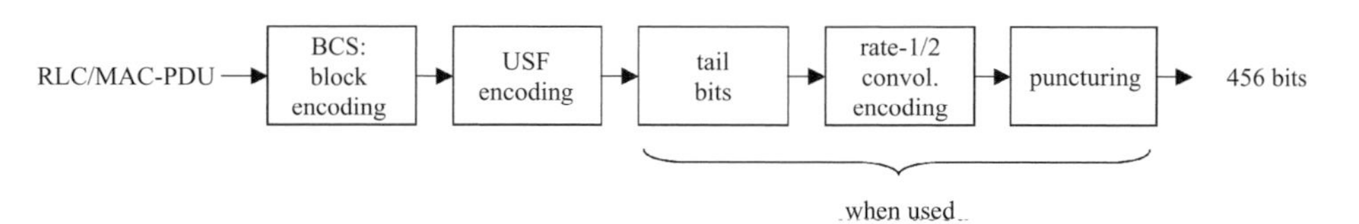

Figure 10.15 Processing of RLC/MAC-PDU

in the figure.) The type of processing carried out is defined by four coding schemes, CS-1 to CS-4, to allow for variations in air interface transmission conditions. CS-1 provides the heaviest coding protection, CS-4 the least. Coding schemes CS-1 to CS-3 all use a constraint length-4, rate-1/2 convolutional encoder for error correction, as shown in Fig. 10.15; CS-4 uses no convolutional encoding. Scheme CS-1 adds 40 bits of block check sequence encoding, BCS, for detecting errors not corrected by convolutional coding; the other schemes use 16 bits for BCS encoding. Schemes CS-2 to CS-4 protect the USF field as well, with USF encoding bits added, as shown in Fig. 10.15. Table 10.6 compares the characteristics of each scheme. Each coding scheme results in the required 456 bits per PDU after processing, as indicated in Fig. 10.15.

The RLC/MAC-PDU size shown in the first column of the table is the resultant size of the PDU allowed before processing, to obtain the necessary 456 bits after processing. This PDU size includes the 3-bit USF, in accordance with Fig. 10.14. Note that the maximum PDU size of 431 bits for CS-4 corresponds to the figure of 54 octets, quoted earlier as the maximum RLC/MAC-PDU size. PDUs thus range in size from 23 to 54 octets, including header. The third column of Table 10.6 indicates the additional bits required to code the USF. (More commonly, figures in the references show the USF encoding box as a "Pre-coding USF" box, with both the USF and its coded bits added at that point (Cai and Goodman, 1997; Bettstetter, 1999). This corresponds to deleting three bits from the first column in Table 10.6 and adding three bits to the USF column. The resultant block of input bits, with the USF bits removed, is referred to as the "payload" of the radio block (Cai and Goodman, 1997).) As noted above, schemes CS-1 to CS-3 all use a rate-1/2 convolutional encoder; CS-4 does not. Four tail bits are added to the encoder, when used, as shown. Schemes CS-2 and CS-3 both produce more than 456 bits at the convolutional encoder output. These extra bits are then "punctured," or deleted, as indicated. The ratio of the convolutional encoder input to the 456 bits out is the effective encoding rate shown as the last column in the table.

Table 10.7 *GPRS throughput rates*
(single time slot per frame; GSM modulation w/o EDGE)

Scheme	Radio block data rate (kbps)	LLC rate (kbps)
CS-1	9.05	8
CS-2	13.4	12
CS-3	15.6	14.4
CS-4	21.4	20

Since each radio block of 456 bits takes an average of 20 msec to be transmitted, scheme CS-4, with a larger PDU size transmitted, provides a substantially higher throughput rate than does CS-1. It is clear from Table 10.6 that the throughput increases with scheme number. Table 10.7 indicates the effective throughput rates for the four schemes both at the RLC/MAC layer (the radio block data rate), and at the LLC layer. The rate shown for the radio block throughput is based on the radio block payload mentioned above column 1 of Table 10.6 less the 3-bit USF (Cai and Goodman, 1997); the LLC throughput is the rounded-off data rate after subtracting the MAC header and 24-bit RLC header from the RLC/MAC-PDU (Bettstetter *et al.*, 1999). Note, however, that the assumption here is that one time slot only per frame is used as the unit on which transmission is based. PDUs could be transmitted using up to as many as eight time slots per frame, increasing a given user throughput by up to that factor. Introducing the EDGE technique for increasing the GSM transmission rate would increase these figures as well, as will be seen in the next paragraph, devoted to a discussion of the EDGE technique.

As noted earlier, GPRS logical channels carrying user data traffic as well as control and signaling information are mapped on to the packet data channels, PDCH. These logical channels are similar to those used in second-generation systems and described earlier in Chapter 8, as well as those used in the other third-generation systems discussed earlier in this chapter. Before discussing the GPRS logical channels and describing their use in GPRS systems, it is useful to indicate briefly how they relate to the discussion just concluded of PDCHs and the radio blocks they utilize. We recall first that user data traffic originates in the higher layers and is carried by LLC-PDUs, as segmented into RLC/MAC-PDUs, as shown in Figs. 10.11 and 10.12. These PDUs serve to carry the user traffic logical channel called the packet data traffic channel, PDTCH. These logical channels are defined for both the uplink and downlink directions.

As seen from Fig. 10.11, the control of the air interface is carried out at the RLC and MAC layers. Signaling and control information for the air interface control functions is thus carried by control RLC/MAC-PDUs. We have seen earlier in this chapter as well as in Chapter 8 that included among these signaling and control functions are such items as mobile registration; downlink broadcast information including base station identification, frequencies of operation, and synchronization signaling; access requests and access grants; paging notification; and others described previously. A number of logical channels defined for GPRS are used to carry messages relating to these various signaling/control functions at the RLC and MAC layers. It is these messages that are, in turn, converted to RLC/MAC-

Table 10.8 *GPRS logical channels*

Group	Channel name	Function	Direction
Packet data traffic channels	PDTCH	Data traffic	Bi-directional
Packet dedicated control channels	PACCH	Associated control	Bi-directional
	PTCCH	Timing advance control	Bi-directional
Packet broadcast control channels	PBCCH	Broadcast control	Downlink
Packet common control channels	PRACH	Random access	Uplink
	PAGCH	Access grant	Downlink
	PPCH	Paging request	Downlink
	PNCH	Multicast notification	Downlink

control PDUs. The GPRS logical channels for both data traffic and control are listed in Table 10.8, grouped by category and function (Cai and Goodman, 1997; Bettstetter *et al.*, 1999). Note how similar they are to logical channels discussed for other cellular systems discussed earlier. In particular, note the similarity with the logical channels described for GSM in Chapter 8.

Consider the logical channels in Table 10.8 one at a time. The packet broadcast control channel PBCCH is used by a base station to broadcast system information to all GPRS mobile terminals in its cell. This function is obviously similar to that carried out by the broadcast control channel BCCH in GSM. Should the PBCCH not be available in a cell, a GPRS terminal would listen to the BCCH to obtain the necessary system information. Coordination between circuit-switched and packet-switched logical channels is thus important (Bettstetter *et al.*, 1999). Similar considerations hold with respect to the packet common control channels and the GSM control channels. The random access channel, PRACH, is used by a mobile, just as in GSM and the other cellular systems already discussed, to indicate to the base station its desire to transmit packet data, hence to request one or more packet data channels, PDTCH. The base station replies using the access grant channel PAGCH to allocate the necessary resources to the requesting mobile. The regular GSM random access channel RACH and access grant channel AGCH could be used as well. A slotted Aloha-type random access technique is again used for requesting access. The associated control channel PACCH is dedicated to a given mobile, being associated with PDTCHs established in either direction for that mobile. It is used to carry packet data acknowledgements, power control information, and resource assignment messages. The paging channel PPCH is used by the base station, as in GSM, to page mobile terminals prior to initiating transfer of waiting downlink packets. GPRS supports multicast transmission of packets to a group of mobiles. The multicast notification channel PNCH is used to inform such a group prior to transferring a group-directed message. Figure 10.16 portrays a possible scenario involving paging of a mobile (Bettstetter *et al.*, 1999). The base station sends a paging request message using the PPCH (or PCH); the mobile responds with a channel access request message using the PRACH (or RACH); a resource assignment

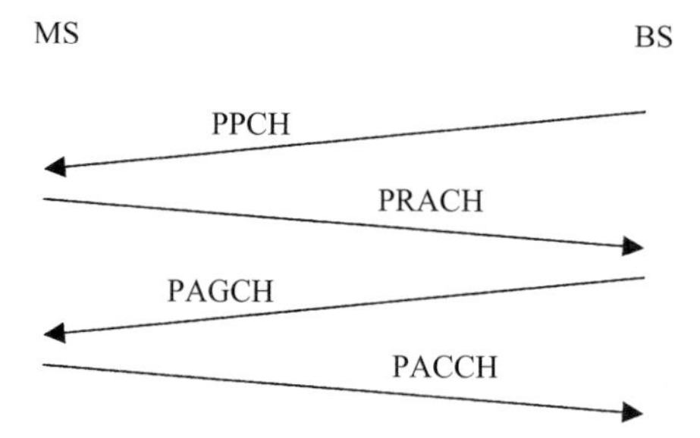

Figure 10.16 Logical channels used in paging

is then sent by the base station using the PAGCH (or AGCH); the mobile acknowledges the page using the PACCH.

EDGE/EGPRS

We now complete our brief discussion of GPRS by describing modifications made to the GSM and GPRS air interface protocols to greatly increase the rates of transmission across the air interface. Two basic changes have been introduced: the use of 8-PSK modulation in place of the GSM use of GMSK modulation and modifications in the radio link control to improve link transmission quality. (See Chapter 5 for a brief discussion of modulation techniques.) The enhanced air interface resulting is referred to as EDGE, for enhanced data rates for global evolution. As noted earlier, EDGE is to be applied to improve the performance of IS-136 as well, with a hope that the various third-generation systems will eventually converge to a smaller number, if not to one universal system. Various papers on the implementation of EDGE in IS-136 systems, as well as to GSM, appear in (IEEE, 1999), already cited. We focus in this discussion on the implementation to GSM/GPRS only. (As noted at the beginning of this section, the G in EDGE has also been interpreted as standing for GSM and GPRS. In keeping with implementations proposed for other 3-G systems, we prefer to use the word global.) EDGE is defined for both packet-switched transmission and circuit-switched transmission. The packet-switched version to be discussed here is referred to as EGPRS; the circuit-switched version is labeled ECSD (Furuskar *et al.*, 1999a).

Moving now to the details of EDGE, consider first the new modulation technique, linear 8-PSK, introduced. Linear 8-PSK extends Quadrature-PSK or QPSK, discussed in Chapter 5, to include eight equally spaced phase positions. It is thus a special case of QAM, also discussed in Chapter 5. (Recall that QPSK, 8-PSK, and QAM are used in cdma2000 as well.) The eight equally spaced phase positions are portrayed in the phase-plane picture of Fig. 10.17. As described in Chapter 5, the introduction of multiple points in the phase plane increases the effective bit rate to be transmitted over a given tranmission bandwidth. Each transmission thus corresponds to a *symbol*, a symbol corresponding to multiple bits of information. In the case of 8-PSK, specifically, three successive bits at the transmitter would be stored and one of the eight possible phase carriers corresponding to a specific set

---

Furuskar, A. *et al.* 1999a. "EDGE: enhanced data rates for GSM and TDMA/136 evolution," *IEEE Personal Communications*, 6, 3 (June), 56–66.

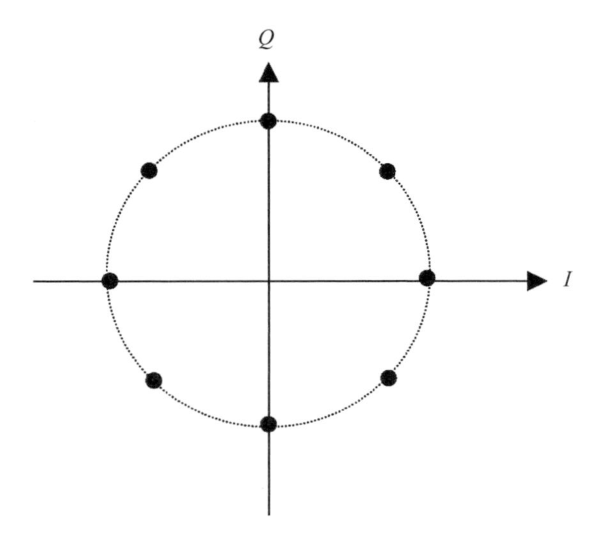

Figure 10.17 8-PSK

of bits would be transmitted every three-bit interval, corresponding to a symbol interval. The bit rate can thus be increased three-fold, as was already noted, in passing, in our discussion of cdma2000 earlier. The downside of such a procedure, of course, is that, as the number of points in the phase plane increases (here to 8 points), the probability of mistaking one point for another (i.e., one phase for another) increases as well. (Recall from Chapter 5 that, with fixed transmitter power, the maximum distance of the points from the origin in the case of QAM cannot increase. Hence, as the number of points increases, they move closer together.) The presence of fading channels and interference from other mobiles and/or base stations means that moving to a modulation scheme such as 8-PSK may increase the probability of packet errors. Hence enhanced error correction procedures and adaptive rates of transmission may have to be used. As we shall see below, EDGE uses both methods.

Now consider the GPRS data rates made possible using EDGE technology (Furuskar *et al.*, 1999a). We note again that an immediate increase by a factor of three is made possible through the use of 8-PSK. The spectral shaping adopted is linearized GMSK (see Chapter 5) to keep the EDGE radio interface parameters as close to those of GSM as possible. To further keep the system parameters as close to those of GSM as possible, the basic time slot or burst format adopted is essentially the same as that of GSM. There is one slight difference: Recall from Fig. 8.2 of Chapter 8 that the time slot in GSM, lasting 0.577 msec, carried 114 data bits. Recall also that a user was assigned one slot in a frame of eight slots, and that 12 frames in every 13 carried a given user's data. This gave rise to a basic data rate per user of $(12/13) \cdot 1/8 \cdot 114/0.577 = 22.8$ kbps. The slot format for EDGE appears in Fig. 10.18. Note first that the word "symbols" appears rather than "bits" as defining each field in the slot. This corresponds to the fact stressed above that one of eight 8-PSK signals is being transmitted every symbol interval, each symbol carrying three bits. Secondly, there are 58 symbols appearing in each of the two data fields for a total of 116 data symbols per slot, rather than the 114 bits per slot in the GSM case.

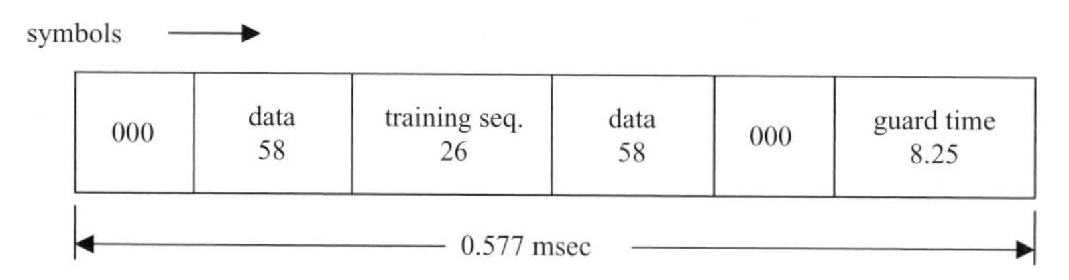

Figure 10.18 Time-slot format, EDGE

Comparing Fig. 10.18 with the GSM case of Fig. 8.2, it is to be noted that the two F bits used in GSM have now been "stolen" to be added to the data field. (The F bits in the GSM case are used to indicate whether data or control traffic was present; no such distinction is made in EDGE. Recall that the RLC/MAC-PDU in GPRS carries its own designation of control or data PDU.) An EDGE time slot thus carries packet data transmission at the rate of $(12/13) \cdot 1/8 \cdot 116/0.577 = 23.2$ ksymbols/sec or 69.6 kbps. The maximum rate of transmission using EDGE is thus eight times this value, or 557 kbps, if a packet data user is assigned all eight time slots for transmission. (Note that the actual rate of transmission per carrier over a given 200 kHz band is 271 ksymbols/sec, no change from that of GSM, except that, with the use of 8-PSK, symbols are transmitted rather than bits.) This maximum rate may not always be possible, however, because of fading, propagation anomalies, and other adverse conditions experienced by cellular systems. We earlier described the QoS attributes defined for GPRS. Included were maximum and average throughput rates. These attributes are carried over to EDGE. In particular, the maximum bit rate established for an EDGE/EGPRS user moving at a speed of 100 km/hr is 384 kbps, despite the theoretical capability of transmitting data at 557 kbps. This is the same maximum value established for W-CDMA, as noted earlier. For mobiles moving at speeds up to 250 km/hr, this maximum allowable rate drops further to 144 kbps (Furuskar *et al.*, 1999a).

We noted, in introducing EDGE and EGPRS, that changes in radio link control have been made in addition to incorporating 8-PSK modulation. Just as in the case of W-CDMA and cdma2000, adaptivity of bit rates has to be built in to accommodate expected changes in link quality. Recall from Table 10.6 that the GPRS, GMSK-based, radio interface provides four coding schemes CS-1–CS-4, used to vary the rate of transmission to adjust for changes in link quality. EDGE/EGPRS follows up on this by adding six additional coding schemes PCS-1–PCS-6 (Furuskar *et al.*, 1999a). The main parameters of each appear in Table 10.9, which lists all ten coding schemes, in order of radio interface bit rate attainable per slot. Rate-1/3 convolutional coding is used for the new coding schemes, all of which use 8-PSK, as shown. Note that these bit rates across the radio interface are higher than the throughput rates shown earlier in Table 10.7, which represented average radio block and LLC rates. The bit rates of Table 10.9 represent the effective "raw" bit rates transmitted across the air interface. These rates could be increased by as much as a factor of eight, as already noted, by assigning additional slots per frame to a given packet data user. The different rates shown in Table 10.9 are obtained by starting with the highest bit rate coding scheme PCS-6, and then successively puncturing more bits per block transmitted, just as is done in moving down from CS-3 to CS-1, described earlier. Recall from Chapter 7 that a rate-1/3 convolutional coder, as used for coding schemes

Table 10.9 *Coding schemes, EGPRS*
(based on Furuskar *et al.*, 1999a, Table 1)

Scheme	Relative code rate	Modulation	Radio interface rate/ time slot (kbps)
CS-1	0.5	GMSK	11.4
CS-2	0.64	GMSK	14.5
CS-3	0.74	GMSK	16.9
CS-4	1	GMSK	22.8
PCS-1	0.33	8-PSK	22.8
PCS-2	0.49	8-PSK	34.3
PCS-3	0.59	8-PSK	41.3
PCS-4	0.74	8-PSK	51.6
PCS-5	0.82	8-PSK	57.4
PCS-6	1	8-PSK	69.6

PCS-1–PCS-6, provides improved performance over a rate-1/2 coder, as used for the GMSK coding schemes CS-1–CS-3.

The particular coding scheme to be used of the ten shown in Table 10.9 depends on radio link quality conditions. EGPRS uses link quality control to adapt to channel conditions. Two examples of link control employed are link adaptation and incremental redundancy (Furuskar *et al.*, 1999a). In the case of link adaptation, the link quality is estimated and the appropriate coding scheme selected to maximize the data transmission for the quality figure found. In the case of incremental redundancy, the highest possible coding is first chosen to transmit a radio block. If the block transmission is unsuccessful, the block is retransmitted using a lower bit rate, obtained, as noted, by puncturing more bits in the retransmission. (ARQ retransmission, described in Chapter 7, is used.) The transmission rate thus adapts automatically and gradually, as required, to lower-numbered coding schemes.

How effective is EDGE/EGPRS in improving QoS parameters such as throughput and packet delay? Results of simulation studies carried out using a regular, large-sized cellular layout, are reported in Furuskar *et al.* (1999a). These simulations focused on downlink Web-type traffic models and incorporated both propagation distance attenuation and log-normal fading. A related simulation study reported in Furuskar *et al.* (1999b), which also incorporated Rayleigh-distributed multipath fading, came up with similar results, albeit with lower packet bit rates found possible, presumably because of the multipath fading model included. We summarize here the results reported in Furuskar *et al.* (1999a) only, for simplicity. Although both papers include simulation results for EDGE applied to IS-136 (TDMA/136) as well, we summarize EGPRS results only, in keeping with the focus of this section.

Furuskar, A. *et al.* 1999b. "Capacity evaluation of the EDGE concept for enhanced data rates in GSM and TDMA/136," *Proc. IEEE Vehicular Technology Conference*, VTC99, 1648–1652.

Table 10.10 *Simulation comparisons, EDGE/EGPRS vs GSM (90 percentile delay results)*

(based on Furuskar *et al.*, 1999a)

	EDGE/EGPRS		GSM		EDGE/GSM (Ratios)	
Load (users/sector)	Through. (kbps)	Spectral effic. (bps/Hz/site)	Through. (kbps)	Spectral effic. (bps/Hz/site)	Through.	Spectr. eff.
30	25	0.14	14.3	0.03	1.7	4.7
40	21.2	0.19	13.3	0.07	1.7	2.7
50	20.2	0.25	12.5	0.11	1.8	2.3
60	16.7	0.28	8.3	0.14	2	2
70	5.9	0.33	3.7	0.17	1.6	1.9

Three-sector macrocells were chosen for the simulation. Load was varied by increasing the number of users per sector. Two basic measures of performance were determined as a function of load, a normalized delay parameter and the spectral efficiency. The latter is defined as the bit rate possible per site for the bandwidth available, in bps/Hz per site, corresponding to a given normalized delay. This parameter thus measures the efficiency of use of the bandwidth available, as is apparent from its designation. Normalized delay is defined as the absolute delay, queuing delay plus transmission time in seconds, normalized to (divided by) packet length in kbits. (Longer packets will clearly experience longer delays. Hence the normalization attempts to reduce the performance dependence on packet length.) The reciprocal of normalized delay is the throughput in kbps. It represents the actual system throughput, as contrasted with the throughput measures noted earlier, such as the radio interface bit rate (Table 10.9), the radio block data rate, and the LLC data rate (Table 10.7). We encountered such a distinction between actual, or effective, throughput rate and data rate earlier in this chapter, in our discussion of IS-95. A summary of some of the throughput and spectral efficiency simulation results taken from Furuskar *et al.* (1999a) is presented in Table 10.10. The results shown are for the 90-percentile of normalized delay. This means that 10% of the users have throughputs less than those shown.

Note that the throughput drops as the load increases, as might be expected. (The equivalent normalized delay increases.) This decrease is relatively gradual until the load of 70 users per sector is reached, at which point congestion clearly begins to set in, and the throughput drops precipitously (the delay increases rapidly). The spectral efficiency continues to increase, however. (Obviously, as the "pipe" begins to fill up, and gaps between transmitted packets begin to disappear, the bandwidth is used more efficiently. This thus provides the typical tradeoff between the performance offered to the users, in delay or equivalent throughput, and the performance desired by the network operator, in spectral efficiency.) The improvement due to the introduction of EDGE is clearly seen from the last two columns, which show the relative performance of EDGE to GSM, as measured by the ratios of both in throughput and spectral efficiency. The throughput ratio, in particular, remains just below a factor of 2:1 over most of the range of loads.

The simulations were also designed to determine the distribution of average bit rate per user provided by EDGE, a measure of system fairness to individual users. Details appear in Furuskar *et al.* (1999a). In particular, just below the onset of congestion indicated in Table 10.10, at the load of 60 users per sector, it was found that 97% of the users were able to attain an average bit rate exceeding 18 kbps using one slot per frame or eight times that, 144 kbps, if all eight slots were used; 30% of the users were able to exceed an average rate of 48 kbps using one slot, 384 kbps using all 8 slots.

This discussion of the effectiveness of EDGE in improving the performance of GSM/GPRS systems concludes our discussion of 2.5G/3G cellular systems. In Chapter 12 we discuss the much higher bit rate family of 802.11 wireless LANs, as well as the small-area high bit rate scheme Bluetooth. These systems, although designed to cover smaller areas than the cellular systems described in this chapter, can be interconnected to each other or to the cellular systems, to provide the extensive coverage needed to handle the bit-rate-intensive multimedia applications we stressed in the introduction to this chapter. In the next chapter we discuss the multi-access issues made necessary by the introduction of all of these newer wireless systems: how are the bandwidth resources to be allocated in effectively transmitting both circuit- and packet-switched information? What are "good" scheduling techniques for handling the combination of voice, video, and data?

Problems

10.1 Explain the statements made in the text that, at a CDMA chip rate of 3.84 Mcps, path differences of 78 meters or 240 ft can be resolved, while, for a chip rate of 1.2288 Mcps, path differences of 240 meters can be resolved.

10.2 **(a)** Explain how multi-code CDMA allows transmission at higher bit rates without reducing the spreading gain. What is meant by the statement that the "codes used in parallel must be orthogonal to one another"? What is meant by the phrase "self-interference at the receiver"? How does othogonality prevent self-interference?

(b) Draw a diagram of multi-code CDMA for the case of tripling the basic bit rate, showing how serial-to-parallel conversion is used to obtain the rate tripling.

(c) Explain why multi-code CDMA results in wider envelope variations. Why would multiple RAKE receivers be required, one for each code?

10.3 Explain the statement that, for a RAKE receiver to be effective, the spread signal bandwidth must be greater than the coherence bandwidth. Why does this result in frequency-selective fading, and why is frequency-selective fading necessary for the resolution of multipath echoes?

10.4 Consider the W-CDMA system.

(a) Explain why the rates of transmission of user data are less than half of the rate of transmission over the air interface. Why would a local-area environment allow a much higher transmission rate than a wide-area environment?

(b) Explain how a bit rate of 5.76 Mps can be attained on the uplink radio channel. Show that the spreading gain is four in this case.

(c) Provide a simple explanation why QPSK, providing twice the bit rate of PSK, is used for downlink transmission in the W-CDMA system while PSK is used in the uplink direction.

10.5 Four QoS traffic types are described in the text. Explain why the real-time conversational class requires low end-to-end delay and limited time variation between successive information entities while the real-time streaming class requires limited time variation between successive information entities only. Why is e-mail traffic considered relatively insensitive to delivery time delays? Provide examples from your own experience of non-real-time interactive traffic, documenting their bursty nature.

10.6 Figure 10.1 shows two W-CDMA physical channels transmitted in parallel. Explain how the two channels can be transmitted simultaneously and separately detected at the receiver.

10.7 Consider the W-CDMA RACH channel.

(a) Explain why transmission over this channel at a 120 kbps rate corresponds to a spreading factor of 32. Explain the statement in the text that power considerations might require the use of a spreading factor of 128. In that case why would the amount of packet data transmitted over the RACH channel be less than 150 bits?

(b) Diagram the access procedure, indicating, in time, the various messages transmitted between mobile and base station. Include the case requiring the retransmission of the RACH preamble a number of times.

10.8 (a) Compare the IS-95B channels with those of IS-95 described in Chapter 8.

(b) Explain how multi-code operation using Walsh codes can be used to increase bit rate transmission.

(c) Consider the two examples of packet data transmission over IS-95B provided in the text. In both cases a 115.2 kbps transmission rate is used. Show why, in the first case of a 1200-byte packet, the effective transmission rate turns out to be 96 kbps, while, in the second case of an 800-byte packet, the effective transmission rate is 80 kbps. Explain the statement that, if the reverse direction access procedure is taken into account, the effective rates of transmission are even less than these calculations indicate.

10.9 Refer to the discussion of cdma2000 in the text.

(a) Table 10.2 provides information on various reverse channel radio configurations. Explain physically why the spreading gains of configuration 3 decrease inversely as the data bit rate. In what kind of environments would you expect the higher bit rate systems to be used as compared with the lower bit rate ones?

(b) Indicate how the data rate of 307.2 kbps in Fig. 10.3 is obtained. What is the actual bit rate as seen at the input to the R-SCH structure shown in that figure?

10.10 Compare the various cdma2000 channels in both directions of transmission with those of IS-95, as described in Chapter 8. Which are the same or similar, which different? A diagram of the various channels defined in each direction would be helpful in making the comparison. Group similar channels together in drawing the diagram. What does this diagram indicate about backward compatibility with IS-95?

10.11 Compare the logical channels, both uplink and downlink, of W-CDMA and cdma2000.

10.12 The text provides a description, in words, of the cdma2000 mobile-powering-up and registration procedure. Provide a diagram of this procedure, indicating, in time, the various signals and messages transmitted in both reverse and forward directions, until the mobile goes into its idle state.

10.13 Table 10.4 shows the various data rates available for transmitting packets in the forward direction in the cdma2000 $1 \times$EV-DV system.

(a) Explain in your own words how the various data rates shown are obtained, expanding on the discussion in the text. Find, in each case, the symbol rate appearing at the outputs of the modulators of Fig. 10.7 (points A and B). What is the purpose of the channel interleaving function shown in Fig. 10.7?

(b) Explain the statement in the text that a much smaller SIR is required at a receiving mobile for 38.4 kbps forward transmission than for the higher bit rate transmissions using 8-PSK and 16-QAM.

10.14 The GPRS layered architecture of Fig. 10.11 is to be applied in two application scenarios. In the first one, a mobile station is connected via GPRS to the Internet, and then, over the Internet, to a web site it is accessing. For this purpose assume the mobile is M1 of Fig. 10.8, with the Internet being the PDN shown. Say the web site, located within the Internet, requires two routing hops over the Internet, similar to those shown in Fig. 10.9, to get to it from the GGSN. Combine the appropriate portions of Figs. 10.8, 10.9, and 10.11 to show an application data unit, an A-PDU, starting at M1 finally arriving for processing at the web site. In the second application scenario, mobile M1 is transmitting packet data to mobile M3. The PDN in this case is again represented by two routers but is only used to transport data units from one GGSN to the other GGSN connected to the PDN. Again combine the appropriate portions of the various figures to show how an application packet starting at M1 finally arrives at M3. In both scenarios indicate which protocol layers at each system along the path taken in each case are invoked in getting the application data unit from one end of the path to the other. Show how and where appropriate headers are added and stripped off, as the case may be. For simplicity, say the A-PDU generated at M1 is short enough not to require any of the segmentation shown in Fig. 10.12.

10.15 **(a)** Compare the GPRS logical channels shown in Table 10.8 to the logical channels of GSM discussed in Chapter 8.

(b) Compare the GPRS paging procedure shown in Fig. 10.16 with that of GSM.

10.16 **(a)** Show that the effective throughput rate of GPRS coding scheme CS-4 is 22.8 kbps if one time slot per frame is used. What is the maximum effective throughput rate? (These are also called radio interface rates. See Table 10.9.)

(b) Show the effective throughput rates of CS-1 to CS-3 are those given in Table 10.9. Compare with the relative effective rates shown as the last column of Table 10.6.

(c) Validate the various GPRS throughput rates shown in Table 10.7.

10.17 Explain, in your own words, how the various bit rates shown in Table 10.9 are obtained.

10.18 Compare the throughput results of Table 10.10 with those of Table 10.9. Why are the Table 10.10 entries so much lower than the maximum value of 69.6 kbps shown in Table 10.9? Why do the Table 10.10 throughput values decrease with load?

Chapter 11

Access and scheduling techniques in cellular systems

We have noted throughout this book, in discussing cellular systems, that the radio channel used for communication between mobile stations and a base station limits the communication possible. In particular, whether using TDMA or CDMA as the underlying multiple access technique, only a limited number of users may be accommodated in a given frequency band. Users must thus be given permission to transmit, using a given frequency and a specified number of time slots or codes, as the case may be. To obtain permission to transmit, a user must first access the base station indicating a desire to use the system. The availability, in turn, of only a limited number of access channels, whether a specified set of frequencies and time slots, or frequencies and codes, may result in interfering access attempts or "collisions" by users attempting access at the same time. Most commonly, as seen in our discussion of second- and third-generation cellular systems in Chapters 8 and 10, respectively, random access strategies based primarily on the slotted-Aloha technique are used to provide the necessary access, including mechanisms for resolving "collisions." In the case of a circuit-switched voice call, once the desired access is made, the user receiving a dedicated time slot or code at a given frequency, as the case may be, this system resource is normally held for the length of the call. The case of packet-switched data in the third-generation systems described in Chapter 10 presents more complex issues.

We noted in that chapter that there are a number of different packet-based data applications with different traffic characteristics that are expected to be carried by these systems. We indicated there that these applications may be usefully categorized as four different types: conversational and streaming real-time traffic; interactive and bursty non-real-time traffic. For the purpose of discussing the access problem at this point, however, it is useful to categorize these various packet-switched data traffic types somewhat differently, in two basic categories. This simplifies the discussion. The two basic categories may be defined as continuous-time traffic and bursty traffic. Examples of continuous-time traffic include voice and video. Handling of this type of traffic, once access is established to a mobile network, will involve the transmission of packets continuously, as the name indicates, at a regular rate, for some time, requiring the dedication of radio channel and network resources, just as in the case of circuit switching. Hence one access attempt, if successful, will, in that case, provide access to the bit rate capacity needed for some interval of time. Bursty data, with each burst consisting of one or a series of consecutive packets may, however, require multiple access attempts. Depending on how the system is designed, access may be granted for one packet only at a time, for packets in a portion of the burst, or for all packets in the burst. Access is, in turn, separately required for each burst. In addition, for each access grant interval, the radio channel capacity or transmission rate actually allocated to a given user, in either direction, uplink or downlink, may vary from frame to frame, as we saw in Chapter 10. The particular allocation of resources during this interval may depend on the characteristics of the data to be transmitted, radio channel conditions, channel resource availability, and quality-of-service (QoS) characteristics specified for that user and type of traffic. Some sort of packet-scheduling mechanism must then be used at the base station to establish the frame-to-frame resource allocations in both directions to ensure equitable treatment of all data users, according to individual traffic type/QoS specifications and channel transmission conditions.

We thus see that packet transmission requires both appropriate access control and the scheduling of packet transmission, once access is granted. Both of these issues are discussed in this chapter. It is of interest to note that both access control and scheduling are properly defined as being operative at the MAC or medium-access sublayer of the data link layer described in our discussion of layering in Chapter 10. We shall see, in the next chapter on wireless LANs and personal area networks, that the systems discussed there are also all operative at the MAC sublayer and the physical layer supporting it below. Specifically then, the basic question to be asked is, what *are* appropriate access strategies? How do different access strategies compare? This is a problem occurring for both circuit-switched and packet-switched systems. The scheduling of data packet transmission, once access has been granted, should be carried out, as noted above, in accordance with the type of traffic and QoS requirements, channel conditions permitting. What are then some appropriate scheduling mechanisms? Both of these design issues are discussed in the sections following.

In Section 11.1, we begin the discussion of access control by focusing on the commonly adopted slotted-Aloha access procedure. We provide a simple analysis of its performance, first in a non-fading environment, and then in the presence of (fast) Rayleigh fading. Some proposals to improve the (random) access performance are then presented and their performance compared to that of slotted-Aloha. One example already noted in Chapter 10, in discussing the W-CDMA access procedure, is CPCH. In Section 11.2 we extend this work by presenting some strategies proposed for appropriately combining two types of packet traffic, with different characteristics, in a TDMA-based frame structure. A common example would be voice and bursty data. These strategies involve aspects of both access control and scheduling. We begin with an early proposal, called PRMA for *packet-radio multiple access*, and continue with improved strategies such as PRMA++ and the related access strategy DQRMA. In Section 11.3 we conclude this chapter by discussing a number of proposed scheduling algorithms for both TDMA and CDMA systems. Some of these scheduling techniques have been proposed for future, all-packet, fourth-generation cellular systems. A number of these have had their genesis, as we shall see, in scheduling algorithms designed for wired packet networks.

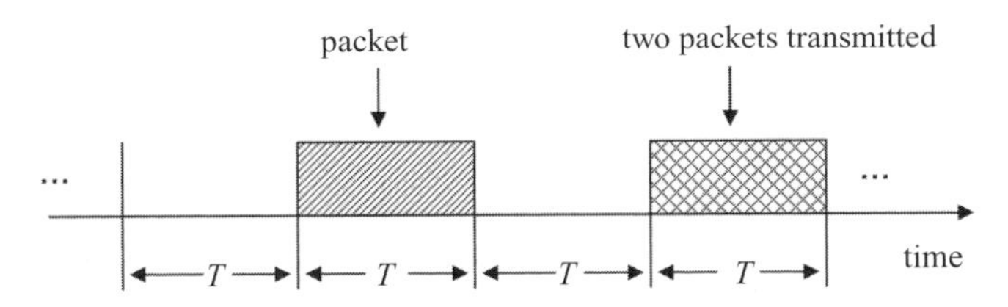

Figure 11.1 Slotted time

11.1 Slotted-Aloha access

As noted above, the slotted-Aloha access technique and variations thereof have been widely adopted in both second- and third-generation cellular systems. We gave examples of the use of the slotted-Aloha technique in both Chapters 8 and 10, in discussing second- and third-generation systems, respectively. We now provide the details of the generic technique, a simple performance evaluation of the technique, the impact of fading channels on its performance, and some variations of the technique designed to improve its performance.

The slotted-Aloha access procedure is one of a class of Aloha-type access methods first proposed and developed by Norman Abramson and his associates at the University of Hawaii in the early 1970s (Abramson, 1973). The basic concept of an Aloha access scheme is a very simple one: transmit at will. If a "collision" is detected, with two or more transmitters attempting to transmit at overlapping times, repeat the transmission attempt, but at a random time later. (The random retry interval is necessary to reduce the probability of a collision at the retransmission attempt.) This simplest form of a *random access* procedure is often called pure Aloha. It turns out that, despite the clear possibility of collisions, this technique provides useful throughput if user transmission attempts are bursty and packets are short. It is thus a viable technique, particularly because of its simplicity. Ethernet local-area networks (LANs) prevalent throughout the world use a modified form of the Aloha access technique, which provides higher throughput capability and, yet, is still quite simple and cheap to implement. A brief summary of the Ethernet technique will be presented in the next chapter. Wireless LANs to be discussed in detail in the next chapter also use an access technique derived from the Aloha procedure.

Now consider the slotted-Aloha access technique. The procedure is essentially the same as that of the pure Aloha technique with the one significant difference that repetitive time slots are made available, and access packets, assumed to occupy a time slot in length, are all constrained to begin transmission at the beginning of a time slot only. The basic picture is portrayed in Fig. 11.1. Slots in that figure are taken to be T-sec long and repeat periodically as shown. An access packet is shown being transmitted in one of the slots; two colliding packets are shown attempting transmission in another slot. The access procedure in this case is, again, transmit at will, but only within a time slot interval. If a transmitting

Abramson, N. 1973. "The Aloha system," Ch. 14, in *Computer Networks*, ed. N. Abramson and F. Kuo, Englewood Cliffs, NJ, Prentice-Hall.

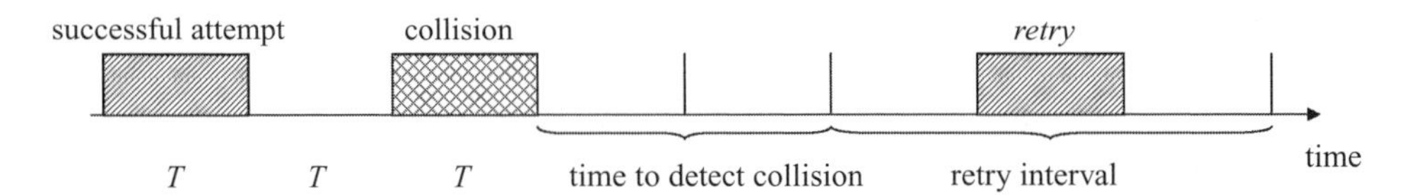

Figure 11.2 Slotted-Aloha procedure

user is alerted of a collision, i.e., more than one user has attempted transmission during the same time slot, that user reschedules the access attempt a random number of time slots later. An example appears in Fig. 11.2. A successful access attempt is shown first. (The time at which the user is notified of success is not shown.) A collision is then shown occurring some time later. The time taken for users with colliding packets to be so notified is shown as well. The interval in which to attempt a retry then follows, with the randomly chosen slot of the retry attempt by one of the users indicated. In the case of cellular systems being considered here, it would be the base station that notifies a user mobile station of a successful access attempt. A collision or unsuccessful access attempt is then detected by the user station when it does not receive an acknowledgement from the base station of its access attempt. The slotted-Aloha schemes adopted by the various cellular systems discussed in Chapters 8 and 10 all build on this basic procedure, with some refinements added. These include, among others, the increase of transmission power on each retry or an increase in the random interval, in units of slots, over which to retry. In the next section we describe another modification labeled p-persistence, in which retries are attempted with a probability p. Another modification might be one in which retries are allowed only once a frame interval, a frame consisting of a specified number of time slots, as in the systems discussed in previous chapters. Alternately, specified time slots only may be allocated for access attempts.

But now consider the basic slotted-Aloha procedure of Fig. 11.1. Note that the maximum possible throughput, or the capacity, of the system, is one packet per T sec. For example, if time slots are 5 msec long, at most 200 packets per second may be transmitted. By system capacity, we mean the total allowable throughput of this radio access channel at a given frequency, i.e., the summed throughput of all user stations using this access channel cannot exceed this value. As the number of potential users increases, the throughput allotted to each user must decrease correspondingly. How well does slotted-Aloha perform compared with this maximum possible throughput or capacity value of one packet per time slot? We shall see, through a simple analysis, that the *maximum-possible slotted-Aloha throughput* is e^{-1}, or 0.368, of the capacity, i.e., $0.368/T$ packets per second. Actually, it turns out that a system must operate significantly below this maximum value. Slotted-Aloha, in common with other random access systems, exhibits instability near the maximum value and must be controlled to operate in a stable form (see references in Schwartz, 1987: p. 430). One simple way of stabilizing such systems is to increase the random retry interval after each (unsuccessful) retry, one of the modifications to the basic procedure noted above. The reduction in allowable throughput using the slotted-Aloha

Schwartz, M. 1987. *Telecommunication Networks: Protocols, Modeling, and Analysis*, Reading, MA, Addison-Wesley.

procedure seems very significant, but is actually quite acceptable in most cases, considering the simplicity of the procedure. Consider the example noted above. With a time slot of 5 msec, the maximum-possible slotted-Aloha throughput is $0.386 \cdot 200$, or 77 access attempts per second. Say the system operates at one-half this rate to ensure stability, then, at most, 38 such attempts per second are allowed. If 100 terminals are associated with a base station, this means that, on the average, each station may make an access attempt every 2.5 seconds. This does not seem very onerous! This is the reason why the Aloha schemes are utilized. So long as the traffic intensity is not too great, they operate very effectively, and are relatively simple to implement. They were originally designed to handle bursty data traffic and operate particularly well for that type of traffic, precisely the type of application for which Ethernet and wireless LANs are best suited. They thus work well for sparsely occurring access attempts.

To derive a simple expression characterizing the performance of slotted-Aloha, we use the following common model (Schwartz, 1987): assume mobile uplink access attempts for a given cell are generated with Poisson statistics, with an average rate of λ attempts per second. Each such attempt corresponds to generating an access packet of length T sec. There are, thus, on the average, λT access attempts in any given time slot of length T sec. The assumption of Poisson statistics implies random and independent occurrence of access attempts from time slot to time slot. For Poisson statistics the probability that K such access attempts are made in a T-sec time interval (a slot time) is given by

$$P(K) = (\lambda T)^K e^{-\lambda T}/K! \tag{11.1}$$

Some of these access attempts will generate collisions, leading to retries at random times later. These retry attempts are clearly connected to the original access attempts. We assume, despite this correlation with access occurrences, that the retry attempts are Poisson occurrences as well. Aside from the obvious need to invoke this assumption to make the analysis feasible, there are several reasons why the assumption is usually considered a reasonable one: the retries may be spaced sufficiently far enough in the future to reduce the correlation; retries are generated at random-chosen intervals later; the ensemble of retries in any one slot some time later is not connected to any one specific user or earlier occurrence. In any given time slot, then, there may be present new, randomly generated access attempts, and retry attempts from earlier time slots. Let the sum of the access attempts, new and retrying, occur at an average rate of $\lambda' > \lambda$ attempts per second. The probability of K such composite attempts in any T-sec slot interval is then also given by (11.1), but with λ' replacing λ.

We now introduce some modified notation commonly used in describing the performance of Aloha-type random access schemes. We let the letter S represent the average number of new access attempts made in a T-sec slot interval, while the letter G represents the average of all access attempts, the sum of new attempts and retries in the same interval. We thus have, respectively, $S \equiv \lambda T$ and $G \equiv \lambda' T$. The symbol S is then a measure of the actual *throughput*, the average of new access attempts per slot interval. (The average rate of new access attempts must equal the average rate of successful attempts, since there is no blocking considered here. All attempts eventually succeed.) Equation 11.1 may now be rewritten using S in place of λT. A similar equation may be written as well using G in place of $\lambda' T$ to determine the probability of K composite access attempts, new plus retries,

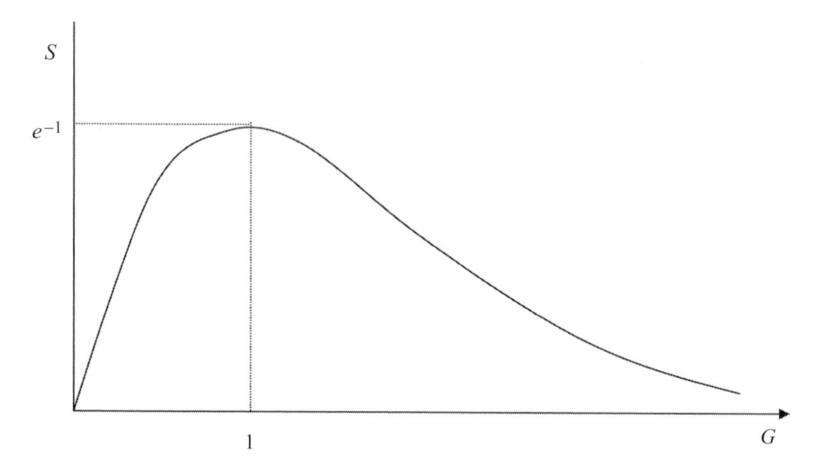

Figure 11.3 Slotted-Apoha performance curve

in a T-sec slot interval. In particular, the probability that no collisions will take place in a T-sec interval is just e^{-G} and the probability that at least one collision will take place is given by $1 - e^{-G}$. The average number of *retry* attempts per T-sec interval is then just $G(1 - e^{-G})$. Since G represents the average of *all* attempts in a T-sec interval, we must have $G = S + G(1 - e^{-G})$, or

$$S = Ge^{-G} \tag{11.2}$$

This expression relating the throughput S, the average number of new (and *successful*) access attempts in a slot interval, to the total number of access attempts in a slot interval is the well-known throughput performance expression for slotted-Aloha (Schwartz, 1987). Equation (11.2) may be derived a somewhat different and even simpler way: over a long interval of time $T' \gg T$, the probability of a successful access attempt must just be the ratio of successful attempts ST' to the total number of attempts GT', or ST'/GT'. But this is just the probability there are no collisions, or e^{-G}. Equating the two expressions, we get (11.2).

Equation (11.2) is plotted in Fig. 11.3. It is easy to determine that the maximum throughput is just $e^{-1} = 0.368$, as noted earlier, and that it is obtained at the point $G = 1$, both as shown in the figure. We note also that increasing G beyond the value of 1 results in a decrease of throughput. This is a sign of an unstable system, also as noted earlier. It is thus clear that one must operate well below the maximum value of S. Interestingly, the pure Aloha system performance derivation is quite similar. In that case, because there are no slots and, hence, no slot boundaries at which a user is constrained to attempt an access, users transmitting at will at any time, users may suffer collisions due to users' access attempts overlapping over an interval of $2T$, T being the access packet length. (The reader is asked to justify this statement.) An analysis using Poisson statistics, with $2T$ replacing the parameter T used in the analysis above, then shows the pure Aloha performance equation to be given by $S = Ge^{-2G}$. Comparing with (11.2), it is clear that the maximum throughput as a fraction of the capacity turns out to be $e^{-1}/2$ or 0.18 at $G = \frac{1}{2}$. Introducing the slot structure doubles the maximum possible throughput. Details are left to the reader.

Effect of fading on access control: "capture effect"

We now consider the effect of a fading medium on the slotted-Aloha access procedure. We shall see here that the maximum possible throughput actually *increases* as compared with the non-fading result obtained above! This is due to a property of random access called the *capture effect*, which has long been known in Aloha operational studies (Abramson, 1977), as well as in cellular system performance analyses (Goodman and Saleh, 1987; Yacoub, 1993). The capture property is due to the fact that a mobile with a sufficiently large transmitting signal may be "heard" over the signals of other users, effectively "capturing" the channel. Mobiles close to base station locations, as compared with those further away in a given cell, provide a simple example of such a possibility. Received power may vary over many decibels due to mobile location, shadowing effects, and more rapidly varying fading phenomena. Power control is thus very important in cellular systems, particularly so, as we have seen in earlier chapters, in CDMA systems. Mobiles normally have two types of power control, a more coarse open-loop control, and a finer closed-loop control. In the case of open-loop control, a mobile can initially estimate the transmitting power it should use based on its measurement of the downlink base station power it receives (Viterbi, 1995). This reduces mobile power differences due to propagation distances and shadow fading. Closed-loop control, involving signals from the base station to a specific mobile to either reduce or increase power, is needed, however, to adjust for the more rapidly varying Rayleigh or Ricean fading.

Consider the access procedure now. The mobile can, obviously, "hear" the base station signals and estimate the power it should use before being recognized by the base station. It can therefore presumably adjust its transmitter power to compensate for propagation effects and the relatively slowly varying shadow fading. But, since it still has not identified itself to the base station (the very reason for the access procedure!), closed-loop control is not possible. Hence differences in mobile power as received at the base station will appear due to rapid fading. A mobile caught in a deep fade will clearly not be "heard" by the base station as well as one in a temporarily better location. These variations in power received at the base station during access attempts may thus give rise to the capture effect noted above. It is the analysis of this effect which we summarize here. We follow the approach of Arbak and Van Blitterwijk, assuming Rayleigh fading (Arbak and Van Blitterwijk, 1987). In particular, we assume a homogeneous situation, with the base station receiving the same average signal power P_0 from each mobile. This thus implies, as noted above, successful operation of the coarse open-loop power control at each mobile. We also assume that the various mobile signals received at the base station add incoherently. (There is, of course, no reason why there should be any phase synchronization among the various signals.) This means attention must be placed on received signal powers, which will add at the base station.

Abramson, N. 1977. "The throughput of packet broadcasting channels," *IEEE Transactions on Communications*, COM-25, 1 (January), 117–128.

Goodman, D. J. and A. A. M. Saleh. 1987. "The near-far effect in local Aloha radio communications," *IEEE Transactions on Vehicular Technology*, VT-36, 1 (February).

Yacoub, M. D. 1993. *Foundations of Mobile Radio Engineering*, Boca Raton, FL, CRC Press.

Viterbi, A. J. 1995. *CDMA, Principles of Spread Spectrum Communication*, Reading, MA, Addison-Wesley.

Arbak, J. C. and W. Van Blitterwijk. 1987. "Capacity of slotted Aloha in Rayleigh fading channels," *IEEE Journal on Selected Areas in Communications*, SAC-5, 2 (February), 261–269.

Focusing on the ith received signal, it will be recalled, from Chapter 2, that, with Rayleigh fading, the received signal power obeys the following exponential probability density function with average value P_0

$$f_{P_i}(P_i) = \frac{1}{P_0} e^{-P_i/P_0} \tag{11.3}$$

Say now, that at a given time slot, n such (interfering) packets are received. Their powers then add to form the random variable T_n

$$T_n = \sum_{i=1}^{n} P_i \tag{11.4}$$

This random variable T_n, with average value nP_0, is readily shown to have as its probability density function the Gamma distribution given by

$$f_{T_n}(T_n) = \frac{1}{P_0} \frac{(T_n/P_0)^{n-1}}{(n-1)!} e^{-T_n/P_0} \tag{11.5}$$

Now consider a test access packet with power P_S transmitted during the same slot interval as these n interfering access packets. We again assume the power P_S is exponentially distributed with average value P_0. This packet will capture the transmission if its power is sufficiently greater than the sum of the powers of the other n access packets. Specifically, capture will occur if the ratio $P_S/T_n \geq z_0$, z_0 some specified threshold ≥ 1. Conversely, the probability of a collision is the probability that $P_s/T_n < z_0$. This threshold z_0, or capture ratio, as it is most often called, depends on system considerations. A typical value of 2, or the equivalent 3 dB, is commonly used in making calculations.

Capture of the transmission by this test access packet is equivalent to a successful transmission. From our discussion of slotted-Aloha performance in the previous paragraph the probability of a successful transmission, or capture in this case, is just S/G. Hence calculating the probability of capture will give us the desired value of the ratio S/G taking Rayleigh fading into account. In particular, call the probability of capture (success) $p_c(z_0)$. The calculation of $p_c(z_0)$ proceeds as follows (Arbak and Van Blitterwijk, 1987). We note first that the n interfering access packets a test access packet will encounter in a given time slot include both new packets and retries, all in this environment which includes Rayleigh fading. The probability that n such access packets are transmitted in a given time slot is just the Poisson probability $P(n)$ defined in (11.1), with the rate parameter now adjusted to include all packets. The probability $P(n)$ is therefore given by $G^n e^{-G}/n!$, G again representing the average number of access attempts in a given time slot, but now in this Rayleigh fading environment. This expression is used in determining the desired probability of capture $p_c(z_0)$. For we note that the probability of capture is just the probability there is no collision, i.e., 1− the probability of a collision. It is thus given by

$$p_c(z_0) = 1 - \sum_{n=1}^{\infty} P(n) \operatorname{Prob}\left(\frac{P_S}{T_n} < z\right) \tag{11.6}$$

Here the expression $\operatorname{Prob}(P_S/T_n < z_0)$ is defined as the probability of a collision in a time slot, conditioned on n interfering packets being transmitted. As a check, let the capture ratio z_0 be very large, approaching ∞. Then $\operatorname{Prob}(P_S/T_n < z_0) = 1$ and $p_c(z_0)$ from (11.6) is given by $1 - \sum_n P(n)$. This is just the probability that 0 interfering packets are transmitted, or e^{-G} from (11.1), using G as the Poisson parameter, as noted above. Hence we get $S/G = e^{-G}$, as expected.

More generally, we have z_0 finite, and we must calculate Prob$(P_S/T_n < z_0)$ to determine the probability of capture $p_c(z_0)$. This is done as follows, using a two-dimensional transformation of random variables similar to that carried out in Chapter 9 (Arbak and Van Blitterwijk, 1987). We first introduce a parameter $z_n \equiv P_S/T_n$. We also write $w \equiv T_n$. We then carry out the transformation of random variables, transforming (P_S, T_n) to the new variables (z_n, w). Integrating over w, we calculate the desired probability density function of z_n from which the probability of capture can then be found. Specifically, we have, just as in Chapter 9

$$\begin{aligned} f_{z_n,w}(z, w) &= f_{P_S,T_n}(P_S, T_n)\left|\frac{\partial(P_S, T_n)}{\partial(z, w)}\right| \\ &= f_{P_S}(zw) f_{T_n}(w) w \end{aligned} \tag{11.7}$$

Replacing P_i in (11.3) with zw gives us the first term in (11.7); the Gamma function of (11.5) with w as the variable gives us the second term. We thus get

$$f_{z_n,w}(z, w) = \frac{w(w/P_0)^{n-1}}{P_0^2(n-1)!} e^{-(wz+w)/P_0} \tag{11.8}$$

The desired probability density function of z_n is now readily found by integrating (11.8) over all values of the variable w from 0 to ∞. It is left for the reader to show that the resultant density function is given by

$$f_{z_n}(z) = n(z+1)^{-n-1} \tag{11.9}$$

But recall that the variable z_n represents, in the transformation above, P_S/T_n, and that it was Prob$(P_S/T_n < z_0)$ we were interested in calculating. This is therefore given by the integral of (11.9) for all $z < z_0$, just the cumulative distribution function of z_n. This is readily found to be given by $1 - (z_0 + 1)^{-n}$. Introducing this result for Prob$(P_S/T_n < z_0)$ in (11.6), carrying out the indicated summation over n, and simplifying the resultant expression, we find, finally, that

$$S = Ge^{-G\frac{z_0}{z_0+1}} \tag{11.10}$$

Comparing with the slotted-Aloha performance expression of (11.2), it is clear that the capture effect due to Rayleigh fading has improved the average throughput somewhat. For note that the maximum value of S is now $e^{-1}(z_0 + 1)/z_0$, instead of the previous value of e^{-1}. As an example, if we take the capture ratio z_0 to be 3 dB or the value 2, as suggested above, this represents an increase in the maximum possible value of S by 50% to 0.55. But recall that we cannot really operate at this value of throughput for stability reasons. Later work by Zorzi and Rao on the stability of capture-slotted-Aloha shows that retransmission control is required to keep the slotted-Aloha access mechanism stable in the presence of Rayleigh fading (Zorzi and Rao, 1994). Their analysis indicates that

---

Zorzi, M. and R. R. Rao. 1994. "Capture and retransmission control in mobile radio," *IEEE Journal on Selected Areas in Communications*, 12, 8 (October), 1289–1298.

stability requires that the average throughput S in new accesses per slot interval be less than $2/\pi\sqrt{z_0}$. For the case of $z_0 = 2$, this value of the maximum S allowed to ensure the access mechanism is a stable one, is $S = 0.45$. This corresponds to a maximum allowable value of G of 0.75. Without the capture effect, operating at this value of G results, from (11.2), in a throughput of $S = 0.35$. There is thus a potential improvement of 29% due to the capture effect if the system operates at this point. But note again, that an increase in the retransmission interval with each additional retry is needed to ensure stability. This potential improvement due to the capture effect reduces as the rate of accesses decreases. The capture effect in GPRS systems, which, in common with all systems discussed in this book, uses slotted-Aloha access, has been shown by simulation to provide a substantial improvement in throughput as compared with a system without the capture effect (Cai and Goodman, 1997). The one basic difference with the analyses discussed in this section, however, is that random access packets are queued while waiting to be transmitted. As a result, the instability described above does not appear, the performance curve not folding over as shown in Fig. 11.3, but saturating at a fixed value. The tradeoff is then added delay time (Cai and Goodman, 1997).

Improved access strategies

A number of procedures have been suggested to provide some improvement in throughput over the basic slotted-Aloha access technique described above. We describe some of them briefly here. Some have been adopted in third-generation systems, as noted in our discussion of these systems in Chapter 10.

Consider, first, the concept of power ramping during the access stage. This procedure is used for second-generation systems, as indicated in our discussion of these systems in Chapter 8. We noted in our discussion of third-generation CDMA systems in Chapter 10 that power ramping during access has been adopted there as well. With this procedure, the power chosen to send an initial access message is generally selected to be substantially below the open-loop transmitted power estimate a mobile might normally use. It is to be noted that such estimates may be inaccurate because of measurement error and fading. To account for the possibility of fading, a large fading margin would normally have to be used, resulting possibly in increased interference. Instead, mobiles begin their initial access with powers below the power suggested by the estimate, and then increase the power by a fixed amount for each access retry required (Olafsson *et al.*, 1999; Moberg *et al.*, 2000). Another question raised in Chapter 10 is whether it is better to send repetitive access preambles until access is granted, followed then by a longer message, or to send the message itself, repeating it if necessary until access is finally granted. In either case, power ramping is used. Processing delay is longer in the latter case, but the message is guaranteed to have been received once access is granted. In the former case, using

Cai, J. and D. J. Goodman. 1997. "General packet radio service in GSM," *IEEE Communications Mag.*, 35, 10 (October), 122–131.

Olafsson, H. *et al.* 1999. "Performance evaluation of different random access power ramping proposals for the WCDMA system," IEEE Personal Indoor Mobile Radio Conference.

Moberg, J. *et al.* 2000. "Throughput of the WCDMA random access channel," Proc. IST Mobile Communication Summit, Galway, Ireland, October.

preambles first to acquire access, processing time is reduced, but the message itself might be lost due to changes in the fading conditions when access is finally granted (Olafsson *et al.*, 1999; Moberg *et al.*, 2000). However, it is also to be noted that the possibility of interference and hence collisions is reduced with the use of short preambles.

CPCH, the uplink common packet channel access procedure, was mentioned in the introduction to this chapter as another technique for improving access throughput. Recall that CPCH was briefly mentioned in our discussion of W-CDMA in Chapter 10 as an alternative option to the RACH access procedure. The distinction between the two procedures arises only after the initial access stage. In the RACH access stage, a mobile selects a RACH sub-channel from those broadcast by the base station, estimates the initial transmission power level to be used, based on its measurement of the downlink power received, and then transmits a brief 1 msec preamble using a randomly selected code and signature. If the base station detects this preamble, it echoes the signature back on the AICH. The terminal then follows by transmitting a 10 or 20 msec portion of the RACH message. Note, however, that other accessing terminals might have used the same signature, and replied to the AICH message as well, resulting in a collision. The CPCH procedure attempts to reduce the possibility of a collision at this point by sending a collision detection (CD) preamble instead of a (longer) portion of the RACH message, using a different, randomly selected signature for this purpose. The base station echoes this new signature back to the mobile on the CD Indication Channel, CDICH. If the echoed signature is correct, the mobile now starts transmission of a longer message. It is the combination of a short CPCH preamble and the newly selected random signature transmitted with it that provide the improved access performance. A simulation study described in Moberg *et al.* (2000) does, in fact, show that significant improvement of the random access procedure throughput at high loads is obtained simply by selecting a new random signature to transmit the message once access is granted by the base station. This second randomization clearly serves to reduce potential "collisions" of multiple message transmissions once access is granted.

11.2 Integrated access: voice and data

The previous section focused on slotted-Aloha random access to an uplink channel. Initial access only was considered. The appropriate handling of channel resources once access was attained was not considered. In this section we do consider the joint problem of access control and channel resource assignment. We also enlarge the problem set to include multiple types of traffic, with differing traffic characteristics and differing quality-of-service (QoS) requirements. As pointed out in the introduction to this chapter, it often suffices to consider two basic types of traffic, continuous-time traffic, such as real-time voice, and bursty data. We therefore focus on the access control and resource assignment of these two types of traffic. We first describe and compare, where possible, a number of proposals that have appeared in the literature for handling voice and bursty data in a TDMA uplink channel. As we shall see, these proposed access strategies all build on slotted-Aloha. These access control procedures are then extended to the CDMA environment.

The first, pioneering, approach to the multi-access control problem was developed in a series of papers by Goodman and associates (Goodman *et al.*, 1989; Goodman and Wei, 1991; Nanda *et al.*, 1991). The basic strategy, called *packet reservation multiple access*, or PRMA for short, requires a slotted-Aloha type of access and reserves time slots in successive frames for voice packets, once access is successfully accomplished. The voice-call reservations are held so long as the voice packets continue to arrive periodically. Time slots are not reserved for data packets, however. Data packets must individually compete for time slots. We describe the access/reservation procedure in more detail first for the case of competing voice calls only, and then add the requirement for data packet access as well. The frame structure chosen for simulation studies of the performance of PRMA envisions a mobile user voice call, once access is granted, as requiring one time slot per frame.

Simulation studies described in the PRMA papers use the following voice-structure example. As is commonly the case (refer to the discussion in Chapter 8), voice samples are assumed taken every 125 μsec. Say a transmission rate of 32 kbps is used. This then corresponds to 4 bits per sample. Say 128 such samples are assembled to form a voice packet of 512 bits. There is therefore one packet assembled every 16 msec. This is then the time chosen for a frame length. Now let the transmission rate be 720 kbps. Note that the third-generation systems described in Chapter 10 do allow a transmission of this value, although the actual system using a rate of this value might be operating indoors. Let a frame encompass 20 slots. With the transmission rate of 720 kbps, 576 bits are available per slot. This implies that 64 bits of control and header information are added to each 512-bit voice packet.

Now consider voice access. As noted above, PRMA adopts a slotted-Aloha access procedure, but with the following modifications. A complete packet is transmitted as part of the access attempt. If successful, the base station replies with an acknowledgement (ack). The slot used by the voice mobile is then considered reserved for that mobile, and is so marked by the base station. Slots in PRMA are thus of two types, those reserved for ongoing voice calls, and those available for random access. Note that the round-trip time to receive an ack is much smaller than a frame time, often even smaller than a slot time: a slot time is, in this example, 0.8 msec. At radio propagation speed, the time to cover a macrocell-size round-trip distance of 30 km is 0.1 msec. Adding some processing time at the base station, there is ample time for an ack to be received before the next frame.

Continuing with the basic PRMA procedure, in the event no ack is received, a voice mobile tries again the next frame with a probability p_v, a design parameter. This PRMA retry procedure thus differs from normal slotted-Aloha in which a random interval later is chosen for the retry attempt. This technique of probabilistically retrying has long been known as *p-persistent-Aloha*. (We shall encounter this retry strategy again in the next chapter in establishing the performance of wireless LANs.) Slot reservation in PRMA is an example as well of a *reservation-Aloha* access strategy, also long-known in the Aloha

---

Goodman, D. J. *et al.* 1989. "Packet reservation multiple access for local wireless communication," *IEEE Transactions on Communications*, 37, 8 (August), 885–890.

Goodman, D. J. and S. X. Wei. 1991. "Efficiency of packet reservation access," *IEEE Transactions on Vehicular Technology*, 40, 1 (February), 170–176.

Nanda, S. *et al.* 1991. "Performance of PRMA: a packet voice protocol for cellular systems," *IEEE Transactions on Vehicular Technology*, 40, 3 (August), 584–598.

literature. In the PRMA case, a voice mobile persists in retrying a maximum of two times only. The packet in question is then dropped. The reason for this is that real-time voice packets must normally be delivered to the destination within 100 msec at most. Otherwise the real-time conversation becomes unacceptable to the parties engaged in conversing. Note that two frames of delay already correspond to 32 msec. Packet dropping does, of course, affect the quality of a voice conversation. Values of packet loss probability in the range of 1–5% have been found tolerable in studies of packet voice in wireline networks. The probability of packet loss is usually included as one of the QoS attributes of packet voice (Schwartz, 1996).

Simulation studies have been carried out for a PRMA system with the example specifications indicated above and are described in the PRMA papers referenced above. We discuss the results shortly. We first note that a talk-spurt model of packet generation during a voice call model is used in these simulations. This model is well-established as representing the characteristics of voice quite accurately (Schwartz, 1996), and was mentioned briefly in Chapter 6 during the discussion of CDMA system performance. According to this model, a close-up view of a typical speaker's characteristics shows the voice utterance alternating randomly between a talkspurt state and a silent state. Figure 11.4(a) shows packets being generated at a periodic rate during a spurt, alternating with intervals of silence. A two-state Markov (memoryless) model representing this voice-packet generation process appears in part b of the figure. According to this model, a speaker in a talk-spurt state has a probability $\beta\Delta t \ll 1$ of leaving that state, and moving to a silent state, in any small time interval Δt. Conversely, there is a probability $\alpha\Delta t \ll 1$ of moving to the talk-spurt state in any small interval Δt. (This represents the continuous-time model. An equivalent discrete-time model is actually more appropriate for our presentation, since we do assume periodic generation of voice packets while in the talk-spurt state, and we impose a periodic frame structure as well. With the numbers with which we deal here, however, the two models are essentially equivalent, and it is simpler to present the continuous-time version.) The probability of being in a talk-spurt state is readily shown to be given by $\alpha/(\alpha + \beta)$, while the probability of being in a silent state is $\beta/(\alpha + \beta)$. (A thoughtful reader will note that this two-state model is a special case of the more general multistate models of Markov chains discussed in Chapter 9.) It is also readily shown that the average length of a talk-spurt is $1/\beta$, while the average silent interval is just $1/\alpha$. Measured talk-spurts typically vary in length from 0.4–1.2 sec, while silent intervals average 0.6–1.8 sec in length. The probability of being in a talk-spurt is generally taken to be about 0.4. This fact that much of the time a speaker is actually silent has long been made use of in telephone networks, with users sharing time slots. The indication is that as many as 2.5 users may, on the average, share each time slot in a large system. This gives rise to DSI, digital speech interpolation, in digital telephone networks, with 2.5 as many users assigned to a given switch. We noted in Chapter 6 that, in CDMA systems, this gain in usage translates into reduced interference, since 60% of the time, on the average, active voice mobiles are silent. PRMA also expects an improvement in performance due to silent intervals of voice.

Schwartz, M. 1996. *Broadband Integrated Networks*, Englewood Cliffs, NJ, Prentice-Hall.

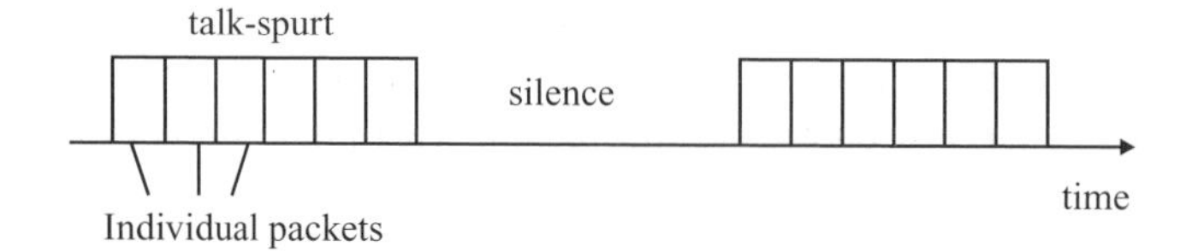

(a) Generation of voice packets during a call

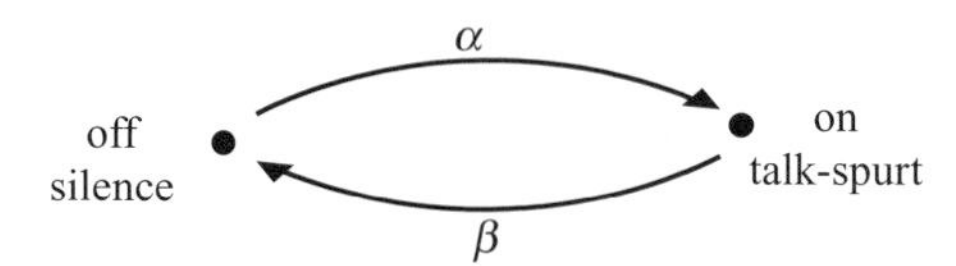

(b) Two-state Markov model

Figure 11.4 Talk-spurt model of speech

Returning now to the PRMA simulations described in the literature cited, it is assumed that a voice mobile gaining access to a time slot in a frame looses that slot once it enters the silent state. It must again contend for access once it begins a talk-spurt. Reservation of a time slot thus lasts for the length of a talk-spurt only. Average talk-spurt lengths used in the PRMA simulations were 0.36 sec, with average silent intervals chosen as 0.64 sec. The probability of being in talk-spurt is then 0.36, with the DSI improvement expected to be about 1/0.36 or 2.8. For a 20-slot frame one might therefore expect to be able to handle up to 56 voice models. This turned out to be too optimistic. The simulations indicated that the best choice of p_v using the p-persistent Aloha strategy was 0.3, with 37 voice mobiles then accommodated with a packet-dropping probability of 1%. Note that this value of users that may be accommodated is not quite 2.8 times the slot number, but it is still almost double the number possible by simply assigning users a slot during a voice call and allowing them to retain the slot for the length of the call. Why a "best" choice of p_v? With p_v too small, terminals wait too long to access the radio channel and too many packets are dropped as a result. With p_v too large, excessive collisions result, and instability sets in.

The simulation studies just summarized focused on the case of voice traffic only. Many subsequent studies reported on in the literature considered the effect of both PRMA voice and data terminals competing for access on the same uplink radio channel. We summarize the results of one of these simulation studies only before moving on to later variations of the PRMA technique. This study, by Grillo and associates, considered the impact of realistic operating conditions, in a multicellular system, with path loss and slow fading included (Grillo *et al.*, 1993). More specifically, the cell structure modeled was hexagonal of 1km radius. Path loss was modeled using a 3.5 power law for attenuation. Slow fading

---

Grillo, D. *et al.* 1993. "A performance analysis of PRMA considering speech, data, co-channel interference and ARQ error recovery," *Mobile and Personal Communications*, IEEE Conference Publication No. 387, December 13–15, 161–171.

was log-normal with 4 dB standard deviation. Fast Rayleigh fading was not explicitly considered. It was assumed that a minimum uplink SIR, as measured at the base station, was required to ensure correct packet receipt at the base station. Two values of minimum SIR, 9 dB and 12 dB, were considered. The capture effect was modeled as well: with two or more packets contending for access, the one with the highest SIR succeeds in gaining access, provided, however, its SIR exceeds the minimum value required.

The PRMA frame structure adopted was the same as that noted above, using 16-msec frames of 20 slots each, again operating at 720 kbps transmission rate. Voice packets, generated once per frame, were again assumed to carry 512 information (speech) bits, with 64 bits added. The one change made was in the talk-spurt/silence parameters. The average talk-spurt interval chosen was 1 sec; the average silence interval was taken to be 1.35 sec. Note that, with these numbers, the probability of a voice station being in talk-spurt is 0.43; the projected DSI gain is 2.33. Voice terminals again retained a given time slot each frame, once access was granted, for the length of a talkspurt. Data terminals, on the other hand, had to access the channel each time a data packet was to be transmitted. There was no data packet reservation from frame to frame. Data packets were one slot in length. Data terminals were taken to be of relatively low bit rate, each transmitting at a rate of 4.8 kbps, with a 50% duty cycle: active for 0.588 sec, not transmitting for the same time. The average bit rate per terminal was thus 2.4 kbps. Unlike voice, with tight constraints on voice-packet delivery time, data packet delay is not normally a QoS objective, as noted in the previous chapter. Hence, in the access procedure adopted for data packets in this simulation study, retransmission times were relatively long. (Note that data packets and voice packets compete for the same time slots. Once a voice source gains access to a time slot, however, that slot is reserved for the source for the length of the talkspurt, reducing the number of slots available for both voice and data packets.) In addition, an ARQ procedure with variable window size was utilized to control data packet flow. In the simulations carried out (Grillo *et al.*, 1993), 30 data terminals were assumed to always be active; traffic was increased by increasing the number of voice stations. The simulation results then showed that the system could support about 29 voice sessions with the required SIR of 9 dB or 25 voice sessions with a required SIR of 12 dB. Both cases had a voice-packet loss of 1%. For these numbers, the average number of data transmission attempts required per successful data packet transmitted ranged from 1.4 to 1.8. Note that the added data packet attempts reduce the effective system capacity. Consider the minimum SIR case of 9 dB as an example. With 1.4 transmission attempts required to transmit each data packet, the average bit rate being used by the 30 data stations is $30 \cdot 1.4 \cdot 2.4$ kbps, or 101 kbps. If the data stations are disabled, and voice only transmitted, the simulations show that, ideally, 37 voice stations can be supported with the packet loss of 1%. This is in agreement with the earlier simulation studies cited above. If, with voice stations only transmitting, a minimum SIR of 9 dB is required for voice-packet reception, the number drops to 35 stations. Requiring a minimum SIR of 12 dB reduces this number to 32 stations. These figures demonstrate the various tradeoffs introduced with the impact of realistic transmission conditions included.

Improvements can, of course, be made on the basic PRMA access procedure. One such improved access strategy proposed by members of the European wireless research community, working early on proposals for third-generation systems, was labeled PRMA++

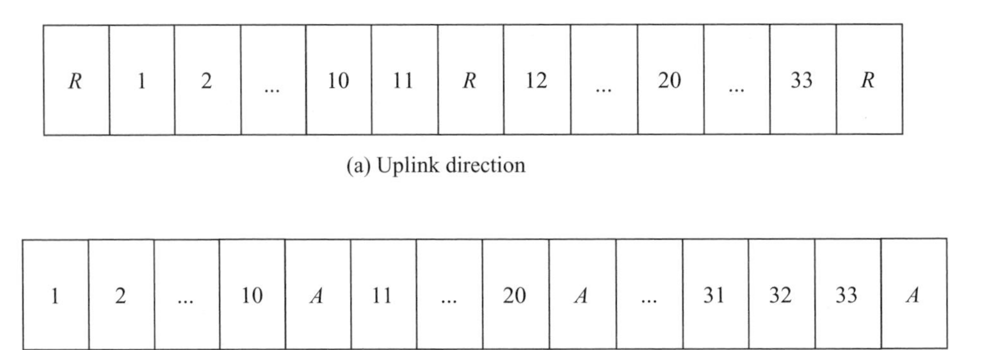

Figure 11.5 Examples of PRMA++ frame structure

(DeVille, 1993). This access procedure differs from PRMA in distinguishing between two types of slots per frame in the uplink direction. Reservation or *R* slots are used exclusively for slotted-Aloha-type access control. Information or *I* slots are used for transmitting information packets, once an access attempt is successful. Figure 11.5(a) portrays an example of this frame structure for a 36-slot frame. Here three *R* slots are specifically identified; the remaining 33 slots are information slots. The downlink slot structure contains two types of slots as well, acknowledgement or *A* slots, acknowledging a successful access attempt, and *I* slots, used to carry downlink information packets. These two types of slots are identified in Fig. 11.5(b) for the same example of Fig. 11.5(a). Additional specialized slots are defined downlink in this proposed access protocol to transmit fast-paging information, with paging-ack slots also set aside uplink. We ignore these added types of slots in this brief discussion. (The idea of introducing special reservation slots, in addition to information slots, was first proposed in Mitrou *et al.* (1990).) In this protocol, a mobile seeking access transmits an access-contention burst on an *R* slot, the burst containing a truncated mobile address and the number of *I* slots required for information transmission. The downlink *A* slots carry the addresses of successfully contending mobile accesses, as well as the uplink *I* slots to be used for transmitting information uplink. These uplink *I* slots then remain allocated to the mobiles in question so long as their transmissions are active. This procedure thus allows multiple slots, hence varying transmission capacity, to be allocated to a given mobile. By using a limited number of reservation, *R*, slots in the uplink direction, the number of acknowledgement, *A*, slots is reduced as well. This results in a significant reduction in the downlink capacity devoted to signaling resolution of access attempts (DeVille, 1993).

Some thought will indicate that, with the number of slots per frame fixed, there is an optimum number of *R* slots per frame in terms of minimizing the access delay. This delay is made up of two components, contention delay and delay waiting for an *I* slot to become free, once access is successful. If a small number of *R* slots is made available,

DeVille, J. M. 1993. "A reservation-based multiple access scheme for a future universal telecommunications system," *Mobile and Personal Communications*, IEEE Conference Publication No. 387, 13–15 December, 210–215.

Mitrou, N. M. *et al.* 1990. "A reservation multiple access protocol for microcellular mobile communication systems," *IEEE Transactions on Vehicular Technology*, 39, 4 (November), 340– 351.

the contention delay is high. Increasing the number reduces the contention delay, but decreases the number of *I* slots as well. Eventually, if the number of *I* slots is small, the wait to receive an *I* slot takes over, and access delay again increases. This optimum number of *R* slots turns out to depend on the number of mobiles in a given system. As an example, simulation results for voice-packet mobile communications described in DeVille (1993) indicate that, with a 72-slot 5-msec frame structure, a minimum access delay of 20 msec (i.e., 4 frames) can be achieved for 128 contending mobiles by assigning 16 of the slots as *R* slots. The average talkspurt in this case was taken to be 1.41 sec, while the average silent interval was chosen as 1.78 sec. The probability of being in talkspurt is thus 0.44 and the projected DSI gain would be 2.27. The actual gain over the circuit-switched mode is 128/72, or 1.78, roughly comparable with the PRMA simulation figures quoted above, for a different frame structure. But note that only 16 *A* slots would be required on the downlink, leaving 56 slots available for information transmission. (We are again ignoring other up/downlink slot usages such as those required for paging.) If the access delay is now allowed to increase to 30 msec (close to the 32-msec maximum value allowed in the PRMA simulations quoted above), 136 mobiles can be accommodated, with the number of *R* slots per frame reduced to 12 (DeVille, 1993).

A number of other extensions of PRMA and PRMA++ have been proposed in the literature. An access protocol called SIR (Service Integration Radio), for example, extends PRMA++ to incorporate bursty data transmission as well as voice-packet transmission (Anasti *et al.*, 1996, 1997). In this proposed protocol uplink *R* slots are divided into two groups, *VR* and *DR* slots. The former are used, as in PRMA++, to contend for access. Voice transmission requests use these slots to request bandwidth allocations as well. The *DR* slots are used to provide information, contention-free, to the base station as to data bandwidth requests: once data mobiles succeed in gaining access, they are periodically polled by the base station, round-robin fashion, to determine their bandwidth requirements. *I* slots are then assigned to the data mobiles by the base station in a round-robin, contention-free, manner. Dedicated data *I* slots may be made available to data mobiles for this purpose. In addition, any voice-assigned *I* slots not being used in a given frame by voice mobiles are assigned to data terminals for this frame only. This procedure for allowing data to access unused voice slots is referred to as a "movable boundary" scheme, and has long been used in wired link assignment strategies. The SIR access protocol thus assigns data mobiles information slots, once access has been accomplished, in a contention-free, round-robin manner without affecting voice information slots.

Another proposed integrated-access protocol, Frame Reservation Multiple Access or FRMA, modifies the basic PRMA protocol two ways: it specifically attempts to maximize the number of slots allocated to data each frame, keeping the voice packet-dropping probability at a specified value of 1%, and has the base station broadcasting one acknowledgement per frame only, to reduce mobile receiver activity and hence enhance mobile battery life (Narasimhan and Yates, 1996). Specifically, the uplink, mobile to base station, frame

Anasti, G. *et al.* 1996. "A bandwidth reservation protocol for speech/data integration in TDMA-based advanced mobile systems," *Proc. IEEE Infocom'96*, San Francisco, CA, March, 722–729.

Anasti, G. *et al.* 1997. "A contention/reservation access protocol for speech and data integration in TDMA-based advanced mobile systems," *Mobile Networks and Applications*, 2, 3–18.

Narasimhan, P. and R. P. Yates. 1996. "A new protocol for the integration of voice and data over PRMA," *IEEE Journal on Selected Areas in Communications*, 14, 4 (May), 621–631.

structure is divided into three slot types: so-called R slots reserved for voice information packets, with an active voice mobile allocated only one such slot each frame, V slots used for voice contention, and data traffic slots used for both data contention and to carry data information packets. (Note the difference in notation as compared with the PRMA++ work cited above: the V slots here are comparable with the R slots in the PRMA++ case; the R slots here are comparable with the I slots there.) A movable-boundary strategy is adopted as well, with unused, dedicated voice slots each frame allocated to data traffic. With N slots in a frame, then, we must have the number of data slots given by $N - R - V$. In this scheme the base station adaptively calculates both V and the voice p-persistent probability p_v required to maximize the data-slot (bandwidth) allocation each frame and broadcasts those values to all voice mobiles for their use each frame. The maximization of data bandwidth is carried out with the constraint that the voice-packet dropping probability be close to 1%, as noted. In carrying out this optimization each frame, the base station knows, at the beginning of each frame, the number M of voice mobiles with ongoing voice calls, and the number R that are in the talk-spurt state, hence expected to transmit a voice-information packet each. (Note that $R \leq M$.) The two parameters V and p_v to be calculated are functions of both R and M.

What are the results of using this access protocol? Simulations carried out show that, because of the optimization adopted, its data-terminal performance far exceeds that of pure PRMA (Narasimhan and Yates, 1996). Specifically, consider a system with parameters similar to those used in the PRMA system summarized above: repetitive frames carry 20 slots and are 16 msec long; the radio channel rate is 720 kbps and voice calls transmit at a 32-kbps rate each; 64 bits of overhead are added to each 512-bit voice packet per slot; the average talk-spurt is 1 sec long and the average silence interval lasts 1.35 sec. (The talk-spurt/silence intervals thus differ from those used in the PRMA simulations referenced earlier.) As in the PRMA example, fading effects are not considered in carrying out the simulation. The results then are as follows: as the number of voice terminals increases, the proportion of the bandwidth (slots) made available to the data terminals decreases linearly with the number of voice terminals, with 20% of the bandwidth or 144 kbps made available to data transmission, on the average, with 32 voice terminals in the system. Note that the voice-only PRMA system allowed 37 voice terminals to be accessing the base station. The integrated PRMA system, carrying both voice and data, cited above (Grillo *et al.*, 1993), showed 72 kbps, or 10% of the bandwidth, was made available to data terminals (30 such terminals carrying an average of 2.4 kbps each), with 30 voice terminals in the system. Consider, as another example, the case of 20 voice mobiles and 20 data terminals in the system. Then, with the FRMA strategy, as much as 50% of the bandwidth or 360 kbps is found to be allocated to data transmission, i.e., as much as 18 kbps per terminal. Even more significantly, the average data packet delay waiting on queue before transmission over the radio channel is begun is found to be significantly reduced with FRMA as compared with PRMA. Thus, consider the average data packet delay in this specific case of 20 voice mobiles and 20 data terminals, as the data rate per terminal is increased. With a data rate of 1000 bps per terminal, the PRMA scheme results in an average data packet delay of about 400 msec. The corresponding FRMA delay is less than 100 msec. The queueing delay for both schemes then increases as the data rate per terminal climbs above this value, following a typical queueing curve. The PRMA scheme saturates at a maximum data rate per terminal of 1400 bps, for which the delay is

800 msec. The FRMA scheme is found to reach its maximum data rate of 18 kbps at a delay of less than 800 msec (Narasimhan and Yates, 1996). It is important to note, however, that a price is paid for these significant improvements in performance provided by the FRMA access procedure: the voice-packet dropping rate is always kept at about 1%, even when the system is lightly loaded. This number is a constraint on, and, hence, built into the optimization procedure. PRMA has the 1% packet dropping rate as a threshold, i.e., the voice-packet dropping rate is less than or equal to 1%. A lightly loaded PRMA system would have a packet dropping rate of much less than 1%.

The final TDMA-based access scheme we discuss (many more have been proposed and studied in the literature!) is DQRUMA (Distributed Queueing Request Update Multiple Access), a multiple-access protocol designed principally for fixed packet transmission over wireless local-area networks (LANs) (Karol *et al.*, 1995a). (In another paper by the same authors, the work is extended to show how DQRUMA may be adapted to large cell-size systems (Karol *et al.*, 1995b).) This protocol attempts to provide more efficient use of system capacity. Uplink frames carry multiple mini-slot request access (RA) channels plus the usual packet-length information slots. The mini-slot RA channels thus use less capacity than would be the case with full-slot contention channels. Mobiles contend for access, transmitting a b-bit short mobile-access id, using one of the RA channels, selected randomly from those available. (The number of bits b used depends on the number of mobiles a base station can support.) The number of RA channels varies dynamically from frame to frame, depending on the number of mobiles requesting access. With traffic light, the base station can allocate more RA channels in a frame, so announcing in a downlink message.

The downlink frames, from base station to mobiles, carry acknowledgement channels, one for each RA-channel uplink. If access contention by a mobile, using one of the RA channels, is successful, the mobile recognizing its id, as transmitted downlink on the corresponding ACK channel, it transmits an information packet on the next slot uplink. (This procedure thus applies to wireless LANs principally, for which the round-trip processing plus propagation delay is relatively small. As noted above, however, other work indicates how the DQRUMA protocol may be adapted to large cell-size systems (Karol *et al.*, 1995b).) If the mobile has additional information packets to transmit (its packet buffer is non-empty), it so signals with a one-bit piggyback request flag attached to its packet transmission. Subsequent packet accesses are thus collision-free. (A CDMA version of DQRUMA, to be described briefly below, incorporates a p-bit piggyback request field, indicating the number of additional packets the mobile proposes to transmit. This field is incorporated as well in the initial access contention using the RA minislot.) A modified slotted-Aloha access procedure, mobiles using an adaptive access attempt probability, is suggested as the contention access procedure (Karol *et al.*, 1995b). Simulations indicate that this dynamic slotted-Aloha access technique as well as another proposed algorithm perform quite well, resulting in a packet delay vs throughput characteristic close to that of an "ideal" access scheme, in which the base station has complete, perfect information of all access requests each slot time. The piggyback request-bit procedure

Karol, M. J. *et al.* 1995a. "An efficient demand-assignment multiple-access protocol for wireless packet (ATM) networks," *Wireless Networks*, 1, 5 (October), 267–279.

Karol, M. J. *et al.* 1995b. "Distributed-queueing request update multiple access (DQRUMA) for wireless packet (ATM) networks," IEEE International Conference on Communications, ICC'95, Seattle, WA, June, 1224–1231.

reduces the random-access requirement as well, since a mobile with a non-empty buffer has no need to use random access beyond the first access. Simulations indicate that the packet delay-throughput performance improves as the mobiles' information packet burst size increases from one to two packets. (Note that this procedure is somewhat similar to that designed into voice access in PRMA, except that there it takes a frame interval for a base station to notice a previously assigned voice-information slot is empty, indicating the voice mobile is no longer in talk-spurt. Here the knowledge as to whether a mobile has additional packets to transmit comes with the current transmission.)

CDMA access strategies

In this section we describe access strategies derived from the original PRMA scheme, but adapted to CDMA systems. These schemes utilize multiple codes for transmission, overlaid on a slotted-frame structure, the procedure adopted for the third-generation CDMA systems discussed in Chapter 10. They are thus directly applicable to those systems.

We begin with a CDMA version of DQRUMA, mentioned in passing in discussing DQRUMA in the previous section (Liu *et al.*, 1996). This proposed multiple-access scheme uses multicode CDMA, the form of multirate CDMA using multiple orthogonal spreading codes in parallel to obtain higher transmission rates, as described in Chapter 10. A procedure called sub-code concatenation is used to obtain the orthogonal spreading codes. In this procedure, each mobile is assigned a primary pseudo-random code. These codes are not completely orthogonal. The orthogonal spreading codes are derived from the primary codes by operating on them with a set of orthogonal Walsh codes (Chin-Lin I and Gitlin, 1995c; Liu *et al.*, 1996).

As with TDMA-based DQRUMA, slotted-Aloha access contention is carried out using RA minislots assigned to each frame. A mobile contending for access uses a unique orthogonal code as well as a b-bit short id. Contention arises when more than one mobile chooses the same orthogonal access code. The access attempt includes a p-bit number indicating the number of packets the mobile wishes to transmit. Multiple request-access receivers, each configured to accept a specified code, can be set up at the base station, allowing more than one mobile access attempt to be acknowledged on the corresponding downlink acknowledgement minislot channel. Multiple mobile access attempts, each using a different code, may therefore be recognized and acknowledged. This approach of allowing multiple simultaneous access attempts to be recognized is characteristic of CDMA systems, as compared with TDMA systems. But we must also recall that CDMA is a much wider-band access technique than TDMA, as first noted in Chapter 6. (The performance results of the various TDMA access schemes described in the previous paragraph were all determined for one frequency channel.) In addition, the price paid for multiple simultaneous accesses is that interference increases as well.

In this multicode CDMA version of DQRUMA, a mobile waits to send information packets until a transmit-permission indication is received from the base station. This indication is sent as one of multiple transmit-permission indications in a manner similar to

Liu, Z. *et al.* 1996. "A demand-assignment access control for multi-code DS-CDMA wireless packet (ATM) networks," *Proc. IEEE Infocom'96*, San Francisco, CA, March, 713–721.

Chih-Lin, I. and R. D. Gitlin. 1995. "Multi-code CDMA wireless personal communication networks," IEEE International Conference on Communications, ICC95, Seattle, WA, June, 1060–1064.

that of the ACK procedure described for TDMA-based DQRUMA, but using a multicode CDMA procedure over a Transmit-Permission minislot channel. Transmit permission directed to a mobile carries an indication of the number of codes, hence rate of transmission, it is permitted to use, a field assigning the primary pseudo-random code to be used for packet transmission in the next time slot, and a field indicating the power level to be used. Since codes are assigned on a slot-by-slot basis, and, hence, are reused, the number of codes required is reduced (Liu *et al.*, 1996).

Simulation performance results for data traffic for this multicode CDMA version of DQRUMA are similar to those found for TDMA-based DQRUMA, as summarized in the previous paragraph (Liu *et al.*, 1996). For packet data bursts carrying multiple packets each, the delay-throughput curve, obtained by increasing the number of bursts per slot, approaches the "ideal access" curve closely. This performance, close to that of ideal, is due to the use of the piggybacking request-access feature of DQRUMA. In fact, simulations for the case of single-packet bursts show that the performance of DQRUMA only approaches that of the ideal curve in the region of heavy traffic, in which a mobile buffer is generally non-empty most of the time, allowing the piggyback request-access feature to become operative.

We turn now to a proposed CDMA access procedure for both voice and data terminals based more closely on PRMA (Brand and Aghvami, 1996). As in the earlier TDMA-based PRMA, terminals contending to transmit a packet do so with probability p_v for voice and p_d for data. These permission probabilities may vary from frame to frame, and from slot to slot within a given frame. They are calculated by the base station for each slot in a frame, and are transmitted downlink by the base station, to be used in the frame following. In the work presented in this paper, p_d is always a fraction of p_v, say 0.1 or 0.2, hence the voice–permission probability only needs be calculated. The objective chosen for the calculation is that of maximizing the throughput with voice-packet loss probability constrained to always be less than some specified value, such as 0.01 or 0.02. This objective turns out to be equivalent to controlling the number of simultaneous "periodic" users in a slot in a current frame. Periodic users are those transmitting one information packet per frame for a number of frames. Voice-packet transmission provides an example: as noted a number of times earlier, voice-packet transmission continues periodically throughout a talk-spurt interval. Data packets, in the model used here, may be periodic or may be random, in which case one packet only is transmitted. As in PRMA, once a periodic user successfully accesses a slot in a frame, that slot is kept reserved for that user, so long as it keeps sending packets. An optimum number with reserve status in a given slot corresponds to the number satisfying the maximum throughput objective. It is thus clear that, in the case of large numbers of users in the reserved state, the permission probability should be low. In fact, if the optimum number of users for a slot has been attained, the permission probability should be zero. As the number of users drops below the optimum value, the permission probability should increase. For calculation purposes in this work, the function relating permission probability to the number of users has been heuristically defined to be made

---

Brand, A. E. and A. H. Aghvami. 1996. "Performance of a joint CDMA/PRMA protocol for mixed voice/data transmission for third generation mobile communications," *IEEE Journal on Selected Areas in Communications*, 14, 9 (December), 1698–1707.

up of two linear segments, an initial shallow curve starting with the highest permission probability for zero users in a slot, and decreasing slowly with the number of users, and then, at a specified breakpoint, dropping much more rapidly to 0 permission probability as the number of users increases.

Simulations of this scheme have been carried out for both the case of a single cell and one with a hexagonal cellular structure (Brand and Aghvami, 1996). 8-kbps voice sources were assumed, as in the cellular systems described in previous chapters. These sources were assumed to have average talk-spurt intervals of 1 sec and average intervals of silence of 1.35 sec, as in the speech model described earlier in (Grillo *et al.*, 1993). This corresponds to a DSI gain of 2.35. A frame length of 20 msec, with 20 slots per frame, was chosen for the simulations. Each voice source thus supplied 160 information bits per frame. These bits plus header bits, after coding, constituted one voice packet to be transmitted in a given slot. Packets were dropped if they could not be transmitted within a frame time of 20 msec. Consider the results of the single cell simulations first. For the case of voice traffic only, the optimum number of periodic sources per slot was found to be 9, with the maximum loss probability in the range of 1 to 2%. The best heuristic permission probability curve was found to be one starting at a permission probability of 0.3 for no users in a given slot (channel), decreasing linearly to 0.25 with six users in a slot, then dropping rapidly to 0 at the optimum value of nine users per slot. With the loss probability constrained to a maximum value of 2%, the resultant number of simultaneous voice calls that could be supported was found to be 379. Reducing the probability of loss constraint to 1% lowered the number of calls that could be supported to 358. Note that these numbers contrast with $9 \cdot 20 \cdot 2.35 = 435$ as the number obtained using the simple DSI advantage calculation. More significantly, if pure random access were used, i.e., setting $p_v = 1$, the number of conversations simultaneously possible with the loss probability constrained to be no more than 2% was found to be 227. So the technique of controlling the permission probability resulted in a performance gain of 67%. The gain was found to be 78% for the tighter loss probability constraint of 1%. Interference patterns plotting the number of users per slot as a function of time show the number fluctuating widely about an average of eight or so in the random access case; the number is constrained quite tightly in the range of seven to nine, in the controlled access case.

The impact of adding single-packet data traffic to the traffic mix was evaluated through simulation by steadily increasing the number of data terminals and determining the number of voice conversations that could be supported with the voice loss constraint of 2% (Brand and Aghvami, 1996). Data terminals were assumed to operate at a bit rate of 3400 bps, the same as the average voice-terminal bit rate (i.e. 8 kbps $\cdot$ 0.43). As the number of data terminals increased the number of voice conversations that could be supported decreased, as would be expected. The net throughput reduced as well, due to the increased burstiness of the traffic. But a considerable improvement in capacity was found as compared with random access (Brand and Aghvami, 1996).

Simulations carried out using a hexagonal-cellular structure provided similar conclusions as to the efficacy of access control. The number of simultaneous conversations that could be supported for a given loss probability constraint was reduced as compared to the single-cell case: inter-cell interference increased the total interference substantially. (Recall that CDMA capacity calculations carried out in Chapter 6 show the inter-cell

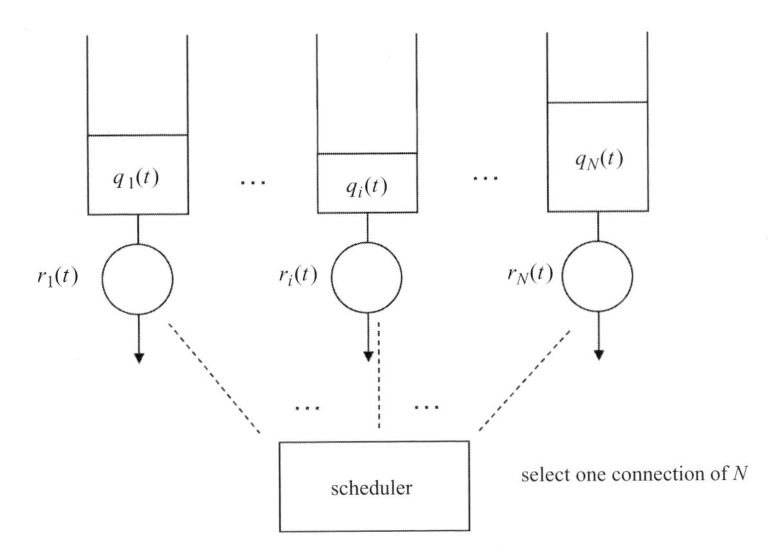

Figure 11.6 Scheduling of multiple data streams

interference can be substantially more than intra-cell interference.) Controlled access still showed a significant improvement over pure random access however (Brand and Aghvami, 1996), the improvement experienced depending on the power-propagation model used. Consider the common $1/d^4$ propagation model described in earlier chapters as an example. With a loss probability constraint of 1%, the single-cell figure of 358 simultaneous conversations possible using controlled access is reduced to about 280. The corresponding number with random access for the cellular model is found to be 160. Controlled access thus provides an improvement of 75%, similar to that found for the single-cell model, as noted above.

11.3 Scheduling in packet-based cellular systems

As noted in the introduction to this chapter, resource allocation in cellular systems designed for packet transmission, such as the third-generation cellular sytems described in Chapter 10, or those being proposed for fourth-generation systems of the future, requires appropriate scheduling of data packets both uplink from each mobile in a given cell, and downlink, to each mobile. The purpose of scheduling is two-fold: to provide the appropriate QoS measures such as maximum packet delay, the related throughput guarantees, packet loss probability, and other QoS performance guarantees to individual users, where possible, propagation conditions allowing, and to ensure full use of the resources, most commonly link bandwidth or capacity. We commented briefly on the need for resource allocation at the end of the paragraph on cdma2000 in Section 10.2 of Chapter 10, discussing the relation between measured SIR and the data bit rate transmission possible for that SIR. As an example of how a QoS requirement affects resource allocation, we indicated, using a reference from the literature, how a change in the data latency requirement could improve overall system throughput.

A graphical picture of the meaning of scheduling appears in Fig. 11.6. This is a rather abstract model, showing N data buffers, each containing a number of packets at time t, indicated, for the ith queue, by the symbol $q_i(t)$. The rate of transmitting or releasing packets from the ith queue is represented by the symbol $r_i(t)$. The data buffers could represent data packets waiting to be transmitted uplink, in the reverse direction, to the base station at each of N data mobiles in the cell controlled by that base station; they could equally well represent individual buffers associated with each data mobile at the base station, holding data packets waiting to be transmitted downlink, in the forward direction, to each of the mobiles. The scheduling problem, both uplink or downlink, is then that of determining which packet or packets to schedule for transmission at time t. The packet transmission rate $r_i(t)$ shown in Fig. 11.6 could be the result of applying the scheduling algorithm; it could be incorporated as part of the solution to the scheduling problem. In our first example below, we shall see that this rate is effectively that resulting from applying the algorithm. In this example, the *desired* rate of data transmission for a given mobile is determined by its measured SIR, i.e., the ability of a mobile to send or receive the waiting data packets. This was the approach adopted in the work cited in Chapter 10. The *actual* rate of transmission may differ, after application of the algorithm.

Scheduling algorithms are often designed to achieve some desired performance objective or objectives. These objectives could include the attainment of a desired QoS for each data traffic type, conditioned on the possibility of adverse propagation conditions, as noted above, and as considered in the example discussed in Section 10.2. They would also include the attainment of maximum possible overall throughput on a given radio link, as stated above. The scheduling algorithm adopted should obviously take the type and characteristics of a given traffic stream into account as well. (Some of the properties of a given traffic type could of course be folded into its desired QoS.) Thus the scheduling of bursty data traffic would presumably take the bursty characteristics into account; the handling of more regular, continuous-type traffic would reflect the characteristics of that type of traffic as well. Wider-band traffic, such as multi-media traffic from a web site that has been accessed by a given mobile and that is to be sent downlink to that mobile, would presumably be handled differently than e-mail packets that require much lower bit rates.

In the context of the system designs we have been considering, scheduling assignments could be varied from time slot to time slot, or from frame to frame. The time t used in Fig. 11.6 reflects whichever choice is made. The actual rate of transmission of packets, in either direction, for a given mobile might be effectuated by appropriately assigning codes in a CDMA system, or varying the number of time slots in a frame in a TDMA system. Examples of both cases were cited in Chapter 10. The scheduling could be done by the base station itself or by a scheduler located in a network controller responsible for control of a number of cells within a given system (WCDM, 2000). In either case, the link control channels described in Chapter 10 would be utilized to transmit the necessary information required to carry out the scheduling algorithm, such as measured power and SIR at each mobile, as well as the characteristics and number of packets waiting to be transmitted on a mobile queue. (Note that the DQRUMA system discussed in the previous section

---

WCDM. 2000. *WCDMA for UMTS*, ed. Harri Holma and Antti Toskala, John Wiley & Sons.

includes mobile-queue length among the attributes transmitted uplink from a mobile to its base station.)

The concept of scheduling packets from multiple streams of traffic to meet specified performance objectives, the abstract picture portrayed in Fig.11.6, has long been discussed in wired packet networks (see Zhang (1995) for an overview of packet network scheduling procedures). Some of the scheduling algorithms adopted or proposed for use in wireless systems are, in fact, variations of those originally proposed for use in the wired world. A compilation of scheduling techniques proposed for use in both TDMA and CDMA wireless networks appears in Fattah and Leung (2002). This compilation includes references to their earlier suggested use in wired networks, where first described there. We discuss in this section only a selected number of scheduling algorithms to simplify the discussion and to provide more focus on the principles of scheduling. We begin our discussion of scheduling algorithms appropriate to wireless systems by describing a specific algorithm introduced for use with the cdma2000 1X high data rate or HDR system mentioned in Chapter 10. This is the *proportionally fair*, PF, scheduling algorithm (Rental *et al.*, 2002). This algorithm is time-slot based and is quite simple to carry out: it chooses for transmission in a given slot packets waiting in the user queue that have the highest (maximum) ratio of requested transmission rate for that queue to the average rate of transmission from that queue. If a tie occurs, it is broken randomly. Specifically, let $DRC^i(n+L)$ be the rate requested by user i for slot $(n+L)$, based on the measured SIR. (As noted above, this choice of rates, based on the measured SIR, was described at the end of the cdma2000 discussion in Section 10.2. These rates were the maximum values possible based on a requirement of a 1% packet error rate (Bender *et al.*, 2000).) Let $R^i_{av}(n+L-1)$ be the average transmission rate received by user i over some long interval, as measured for the previous slot. The user whose packets are chosen for transmission in slot $(n+L)$ is then the one for which its ratio $DRC(n+L)/R_{av}(n+L-1)$ is maximum. Specifically, the PF algorithm selects user j for transmission in time slot $(n + \mathrm{L})$ if j has the maximum value indicated in (11.11):

$$\text{PF:} \qquad \max_i \left[DRC^i(n+L)/R^i_{av}(n+L-1) \right] \quad 1 \le i \le N \tag{11.11}$$

Note that this algorithm attempts to satisfy the need of the user with the highest requested rate of transmission, but introduces an element of fairness by normalizing the requested rate to the average rate of transmission. Hence, the priority of a user with a high average rate of transmission is reduced to accommodate users with lower average rates of transmission. The algorithm attempts to maximize the channel throughput (the choice of the user with the maximum requested rate would clearly accomplish this), but introduces a measure of fairness by normalizing to the average rate, preventing users from "hogging" the channel.

Zhang, H. 1995. "Service disciplines for guaranteed performance service in packet-switching networks," *Proc. IEEE*, 53 (October), 1374–1396.

Fattah, H. and C. Leung. 2002. "An overview of scheduling algorithms in wireless multimedia networks," *IEEE Wireless Communications*, 9, 5 (October), 76–83.

Rental, C. H. *et al.* 2002. "Comparative forward link traffic channel performance evaluation of HDR and 1EXTREME systems," Proc. 55th IEEE Vehicular Technology Conference, VTC Spring, 160–164.

Bender, P. *et al.* 2000. "CDMA/HDR: a bandwidth-efficient high-speed service for nomadic users," *IEEE Communications Magazine*, 38, 7 (July), 70–77.

Any averaging algorithm could be used to carry out the averaging of rates. The one proposed by Rental *et al.* (2002) uses first-order recursive updating

$$R^i_{av}(n) = \left(1 - \frac{1}{t_c}\right) R^i_{av}(n-1) + \frac{1}{t_c} r_i(n-1) \tag{11.12}$$

The term $r_i(n-1)$ is the actual transmission rate received by user i in the prior slot $n-i$. This is the transmission rate shown in Fig. 11.6. The number t_c is the number of slots over which the first-order averaging is carried out. In the simulations reported on in Rental *et al.* (2002), 1000 slots were used as the averaging time. Note from (11.12), that the average transmission rate is expected to change very slowly with time.

Example

A simple two-user example of the application of this algorithm, taken from Andrews *et al.* (2001), shows its effectiveness in increasing user and overall throughput, as compared with a simple round-robin allocation of system resources. In the round-robin approach, users are allocated a transmission opportunity one after another, without any priority. In this example, say mobile user 2 experiences better channel quality than does user 1, on the average. Say it requests transmission, as a result of measuring its SIR, at either of two transmission rates, 153.6 kbps or 307.2 kbps. Assume these requests are each made, on the average, 50% of the time. User 1, experiencing poorer channel quality, is assumed to be able to transmit only at the rates of 76.8 kbps and 153.6 kbps, each again occurring 50% of the time. (Note that these represent transmission rates offered by cdma2000 1X, as shown in Tables 10.2 and 10.3 in Chapter 10.) It is left for the reader to show that the average round-robin service throughput, i.e., the average channel throughput under round-robin scheduling, is then 172.8 kbps. Of this average throughput, user 1 is allocated an average throughput of 57.6 kbps; user 2 has an average throughput of 115.7 kbps.

Now introduce proportionally fair scheduling. The average rates R_{av} to be used for normalization as in (11.11) are obtained in this simple example by averaging over the two possible rates of transmission for each user, and are given, respectively, by $R^1_{av} = 115.2$ kbps and $R^2_{av} = 230.4$ kbps. (There is clearly no need to invoke the recursive relation (11.12) in this simple case. Steady-state averages suffice.) There are four possible, equally likely, events that can occur in invoking the proportionally fair algorithm in this two-user example: there are two tie-breaker cases, $76.8/115.2 = 153.6/230.4 = 0.67$ and $153.6/115.2 = 307.2/230.4 = 1.33$. The first case produces the average transmission rate $0.5(76.8 + 153.6) = 115.2$ kbps. The second tie-breaker case results in an average double that value, 230.4 kbps. A third possible event has user 2 transmitting at the rate 307.2 kbps winning out, i.e., $307.2/230.4 > 76.8/115.2$. The fourth possibility has user 1 transmitting at the rate 153.6 kbps winning out, i.e., $153.6/115.2 > 153.6/230.4$. Averaging over all four possible events, we find the average overall channel throughput to be 201.6 kbps, clearly larger than the round-robin result of 172.8 kbps. The individual average throughput values are larger as well, being 67.2 kbps and 134.4 kbps, respectively. Details are left for the reader to check.

Large-scale simulations of an HDR system described in Rental *et al.* (2002) show proportionally fair scheduling provides substantial improvement in overall throughput in a pedestrian-based three-sector cell environment as the number of mobile users increases. Path loss, shadow fading, and Rayleigh fading were all included in the simulations. The average throughput per user with 16 users present per sector was measured to be 100 kbps, with a standard deviation of 55.6 kbps. The standard deviation provided a measure of the fairness of the algorithm. In a second set of simulations involving a vehicular cellular environment, with vehicles traveling at 60 km/hr, a marginal improvement only in performance using the PF algorithm was observed (Rental *et al.*, 2002).

Proportionally fair scheduling is a special case of a more general scheduling algorithm proposed by Andrews *et al.* (2001). Their more general algorithm, termed *modified largest weighted delay first* or *M-LWDF* scheduling, has the property that all queues of all users are guaranteed to be kept stable, i.e. no queues "blow up." (The assumption here is that queue sizes are unlimited in length. Introducing a finite limit ensures the queues will not blow up, but, in that case, the equivalent of instability is that a queue will saturate at its maximum value and remain there, with all new packets attempting to enter the queue blocked from doing so.) This algorithm schedules packets from queue j at time slot t if, for j, the following quantity is maximum (Andrews *et al.*, 2001):

$$\text{M-LWDF:} \qquad \max_i[\gamma_i W_i(t) r_i(t)] \qquad 1 \leq i \leq N \tag{11.13}$$

Here γ_i is an arbitrary constant to be discussed further below; $W_i(t)$ is the head-of-the-line waiting time in queue i, and $r_i(t)$ is the rate of transmission of packets at queue i, as shown in Fig. 11.6 and discussed earlier. The waiting time $W_i(t)$ may be replaced as well for some or all of the users by the queue length $q_i(t)$ shown in Fig. 11.6 (Andrews *et al.*, 2001). How does one select the constant γ_i? This depends on the type of traffic and the QoS objective desired. Andrews *et al.* provide an example for delay-constrained traffic. Real-time traffic such as voice or video would be examples. For this type of traffic, the delay constraint at queue i might be presented as

$$\text{Prob}[W_i > T_i] \leq \varepsilon_i \tag{11.14}$$

Here T_i is the delay constraint; ε_i is a specified probability. For example, for real-time voice, we might have $T_i = 50$ msec and $\varepsilon_i = 0.01$. There is thus a 99% chance in this case that the delay constraint of 50 msec is satisfied. Given the delay constraint (11.14), Andrews *et al.* suggest the constant γ_i be written as a_i/R^i_{av}, R^i_{av} just the average transmission rate for user *i*s traffic, as used in the PF algorithm, and the parameter a_i defined as $(-\log\varepsilon_i)/T_i$. For example, if, for user i, we pick ε_i to be 0.01 and $T_i = 50$ msec, the example just given, we have $a_i = 0.04$. If ε_i is increased to 0.1, suggesting a looser constraint, a_i is reduced to 0.02. The priority of user i is thus reduced correspondingly. Note therefore that the transmission priority of individual mobile users in a given cell may be adjusted by appropriate choice of the parameter a_i. Using the relation $\gamma_i = a_i/R^i_{av}$, the M-LWDF scheduling algorithm may then be defined as one which chooses user j for transmission in slot t if, for user j, the following quantity is the largest among the N mobile users

$$\text{M-LWDF:} \qquad \max_i\left[a_i W_i(t) r_i(t)/R^i_{av}\right] \qquad 1 \leq i \leq N \tag{11.15}$$

Note how similar the M-LWDF scheduling algorithm in this form is to the PF algorithm described by (11.11). Here there are three sets of parameters that determine the choice of user to serve at time t: a_i and $W_i(t)$ can be visualized as providing the QoS delay-constraint requirement inputs, the ratio $r_i(t)/R^i_{av}$ providing the ability to maximize the overall channel throughput, yet without unduly penalizing lower average-rate users, just as

---

Andrews, M. *et al.* 2001. "Providing quality of service over a shared wireless link," *IEEE Communications Magazine*, 39, 2 (February), 150–153.

in the PF algorithm. Although we discussed the M-LWDF algorithm above for a real-time traffic example, it is apparent that the algorithm is readily adapted to non-real-time traffic as well. In that case, one might ignore the parameter a_i. (Choosing it to be a constant value for all mobile users effectively does this.) The wait-time quantity $W_i(t)$ or its equivalent queue length $q_i(t)$ then provides a means for ensuring that packets in individual queues are not unduly delayed.

The M-LWDF algorithm incorporates, as one of its elements, the wait time $W_i(t)$ or its equivalent queue length $q_i(t)$. Priority is thus given to packets that have been waiting longer than others to be transmitted. This strategy of favoring packets waiting the longest is related to a class of scheduling algorithms labeled generically "Earliest-Deadline-First" scheduling algorithms. This class of algorithms, and variations thereof, have long been studied in the context of wired networks. In its simplest version, the "Earliest-Deadline-First" (EDF) algorithm involves choosing for transmission in each time slot the packet at the head of a queue which is closest to its threshold delay for transmission. In terms of the notation adopted thus far, and referring to Fig. 11.6, this means choosing for transmission the head-of-the-line packet in queue i if $(T_i - W_i)$ is the smallest value among all N queues. Here T_i is the delay threshold defined in (11.14). Various versions of this algorithm, adapted to make them appropriate to a CDMA mobile environment, have been studied and compared in Varsou and Poor (2002) and Solana *et al.* (2002). This latter reference uses the time stamping of packets when they first arrive at the transmission queue on which to base the waiting time measured once they reach the head of the queue. The time-stamping procedure has been proposed by other investigators studying scheduling algorithms based on packet waiting time.

The earliest-deadline-first (EDF) class of scheduling algorithms has been shown, in an unpublished September 1999 Bell Laboratories' memo referenced in Varsou and Poor (2002), to provide the best performance in a comparison made among a number of down-link scheduling schemes for bursty data users in a CDMA environment and for certain QoS requirements. Varsou and Poor extend this unpublished work as follows: in the original work the downlink packet at the head of a queue with the closest deadline is selected for transmission each time slot. All available power is assigned to that single transmission. The extension to that work described in Varsou and Poor allows the next earliest deadline user to be served each time slot should there be power left over to do so. This scheduling strategy is labeled "powered earliest-deadline-first," PEDF, scheduling. A further variation, called "powered earliest deadline first fair" or PEDFF, shares any remaining power among all the other active users. This introduces an element of fairness into the scheduling algorithm. Both extensions of the EDF algorithm are shown to provide improvements in performance, using power more efficiently (Varsou and Poor, 2002). The PEDFF algorithm in turn is found to outperform the PEDF algorithm for bursty traffic and mixed (heterogeneous) traffic. Further performance improvement is shown to be obtained by using a different algorithm favoring short-packet transmission, as well as penalizing packets that have missed the deadline threshold (Varsou and Poor, 2002).

Varsou, A. C. and H. V. Poor. 2002. "Scheduling algorithms for downlink rate allocation in heterogeneous CDMA networks," *Journal of Communications and Networks*, 4, 3 (September), 199–208.

Solana, A. H. *et al.* 2002. "Performance analysis of packet scheduling strategies for multimedia traffic in WCDMA," *Proc. 55th IEEE Vehicular Technology Conference*, VTC 2002, Spring, 155–159.

We conclude this section on scheduling by focusing on a CDMA fair-scheduling scheme adapted from a similar strategy studied quite extensively in the wired networking literature. We describe the wired-networking version first. The strategy, generalized processor sharing, GPS, was first proposed and analyzed by Parekh and Gallager (1993). A summary of this work appears in Schwartz (1996). (The GPS algorithm is an idealized one, not being implementable in practice. A realizable extension called packet GPS, PGPS, also known as weighted fair queueing, is discussed in Parekh and Gallager (1993) as well. We describe GPS only to simplify the discussion.) GPS is essentially a modified round-robin strategy, designed to allocate link capacity (bandwidth) fairly among multiple users. It accomplishes this objective by weighting the capacity allocated to each user as it carries out the round-robin scheduling. Specifically, consider the N queues of Fig. 11.6. Let these queues share a common link of capacity C. In the GPS literature, each queue and associated transmission element is designated a *connection*. Visualize each connection transmitting in round-robin fashion. Let the flow of packets from connection i be assigned a weight ϕ_i so that it is guaranteed service at a rate g_i given by

$$g_i = \phi_i C \bigg/ \sum_{j=1}^{N} \phi_j \tag{11.16}$$

The diagram of Fig. 11.7(b) provides a conceptual view of the GPS scheduler. (The regulators shown will be described shortly.) The GPS scheme is defined to be *work conserving* as well. This means that, at every opportunity for a packet transmission, one should take place. As a result, if a particular connection has no traffic to transmit when its turn to do so comes around, its allocation is apportioned to a connection with traffic to send. As a result, the actual connection-i traffic served in an interval of length T designated as S_i obeys the following inequality

$$S_i/S_j \geq \phi_i/\phi_j, \qquad j = 1 \ldots N \tag{11.17}$$

It may then be shown that GPS provides the following two guarantees (Parekh and Gallager, 1993):

1 Let R^i_{av} be the average connection-i service (transmission) rate. So long as $R^i_{av} \leq g_i$, service rate g_i is guaranteed, independent of other connections.
2 The maximum delay of traffic on any connection can be bounded, based on its own characteristics, independent of traffic on other connections.

A specific bound on the GPS delay performance may be attained as follows. Let the traffic flow on each connection be smoothed or regulated by a so-called "leaky bucket" regulator (Schwartz, 1996). This regulator is designed to limit the rate and burstiness of arriving traffic. It does so by setting two parameters σ and ρ. A leaky bucket implementation using a settable counter indicates how the rate and burstiness is regulated. The counter is set to increment by 1 every $1/\rho$ seconds, to a maximum value of σ. It is decremented by 1 each time a packet is received and transmitted. Once the counter reaches 0, packets cannot be transmitted. Some thought will indicate that the leaky bucket controller has constrained the rate of packets arriving to be at most the parameter ρ, while the maximum

Parekh, A. K. and R. G. Gallager. 1993. "A generalized processor-sharing approach to flow control in integrated services networks: the single-node case," *IEEE/ACM Transactions on Networking*, 1, 3 (June), 344–357.

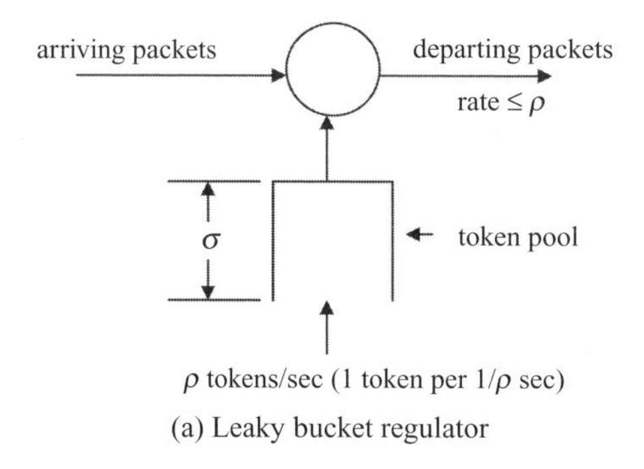

(a) Leaky bucket regulator

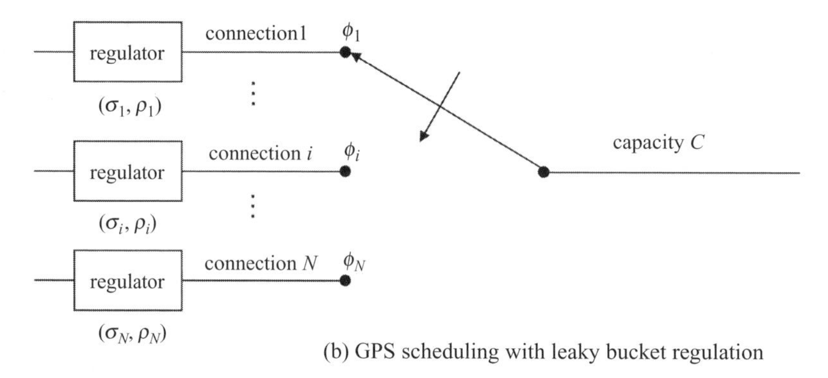

(b) GPS scheduling with leaky bucket regulation

Figure 11.7 GPS and leaky bucket regulation

number of packets arriving as a burst and being transmitted as such, i.e., the maximum burst size, is limited to the parameter σ. An alternate conceptual implementation involves a token buffer holding at most σ tokens (Schwartz, 1996). This token buffer-based leaky bucket controller is depicted in Fig. 11.7(a). Its operation is explained as follows: the token buffer is incremented by 1 to its maximum value of σ once every $1/\rho$ seconds. An arriving packet must take a token to be transmitted. Now let each GPS connection have a leaky bucket regulator with parameters ρ_i and σ_i chosen independently for each connection. The resultant N-connection GPS scheduler appears in Fig. 11.7(b). The only stipulation here is that the leaky bucket-controlled packet arrival rate be less than or equal to the connection service rate g_i, i.e., $\rho_i \leq g_i$. It may then be shown that this inclusion of a leaky bucket regulator on each connection results in a bound, given by the ratio σ_i/g_i, on the maximum packet delay on each connection (Parekh and Gallager, 1993; Schwartz, 1996).

This description of the GPS fair scheduling algorithm, and summary of its properties, has assumed, as noted above, a wired networking environment. Its adaptation to the link scheduling of heterogeneous traffic in a CDMA-type cellular wireless system has been proposed by a number of investigators, including Arad and Leon-Garcia (1998) and Xu *et al.* (2002). We describe here the implementation proposed by Xu *et al.* for

Arad, M. A. and A. Leon-Garcia. 1998. "A generalized processor sharing approach to time scheduling in hybrid CDMA/TDMA," *Proc. IEEE Infocom'98*, San Francisco, CA, March, 1164–1171.

Xu, L. *et al.* 2002. "Dynamic bandwidth allocation with fair scheduling for WCDMA systems," *IEEE Wireless Communications*, 9, 2 (April), 26–32.

multicode CDMA cellular systems. Uplink scheduling only is studied specifically in this work, although it is clear that GPS fair scheduling may be applied as well to downlink scheduling. The strategy invoked by Xu *et al.* is a rate-based one, dynamically allocating transmission rates to each user each time slot. The rate of each uplink user is varied by assigning different spreading factors and/or differing numbers of orthogonal code channels. The scheduling procedure used is termed *code-division GPS fair scheduling* or CDGPS. In this proposal each mobile transmits to the base station a short message at the end of each time slot reporting its buffer state (number of backlogged packets) and the amount of traffic that arrived during that time slot. The random access channel or a dedicated uplink channel could be used to carry this message. The scheduler, whether at the base station or at a more central location, uses this uplink information gathered from all the active mobiles, plus the known QoS requirements for each mobile, to calculate the channel transmission rate allocated to each mobile for its uplink transmission the next time slot. This channel rate is translated into the appropriate spreading factor and number of codes to be used in the next time slot, and this information is sent downlink by the base station to each mobile in a resource allocation message.

The specific computational procedure proposed by Xu *et al.* (2002) for their CDGPS scheduling scheme may be written as follows :

(*Note*: The symbols S_i, C, and ϕ_i appearing below are the same as those used above in (11.16) and (11.17) in defining the GPS procedure.)

CDGPS: Objective – to determine capacity allocation $C_i(k)$ for mobile i, time slot k

Let $q_i(k-1)$ be the packet backlog, mobile i, end of slot $k-1$. Let $r_i(k)$ be the estimated traffic arrival rate during time slot k. (This rate may be estimated directly from the traffic arriving during time slot $k-1$ or by using some type of averaging scheme.) Time slot length is T. Estimated backlog, time slot k, is $B_i(k)$.

Steps in procedure

1
$$B_i(k) = q_i(k-1) + r_i(k)T \tag{11.18}$$

2 Calculation of $S_i(k)$, traffic expected to be received by mobile i (see (11.17))

$$\begin{aligned} &\text{If } B_i(k) = 0, \quad \text{then } S_i(k) = 0 \\ &\text{If } B_i(k) > 0, \quad \text{then } S_i(k) = g_i T \end{aligned} \tag{11.19}$$

g_i given by (11.16)

$$\text{If } \sum_i S_i(k) < CT$$

distribute remaining capacity in accordance with ϕ_i

3 Capacity allocated to user i:

$$C_i(k) = S_i(k)/T \tag{11.20}$$

How well does CDGPS, this version of GPS modified for use with CDMA systems, perform? Simulations for different types of traffic have been carried out and results are reported in Xu *et al.* (2002). In addition, assuming leaky bucket regulators are used as shown in Fig. 11.7(b), maximum delay bounds very similar to those quoted above for GPS may be established as well. Specifically, the maximum delay bound for mobile i turns out to be almost the same as the GPS quoted earlier, with the addition of the scheduling (time slot) interval T (Xu *et al.*, 2002):

$$\text{Delay}_{\max,i} = \sigma_i / g_i + T \qquad (11.21)$$

For small time slot values the maximum delay bound for CDGPS is obviously close to that of GPS.

Consider now the simulation results reported by Xu *et al.* (2002). Specifically, the system simulated had CDMA uplink channels of total capacity $C = 2$ Mbps. The time slots were 0.1 sec long. The scheduler has to not only send rate information, but power control information as well to control intra-cell and inter-cell interference. In these simulations, perfect power control and error-free transmission channels were assumed. Channel fading was not modeled. The first case simulated was that of four homogeneous, best effort, packet flows, with packets $L = 5120$ bits long, shaped by a leaky bucket regulator with parameters $\sigma = 4L$ and $\rho = C/4$. The traffic flows were modeled as Poisson processes with average arrival rates, the offered load for the system, increasing from low load to a load approaching the maximum possible. The maximum delay bound from (11.21), setting $g_i = \rho$, turns out to be 0.14 sec. The actual maximum delay at the maximum load simulated turns out to be 0.13 sec, below the delay bound, as expected. A static allocation scheme, with each flow assigned the same fixed transmission rate, results in substantially higher maximum delay at the high loads. A similar result is found for average delay. At low load values, the static and dynamic allocation techniques do not differ very much in performance, as might be expected, since, with relatively little traffic, dynamic allocation is clearly not needed. Simulation of a heterogeneous traffic situation involving a combination of ten voice flows, three video flows, and four best-effort data flows, was carried out as well. The results obtained using CDGPS scheduling again had the maximum delay for traffic source below the calculated maximum delay bound for its particular class of traffic. Both the measured maximum delays and average delays within each class were found to have relatively little variation among the various flows in a class. Measured throughputs among the video and data traffic sources showed relatively little variation as well, indicating fair scheduling was being accomplished for these flows. Throughputs among the ten voice sources did vary significantly, however.

Problems

11.1 It was noted in Chapter 8 that the GSM system uses a slotted-Aloha-type random access procedure. In that procedure, uplink RACH messages may be sent using the "0" time slot each frame of selected frequency channels. GSM frames are 4.615 msec long. For the purpose of this problem, assume one frequency channel only is used to carry RACH messages.

(a) What is the maximum number of RACH messages per second that may be sent if the GSM random access procedure is modeled as using slotted-Aloha as described in this chapter? What would a reasonable value for the number of access attempts/sec be if stability is to be ensured?

(b) Say the Erlang load in a given cell is limited to 300 Erlangs, based on call-blocking considerations. Comment on the possibility of a random-access "collision" being experienced by a typical mobile terminal in this cell. In studying this issue, try different values of user call attempts per unit time, as well as user call-holding times. (See Chapter 3 for a discussion of Erlang load.)

11.2 Repeat problem 11.1 for the case of IS-136 (D-AMPS). Recall, from Chapter 8, that this system assigns slots 1 and 4 from frames in a selected set of frequency channels for use by the random-access procedure adopted for the system. Frames are 40 msec long.

11.3 Repeat the slotted-Aloha analysis carried out in the text for the case of pure Aloha, without any slot boundaries.

(a) Show that collisions are due to access attempts over an interval of 2T.

(b) Show the pure-Aloha throughput equation is given by $S = Ge^{-2G}$. Show the maximum possible normalized throughput (fraction of capacity) is $e^{-1}/2 \approx 0.18$. Re-do the example of 11.1 for this case.

11.4 **(a)** Verify that the Gamma distribution of (11.5) is properly normalized and that its average value is nP_0.

(b) Carrying out the integration over w of (11.8) as indicated in the text, show that (11.9) results.

(c) Complete the analysis leading to (11.10). Superimpose a sketch of (11.10) for a capture ratio of 2 on a sketch of the slotted-Aloha throughput equation (11.2) and compare the two throughput results.

11.5 Refer to Problem 11.1. Repeat the calculations if the capture effect is taken into account.

11.6 A number of improved random-access strategies are described in the text. One such strategy uses power-ramping; another method starts with short preambles only; another method proposes repetition of preambles.

(a) Explain why each of these methods might provide improved access.

(b) Explain why the CPCH access procedure might be expected to provide improved access performance.

11.7 Figure 11.4 describes the two-state model of speech.

(a) Show the probability of being in a talk-spurt state is $\alpha/(\alpha + \beta)$, while the probability of being in a silent state is $\beta/(\alpha + \beta)$.

(b) Show the average talk-spurt length is $1/\beta$, while the average silent interval is $1/\alpha$.

(c) Calculate the two probabilities of (a) above for (1) average talk-spurt length of 0.4 sec, average silent interval of 0.6 sec; (2) average talk-spurt length of 1.2 sec, average silent interval of 1.8 sec.

(d) Explain the statement in the text that, with users each having a probability of 0.4 of being in a talk-spurt, "as many as 2.5 users may, on the average, share each time slot in a large system." Why the focus on *large* systems?

(e) Referring back to the discussion of CDMA capacity in Chapter 6, explain how an improvement in capacity of 2.5 is obtained by use of the two-state talk-spurt model.

11.8 **(a)** Prepare a finite-state machine (FSM) diagram of the PRMA procedure for voice access described in the text.

(b) Consider the PRMA system described in the text. Show why, with a frame length of 16 msec and 20 slots per frame, the system might be expected to handle up to 56 PRMA users. Why do the simulation results show this number to be too optimistic?

11.9 Consider the PRMA++ access procedure described in the text.

(a) A given mobile is engaged in a voice call. Compare the procedure using PRMA with that using PRMA++.

(b) Refer to Fig. 11.5. Redraw the full figure, using a common time scale for uplink and downlink transmissions, but allowing for some packet propagation delay in either direction. Show, over time, how a number of mobiles involved in voice communication might use PRMA++ in carrying out their respective transmissions. (Each mobile should be separately identified by number. Use a two-state voice model for each mobile, taking care, of course, that these models are all independent of one another.)

11.10 This problem refers to the two-user example of proportionally fair scheduling, as described in the text. Mobile user 1 experiences poor channel quality, and may transmit at either 76.8 kbps or 153.6 kbps. Assume it makes requests to the base station, on the average, to be scheduled 50% of the time at each transmission rate. User 2 experiences higher channel quality and may, as a result, transmit at double the rate of user 1, 153.6 kbps or 307.2 kbps. Its scheduling requests are also made, on the average, 50% of the time for each rate.

(a) Say the base station invokes round-robin scheduling. Show user 1 is allocated an average of 57.6 kbps, while user 2 is allocated an average of 115.7 kbps. Show the average system throughput is 172.8 kbps.

(b) Repeat the calculation with proportionally fair scheduling used. Show the individual throughputs are now improved considerably, as is the overall system throughput.

11.11 Round-robin scheduling and PF scheduling of data packets are to be compared, using simulation, for a mobile system of at least five mobiles. Say mobiles may transmit at one of three rates, 76.8 kbps, 153.6 kbps, or 307.2 kbps. A packet at this last, highest, rate takes one time slot to be transmitted. These three rates are chosen on the basis of measured SIR. Let the SIR measured by each mobile randomly change every six time slots, taking on one of three equally likely values, such as 10 dB, 15 dB, and 20 dB. The transmission rate requested by a mobile then corresponds to its measured SIR. Design a simulation for this example allowing the use of either round-robin or PF scheduling. Make any other assumptions you need to carry out the simulation. Run the simulation and compare the two algorithms on the basis of (1) average system throughput, (2) average throughput per user.

Chapter 12

Wireless LANs and personal-area networks

Previous chapters in this book have focused on cellular wireless networks. These were originally developed for voice use almost exclusively, as extensions of the wire-based telephone networks worldwide. Second-generation systems have been tremendously successful in this regard, with the number of cellular users expanding at a remarkable rate throughout the world. As noted in Chapter 10, data traffic on these cellular networks has been slower to develop. Higher-capacity third-generation systems discussed in that chapter have been designed specifically to handle packet-switched data. Wireless local-area networks, WLANs, have been developed as well to handle data traffic and have been proliferating worldwide. It is these networks and their even-smaller relatives, personal-area networks, that we discuss in this last chapter.

More specifically, we discuss in Section 12.1 the widely popular IEEE 802.11 and 802.11b wireless LAN standards, running at transmission rates of up to 11 megabits per second (11 Mbps) over the 2.4 GHz unlicensed frequency band. We treat briefly as well the newer very-high bit rate WLAN standards IEEE 802.11g and 802.11a running at rates up to 54 Mbps over the 2.4 GHz and 5 GHz unlicensed bands, respectively. The 802.11g standard is specifically designed to support extension of the 802.11b LAN

IEEE 802.11. IEEE Standard for Information technology-Telecommunications and information exchange between systems-Local and metropolitan area networks-Specific requirements-Part 11: Wireless LAN Medium Access Control (MAC) and Physical Layer (PHY) Specifications, ANSI/IEEE Std 802.11, 1999 Edition, IEEE Standards Board, Piscataway, NJ.

IEEE 802.11b. Supplement to IEEE Standard for Information Technology-Telecommunications and information exchange between systems-Local and metropolitan are networks-Specific requirements-Part 11: Wireless LAN Medium Access Control (MAC) and Physical Layer (PHY) specifications: Higher-Speed Physical Layer Extension in the 2.4 GHz Band, IEEE Std 802.11b-1999, IEEE, January 2000.

IEEE 802.11g. IEEE Standard for Information technology-Telecommunications and information exchange between systems-Local and metropolitan area networks-Specific requirements-Part 11: Wireless LAN Medium Access Control (MAC) and Physical Layer (PHY) specifications; Amendment 4: Further Data Rate Extension in the 2.4 GHz Band, IEEE Std 802.11g-2003, IEEE, New York, NY, 25 June 2003.

IEEE 802.11a. Supplement to IEEE Standard for Information technology-Telecommunications and information exchange between systems-Local and metropolitan area networks-Specific requirements-Part 11: Wireless LAN Medium Access Control (MAC) and Physical Layer (PHY) specifications: High-Speed Physical Layer in the 5 GHz Band, IEEE Std 802.11a-1999, IEEE, New York, NY, 16 September 1999.

standard to much higher bit rates over the same band. In Section 12.2 we then discuss wireless personal-area networks, focusing on the Bluetooth system, standardized as IEEE 802.15.1. The IEEE 802.11b standard has also been dubbed "Wi-FI" and is frequently referred to as the wireless Ethernet.

12.1 IEEE 802.11 wireless LANs

Local-area Networks, or LANs, as they are most frequently called, were first developed in the late 1970s and early 1980s to interconnect data users via wire or cable in locally confined campuses or areas of at most a few kilometers in extent. Optical fibre installations followed once fibre began to become commercially available at low enough prices and costs. The IEEE, recognizing the need to develop standards for this newly arisen mode of data communications, established the IEEE 802 standards committee to provide some standardization to an increasingly large number of *ad hoc* solutions to the problem of communicating over small areas. A whole IEEE 802 family of standards has resulted from work over the years. Standards in this family are confined to the physical and data link layers of the protocol stack described in Chapter 10. The most popular of these standards is 802.3 or Ethernet, which has come to be installed ubiquitously in most offices, businesses, buildings, and campuses throughout the world. This standard covers the medium access control, or MAC sublayer of the data link layer, as well as the physical layer below. The 802.11 wireless subset of 802 LANs to be discussed in this section covers the same two layers. All 802 LAN standards are designed to be interconnected through a common sublayer of the data link layer above the MAC sublayer, 802.2, logical link control. The combination of logical link control and medium access control below it comprises the full data link layer of the LAN standards.

We recall from Section 10.3 of Chapter 10 that the units of data for each layer in a layered architecture are referred to as protocol data units or PDUs for that layer. Thus we had T-PDU referring to the transport layer PDU and N-PDU referring to the network layer PDU. It is the latter that is usually more commonly referred to a *packet*. The data link layer PDU or DL-PDU and the MAC sublayer PDU, MAC-PDU, are most commonly referred to as *frames*. This is the terminology used in the IEEE 802 standards and the one we shall be using throughout this section as well, although we shall occasionally find ourselves using the term packet to denote the block of data. (We noted in Chapter 10 that, despite the attempt to standardize the data unit terms, the word packet is often used informally to represent the PDUs at layers other than the network layer, for which it is the standard definition.) We begin this section with a brief summary of the Ethernet access control procedure and then proceed with a discussion of IEEE 802.11 and 802.11b. We follow that discussion with an introduction to the IEEE 802.11g and 802.11a standards, which use the OFDM technique described in Chapter 5.

Ethernet

Consider, first, therefore, the wired LAN standard, IEEE 802.3 or Ethernet. As we shall see, the access method of the 802.11 wireless standard to be discussed in this section has similarities to the earlier 802.3, although there are substantial differences because of the difference in media over which the two operate. Ethernet utilizes a random access protocol arising out of, and extending, the Aloha-type protocols discussed in Chapter 11. This access protocol appears as part of the MAC sublayer of the 802.3 standard. It is a distributed protocol with users assumed connecting to, and intercommunicating over, the same physical link. (This is a simplification only. We ignore here the use of hubs, bridges to interconnect LANs, and virtual LANs designed to effectively increase the size of the LAN.) The protocol assumes that users are located close enough to individually recognize "collisions" when they occur. Because of this ability to recognize collisions, it turns out that the normalized transmission capacity of Ethernet can exceed the relatively low maximum value of e^{-1} or 0.368 packet per packet slot that we showed characterized slotted-Aloha in Section 11.1. (Note the use of the word packet rather than the more-correct word frame here!) For the Ethernet protocol to attain these higher throughput values a collision detection mechanism labeled CSMA/CD (*carrier sense multiple access/collision detection*) is used. The basic mechanism operates as follows: a data station wishing to send a message "listens" to the link and transmits only if it detects no energy on the link. This is the *carrier sense* portion of the access strategy. It is apparent that "collisions" may still occur, however, since one or more other data terminals using the same access strategy might have started transmitting before the data station in question, or might have begun transmitting afterwards, with those transmissions not having had time to propagate to the location of the station of interest. This data station, and all other Ethernet stations transmitting on to the link continue to monitor the link after transmitting, however, and will detect signals on the link other than the one they transmitted. Once a station detects a collision (*collision detection*), it stops transmitting, transmits a special *jam* signal to notify other users active on the link to stop transmitting, and defers retransmission for a random time interval (as in the slotted-Aloha protocol) to reduce the possibility of users again colliding. If successive retries again result in collisions, the random retry interval is doubled after each collision. This retry procedure is referred to as an *exponential backoff* procedure. Note that, for this random access strategy to be effective, stations must be close enough together to sense another station's transmission before it is completed. This implies that the length of the message transmitted on to the link by the physical layer (the MAC-PDU or Ethernet *frame*) in units of time has to be longer than the end-to-end propagation delay on the link; i.e., the frame has to physically cover the entire link length as it propagates. This length in time depends on the transmission rate: the higher the rate, the shorter the message length. Hence, as Ethernets increase in their transmission bit rate, initially into the tens of megabits per second, now into the gigabits per second range, the link length allowable decreases correspondingly. Depending on the length of the frame transmitted and the length of the Ethernet link, normalized capacities substantially above that of slotted-Aloha may be attained. Further details of the operation of Ethernet/802.3, including a simple performance analysis, may be found in Schwartz (1987).

IEEE 802.11 access mechanisms

Now consider the wireless Ethernet IEEE 802.11 standard, the subject proper of this section. This wireless LAN standard maintains some of the characteristics of the 802.3

Schwartz, M. 1987. *Telecommunication Networks: Protocols, Modeling, and Analysis*, Reading, MA, Addison-Wesley.

Ethernet standard, as noted above, but has some significant differences because of the wireless medium over which it is designed to operate. This medium is clearly much less reliable than wired media. Its propagation properties can be time-varying and asymmetric. The medium possesses no readily observable boundaries; it lacks full connectivity, and user stations may not always be able to hear each other. In addition, stations may be mobile, leading to dynamic topologies. These differences in the medium and user station characteristics between wired and wireless LANs lead to the requirement in the case of 802.11 for a different random access strategy than that of 802.3 or Ethernet. The 802.11 access method we shall be discussing is also defined at the MAC sublayer, and is referred to as the *distributed coordination function* DCF, with the specific access strategy to be used called *carrier sense multiple access with collision avoidance*, CSMA/CA, rather than the CSMA/CD of Ethernet. The concept of DCF means that all stations capable of communicating with one another in a given area use the same access strategy. The critical change in access strategy from that of Ethernet is then one of *avoiding*, rather than *detecting* a collision. The CSMA/CA access protocol also appears as part of the MAC sublayer (IEEE 802.11). All 802.11 standards use the same access mechanism. They differ as to the transmission bit rates possible and as to the physical layer mechanisms defined to implement the various bit rates adopted. The original 802.11 standard is defined to operate at 1 and 2 Mbps; Wi-Fi, the 802.11b standard, operates at bit rates of up to 11 Mbps; the newer 802.11g standard is designed to take the operating bit rate as high as 54 Mbps. All three of these operate over the unlicensed 2.4 GHz band. The 802.11a standard is similar to 802.11g, but operates over the 5 GHz unlicensed band. We discuss the physical layer mechanisms for each of these different systems briefly in the paragraphs following this one on access mechanisms. The very high bit rate systems, 802.11g and a, in particular, use the OFDM technique discussed in Chapter 5 to achieve their very high bit rates.

Before discussing this access protocol, we describe the 802.11 architecture briefly (IEEE 802.11). The IEEE 802.11 wireless LAN standard defines a cell-like structure called a basic service set (BSS) containing at least two stations in communication with one another. Two types of BSS are supported. The first, most basic, type, called an independent BSS (IBSS), involves stations communicating directly with one another, with no administrative pre-planning required as to structure or organization. Stations move in and out of the BSS and establish communication with other stations in the BSS in an independent manner. The second, more common, type of BSS is closer to the concept of a cell as used throughout this book in referring to cellular systems. In this structure multiple BSSs may combine to form an extended network, or may be connected to external wired LANs such as Ethernets. Communication between BSSs or to external wired networks is made through a special station associated with each BSS and called an *access point* or AP. Stations within each BSS communicate with the AP only. The AP thus serves to relay messages to and from external wired networks or other BSSs. In essence, then, we may visualize the BSS to be similar to a cell in a cellular network, with the AP playing a role somewhat similar to a base station. This second type of BSS is particularly appropriate in handling messages between stations that may not be able to "hear" one another. This problem of providing communication between stations not in direct radio contact with one another is often referred to as the "hidden terminal" problem.

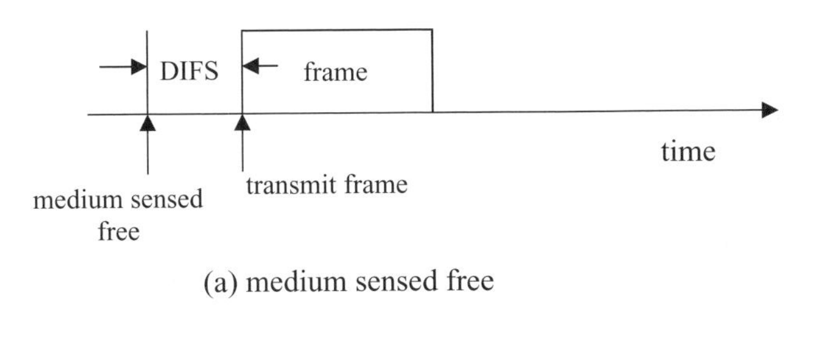

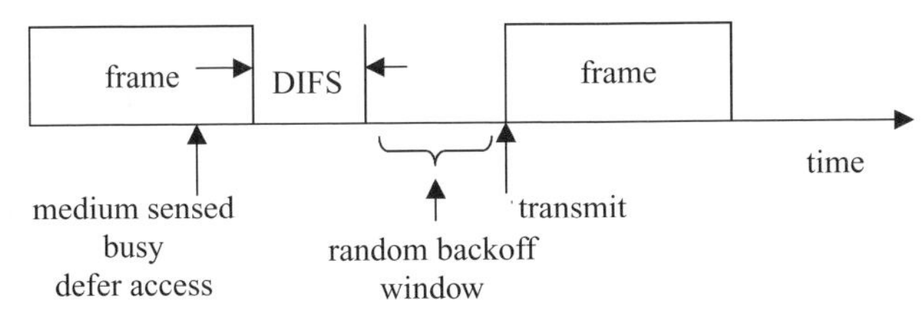

Figure 12.1 Basic access mechanism

With the two forms of wireless LAN architecture supported by IEEE 802.11 described, we are now in a position to return to a more detailed discussion of the MAC-sublayer-based carrier sense–collision avoidance, or CSMA/CA, access strategy utilized by this standard. The use of the terms carrier sense and collision avoidance means that a station wishing to transmit first senses the medium. If the medium is idle, it may transmit, but only after a prescribed minimum time after the last transmission sensed has elapsed to reduce the chance that a message has been transmitted by another station, and has not yet been sensed by the one in question. If the medium is sensed busy, access is again deferred until a specified minimum time after the medium is sensed idle, and then only after a randomly chosen interval has elapsed. This procedure reduces the possibility that two or more stations all wanting to transmit do so at the same time. There are two methods defined for carrying out carrier sensing. The first, basic method, also referred to as a *physical carrier-sense mechanism* (IEEE 802.11), is the one mentioned above of sensing and then waiting a prescribed length of time before transmitting if nothing is sensed on the medium. The second method, referred to as a *virtual carrier-sense mechanism*, involves the exchange of short, reservation-type, frames before attempting to send a data frame. This latter mechanism is particularly significant in exchanging information between stations not able to "hear" one another. (These are the "hidden terminals" mentioned above.) We discuss each of these techniques in more detail beginning with the basic sensing mechanism.

The basic mechanism is portrayed graphically in Fig. 12.1. Two cases are shown there. In part (a) of the figure, a station senses the medium to be free when desiring to transmit a frame; it then waits a specified time DIFS (DCI interframe space) and, sensing no transmission over the medium during that time, begins transmission of its pending frame at the end of that interval. Part (b) of the figure portrays the case in which the medium is

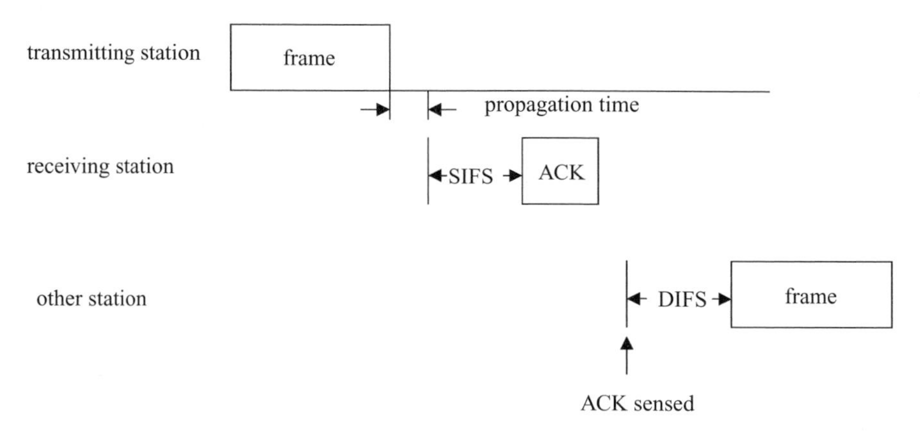

Figure 12.2 Use of SIFS/ACK procedure in basic access

sensed busy. The station desiring to transmit again waits an interval DIFS, then selects a backoff time randomly from a specified contention window length, the collision avoidance part of the CSMA/CA procedure noted above, starts counting down for that backoff time, and transmits at the end of the interval, if no transmission on the medium is sensed during that time. This choice of a random time to wait before attempting to transmit reduces the possibility that more than one station will attempt to transmit at the end of the busy interval, i.e., that a "collision" will take place. Despite this randomization, collisions may still occur due to contention. In this case, contending stations repeat the same collision avoidance procedure, using an exponential backoff procedure similar to the one used in the Ethernet CSMA/CD procedure, as noted earlier: the contention window, from which the random time to transmit is chosen, is doubled each time a collision occurs, until a maximum contention window size is reached. Subsequent retransmission attempts then use that maximum window length from which to choose the random time for a transmission attempt.

Correct reception of each frame transmitted is acknowledged by the recipient station by transmission of an ACK frame. The recipient station transmits the ACK frame after a prescribed time interval SIFS (short interframe space). This procedure is portrayed in Fig. 12.2. The interval SIFS interval accounts for the time for the receiving station to process the frame it has received, up to, and including, the MAC sublayer. This interval plus the time for the ACK frame to propagate throughout the BSS will be less than the DIFS interval, so that other stations wishing to transmit frames are precluded from doing so until after the ACK has been received by the original transmitting station. Even then, stations wishing to transmit must wait an interval DIFS after sensing the ACK as it propagates throughout the BSS. This procedure is also shown in Fig. 12.2. Its use thus provides implicit priority to the transmitting station for use of the medium until it has successfully received the ACK. (We shall see the SIFS interval used in the virtual mechanism as well.) The use of the SIFS and DIFS intervals thus serve to prevent other stations from accessing the medium while transmission and acknowledgement of a frame are in progress.

Timeouts are used by stations transmitting frames to ensure correct reception of data frames: A station transmitting a frame sets a timer and waits to receive the ACK for a

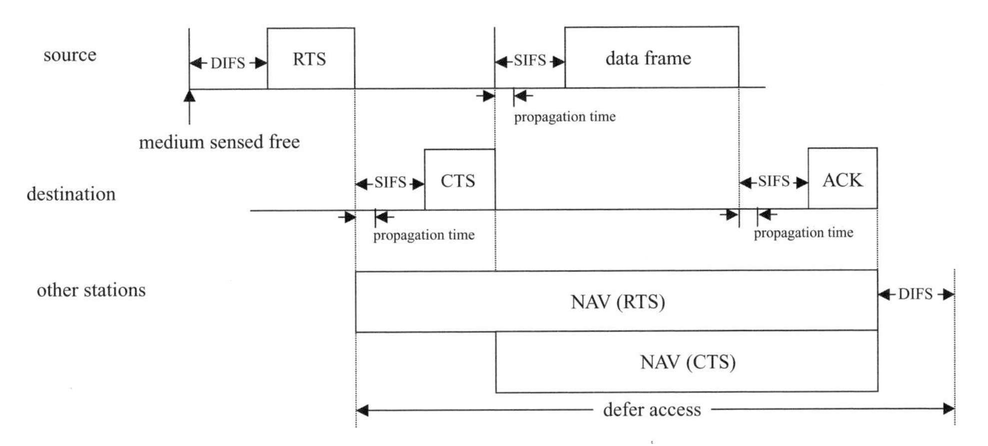

Figure 12.3 Virtual carrier-sense access mechanism

prescribed timeout interval. If the timeout interval expires without the receipt of the ACK, the station invokes the random backoff interval noted above before attempting to retransmit the original frame. The random backoff interval utilized in the 802.11 standard is defined as multiples of a specific unit-interval called a SlotTime, up to a specified maximum number of such units constituting the contention window. This SlotTime effectively measures the time for a station to detect the transmission of a frame from any other station, being made up of the sum of the time required to physically sense the medium and declare the channel "clear" (the so-called "clear channel assessment function"), MAC processing delay, air propagation time, and "receiver/transmitter turnaround time," the time required for the physical layer to change from receiving to transmitting the start of the first bit (IEEE 802.11). The value prescribed by the 802.11 standard for SlotTime is 20 μsec. (As a point of comparison, the value prescribed for SIFS is 10 μsec. The DIFS interval value is prescribed as SIFS plus 2 · SlotTime, or 50 μsec.) The contention window length from which a given station randomly chooses its backoff interval begins with an initial value of eight slot times and is doubled each subsequent retry without success until a maximum window length of 256 slot times is reached. With the backoff time selected, a station begins to count down, in units of SlotTime, to determine when it should attempt transmission. Should a station sense transmission on the channel during a particular SlotTime, while counting down, it suspends the backoff procedure. It again resumes countdown at the same slot number once it again detects a clear channel, but only again after waiting the interval DIFS.

We noted earlier that 802.11 provides two mechanisms for carrying out carrier sensing. The first, the basic, physical carrier-sense, mechanism, has just been described. The second mechanism, referred to as the virtual carrier-sense mechanism, involves the exchange of special short frames between a source and the intended destination, alerting other user stations that packet transmission for a specified time will be taking place. In essence, this is an announcement to other users that the given source wishes to reserve the transmission channel. Collisions will then, hopefully, be avoided during that time. This virtual carrier-sensing procedure is illustrated in Fig. 12.3 (IEEE 802.11). A station wishing to send data

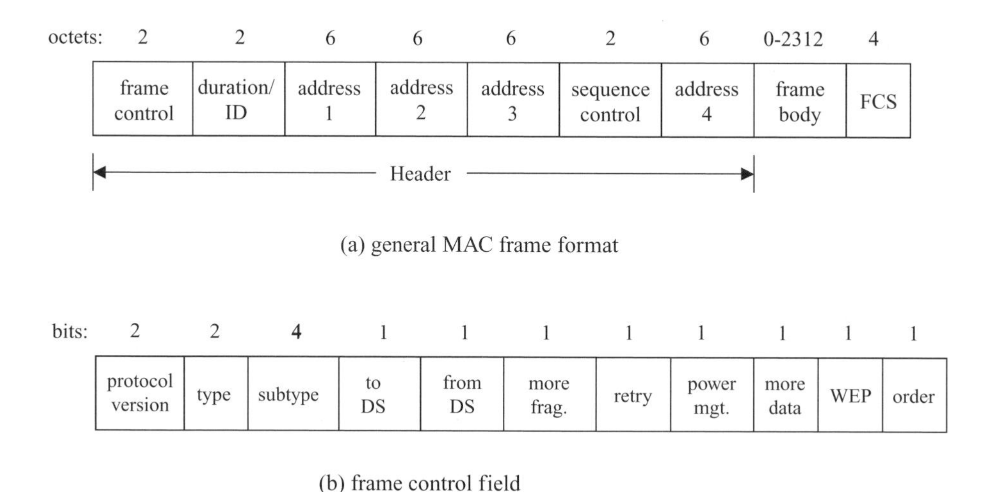

Figure 12.4 MAC frame format, 802.11 (from IEEE 802.11)

to a specific destination on the LAN first sends a special RTS, request-to-send, frame. This frame includes, as we shall see below, the duration, in microseconds, of the data the source station wishes to send. Note that the RTS frame is only sent after the medium has been sensed free for an interval DIFS, the same procedure used in the basic mechanism. The destination then replies, after time SIFS, with a clear-to-send CTS frame. This frame also includes duration information. The source receiving the CTS frame then proceeds to transmit the desired data frame, after an interval SIFS has elapsed. The destination acknowledges correct receipt of the data frame after the interval SIFS. There are thus two exchanges of frames between a source and the intended destination, as compared with the one exchange in the basic access case. Now note that other stations listening on the medium may sense the RTS frame as well. Each such station uses the time duration information carried by the RTS to update an internal indicator called the network allocation vector, NAV, which indicates the time period over which the station should not initiate any transmission, even if the medium is sensed free. The update is made only if the new NAV value is greater than its current value. (Earlier exchanges may have previously "reserved" the channel as well.) On sensing the CTS frame sent by the destination as a reply to the source station, each station may update its NAV with the time duration information carried by the CTS frame, again only if the duration is greater than that already carried by its NAV. Note that not all stations may have sensed the RTS frame. Sensing the CTS then provides them with a further means to update the NAV. The lower NAV(CTS) indication in Fig. 12.3 refers to this case. All stations other than the source-destination stations now defer possible access until a time DIFS after the acknowledgement frame from the destination has been received, as shown in Fig. 12.3.

Note that the two access procedures described require the sending or exchange of a number of MAC frame types. The formats of these various frames, including the data frame, are now described. Note, first, that all 802.11 MAC frames obey the general frame format shown in Fig. 12.4(a) (IEEE 802.11). Consider the 2-octet, 16-bit frame control field first. Its field breakdown appears in Fig. 12.4(b). The Protocol Version field indicated simply identifies the current version of the standard. As of this writing, this field is set at 0. The Type field identifies which of three frame types, control, data, and management, a particular frame represents. Within each of these three categories the particular frame in question is identified by the 4-bit Subtype field. Management frames, not to be discussed in this brief section on 802.11, are identified by the Type field 00, control frames by 01, and data frames by 10. Among the various control frame types mentioned in our discussion of the two access mechanisms above, the RTS frame has the subtype value 1011, the CTS frame the value 1100, and the ACK frame the value 1101. Consider now the two 1-bit DS fields in the frame control field of Fig. 12.4(b). DS stands for *distribution system*, an 802.11 architectural concept designed to provide flexibility to 802.11 LANs. The distribution system concept allows multiple BSSs (basic service sets) to be interconnected. This is particularly useful where the normal allowable physical distance between stations is insufficient to prove the desired network coverage (IEEE 802.11). The 1-bit To DS field in the frame control field is set to 1 in data frames destined for the DS; it is set to 0 in all other frames. Similarly, the From DS field is set to 1 in all data frames exiting the DS, and to 0 in all other frames. Continuing our discussion of the Frame Control field of Fig. 12.4(b), the More Fragments field is set to 1 in cases where a data or management frame has been fragmented into multiple frames, indicating another fragment of a current frame will follow. The Retry field is used to designate a retransmission of a previous data or management frame; the More Data field is used to save station power, indicating to a station that more frames destined for that station are buffered at the access point (AP); the WEP and Order fields are beyond the scope of this brief introduction to 802.11, being used to designate the use of a specialized algorithm and service class, respectively.

Now return to the general frame format of Fig. 12.4(a). The various frames we shall be discussing use different, specialized versions of this general format. All frames carry the end-of-frame 4-octet frame check sequence, FCS, however. This 32-bit cyclic-polynomial code is used to check for frame-transmission errors, and is calculated over all the preceding fields of a frame, including the frame body, if it is present. (See Chapter 7 for a discussion of cyclic polynomial codes. The specific 32-degree generator polynomial used in calculating the FCS field of 802.11 frames and a description of how the FCS calculation is carried out appear in IEEE 802.11.) Finally, note that there are four address fields shown. Not all frames use all of these, however.

Consider data frames as the first example. Multiple data frame types are defined for 802.11. To simplify the discussion here, we take the simplest, most common case only, with stations confined to a given basic service set (BSS) in a given service area only, sending data to one another. For this case, from the brief discussion earlier, the To DS and From DS fields in the Frame Control field, Fig. 12.4(b), are both set to 0. (Other types of data-frame types are discussed in IEEE 802.11.) The Sub-type field of Fig. 12.4(b) is set to 0000. For these data frames, three 6-octet (48 bit) addresses only are used. (The fourth address field indicated in Fig. 12.4(a) does not appear.) These consist, respectively, of the destination station MAC address, the source station MAC address, and the id of the basic service set, i.e., the id of the cell in which this communication is taking place. The "frame body" carries the information or data portion of the data frame, the MAC service data unit or MSDU, or a portion thereof. The maximum length of this field is 2312 octets, as

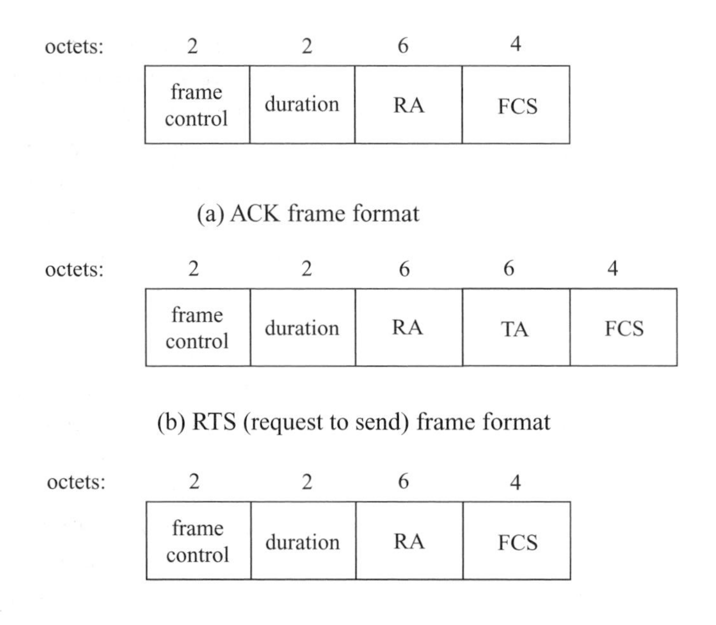

Figure 12.5 Frame formats for basic and virtual access (from IEEE 802.11)

indicated in Fig. 12.4(a). The MSDU contains the MAC-layer data sent from the source station to the destination. The 16-bit Sequence Control field shown in Fig. 12.4(a) has two parts for a data frame: a 12-bit Sequence Number field, followed by a 4-bit Fragment Number field. Each MSDU transmitted by a given station carries a sequence number, starting at 0 and increasing by 1 until a maximum of 4095 is reached. It then recycles back to 0. The Fragment Number is set to 0 in the first or only fragment of an MSDU, and, in the case of an MSDU transmitted as multiple fragments, is incremented by 1 for each successive fragment. (Note again, from Fig. 12.4(b), that the More Fragments bit in the Frame Control field indicates whether additional MSDU fragments are to be transmitted.) Coming now to the Duration field of Fig. 12.4(a) for the data frame, the setting of this field for the data-frame type we are considering here depends on the setting of the More Fragments bit. If that bit is set to 0, the duration value is the time in microseconds to transmit one ACK frame, plus one SIFS interval. If the bit is set to 1, meaning more fragments are coming, the duration interval, in microseconds, is given by the time to transmit the next fragment of this data frame, plus two ACK frames, plus three SIFS intervals.

We continue now with the other frame types described in our discussion of the basic and virtual access mechanisms. These were the ACK, RTS, and CTS frames shown being sent in Figs. 12.2 and 12.3. They are all in the category of control frames, as noted above. Their frame formats, all special cases of the general frame format of Fig. 12.4(a), appear in Fig. 12.5 (IEEE 802.11). Consider the ACK frame format of Fig. 12.5(a) first. This carries one address only, indicated as RA (receiver address). Recall from the basic access data frame acknowledgement procedure described pictorially in Fig. 12.2 that the ACK is always returned by the destination station to the source station on reception of a valid frame. The RA field thus carries the source station address, and is copied from the second address field, the source address, in the received data frame. If that data frame carries a 0 More Fragment bit in its Frame Control field, the ACK frame Duration field is set to 0. If the data frame More Fragment bit is set to 1, the ACK Duration field is set to the value contained in the data frame Duration field less the time, in microseconds, required to transmit the ACK, plus an SIFS interval (IEEE 802.11). The other two control frames whose formats appear in Fig. 12.5 are used in carrying out the virtual access procedure described above (Fig. 12.3). In particular, consider first the RTS frame format shown in Fig. 12.5(b). Note that this frame carries two addresses. The first, RA, address indicates the address of the destination station to which both this frame and the waiting data frame are to be sent (see Fig. 12.3). The second, TA, address is the address of the source station sending the RTS frame. The Duration field value carries the information to be used, as noted earlier, by other stations sensing the propagating RTS frame to update their NAV vector, as part of the virtual access mechanism. This information consists of the time, in microseconds, required to transmit the data frame waiting at the source station, plus the time to transmit one CTS frame, one ACK, and three STS time intervals, just the total time required to transmit a data frame and receive an acknowledgement. Note from Fig. 12.3 that this is just the NAV(RTS) time indicated there. The CTS frame format of Fig. 12.5(c) is similar to that of the RTS frame, except that one address only is used. That address, indicated as RA in Fig. 12.5(c), is copied from the TA field of the preceding RTS frame to which this represents a reply. The duration value is also that carried in the Duration field of the RTS frame, less the time to transmit the CTS frame and an SIFS interval (see Fig. 12.3).

Physical layer specifications, 802.11b and 802.11

The paragraph above has focused on the IEEE 802.11 access mechanisms. In this paragraph we provide a brief description of the 802.11 physical-layer specifications for the original 1 and 2 Mbps 802.11 standard, as well as the widely-used 802.11b, Wi-Fi, standard that provides 5.5 and 11 Mbps data transmission capability. Both use direct-sequence spread spectrum (DSSS) coding techniques followed by some form of PSK to obtain the desired processing gain for transmission over the wireless channel. Both use the unlicensed frequency band at 2.4 GHz. In the next paragraph we provide a brief description of the newer 802.11 standards that provide up to 54 Mbps transmission capability. All these versions of 802.11 use essentially the same access procedure, the one just described in the previous paragraph. Details of the 802.11b standard appear in IEEE 802.11b. The original lower bit-rate 802.11 standard is described in IEEE 802.11.

The physical layer maps the MAC-PDUs, MAC sublayer protocol data units, or MAC frames, described in the preceding paragraph into the format necessary for sending and receiving user information over the wireless medium. In all versions of 802.11, the physical layer appends a 144-bit physical-layer preamble followed by a 48-bit physical-layer header to the MAC-PDU being transmitted. The resultant protocol data unit is called the PPDU, or physical layer-PDU (IEEE 802.11; IEEE 802.11b). There are two different sets of physical-layer preambles and headers defined for 802.11b, one mandatory, the other optional. In the mandatory case, the preamble-header combination is transmitted in 192 μsec using 1 Mbps differential PSK, DPSK, modulation. The data portion of the PPDU is then sent at the rate selected, 11 or 5.5 Mbps. The resultant PPDU in the 802.11b system is referred to as having the *long* PPDU format (IEEE 802.11b). A *short*, optional, PPDU format for the 802.11b system has the preamble-header combination transmitted in 96 μsec (IEEE 802.11b). This, as we shall see later, improves the performance of the 802.11b system.

The original 1 or 2 Mbps version of 802.11, designed, as noted above, to operate over the 2.4 GHz unlicensed transmission band, uses the 11-chip Barker code sequence, $+1, -1, +1, +1, -1, +1, +1, +1, -1, -1, -1$ to spread each bit in the MAC frames being transmitted. The 1 Mbps system uses DPSK for the transmission, while the 2 Mbps system uses DQPSK, differential QPSK (IEEE802.11). The higher-speed IEEE 802.11b system operates over the same 2.4 GHz band as noted above, but uses Complementary Code Keying, CCK, followed by DQPSK modulation, as the overall modulation scheme in place of the Barker code of 802.11 to attain the higher data bit rates of 5.5 and 11 Mbps (IEEE 802.11b). In this scheme eight bits at a time are operated on to attain 8-bit codewords. Call the successive data bits in each group of eight $d_0, d_1, d_2, \ldots, d_7$. We focus specifically on the 11 Mbps system in the material that follows. In this system, pairs of successive bits, (d_i, d_{i+1}), are grouped together and are each encoded into one of four phase angles $0, \pi, \pi/2$, and $-\pi/2$. The four possible values of the bit pairs and the phases they encode are $00 \rightarrow 0$, $01 \rightarrow \pi/2$, $10 \rightarrow \pi$, and $11 \rightarrow -\pi/2$. Now let φ_1 represent the resultant phase angle of the pair (d_0, d_1), let φ_2 be the resultant phase of the pair (d_2, d_3), φ_3 be the phase determined by the pair (d_4, d_5), and φ_4 be the phase given by the pair (d_6, d_7). These four different phases are then used to derive the following 8-bit CCK codeword $\boldsymbol{c}$, written in complex-vector form, which serves as an 11-chips/sec spreading code (IEEE 802.11b)

$$\begin{aligned}\boldsymbol{c} = \big[&e^{j(\varphi_1+\varphi_2+\varphi_3+\varphi_4)}, e^{j(\varphi_1+\varphi_3+\varphi_4)}, e^{j(\varphi_1+\varphi_2+\varphi_4)},\\ &-e^{j(\varphi_1+\varphi_4)}, e^{j(\varphi_1+\varphi_2+\varphi_3)}, e^{j(\varphi_1+\varphi_3)}, -e^{j(\varphi_1+\varphi_2)}, e^{j\varphi_1}\big]\end{aligned} \tag{12.1}$$

The 8-bit codewords are clocked at a 1.375 MHz rate, with serial-to-parallel conversion used to convert the incoming 11 Mbps data stream to the 8-bit codewords. The resultant signal occupies the same bandwidth as the 2 Mbps 802.11 Barker code-based system. As an example, say a group of eight bits is given by 00101011. We then have $\varphi_1 = 0$, $\varphi_2 = \pi$, $\varphi_3 = \pi$, and $\varphi_4 = -\pi/2$. The corresponding codeword $\boldsymbol{c}$ is given by $[-j, j, j, j, 1, -1, 1, 1]$.

The manner in which this complex-vector codeword is used to provide the desired spread signal for the 11 Mbps system is defined in IEEE 802.11b. The 5.5 Mbps system also uses (12.1) to derive the 8-bit codewords. Four data-bit symbols are used, however, and the DQPSK encoding as well as the definition of the four phase angles differ from that of the 11 Mbps system (IEEE 802.11b). A tutorial discussing CCK codes and their application to the IEEE 802.11b system appears in Pearson (2001). Included are block diagrams of possible modulator and receiver implementations. The bit and packet error rate performance as a function of the received signal-to-noise ratio

---

Pearson, R. 2001. "Complementary code keying made simple," Application Note AN9850.2, Intersil Americas Inc., November.

of both the 5.5 Mbps and the 11 Mbps systems is discussed by Heegard *et al.* (2001). An additive white gaussian noise environment is assumed for the comparative study. Comparisons are made between these systems and the lower bit rate Barker code-based systems, as well as with a higher performance mode for both 5.5 Mbps and 11 Mbps, proposed as an option in the 802.11b standard. This optional mode is termed *packet binary convolutional coding* or PBCC (IEEE 802.11b; Heegard *et al.*, 2001). The performance curves in Heegard *et al.* indicate that the 5.5 Mbps CCK scheme requires 2 dB more in signal-to-noise ratio than the 2 Mbps Barker code system. The 11 Mbps CCK scheme requires an additional 3 dB in signal power as compared with the 5.5 Mbps scheme.

Physical layer specifications, 802.11g and 802.11a

As noted earlier, the two newest additions to the 802.11 WLAN family are the 802.11g and the 802.11a standards. Both use orthogonal frequency-division multiplexing (OFDM) technology to provide up to 54 Mbps transmission capability. The 802.11g scheme was standardized in 2003 and operates over the same 2.4 GHz unlicensed band as do the original 802.11 scheme and the 802.11b, Wi-Fi, scheme just described. It is, in fact. designed to be compatible with the 802.11b scheme. The 802.11a scheme operates over a different unlicensed band, one at 5 GHz. Both schemes use basically the same MAC protocol as do the earlier developed schemes. Details of these two standards appear in IEEE 802.11g and IEEE 802.11a, respectively. Note that all the 802.11 standards, the original 802.11 standard and the ones following on 802.11b, g, and a, are tied together, newer standards making reference to, or modifying, specific sections and paragraphs appearing in the earlier ones.

The physical layer specifications for these two higher bit rate standards are essentially the same, the prime difference being the frequency band over which they are designed to operate. The later 802.11g standard, therefore, draws on the techniques prescribed in the earlier 802.11a standard. Operation is prescribed to take place within 20-MHz wide bands within the overall frequency band used. Subcarriers spaced 312.5 kHz apart are defined as providing the desired OFDM operation, with the equivalent inverse Fourier Transform technique discussed in Chapter 5 actually being used to carry out the orthogonal frequency-division multiplexing. Fast Fourier Transform techniques are used for this purpose. The carrier spacing of 312.5 kHz over the 20-MHz band allows 64 subcarriers to be transmitted simultaneously. In practice, a bandwidth of 16.6 MHz within the center of the 20-MHz band is actually used to define the OFDM operation. Fifty-two subcarriers evenly spaced from each side of the band in to the band center, as shown in Fig. 12.6, are used to cover the 16.6 MHz band. The numbers shown in the figure represent the carrier designations counting up (+) and down (−) from the center of the band, designated as 0. We note that the center of the band covering 625 kHz is left free. Of these 52 carriers, 48 are used to carry data; four are used to carry pilot information (IEEE 802.11a). As we shall see below, OFDM symbols are defined as 48 complex numbers, each representing one of the data subcarriers. The OFDM data rates supported by these standards are 6, 9, 12, 18, 24, 36, 48,

---

Heegard, C. *et al.* 2001. "High-performance wireless internet," *IEEE Communications Magazine*, 39, 11 (November), 64–73.

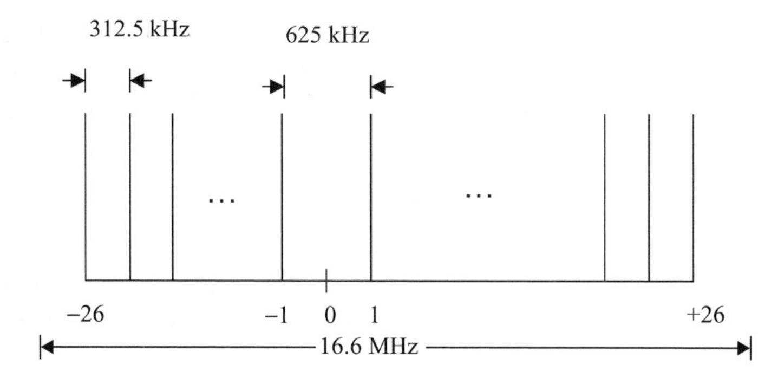

Figure 12.6 802.11a and g OFDM carrier assignments

and 54 Mbps. In addition, the 802.11g standard supports the 1, 2, 5.5, and 11 Mbps data rates of the 802.11 and 802.11b (Wi-Fi) standards described in the previous paragraph, using the DSSS and CCK techniques discussed there. It also allows two optional data rates of 22 and 33 Mbps to be used (IEEE 802.11g). For simplicity of presentation, we focus here on the eight data rates from 6 to 54 Mbps supported by both of the two higher bit rate standards.

These eight data rates are obtained by using various combinations of QAM with OFDM, exactly the procedure described in our discussion of OFDM earlier, in Chapter 5. In particular, the 6 and 9 Mbps rates use PSK; the 12 and 18 Mbps data rates use QPSK; the 24 and 36 Mbps rates use 16-QAM; and the highest rates, 48 and 54 Mbps, both use 64-QAM. Details appear below.

Consider now the OFDM operation. Recall from the OFDM discussion of Chapter 5 that, for orthogonal-frequency operation, the sub-carrier spacing and bandwidth must be $1/T_S$, T_S being the OFDM symbol interval. With a sub-carrier spacing here of 312.5 kHz, we have $T_S = 3.2$ µsec. From our discussion of frequency-selective fading in Chapter 2 (see equation (2.48a)), we note that this value of the symbol interval suggests that any fading encountered during data transmission over a local-area network would be non-frequency-selective; i.e., choosing these OFDM bandwidths should result in flat fading for a local-area network. (Why is this so?) This demonstrates why OFDM was chosen for use with these high bit rate systems; for note that a 64-QAM system operating at a 54-Mbps data rate transmits at a rate of 9 Msymbols/sec. Without OFDM, frequency-selective fading would be encountered for delay spreads of 0.03 µsec or more. The use of OFDM increases that number substantially. The calculation of the inverse Fast Fourier Transform technique (IFFT) used to implement the OFDM processing is carried out every 4 µsec, a guard interval in time of 0.8 µsec being added to the 3.2 µsec OFDM symbol interval. The total interval of 4 µsec is referred to in the standard as the symbol interval, the 3.2 µsec interval as the IFFT/FFT interval (IEEE 802.11a).

Figure 12.7 portrays a simplified version of the 802.11a and g OFDM transmitter using the IFFT calculation (IEEE 802.11a). The input bit rate R is one of the eight bit rates noted above. Forward-error correction using a $K = 7$, rate-$1/2$ convolutional encoder is then carried out. The output bit stream of this encoder is then punctured (some bits omitted, the number depending on the input bit rate) to attain the desired output bit rate.

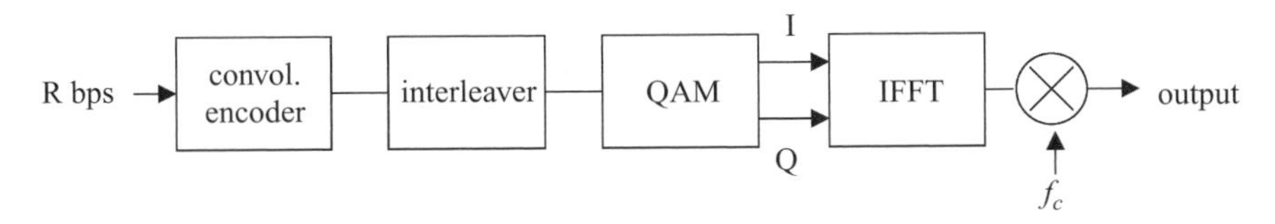

Figure 12.7 Simplified OFDM transmitter, IEEE 801.11a and g

(Examples of puncturing will be discussed below.) An interleaver is then used to spread out the punctured output stream, and the resultant bit stream stored the appropriate number of bits to provide a QAM symbol. Each QAM-symbol interval the QAM-constellation complex number corresponding to the set of input bits received in the QAM interval is stored until 48 such complex numbers, represented by a sequence of I (inphase) and Q (quadrature) numbers in a 3.2 μsec interval, are accumulated for input to the IFFT calculator. These complex numbers are represented by the symbols I and Q in Fig. 12.7. These 48 numbers, corresponding to the data to be transmitted, are augmented by four numbers corresponding to the four pilot signals. A 64-point IFFT is used, with 16 of the IFFT input coefficients set equal to zero. The IFFT output is then used to modulate the system carrier as shown.

Two examples suffice to describe the operation of this transmitter:

Example 1

Consider the 6 Mbps case first. At this bit rate 24 bits are accumulated every 4 μsec. The rate-1/2 encoder produces 48 bits in this same interval. There is no puncturing in this case. With PSK modulation used, 48 binary numbers, either +1 or −1, are inputted to the IFFT calculator every 4 μsec.

Example 2

Now consider the maximum bit rate case of 54 Mbps. Here 216 bits appear at the encoder input every 4 μsec. For this bit rate, one-third of the encoder output bits are punctured or dropped, with 288 bits appearing at the output every 4 μsec. The effective encoder rate is thus $\frac{3}{4}$. 64-QAM is used in this case, as noted earlier, with the interleaved bits, taken six at a time, used to determine which one of 64 possible complex numbers is to be stored for IFFT processing. Note that 48 such complex numbers (288/6) are stored every 4 μsec.

Other input bit rates are processed in a similar manner, all producing the required 48 complex numbers to be inputted to the IFFT calculator every 4 μsec. It is left to the reader to show that the QAM values and effective encoder rates shown in Table 12.1 all produce the 48 complex numbers required to carry out the IFFT calculation every 4 μsec.

Throughput performance analysis

We have described the physical-layer protocols for the various 802.11 implementations in the preceding two paragraphs. We also noted that the MAC-layer 802.11 CSMA/CA access procedures described earlier in this section are essentially the same for all the 802.11 schemes. In this paragraph we return to the access procedures and describe a throughput performance analysis applicable to all the implementations. We provide numerical examples of the analysis for the original 1 Mbps 802.11 scheme and the 802.11b, Wi-Fi, scheme only, however.

Table 12.1 *Data rate parameters, IEEE 802.11a/g*

Data rate (Mbps)	QAM type	Encoder rate
6	PSK	1/2
9	PSK	3/4
12	QPSK	1/2
18	QPSK	3/4
24	16-QAM	1/2
36	16-QAM	3/4
48	64-QAM	2/3
54	64-QAM	3/4

Recall that there are two mechanisms used to carry out the MAC-layer CSMA/CA procedure. These are the basic access and virtual access techniques summarized in Figs. 12.1 and 12.3 respectively. The throughput performance of each technique is measured by how effectively it utilizes the wireless channel capacity. To answer this question we turn to analytical work, backed by simulation, carried out by Bianchi (2000). He has determined the maximum channel utilization obtainable for both access mechanisms as a function of the number of users, for various maximum contention window values. Interestingly, he models the access attempts using a p-persistent strategy similar to the strategies adopted for the various PRMA-type schemes described in the previous chapter. Related work in determining the throughput performance of 802.11 systems appears in Ho and Chen (1996) and Cali *et al.* (1998), as well as other papers cited in all of these references.

The channel utilization or normalized throughput obtainable using the two CSMA/CA access procedures is defined as the fraction of time the channel is used to successfully transmit the information portion of data frames, assuming each user station always has data packets ready to be transmitted. The normalized throughput to be determined is then referred to as the *saturation throughput*, the maximum possible system throughput under stable conditions (Bianchi, 2000). To determine this performance measure, begin at some randomly chosen slot time and consider a long interval of time following, long enough to cover many data frames being transmitted, as well as collisions incurred in attempting to transmit. The length of time in this long interval used to actually transmit information divided by the full interval, i.e., the fraction of time in which information is transmitted, is then the desired channel utilization. Equivalently, the average length of time following a randomly chosen slot time used to transmit user information divided by the average length of time required to transmit that information provides the desired channel utilization. This is the approach, following Bianchi, to be used here (Bianchi, 2000). Denote the information

Bianchi, G. 2000. "Performance analysis of the IEEE 802.11 distributed coordination function," *IEEE Journal on Selected Areas in Communications*, 18, 3 (March), 535–547.

Ho, T.-S. and K.-C. Chen. 1996. "Performance analysis of 802.11 CSMA/CA medium access control protocol," IEEE PRMC, Taipai, Taiwan, October, 407–411.

Cali, F. *et al.* 1998. "IEEE 802.11 wireless LAN: capacity analysis and protocol enhancement," IEEE Infocom'98, San Francisco, CA, March.

portion of a data frame by P and the header portion, including error-correction bits, by H, both measured in units of time. The average information length of a data frame is then denoted $E(P)$. (In the example calculations following, we shall always use a constant value for P to simplify the discussion.) Let T_S be the average length of time the channel is sensed busy while a successful transmission is underway. Let T_C be the average length of time the channel is sensed busy during a collision. Both T_S and T_C depend on the particular access technique used, and their evaluation for each of the two access techniques used in 802.11 will be described below. Using these definitions of times involved in transmitting data frames, the saturation channel utilization S may then be written, following Bianchi (2000)

$$S = \frac{P_{tr} P_s E(P)}{(1 - P_{tr})\sigma + P_{tr} P_s T_S + P_{tr}(1 - P_s)T_C} \tag{12.2}$$

Here σ is the length of a slot, P_{tr} is the probability that there is at least one transmission attempt at the randomly chosen slot time, and P_s is the probability of a successful transmission, i.e., the probability that exactly one station transmits, given at least one transmission takes place. Note, therefore, that $P_{tr}P_s$ is just the probability that one station only transmits at the slot interval chosen. The numerator thus clearly represents the desired average length of time required to transmit the information portion of a data frame, at a randomly chosen slot time. The denominator term represents the three different time intervals that are, on the average, required to successfully transmit the desired information. The first term is the case of an empty time slot; the second provides the average time required to successfully transmit a data packet in the case of no collisions; the third term represents the average time to successfully transmit a frame in the face of collisions on the channel.

It now remains to calculate each of the quantities appearing in (12.2). Consider the probability P_{tr} first. Say there are n user stations in the system. Let p_d be the probability that any one of these stations transmits a packet in a randomly chosen slot interval. This is the quantity analogous to the p-persistent probability described in Chapter 11. The probability that none of the stations transmits is then clearly $(1- p_d)^n$. We thus have P_{tr}, the probability there is at least one transmission attempt at a randomly chosen slot interval, immediately written as

$$P_{tr} = 1 - (1 - p_d)^n \tag{12.3}$$

To find P_s, the probability of a successful transmission, we use the expression noted above that $P_{tr}P_s$ is the probability that one station only transmits in a given slot interval. This is just $np_d(1-p_d)^{n-1}$, i.e., the probability that one transmits and $(n - 1)$ do not. We thus have

$$P_{tr} P_s = np_d(1 - p_d)^{n-1} \tag{12.4}$$

from which P_s may be readily found, using (12.3).

The main problem now is to determine p_d, the probability a station transmits in a given, randomly chosen slot interval. To do this, one must introduce the concept of back-off time before attempting to transmit, the procedure critical to the CSMA/CA technique being modeled here. Recall from our discussion of the CSMA/CA procedure in an earlier paragraph of this section on 802.11 that a station sensing the medium busy defers access for an interval DIFS, and then transmits at a slot interval chosen randomly from a contention window w slots long. Figure 12.1(b) provides an example for the basic access mechanism. Recall also that, with exponential backoff, the window size is doubled after each collision until a maximum window size is reached. We thus have $w = 2^i\ W, 0 \le i \le m$, with m the maximum number of backoff attempts, beyond which the window length remains constant. The parameter W is the initial length of the window in units of slot times. Bianchi now assumes that there is a fixed probability p a data frame will incur a collision on a transmission attempt, whether on the first attempt, or on any subsequent retransmission attempts. (This assumption can only be justified on comparing the results of this throughput analysis with simulation results.) If this collision probability is independent of transmission attempt, it must satisfy the following equation, relating it to the probability p_d of packet transmission

$$p = 1 - (1 - p_d)^{n-1} \tag{12.5}$$

(This equation simply says that at least one of the remaining $n - 1$ stations must be transmitting for a collision to occur.) This puts the burden of finding the desired probability p_d on to finding p. However, Bianchi then shows, using a Markov chain analysis we shall not reproduce here, that the two probabilities are further related through a second equation which we do reproduce in (12.6) following

$$p_d = \frac{2(1 - 2p)}{(1 - 2p)(W + 1) + pW(1 - (2p)^m)} \tag{12.6}$$

The two nonlinear equations (12.5) and (12.6) may be jointly solved for the two desired quantities p and p_d. Note, specifically, that, if exponential backoff is not used, so that $m = 0$, we have, from (12.6), $p_d = 2/(W + 1)$. This is also the value of p_d for zero or very small p. It is also readily shown that p_d decreases monotonically as p increases. Thus we have $p_d\ (0) = 2/(W + 1) > p_d(p) > p_d(1) = 2/[W + 1 + W\ (2^m - 1)]$. For example, if the initial window size $W > 10$ slot times, say, p_d is always less than 0.2. Using (12.5) and (12.6) to determine p_d, and from this calculating P_{tr} and P_s from (12.3) and (12.4), respectively, one may then calculate the channel utilization S for either of the two access mechanisms we have described by introducing the corresponding values of T_S and T_C in (12.2). We shall, in fact, shortly carry out such a calculation. But, first, it is to be noted that the channel utilization lends itself to a maximization by an appropriate choice of p_d (Bianchi, 2000).

In particular, an examination of (12.2) shows that S is maximized by maximizing $P_s/[(T_C/\sigma) + (1 - P_{tr}/P_{tr})]$. Details are left for the reader. Defining $T_C^* \equiv T_C/\sigma$, analysis then indicates (Bianchi, 2000) that for both $n \gg 1$ and $T_C^* \gg 1$, the optimum choice of p_d to maximize S is given approximately by

$$p_{d,opt} \approx 1/n\sqrt{T_C^*/2} \tag{12.7}$$

The optimum choice of p_d thus depends only on the average length of time the channel is sensed busy during a collision interval, in units of slot time, as well as the number of user stations attempting to transmit over the channel. As an example, say $T_C = 1007$ μsec, the value we shall show below is the one appropriate to 1023-octet-long MAC PDUs in an 802.11b 11 Mbps wireless LAN, using the basic access scheme. The slot interval defined is 20 μsec, as noted earlier. We then have, for the optimum choice of p_d, $p_{d,opt} = 1/5n$.

To carry out the analysis, it now remains to determine the two times T_C and T_S, the average time, respectively, that the channel is sensed busy during a collision and the time the channel is sensed busy during a successful transmission. These two times may be found for the two cases of basic access and virtual access by consideration of Figs. 12.1 to 12.3 discussed earlier. In particular, it is left for the reader to show that, with the basic access mechanism and a fixed-length data packet information field of value P, one has

$$\textit{Basic access:} \quad T_C = P + H + \text{DIFS} + \delta \tag{12.8}$$

and

$$\textit{Basic access:} \quad T_S = P + H + \text{SIFS} + \text{ACK} + \text{DIFS} + 2\delta \tag{12.9}$$

All quantities are given in units of time. The term H is the total data-frame header length, the sum of the time required to transmit both the physical-layer signals (192 μsec in the long-format mode) and the MAC-layer header of 28 octets, including the FCS field. The term δ is the propagation time, given as 1 μsec in the IEEE 802.11 standards. We have already indicated earlier that the value of SIFS, as defined by the standard, is 10 μsec, while DIFS is 50 μsec (IEEE 802.11b). The ACK frame format appeared in Fig. 12.5a as well. (But note that the physical preamble/header combination, in time, must be added to the frame length, in calculating the time to transmit the ACK frame.) One can therefore readily calculate T_C and T_S for the basic access mode, using (12.8) and (12.9). As an example, one readily finds T_S to be 1221 μsec and T_C to be 1007 μsec, the figure quoted above, for the 11 Mbps system and 1023-octet-long MAC-PDUs. This assumes the long preamble and header mode of operation. If we now use the standard-assigned value of 20 μsec for the slot time σ (IEEE 802.11), we find $T_C^* = 50.3$ and $p_{d,opt} = 1/5n$, using (12.7). Recall that the assumption made there was that $T_C \gg 1$, clearly the case here, and that $n \gg 1$. So this value of $p_{d,opt}$ should be valid for $n \geq 10$.

Consider specifically the case of $n = 10$ user stations. Then $p_d = 0.02$, $P_{tr} = 0.183$, and $P_s = 0.895$. From (12.2), one finds the maximum system capacity or normalized throughput is $S = 0.49$. Details are left to the reader. (Note, however, that these calculations assume a MAC-PDU frame size of 1023 octets. Results will be somewhat different for other size data frames.) This says that the MAC-PDU frames can be delivered at an average rate of 0.49*11 Mbps = 5.4 Mbps, if data frames are always made available at each of ten user stations. This is quite a significant reduction in the 11 Mbps potential throughput, due principally to the relatively high overhead introduced with the use of the 192 μsec-long preamble/header combination. (Note that the 192 μsec physical header appears twice in the calculation of T_S, preceding both the information-carrying MAC-PDU and the ACK frame. Even if one were to neglect the average time T_C spent in resolving collisions, the impact of the time required to transmit the physical header would result in a normalized capacity of $P/T_S = 0.61$, still a relatively low value.) Much higher normalized throughputs are obtained for delivery of the same size MAC-PDU and the same number of user stations, using the 1 Mbps 802.11 system (Bianchi, 2000). But note that delivery of the data information, the MAC-PDUs, in that case is at the 1 Mbps rate, rather than the 11 Mbps rate assumed here. The critical quantity P, the MAC-PDU length measured in units of time, is, in the 1 Mbps case, 11 times the length in the 11 Mbps case.

The impact of the physical preamble/header on the overhead is thus correspondingly less. Normalized throughputs in the range of 0.8 are then possible (Bianchi, 2000). But although this lower bit-rate system is much more efficient, the actual throughput rate is in the order of 800 kbps, a considerable reduction from the 5.4-Mbps rate made available in this case of basic access at the 11 Mbps transmission rate. (It is to be noted that the system parameters such as slot time, SIFS, and DIFS used in the Bianchi paper are somewhat different than those used here, but the relative throughput results at the 1 Mbps rate using the parameters appearing here would not differ very much. Details are left to the reader.)

Consider now the effect of reducing the physical-layer overhead by use of the optional short preamble and header, transmitted in 96 μsec instead of the long-format 192 μsec. Let MAC-PDU frame lengths remain the same, as does the number of user data stations assumed in making the calculation. The values of both T_S and T_C are then both reduced considerably, the former decreasing by 2×96 μsec to 1029 μsec; the latter by 96 μsec to 911 μsec. The resultant normalized capacity at the 11 Mbps bit rate then increases to 0.6 from the 0.49 value calculated above, using the long preamble and physical header combination. Details are left to the reader. Some improvement in efficiency of transmission is thus possible by using this optional preamble-header combination. The downside is that 1 or 2 Mbps transmission is not possible on the same system: the same 192 μsec long preamble-header combination is standardized for use on the lower as well as higher bit rate 802.11 systems.

We now turn to the virtual access mechanism, calculating the normalized capacity in this case, again at the 11 Mbps transmission rate. The capacity equation (12.2) still applies in this case, as do the equations (12.3)–(12.7) following. The only difference between this calculation and that for the basic access mechanism just described is a change in the calculation of the two critical times T_C and T_S. Recall that the virtual access mechanism was described in the paragraph on access mechanisms appearing earlier in this section, with the procedure summarized in Fig. 12.3. Using that description, it is left for the reader to show that the two times in question may be expressed as shown in (12.10) and (12.11) following (see also Bianchi, 2000):

$$\textit{Virtual access:} \quad T_C = \text{RTS} + \text{DIFS} + \delta \tag{12.10}$$

and

$$\textit{Virtual access:} \quad T_S = \text{RTS} + 3\text{SIFS} + 4\delta + \text{DIFS} + \text{ACK} + \text{CTS} + P + H \tag{12.11}$$

Note that each of the MAC-layer frames transmitted using the virtual access procedure and appearing in (12.10) and (12.11), the MAC-PDU P plus its header H, the ACK, and the two control frames, the RTS and the CTS, must be preceded during transmission by the physical-layer preamble/header combination lasting 192 μsec in the long format case and 96 μsec in the optional short format case. As was the case with the basic access mechanism, these additions to the overhead would be expected to reduce the system efficiency considerably at the 11 Mbps transmission rate. For note, as an example, that the 20-octet RTS frame (Fig. 12.5b) is transmitted in 14.5 μsec in the 11 Mbps system, a fraction of the 192 μsec time required to transmit the physical-layer preamble/header combination using long-format transmission. The 14-octet CTS frame is transmitted in

10.2 μsec (Fig. 12.5c), again a fraction of the physical-layer symbols preceding it. The same is true of the 14-octet ACK frame, already mentioned above in connection with the basic-access capacity calculation.

Consider now the same example worked out above for the basic access case. We again assume ten user stations are in operation, each transmitting 1023-octet data frames at the 11-Mbps rate, with frames always available for transmission. We assume the standard long-format physical layer procedure is utilized. From (12.11) and (12.12) we find quite readily that the corresponding values of the two times are, respectively, $T_C = 258$ μsec and $T_S = 1651$ μsec. Note that the value of T_C, the average time the channel is sensed busy during a collision, has been reduced considerably from the value of 1007 μsec found previously for the basic access procedure. This is already apparent by comparing (12.10) with (12.8). Invoking the virtual access procedure reduces the effect of collisions. The value of T_S, however, the time the channel is sensed busy during a successful transmission, has been increased from the basic-access value of 1221 μsec. This is due to the need to transmit the added control messages, each preceded by the 192 μsec physical-layer preamble and header, before finally transmitting the data frame. It is left for the reader to show that the three probabilities required to calculate the normalized throughput S in this case are given by $p_{d,opt} = 0.04$, $P_{tr} = 0.34$, and $P_s = 0.81$. From (12.2), then, we find $S = 0.42$. This is obviously a low normalized throughout, but note again that data frames are delivered at an 11-Mbps transmission rate. The actual optimum throughput value is $0.42 \cdot 11$ Mbps or 4.6 Mbps. This is to be compared with results for the 1 Mbps 802.11 system, for which the virtual access mechanism is found to provide a normalized maximum throughput of 0.84 (Bianchi, 2000). But note again that the actual data frame throughput in this case is 840 kbps, considerably less than the 4.6 Mbps throughput made available by the 802.11b system.

12.2 Wireless personal-area networks: Bluetooth/IEEE 802.15.1

The IEEE 802.11 wireless LANs discussed in the previous section have been designed principally to provide high-speed wireless transmission over distances of 100m or so. Their deployment, particularly that of 802.11b or Wi-Fi, has been spectacular. The coupling of multiple such WLANs has, in fact, been used to provide high-speed data transmission over much larger distances than that of individual WLANs alone. The next mode of shorter-distance wireless data communication to become standardized is that of the *personal-area network*, or PAN. The objective here is to provide wireless interconnectivity within a small area covering at most a 10m distance between various personal devices, such as notebook computers, personal data assistants (PDAs), digital cameras, etc. The Bluetooth technology, introduced in 1998 by a group of companies and subsequently expanded to include hundreds of organizations, is intended to meet this objective (Bluetooth, 1999; Bisdikian, 2001; Miller and Bisdikian, 2002). (The name Bluetooth comes from the Danish

---

Bluetooth. 1999. "Specification of the Bluetooth System," Specification Volume 1, Core, v1.0.B, www.bluetooth.com, December 1.

Bisdikian, C. 2001. "An overview of the Bluetooth wireless technology," *IEEE Communications Magazine*, 39, 12 (December), 86–94.

Miller, B. A. and C. Bisdikian. 2002. *Bluetooth Revealed*, 2nd edn, Upper Saddle River, NJ, Prentice-Hall PTR.

king Harald Blatand (Bluetooth), who, in the tenth century, united the Scandinavian people (Bisdikian, 2001).) In March 1999 the IEEE 802.15 Standards Working Group for wireless PANs (WPANs) was created and subsequently adopted the Bluetooth specifications as the basis for the IEEE 802.15.1 standard for 1-Mbps WPANs, the transmission rate chosen for Bluetooth (Bluetooth, 1999; IEEE 802.15). The Bluetooth/IEEE 802.15.1 standard calls for transmission in the 2.4 GHz unlicensed band, the same band utilized, as we have seen, for 802.11 WLAN transmission. Other task groups within the IEEE 802.15 Working Group are providing specifications for both higher rate and lower rate transmission within the same band. These include the IEEE 802.15.3 Task Group, working on higher rates of 20–110 Mbps, suitable for portable consumer digital imaging and multimedia applications, and the IEEE 802.15.4 Task Group, devoted to developing specifications for lower rate transmission of 20–250 kbps, designed for such applications as sensors, smart badges, remote controls, etc. We discuss in this section the 1-Mbps Bluetooth/IEEE 802.15.1 specifications only.

As noted above, the IEEE 802.15.1 Bluetooth system operates in the 2.4 GHz unlicensed band. It uses spread-spectrum technology with each device, when transmitting, hopping pseudo-randomly in frequency at a 1600 hops/sec rate among 79 equally spaced 1-MHz transmission channels in the range 2402–2480 MHz. The modulation used is binary gaussian-shaped frequency-shift keying (GFSK), with transmission at the 1 Mbps rate noted above. Typical output power is in the range of 1 mW, allowing the use of battery-enabled portable devices. (Radios of up to 100 mW in power are allowed, however, to cover larger distances.) This low power operation and the resultant small area covered contrast with the 802.11 WLAN operation, where devices might be connected to a wall power plug, moving only relatively infrequently (IEEE 802.15). Bluetooth devices within the coverage range of 10m or so organize themselves into *piconets*, each such piconet consisting of a *master* and at most seven *slaves*, operating under the control of, and communicating with, the master (Bisdikian, 2001; Miller and Bisdikian, 2002). There is thus no contention within the system. Piconets can operate independently of one another, and may change their constituency with time and space. The position of master and slave may vary as well, devices changing from one category to the other. Any devices desiring to enter a piconet already containing the maximum number of eight devices may register with the master, and are listed as *parked*, with the possibility of joining the piconet at a later time. Other devices not associated with a piconet, but within the general vicinity, are said to be in *standby* mode. A single device may be a member of several overlapping piconets, comprising a *scatternet*, thus providing the capability of communications beyond just one piconet. A device can, however, serve as master of one piconet only. If associated with other piconets forming a scatternet, that device serves as a slave in the other piconets. Figure 12.8 provides an example of two overlapping piconets with one device shown serving as a slave in both. The two piconets thus form a scatternet. The symbol S represents slave, M is master, P corresponds to a parked device, and Sb to a standby device (Bisdikian, 2001).

IEEE 802.15. IEEE Standard for Information technology-Telecommunications and Information exchange between systems-Local and metropolitan area networks-Specific requirements, Part 15.1: Wireless Medium Access Control (MAC) and Physical Layer (PHY) Specifications for Wireless Personal Area Networks (WPANs), IEEE Std 802.15.1, IEEE-SA Standards Board, 15 April 2000.

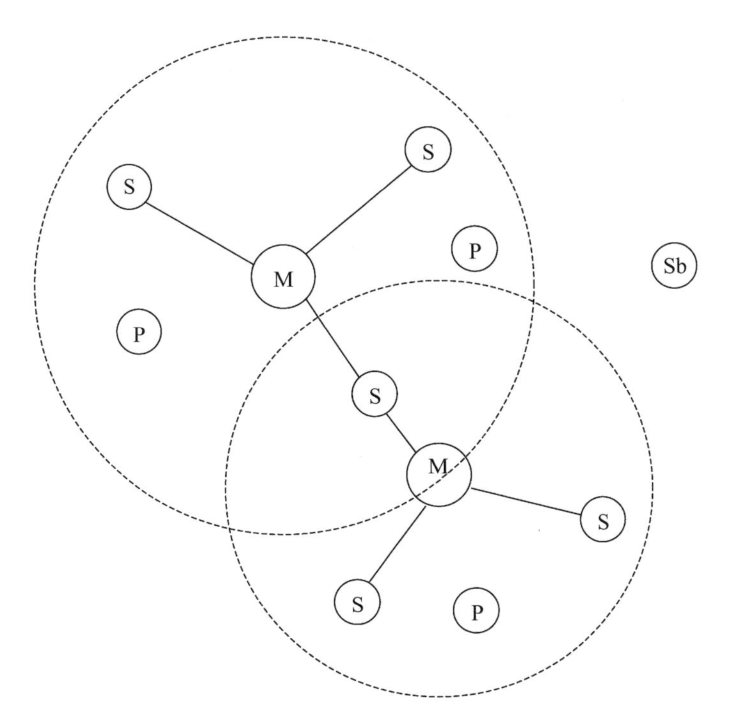

Figure 12.8 Piconet/Scatternet
M-master; S-slave; P-parked; Sb-standby

The Bluetooth specification is defined in terms of the typical set of layered protocols. This Bluetooth/IEEE 802.15.1 protocol stack is represented in Fig. 12.9 (Bisdikian, 2001; IEEE 802.15; Miller and Bisdikian, 2002). The lowest layer shown, the Bluetooth-defined Radio layer, corresponds to the IEEE 802 Physical layer, as indicated in the figure. It is this layer that covers the Bluetooth radio characteristics noted above: the 1600 hops/sec frequency-hopping spread spectrum signal, transmitted using GFSK modulation. The layers from Baseband up to, and including, the layer designated L2CAP, comprise the 802 MAC layer, as indicated. The protocols comprising the composite MAC and Physical layers shown in Fig. 12.9 correspond to the Transport protocol in the Bluetooth specification (Bisdikian, 2001; Miller and Bisdikian, 2002). These protocols appear in every Bluetooth implementation. The layer above the MAC layer consists of middleware protocols, some of which are specific to Bluetooth, others providing interfaces to other systems. The acronym L2CAP stands for Logical Link Control and Adaptation Protocol, a Bluetooth-specific protocol, and provides the packet interface to the upper layers. The acronym LLC at the middleware level stands for the 802-defined Logical Link Control, allowing connection to LANs and other 802-defined networks.

The organization of communications among members of a piconet, including the establishment of Bluetooth communications and the delivery of packets over these links, is defined and specified at the Baseband layer. The basic communication link consists of a continuous set of time slots, each 625 μsec long, the nominal duration of a hop within the 1600 hops/sec frequency-hopping sequence. Packet transmission generally takes place

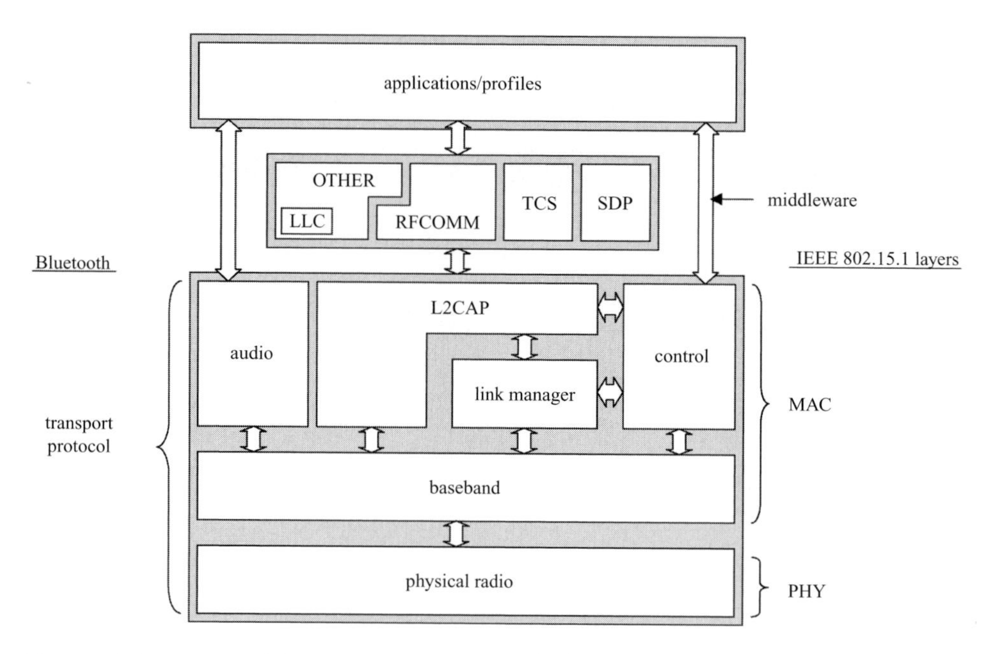

Figure 12.9 Bluetooth IEEE 802.15.1 protocol stack (from IEEE 802.15, Fig. 6)

within the confines of one slot, with the exception of three- and five-slot packets, to be described briefly later. The transmission frequency used changes with each packet transmission, selected pseudo-randomly, as noted above, from the set of 79 frequencies within the transmission band. For single-slot packets, then, frequency hopping is aligned with slot boundaries. For multiple-slot packets, hopping is suspended during packet transmission, but is picked up again with the transmission of a new packet as if the equivalent number of single-slot packets had been transmitted. The necessary slot synchronization and choice of transmission frequency to be used are controlled by the master device of the piconet. For this purpose each Bluetooth device carries a free-running 28-bit clock that ticks every 312.5 μsec, half of a slot interval. In addition, each device carries a unique IEEE-type 48-bit address that is assigned on manufacture. Devices communicating in a piconet acquire the other device addresses. Slaves also acquire the clock of the master, and use this information to establish time-slot synchronization. The frequency hop sequence is determined by the master's address; the phase of the sequence, i.e., the frequency corresponding to a given slot, is determined from the master clock. Time-division duplexing, TDD, is used to transmit and receive information, with the master and slaves alternating in transmission: slave devices transmit only in response to the master device, the master transmitting in even-numbered slots, a slave, in its response, transmitting in the following odd-numbered slots.

Piconets are organized in an *ad hoc* fashion. A two-phase inquiry/page procedure is used for this purpose (Bisdikian, 2001). Consider a Bluetooth radio in the standby mode noted above. Say a user-controlled application in the device to which it is attached decides to enter into communication with one or other Bluetooth devices in its vicinity. It will signal the radio to either begin the inquiry phase or listen for page messages. Focus

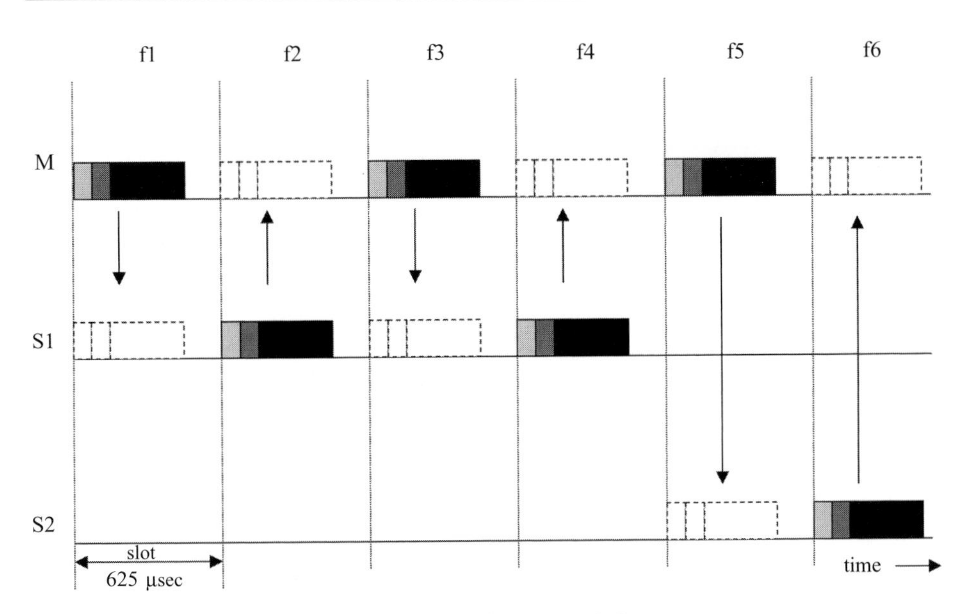

Figure 12.10 Piconet channel, single-slot packet transmission

on the inquiry phase. That phase is a device discovery phase used to determine which devices are in range, as well as the services they can provide. The device initiating that phase transmits inquiry messages inviting other devices in its vicinity to join it in a piconet. Devices respond with messages carrying their address id. Once communication with other devices is established, the initiating device *de facto* becomes the piconet master.

All transmission in a piconet is controlled by the master, using the time-division duplex procedure noted above. It is always the master that initiates transmission, sending a message to a specific device, that device then responding, if a response is required or desired, to the master. (It is to be noted again that the master and a slave may interchange roles at any time.) Two types of transmission are supported by Bluetooth/IEEE 802.15.1: asynchronous, best-effort data transmission, and synchronous transmission providing for 64 kbps periodic transmission, such as voice. With asynchronous transmission, an *asynchronous connectionless* (ACL) link is established between master and slave. With synchronous transmission, a *synchronous connection-oriented* (SCO) link is established. At most three SCO links may exist at any one time in a given piconet. An ACL link uses packet retransmission and sequence numbering, as well as forward-error correction (FEC) procedures, if necessary, to detect and correct errors. Packet retransmission is not used with an SCO link, but forward-error correction may be used, if desired.

Figures 12.10–12.12 provide examples of the master-controlled time-division duplexing procedure defined for Bluetooth.[1] A simple piconet consisting of three devices, a master and two slaves, is used in these examples. Figure 12.10 provides an example of the basic piconet channel, for the case of single-slot packet transmission. It demonstrates the use of frequency-hopping as well as TDD transmission. Transmission in this example

[1] These figures are taken from the IEEE Infocom 2001 tutorial on Bluetooth, presented by Dr. Mahmoud Nagshineh of IBM. The author is indebted to Dr. Naghshineh for permission to use these figures.

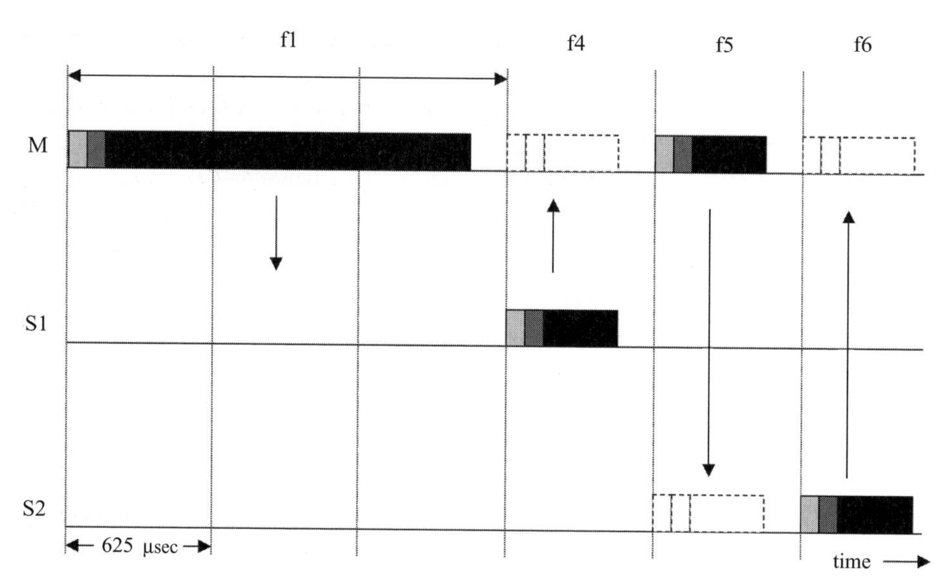

Figure 12.11 Multi-slot packet transmission

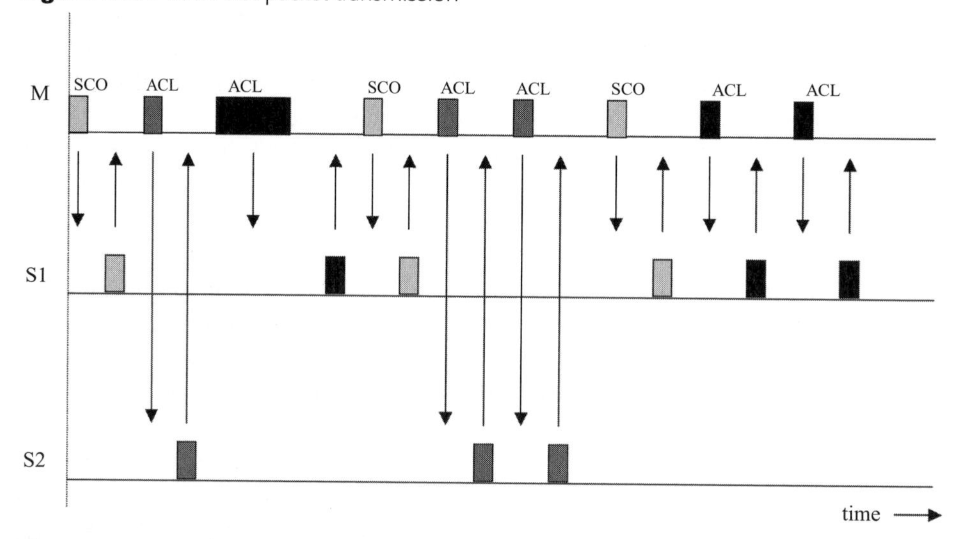

Figure 12.12 Synchronous connection oriented (SCO) and asynchronous connectionless (ACL) links

begins with the master device, labeled M, sending a message to device S1, using frequency f1. That device receives the message and replies during the next slot, 625 μsec later. Its reply packet is transmitted at frequency f2. Transmission to and from device S1 is then repeated during the next two slots, using frequencies f3 and f4, as indicated. During the fifth slot, the master device transmits a packet to device S2, using frequency f5, as shown; S2 replies during the sixth slot, using frequency f6. Note, as explained earlier, that the slot choice of 625 μsec corresponds to a frequency-hopping rate of 1600 hops/sec. The selection of each successive transmission frequency, f1, f2, . . . is made using a pre-determined pseudo-random sequence based on the master device's address.

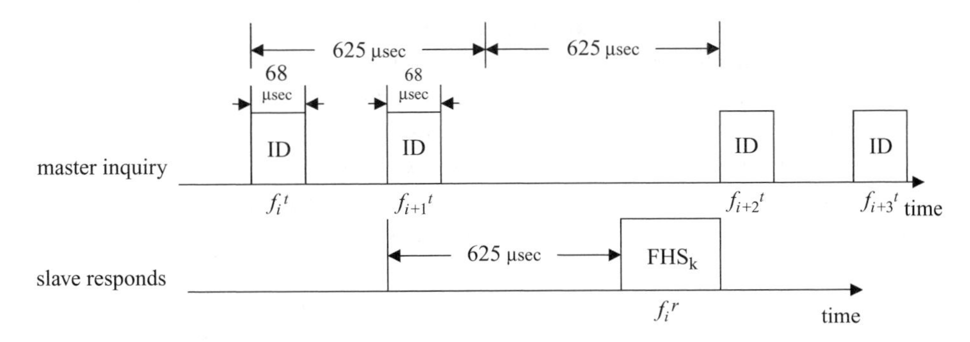

Figure 12.13 Inquiry procedure

Figure 12.11 provides a similar example for the case of multi-slot packet transmission. In this example, master device M initiates transmission, sending a three-slot long packet to device S1, using frequency f1 as shown. Device S1 replies with a single-slot packet in slot 4, using frequency f4, as if three separate packets had been transmitted during the previous three slots. Pseudo-random frequency hopping thus always continues at the 1600 hops/sec rate, even if frequencies during a multi-slot transmission are not used. Finally, Fig. 12.12 provides an example of data transmission for combined asynchronous connectionless (ACL) and synchronous connection-oriented (SCO) links. Note that the SCO link provides periodic transmission: the master device periodically polls device S1 and S1 replies during the next slot. ACL link transmission is used during other slot times.

How are piconets organized in the first place? It was noted above that a two-phase inquiry/page procedure is used for this purpose. We now elaborate on this process. A piconet may be set up by any device wishing to initiate communication with one or more other devices. An example might be a Bluetooth device searching for a near-by printer or Bluetooth-enabled Fax machine; it might want to connect to a cell phone or to a LAN, or possibly to a combination of a number of these devices (IEEE 802.15). In the event it knows of no such device in its neighborhood, it initiates the *inquiry procedure*, in which it sends periodic inquiry, ID, messages to locate devices in its vicinity. These devices would be in the standby mode mentioned earlier, as indicated by the symbol Sb in the simple example of Fig. 12.8. Once the inquiry-initiating device has "collected" the devices to which it may want to connect, it will follow this inquiry phase by initiating the *page procedure* to actually make the connections.

Consider now a device in standby mode receiving the inquiry messages and desiring to enter into communication with the inquiring device. It will, in turn, respond with an inquiry response message. The inquiry procedure is diagrammed in Fig. 12.13 (Miller and Bisdikian, 2002; IEEE 802.15). Note that the device initiating the inquiry procedure automatically becomes the master of the resultant piconet; any responding devices become the slaves. The inquiry messages, labeled ID in Fig, 12.13, are 68 bits in length and carry an *inquiry access code* known to all devices capable of joining the piconet. (There are two types of inquiry access codes, a general inquiry access code, GIAC, and a dedicated inquiry access code, DIAC, specific to a dedicated group of Bluetooth devices sharing a common characteristic (IEEE 802.15). These inquiry or ID messages are sent two per 625 μsec slot interval, spaced 312.5 μsec apart, as shown in the figure. (The 68-bit ID packets are transmitted in 68 μsec at the nominal Bluetooth rate of 1 Mbps.) The inquiring device, the master, waits one slot interval for a possible response as shown, and then transmits two more inquiry messages in the next slot interval, repeating this process of transmitting and waiting for a response until it has located as many devices as deemed necessary by its link manager, or until an inquiry timeout has been reached (IEEE 802.15). Each successive inquiry message is sent at a different hop frequency, as shown, pseudo-randomly selected from a group of 32. The frequencies used by the master in sending its inquiry message and a slave in responding will generally differ. The frequencies chosen by the inquiring device change randomly every 312.5 μsec, as indicated in Fig. 12.13. The responding devices change their *listening* frequencies much more slowly, selecting a new listening frequency every 1.28 sec. It is therefore very likely that a listening device will hear the inquiring device's inquiry message, since the frequencies it uses are changing rapidly. The specific choice of frequencies to use is based on the inquiry access code. Devices available for joining in this piconet thus know to which frequencies to listen, and will eventually receive an inquiry message, as noted. The frequencies shown in Fig. 12.13 represent the master's transmitting frequency, denoted as f_i^t, and the frequency at which a slave responds, denoted by f_i^r. The superscript in each case denotes the transmitting device, t for the inquiring (master) device, r for the responding device. The subscript denotes the particular choice of frequency, changing randomly every 312.5 μsec, as indicated. Figure 12.13 shows as well that a responding device, which automatically becomes a slave, responds to the inquiry message with a special FHS packet 625 μsec, a slot time, after receiving the inquiry message, using its own frequency at which to respond. This packet contains, among other fields, the responding device's 48-bit IEEE address, thus identifying it to the master device. In the example of Fig. 12.13, the responding device address is indicated by the letter k. To avoid possible collisions between responses from different devices that might respond at the same time, the time at which a device responds after receiving an inquiry message is randomly selected from a window of slot times. In the example of Fig. 12.13 the responding device is shown responding to the second ID message. A device could also have responded to the first ID message, responding with an FHS packet 625 μsec after receiving that inquiry message. Devices responding to an inquiry as prospective members of a piconet thus respond in a slot interval in which the device initiating the inquiry, the master, is silent and is listening.

It was noted above that a master having completed the inquiry procedure will normally follow with the paging procedure to actually make the device connections it desires. The paging procedure is similar to the inquiry procedure and is diagrammed in Fig. 12.14 (IEEE 802.15; Miller and Bisdikian, 2002). The difference is that the master now knows the address of the device it pages. A 32-hop set of frequencies determined by the paged device address is used to carry out the paging process and 68-bit ID packets carry a *device access code* based on the 625 μsec slot interval, spaced half a slot apart, as shown in Fig. 12.14. The master transmission frequencies again change randomly every 312.5 μsec, while the frequencies at which the paged devices listen again change much more slowly, every 1.28 sec. The paged device knows the specific ID packet carrying its device access code to which to respond. As an example, the letter k appearing in Fig. 12.14 indicates device k is being paged by the master. Note again that the master device alternates between a transmitting phase and a listening phase, each a 625 μsec-slot long. A paged device may

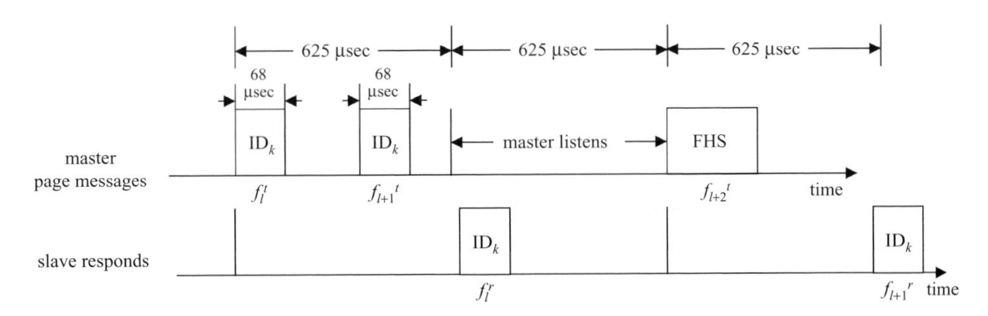

Figure 12.14 Page procedure

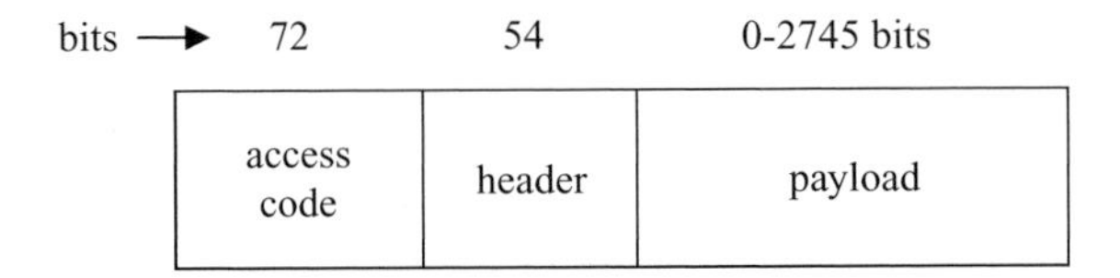

Figure 12.15 Bluetooth packet format (Baseband layer)

respond a slot time later with its own ID packet to either of the two ID packets carrying its device access code. In the example of Fig. 12.14, the paged device *k*, the slave, is shown responding at frequency f_l^r to the first of the two ID packets carrying its device access code and transmitted at frequency f_l^t a slot time earlier. The slave, after responding, then waits for an FHS packet from the master device. The master, on receiving the ID packet with the appropriate device access code from the slave addressed, in turn acknowledges its receipt of the packet by transmitting the FHS packet. This packet contains a 3-bit id by which the slave will be identified in the communication phase following, as well as the master address and clock information from which the slave will determine the frequency-hopping sequence from the set of 79 and the *channel access code* (piconet id) to be used in communicating with the master device. The slave, on receiving the FHS packet, acknowledges its receipt by sending an ID packet with its device access code a slot time later. It then awaits receipt of a special POLL packet, not shown in Fig. 12.14, used by the master to begin the communication process. The POLL packet is transmitted at the appropriate frequency in the sequence of 79 pseudo-randomly hopping frequencies based on the master device's address, as noted earlier in discussing the example of Fig. 12.10. The slave replies to the POLL packet with any type of packet using the next time slot and transmitting at the next frequency of transmission in the 79-frequency pseudo-random hopping sequence.

A number of Baseband-layer control-packet types have been described above. The format of these control packets, as well as data packets at the Baseband layer used for both asynchronous (ACL) and synchronous (SCO) transmission, follows a standard Bluetooth form as diagrammed in Fig. 12.15 (IEEE 802.15). Packet formats begin with a 72-bit access code (the 68-bit device access code and inquiry access code ID packets described above provide the exceptions), followed by a 54-bit header and any payload to be transmitted. (The payload may range from 0 to a maximum of 2745 bits, as indicated in Fig. 12.15.)

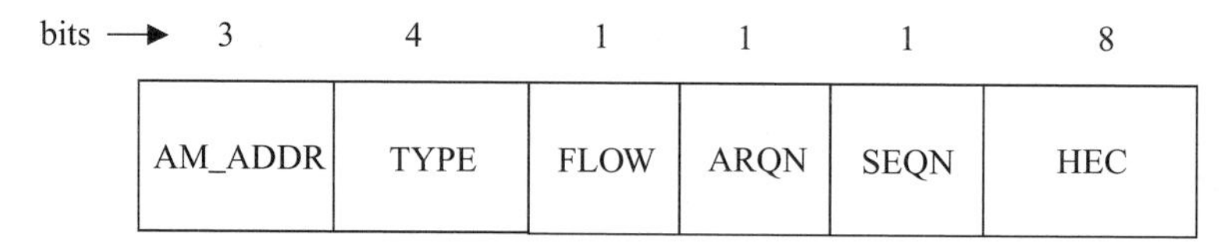

Figure 12.16 Header format

The access code consists of a 4-bit alternating 0-1 preamble, a 64-bit sync word derived from a 24-bit portion of an address, and a 4-bit alternating 0–1 trailer. The various 68-bit ID packets described above omit the trailer. They carry the preamble and sync words only. As noted earlier, the address used to derive the sync word portion of the 68-bit ID packets used in the page procedure (Fig. 12.14) is based on 24 bits of the paged device 48-bit address; the sync word used in the inquiry procedure (Fig. 12.13) is based on either the general inquiry access code (GIAC) or the dedicated access code (DIAC). FHS packets used in the inquiry procedure carry a 72-bit access code, with the sync word derived from the responding device 48-bit address (Fig. 12.13); FHS packets sent during the page procedure carry 72-bit channel access codes as well, with the sync word derived from the master device's 48-bit address.

Now consider the 54-bit header portion of Bluetooth packets, as indicated in Fig. 12.15. Eighteen bits are actually assigned to the header, but each bit is transmitted three times in succession, to ensure reliable transmission. This three-fold repetition procedure is defined as a one-third forward-error correcting code, FEC, procedure. (It is thus written as 1/3FEC.) The 18-bit header format appears in Fig. 12.16 (IEEE 802.15). The first three bits, labeled AM_ADDR, represent the piconet membership id assigned during the page process as indicated above. (Recall that a piconet can contain at most seven slave devices. The all-zero AM_ADDR is used by the master device, in packets other than FHS packets, to broadcast information to all slaves in a piconet.) The 4-bit TYPE field is used to identify one of 16 different possible packet types that might be transmitted over Bluetooth piconets. Some of the packet types that have been defined for Bluetooth are described below. The three 1-bit flag fields following are used, respectively, for ACL link flow control, positive or negative acknowledgement of packet transmission, and sequence numbering in a data packet stream. Finally, the 8-bit HEC or Header-Error-Check field is generated using a specified code polynomial and is used by the receiver to verify correct transmission of the header. The receiver calculating an HEC not in agreement with the value received discards the entire packet.

Consider now the various Bluetooth packets that may be transmitted. These are designated, as indicated above, by the 4-bit TYPE field in the 18-bit header of Fig. 12.16. The packets may be grouped in two types: control packets and data packets. The control packets already mentioned above are the FHS and POLL packets. (The ID packets are control packets as well, but are all 68-bit packets, as already noted, and are thus not included in this list of packet types.) We have already seen the use of the FHS packet in the inquiry procedure as a response from a potential piconet slave device to the master device initiating the inquiry (Fig. 12.13) and in the page procedure as a message from the master device to the slave device in getting ready to initiate communication (Fig. 12.14). It is also used to carry out a master-slave switch. The FHS packet payload (Fig. 12.15) contains

144 information bits plus a 16-bit cyclic redundancy check (CRC) field to protect the information bits. These 160 bits are in turn further protected by 80 bits calculated using a (15, 10) type Hamming code to provide what is called 2/3 forward-error correction or 2/3 FEC. The actual payload is thus 240 bits long, and the full FHS packet is 366 bits long. Note that this covers about 59% of a slot time at the nominal transmission rate of 1 Mbps, allowing the receiving device time to synchronize on the received packet and turn around, if a reply is necessary the next slot time. The 144 information bits are divided into 11 different fields, which include, among others, the 48-bit address and a 26-bit clock field of the device sending the packet (i.e., master in the page procedure; responding slave in the inquiry procedure), as well as the 3-bit address assigned to the slave in the page procedure. (The 3-bit address is set to 0 in the inquiry-procedure case.) The POLL packet carries no payload, consisting only of the 72-bit channel access code identifying the piconet and the 54-bit header. As noted above, this packet is used to initiate communication with a particular slave identified by the 3-bit address in the header. It is also used to address any slave in the piconet once the communication procedure is underway. This packet requires a response from the slave being polled. A broadcast POLL packet, addressed to all slaves in the piconet, is designated by the all-zero 3-bit address. A third control packet is the NULL packet, which, like the POLL packet, carries no payload, consisting only of the 72-bit channel access code followed by the 54-bit header. This packet is used by a slave to acknowledge receipt of a query by the master when it has nothing to send, or to send flow-control information to the master.

A number of data packet types have been defined for Bluetooth (IEEE 802.15). These are used for asynchronous data transmission over an ACL link and synchronous data transmission over an SCO link, the two Bluetooth physical link types mentioned earlier (Fig. 12.12). Consider the SCO packets first. Recall again that no more than three SCO links can be established for a given piconet. SCO packets are never repeated if found in error. There are four SCO packet types, defined to accommodate 64-kbps voice transmission primarily. Three SCO packet types carry SCO traffic only, and, like the FHS packet, are all 366 bits long, with 240-bit payload fields. These three packet types are designated HV1, HV2, and HV3, in order of decreasing error protection. Thus HV1 packets use 1/3 FEC, carrying, in their payload field (Fig. 12.15), 80 information bits of 64-kbps voice, repeated three times, for the full payload field of 240 bits. These packets would be transmitted periodically in alternate time slots assigned to their SCO link. Note that the transmission of 80 information bits every two time slots, or 1.25 msec, corresponds to the desired data transmission rate of 64 kbps. HV2 packets use 2/3 FEC, carrying, in the 240-bit payload field, 160 information bits plus 80 protection bits generated by a (15, 10) Hamming code. These packets are transmitted every four time slots, again corresponding to 64-kbps voice. Finally, HV3 packets carry a full set of 240 information bits in the payload field, and have no FEC protection. These packets are transmitted every six time slots, again corresponding to 64-kbps transmission. The results of published simulation studies designed to measure the relative effectiveness of using these three different SCO packet types are summarized briefly in the paragraph following. The fourth SCO packet defined, the DV packet, is designed to carry both synchronous 64-kbps voice and data. DV packets on a given SCO link are sent periodically, as required for SCO links. The DV packet payload consists of an unprotected 80-bit voice field and a data field of at most

150 bits including a 16-bit CRC for error checking and 2/3 FEC encoding. The voice and data portions are treated separately, the voice field never repeated, the data field checked for errors, and repeated if errors are found.

Seven ACL packet types have been defined (IEEE 802.15). We describe six of them only here. The payload fields of all six are protected with a 16-bit CRC code, and packets for which no acknowledgement of correct reception is received are retransmitted. The six packet types are grouped into two categories, transmission at *medium* or *high* data rate, designated with the letters M and H respectively; within each category packets may be transmitted in single slots, or over three and five slots, as noted earlier. They are thus designated DM1, DM3, DM5, and DH1, DH3, DH5. The DM1 packet payload field may carry a maximum user payload of 17 bytes (136 bits), preceded by a one-byte header, both protected by the 16-bit CRC. The resultant set of bits is then further protected using 2/3 FEC. Since such packets may be transmitted at most every second time slot, the maximum DM1 user data transmission rate is readily calculated to be 108.8 kbps. Now consider the DH1 packet. This packet also carries a one-byte header in its payload field, but the maximum length of the user payload within the payload field is now increased to 27 bytes or 216 bits. The header and user payload are protected by the 16-bit CRC, but further FEC protection is not used. It is left to the reader to show that the maximum user data transmission rate is now increased from the DM1 rate of 108.8 kbps to 172.8 kbps. There is thus an obvious tradeoff between error-correction capability and transmission rate. Similar tradeoffs appear in the choice of moving to multiple-slot data packet transmission with either medium or high data-rate transmission. Consider the three-slot packet case first. The three-slot data packet types DM3 and DH3 can carry a maximum of 121 and 183 user payload bytes within the payload fields, respectively, covering three slots each. In addition to the user payloads, both have two-byte payload headers, all protected by the 16-bit CRC. The DM3 packet type is further protected by 2/3 FEC, while the DH3 packet has no such error correction. This is thus similar to the single-slot packet case. Calculation of the maximum user payload transfer rate is somewhat more complex, however.

Two possible situations appear in calculating the maximum user data transfer rates of these three-slot ACL packets: In one case, labeled the asymmetric case (IEEE 802.15), DH3 or DM3 packets are transmitted in one direction three of every four slots, with corresponding DH1 or DM1 packets transmitted in the other direction every fourth slot. In the other case, called the symmetric case, three-slot packets are transmitted in each direction three of every six slots, while three-slot packets are transmitted in the reverse direction the other three slots. (It is left for the reader to verify and explain these two cases.) Calculation then shows that the maximum user data transfer rates in the asymmetric case are 387.2 kbps and 585.6 kbps in one direction in the DM3 and DH3 cases, respectively, while user data rates in the other direction are 54.4 kbps and 86.4 kbps. (It is to be noted that these are just one-half of the DM1 and DH1 rates quoted above. Why should one expect these values?) Similar calculations for the symmetric case provide the two results, 258.1 kbps for DM3 packets and 390.4 kbps for DH3 packets. Finally, consider the DM5 and DH5 packet types. These packet types both have 2-byte headers within the payload fields, as in the three-slot packet cases. Both have the 16-bit CRC protecting the header and user payload fields. The DM5 packet is further protected by 2/3 FEC, while the

DH5 packet does not have this protection. (Note, therefore, that all DM-type packets have 2/3 FEC protection, while the DH-type packets do not.) The maximum user payloads in the two cases are given as 224 and 339 bytes, respectively. The maximum user data transmission rates must again be considered in the two cases of asymmetric and symmetric transmission. The symmetric-case results now turn out to be 286.7 kbps for DM5 packets and 433.9 kbps for DH5 packets, while the corresponding asymmetric-case results are 477.8 kbps/36.3 kbps and 723.2 kbps/57.6 kbps, respectively. It is apparent that using DH5-type packets in the asymmetric mode can be very useful in handling asymmetric traffic situations in which relatively large files are to be transferred in one direction, with relatively low bit rates acceptable in the other direction. The results of simulation studies designed to measure the performance of these various packet types in the presence of interference are summarized briefly in the next paragraph. Interestingly, the various throughputs, the data transmission rates calculated above, have been verified in these simulations.

The ACL packets are used to carry data traffic and control information to and from the L2CAP and Link Manager layers, respectively, as may be seen by noting from the Bluetooth layered architecture of Fig. 12.9 that the Baseband layer provides the necessary support to both these layers. The one- or two-byte headers within the ACL-packet payload fields, to which reference is made above, carry a two-bit field used to distinguish between packets carrying L2CAP data, or Link Manager control information used in configuring the Bluetooth link. These headers also carry a length field indicating the length of the user payload, the L2CAP data or Link Manager information, following. Link Manager information is normally carried by single-slot ACL packets, although the DV packets can be used for this purpose as well. The L2CAP layer defines a simple data link protocol shielding higher-layer data sources from the Bluetooth Baseband layer. The concepts of master-slave operation and piconets no longer appear at the L2CAP layer. That layer provides a set of logical channels, both connection-oriented and connectionless, over which L2CAP traffic may flow. QoS information may be exchanged over the L2CAP layer. L2CAP-PDUs defined for that layer and carrying asynchronous L2CAP traffic may be much longer than ACL packets: The L2CAP layer then provides the necessary segmentation for the transmitting Bluetooth device; the corresponding layer at the receiving device does the necessary reassembly. Details of the operation of the L2CAP and Link Manager layers appear in IEEE 802.15 and Miller and Bisdikian (2002).

Bluetooth performance

A number of papers have appeared in the literature summarizing the results of simulation performance studies of Bluetooth. We describe some of the findings in these papers here. They include studies of the radio network performance and comparisons of simple polling and packet-scheduling strategies designed to more efficiently transmit ACL data packets.

The radio network performance studies were concerned with the possible interference introduced by multiple piconets operating simultaneously (Zürbes *et al.*, 2000). This

---

Zürbes, S. *et al.* 2000. "Radio network performance of Bluetooth," IEEE International Conference on Communications, ICC, 1563–1567.

interference is, of course, to be mitigated by the use of the 79-frequency random hopping sequence. The results of these studies were, on the whole, quite positive (Zürbes *et al.*, 2000): Using a Worldwide Web (WWW) data traffic model and modeling up to 100 simultaneous piconets operating within a simulated room of size 10m $\times$ 20m, the simulations indicated that the expected throughputs for each of the six ACL packet types, DM1 to DH5, were degraded by no more than 6%. These expected throughputs are just those we calculated in the paragraph above. The simulations incorporated log-normally distributed shadow fading and Ricean-type multipath fading. Piconets were each limited to a master and one slave. Both ACL and SCO links were simulated. WWW traffic over the ACL links was assumed to flow downlink only, in the master to slave direction; NULL packets were used by the slaves to acknowledge receipt of ACL data packets. The WWW session model was taken to consist of geometrically distributed downlink datagrams, averaging 10 datagrams per session. The datagrams were log-normally distributed in length, with average length of 4.1 kbytes (32.8 kbits) and 30 kbytes (240 kbits) standard deviation. These datagrams were then segmented into ACL packets carrying the user payloads discussed in the previous paragraph. Measurements were made, for each ACL packet type, of burst failure probabilities resulting from interference and noise in the fading environment simulated. These failure probabilities were found to increase proportionally with system load, as given by the number of piconets operating simultaneously. For example, the measured forward-link burst failure probability was found to be 6×10^{-3} for 10 piconets in operation and 6×10^{-2}, or 6%, for 100 piconets. (The system model chosen assumed an SIR for the system selected to achieve an uncoded bit error rate of 10^{-3}.) Interestingly, the measured forward-link burst failure probabilities were found to be about the same for all ACL packet types. It is to be noted that the use of longer packet types such as DH5 generates less interference. Hence the fact that the longer, uncoded packet types are more sensitive to interference appears to be compensated for by reduced interference. The measured link throughput for each of the six traffic types was found to be close to that calculated in the previous paragraph.

Calls on the SCO links were modeled as being exponentially distributed, with an average duration of 10 sec to reduce simulation time (Zürbes *et al.*, 2000). The performance measures used in determining SCO-link performance were different than those for the ACL links, since retransmissions are not used in the SCO-traffic case. Instead, an erasure mechanism was adopted: the payload of a burst received with an error in the access code or header was replaced by a specified bit pattern. The probability of an erased burst, labeled the frame erasure rate, FER, was one of the performance parameters measured; a second was the residual bit error rate, RBER, the undetected bit errors remaining in the SCO packets containing no erasures. Results for these two performance measures were roughly proportional to the number of simultaneous SCO sessions. The results did vary with SCO packet type, however. Recall that HV1 packets are transmitted in alternate slots while HV3 packets are transmitted every six slots. The HV1 packets thus create more interference, resulting in more erasures for the same number of sessions, despite the greater protection they receive. Thus, for ten sessions, the measured FER using HV1 packets was 10^{-2}, 0.5×10^{-2} for HV2 packets, and 0.3×10^{-2} for HV3 packets. These numbers increased by a factor of 10 for 100 sessions: roughly 0.1 for HV1 packets, 0.05 for HV2 packets, and 0.03 for HV3 packets. Measurements of the residual bit error rate did, however, show the combined effects of improved protection provided by FEC and increased interference. Thus, for 100 sessions, the measured RBER was about 3.5×10^{-3} for HV1 packets using 1/3 FEC, 7.5×10^{-3} for HV2 packets using 2/3 FEC, and 5.5×10^{-3} for the unprotected HV3 packets. FEC thus appears to be effective in reducing bit errors in the case of HV1 packets, despite the increase in interference; HV2 packets suffer the largest RBER, although not by very much, with the 2/3 FEC used in that case not being effective enough to overcome the increased interference their use introduces. Reducing the number of sessions reduces the RBER proportionately in all cases.

The performance studies summarized above have been devoted to determining the basic performance of the ACL data transmission strategy as well as SCO-link performance in transmitting voice calls. A set of papers has also been published comparing a variety of scheduling techniques at the Bluetooth MAC layer. We recall from the previous Chapter 11 that scheduling of packet transmission plays a critical role in wireless systems. Proper scheduling can improve throughput performance and reduce delays in transmitting packets. Fairness in providing equal opportunities for mobile devices to transmit and receive data must be considered as well. Appropriate scheduling of packet transmission is an important issue in Bluetooth as well. The scheduling problem is compounded in Bluetooth by the strategy adopted of having master and slave alternate transmission in defined slots only, the master transmitting packets to slaves in odd-numbered slots, a given slave, having been addressed, transmitting a response to the master in the following even-numbered slot. The master could be transmitting a data or control packet to a given slave or could just be polling that slave, with no data transmitted, requesting the slave to respond with data packet(s), if it has any to transmit. The basic question then is, how does the master determine which slaves to poll in any given slot? This question is then, obviously, similar to the scheduling problem addressed in Chapter 11: what schedule should the master adopt in polling the piconet slaves? The problem here is that the master could poll slaves when they have nothing to transmit, leading to significant wastage of the transmission channel, with attendant inefficiency in transmission. The simplest scheduling technique which presumably provides inefficient transmission in the heterogeneous traffic case is that of round-robin scheduling: the master simply polls each slave cyclically, one after the other, giving each slave an opportunity to transmit one packet in the next slot, if it has a packet to transmit. The master also uses its polling slot to transmit a packet to the slave polled, providing it has a packet in queue for that slave. This scheme is fair in giving all mobile devices the same opportunity to receive and transmit packets each cycle. It may, however, lead to inefficient use of the channel, as noted. The objective, then, in the Bluetooth case, is to design scheduling procedures that retain the fairness of round-robin scheduling, yet provide service only to those mobiles requiring it. This issue has been addressed in papers by Kalia *et al.* (2000); Capone *et al.* (2001), and Lee *et al.* (2002), among others. We provide a brief overview here of the

---

Kalia, M. *et al.* 2000. "Data scheduling and SAR for Bluetooth MAC," IEEE Vehicular Technology Conference, Spring, Tokyo, 716–720.

Capone, A. *et al.* 2001. "Efficient polling schemes for Bluetooth picocells," IEEE International Conference on Communications, ICC2001, Helsinki, Finland, 1990–1994.

Lee, Y.-Z. *et al.* 2002. "An efficient and fair polling scheme for Bluetooth," *Proc. IEEE Milcom*, 1062–1068.

scheduling procedures proposed in these papers, each of which builds on the one appearing before.

Consider Kalia *et al.* (2000) first. It is assumed in this paper that a backlogged slave, i.e., one with further data to send, so indicates to the master by setting a bit in its reply packet when polled by the master. (Note that this approach is somewhat similar to the one adopted for DQRUMA, described in Chapter 11.) The master can then distinguish between backlogged slaves and those with nothing to send. There are clearly four categories of master–slave pairs now resulting: pairs with both master and slave waiting to transmit a packet to the other, referred to as the 1–1 state; pairs with the master only ready to transmit a packet to the slave, the slave having nothing to transmit, the 1–0 state; pairs with the reverse – the master has no data to send to the slave, while the slave is ready to transmit, but can only do so on being polled, the 0–1 state; and the final category with neither device having anything to transmit, 0–0. Kalia *et al.* propose two scheduling policies that use this information, and compare them with pure round-robin scheduling. The first policy is called a Priority Policy, PP, in which the 1–1 state, with both master and slave backlogged, is given priority over the other states, using a variable priority parameter; slaves in the 0–0 state, with neither the master nor the slave having anything to transmit, are skipped over and not polled at all. The second policy, called a K-Fairness Policy, KFP, carries out round-robin scheduling over the three states 1–1, 1–0, and 0–1, allowing the devices in the 1–1 state to receive longer service, but not more than K slots, K a parameter. Their simulations then show that the KFP procedure provides a higher throughput than the PP procedure and provides a better measure of fairness. Both procedures perform better than the pure round-robin one.

Capone *et al.* (2001) builds on Kalia *et al.* (2000), pointing out that a master's knowledge of the backlogged state of a given slave is not always accurate, a slave having indicated it had nothing to send when last polled, having perhaps become backlogged in the interim. They then propose a number of procedures that do not rely on knowledge of slave queues, comparing them with idealized strategies in which the master is assumed to have knowledge of slave queues. These procedures are based on optimum polling strategies proposed previously in the literature. One such scheme is exhaustive round robin, ERR, in which all packets queued at a given device are transmitted before moving on to the next device in the round-robin cycle. This strategy is obviously not fair, since a master–slave pair in this case could capture the channel. To avoid this problem and introduce a measure of fairness, the authors propose a modified procedure, Limited Round Robin, LRR, with a given pair limited in the number of packets they can transmit. A modified version of this polling strategy, Limited and Weighted Round Robin, LWRR, applies a priority weight to a station, depending on the observed queue status. A station is first given the maximum weight, MP. This weight is reduced by 1 each time the station is polled and no data are exchanged. Once the weight reaches the lowest value of 1, it is not polled again for MP-1 cycles, potentially leading to long waits. Any time data are exchanged during a poll, however, the weight is increased again to MP. The focus in comparatively evaluating these various algorithms was packet delay. Simulation studies indicated that the exhaustive round-robin procedure performed almost as well as an idealized scheme in which the master, assumed to know the queue lengths at all slave stations, serves slaves exhaustively in order of the summed length of master to slave and slave queues. It was also found to provide better performance than the Priority Policy (PP) scheme of Kalia *et al.* It clearly lacked fairness, however. The LRR procedure worked well and was fair as well. The LWRR scheme was found to work well in all kinds of traffic, performing almost as well as exhaustive round robin, as well as providing a measure of fairness. Two later changes to the LWRR strategy were found to improve throughput and delay performance, as well as provide further improvement in fairness (Lee *et al.*, 2002). In one change, suspension of polling was reduced to a specified number of slots rather than MP-1 cycles. This clearly reduces unfairness. The second change, again reducing unfairness, was to introduce a pseudo-random polling cycle. As noted, these two changes resulted in performance improvement in both throughput and delay as well.

Problems

12.1 The IEEE 802.11 set of protocols assumes a local-area network environment. The propagation time chosen for the standard is 1 μsec. What size LAN does this value represent?

12.2 **(a)** Consider the transmission of a data frame using the basic access mechanism of 802.11. Explain why, with the More Fragments bit of the Frame Control field of Fig. 12.4 set to 0, the Duration field value represents the time to send one ACK frame plus one SIFS interval, while with that bit set to 1, indicating more fragments coming, the Duration field value represents the time to transmit the next fragment plus two ACK frames and three SIFS intervals. Comment on the explanation given in the text on the Duration field values of the ACK frame format.

(b) Provide, in your own words, explanations for the Duration-field values carried by the RTS and CTS frames involved in the virtual access mechanism.

12.3 Say a BSS contains three stations transmitting data frames to one another. Draw pictures similar to those of Figs. 12.2 and 12.3 for the basic and virtual access mechanisms showing various scenarios involving the exchange of information among the three stations. Compare the use of the two mechanisms. Note that there are two possible cases to be considered: a station might decide, at some randomly chosen time, to transmit data to another station; a station receiving a data frame attempts to reply to the sending station. Include the possibility of collisions occurring.

12.4 Refer to the discussion of 8-bit codeword generation for the 802.11b system in the text. Verify that, for an 8-bit group given by 00101011, the corresponding codeword is given by [−j, j, j, j, 1, −1, 1, 1]. Repeat this calculation for some other 8-bit groups of your own choice.

12.5 Show why, and how much, the use of OFDM in the high bit rate WLAN standards IEEE 802.11g and a reduces the possibility of inter-symbol interference. Why would a local-area environment be expected to provide flat fading for the OFDM bandwidths adopted?

12.6 Consider the various entries in Table 12.1. Show that each one of them does provide the appropriate set of 48 complex numbers every 4 μsec, as required for the OFDM/IFFT calculation.

12.7 **(a)** Verify that the saturation channel utilization S is given by (12.2). In particular, explain each of the three terms in the denominator and show that they collectively cover the average time required to successfully transmit a frame.

(b) Show S is maximized by maximizing $P_S/[(T_C/\sigma) + (1 - P_{tr})/P_{tr}]$.

12.8 **(a)** Show T_C and T_S for the basic access mechanism are given, respectively, by (12.8) and (12.9).

(b) Show the values for these two times for the virtual access mechanism are given, respectively, by (12.10) and (12.11).

12.9 **(a)** Consider the analysis of the saturation throughput of the 11 Mbps 802.11b basic access mechanism discussed in the text. Let the MAC-PDU frame size be 1023 octets, the example selected there. Show that, for ten stations, the maximum normalized throughput is 0.49 with the standard preamble and header. Repeat the calculation for the optional short preamble/header and show the maximum normalized throughput increases to 0.6.

(b) Try varying the frame size and the number of user stations, and calculate the resulting normalized throughput values. Determine the sensitivity of the throughput to both these quantities.

12.10 Repeat problem 12.9 for the virtual access mechanism and compare with the results for the basic access mechanism.

12.11 Figures 12.10 and 12.11 provide examples of Bluetooth master–slave asynchronous transmission for single-slot and multi-slot packet transmission, respectively. Provide timing diagrams for both cases for other examples involving three slave stations. Let the packet lengths in the multi-packet example vary between one and three slots in length. Include in the diagrams the sequence of frequencies used, as in Figs. 12.10 and 12.11.

12.12 **(a)** Expand the Bluetooth Inquiry procedure of Fig. 12.13 by showing how three slaves join the piconet.

(b) Expand the Page procedure of Fig. 12.14 for the three slaves of (a) joining the piconet.

12.13 The text describes, in words, the formats of the 68-bit ID packets and the 366-bit FHS packets. Show the complete format for each in an expanded form of Fig. 12.15, indicating the length and contents of each of the fields and subfields used. (The Access code, for example, consists of a number of subfields. In the case of the FHS payload field, include the subfields as well.)

12.14 **(a)** Consider synchronous data transmission over an SCO link. Use timing diagrams for a master and a slave to demonstrate the transmission of each of the three SCO packet types.

(b) Diagram the payload field of each of the three SCO packet types.
(c) Explain why transmission of each of these packet types corresponds to 64-kbps transmission.

12.15 Verify the statements in the text that the maximum DM1 user data transmission rate is 108.8 kbps, while that for DH1 packets is 172.8 kbps.

12.16 Consider three-slot ACL-packet transmission.
(a) Diagram transmission between a master and a slave for both the asymmetric and symmetric transmission cases.
(b) Verify the various maximum user data transfer rates cited in the text in both directions of transmission for both asymmetric and symmetric transmission. Explain why, in the asymmetric case, the DM1 and DH1 transmission rates are one-half those found in problem 12.15.

12.17 Provide diagrams of the payload fields of each of the six ACL data packets described in the text.

12.18 Verify the maximum user data transmission rates cited in the text for both symmetric and asymmetric transmission of five-slot ACL packets.

12.19 A number of proposed Bluetooth scheduling techniques are described in the text. These begin with simple round-robin polling and add complexity in order to improve throughput and delay performance and maintain a measure of fairness.
(a) Describe these various techniques in your own words, using master-slave timing diagrams to explain the procedures where possible. What is meant by "fairness"? Why might one technique be "fairer" than another?
(b) Note, as explained in the text, that there are four different cases to be considered in Bluetooth scheduling, involving whether a master or a slave or both have a packet waiting to transmit or have empty buffers. Come up with scenarios involving all of these cases and indicate how round-robin polling and a simple scheduling algorithm of your own choice might compare in each scenario.

References

Abramson, N. 1973. "The Aloha system," Ch. 14, in *Computer Networks*, ed. N. Abramson and F. Kuo, Englewood Cliffs, NJ, Prentice-Hall.

Abramson, N. 1977. "The throughput of packet broadcasting channels," *IEEE Transactions on Communications*, COM-25, 1 (January), 117–128.

Aein, J. M. 1973. "Power balancing in systems employing frequency reuse," *Comsat Technology Review*, 3, 2 (Fall).

Akyildiz, I. F. and J. S. M. Ho. 1995. "Dynamic mobile user location update for wireless PCS networks," *Wireless Networks*, 1, 2 (July), 187–196.

Akyildiz, I. F. *et al.* 1998. "Mobility management in current and future communication networks," *IEEE Network*, 12, 4 (July/August), 39–49.

Anasti, G. *et al.* 1996. "A bandwidth reservation protocol for speech/data integration in TDMA-based advanced mobile systems," *Proc. IEEE Infocom'96*, San Francisco, CA, March, 722–729.

Anasti, G. *et al.* 1997. "A contention/reservation access protocol for speech and data integration in TDMA-based advanced mobile systems," *Mobile Networks and Applications*, 2, 3–18.

Andrews, M. *et al.* 2001. "Providing quality of service over a shared wireless link," *IEEE Communications Magazine*, 39, 2 (February), 150–153.

Arad, M. A. and A. Leon-Garcia. 1998. "A generalized processor sharing approach to time scheduling in hybrid CDMA/TDMA," *Proc. IEEE Infocom'98*, San Francisco, CA, March, 1164–1171.

Arbak, J. C. and W. Van Blitterwijk. 1987. "Capacity of slotted Aloha in Rayleigh fading channels," *IEEE Journal on Selected Areas in Communications*, SAC-5, 2 (February), 261–269.

Atal, B. S. and S. L. Hanauer. 1971. "Speech analysis and synthesis by linear prediction of the speech wave," *Journal of the Acoustic Society of America*, 50, 2 (August), 637–655.

Bar-Noy, A. *et al.* 1995. "Mobile users: to update or not to update?" *Wireless Networks*, 1, 2 (July), 175–185.

Bello, P. A. 1963. "Characterization of randomly time-variant linear channels," *IEEE Transactions on Communication Systems*, 12, CS-11 (December), 360–393.

Bello, P. A. 1985. *A History of Engineering and Science in the Bell System: Transmission Technology (1925–1975)*, ed. E. F. O'Neill, AT&T Bell Laboratories.

Bender, P. *et al.* 2000. "CDMA/HDR: a bandwidth-efficient high-speed service for nomadic users," *IEEE Communications Magazine*, 38, 7 (July), 70–77.

Berrou, C., A. Glavieux, and P. Thitimajshima. 1993. "Near Shannon-limit error-correcting coding and decoding: turbo codes," IEEE International Conference on Communications, ICC93, Geneva, Switzerland, May, 1064–1070.

Berrou, C. and A. Glavieux. 1996. "Near optimum error correcting coding and decoding: turbo codes," *IEEE Transactions on Communications*, 44, 10 (October), 1261–1271.

Bettstetter, C. *et al.* 1999. "GSM Phase 2+ general packet radio service GPRS: architecture, protocols, and air interface," *IEEE Communication Surveys*, 3rd 8tr.

Bianchi, G. 2000. "Performance analysis of the IEEE 802.11 distributed coordination function," *IEEE Journal on Selected Areas in Communications*, 18, 3 (March), 535–547.

Bisdikian, C. 2001. "An overview of the Bluetooth wireless technology," *IEEE Communications Magazine*, 39, 12 (December), 86–94.

Black, D. M. and D. O. Reudink. 1972. "Some characteristics of radio propagation at 800 MHz in the Philadelphia area," *IEEE Transactions on Vehicular Technology*, 21 (May), 45–51.

Blom, J. and F. Gunnarson. 1998. "Power control in cellular systems, Linköping studies in science and technology," Thesis No. 706, Department of Electrical Engineering, Linköpings Universitet, Linköping, Sweden.

Bluetooth. 1999. "Specification of the Bluetooth System," Specification Volume 1, Core, v1.0.B, www.bluetooth.com, December 1.

Brand, A. E. and A. H. Aghvami. 1996. "Performance of a joint CDMA/PRMA protocol for mixed voice/data transmission for third generation mobile communications," *IEEE Journal on Selected Areas in Communications*, 14, 9 (December), 1698–1707.

Brenner, D. G. 1959. "Linear diversity combining techniques," *Proc. IRE*, 47, 6 (June), 1075–1102.

Le Bris, L. and W. Robison. 1999. "Dynamic channel assignment in GSM networks," *Proc. VTC'99*, IEEE Vehicular Technology Conference, Amsterdam, September, 2339–2342.

Cai, J. and D. J. Goodman. 1997. "General packet radio service in GSM," *IEEE Communications Magazine*, 35, 10 (October), 122–131.

Cali, F. *et al.* 1998. "IEEE 802.11 wireless LAN: capacity analysis and protocol enhancement," IEEE Infocom'98, San Francisco, CA, March.

Capone, A. *et al.* 2001. "Efficient polling schemes for Bluetooth picocells," IEEE International Conference on Communications, ICC2001, Helsinki, Finland, 1990–1994.

cdma 2001. cdma2000 high rate packet data air interface specification, 3GPP2 C.S0024, Version 3.0, 3rd Generation Partnership Project 2, 3GPP2, December, 5, http://www.3gpp2.org/public_html/specs/

cdma 2002. Physical layer standard for ccdma2000 spread spectrum systems, Release A, 3GPP2 C.S0002-A, Version 6.0, 3rd Generation Partnership Project 2, 3GPP2, February, http://www.3gpp2.org/public_html/specs/

Chao, C. and W. Chen. 1997. "Connection admission control for mobile multiple-class personal communication networks," *IEEE Journal on Selected Areas in Comunications*, 15, 10 (October), 1618–1626.

Chen, J.-H. *et al.* 1992. "A low-delay CELP coder for the CCITT 16 kb/s Speech Coding Standard," *IEEE Journal on Selected Areas in Communications*, 10, 5 (June), 830–849.

Chih-Lin I and K. Sabnani. 1995a. "Variable speading gain CDMA with adaptive control for true packet switching wireless network," IEEE International Conference on Communications, ICC95, Seattle, WA, June, 725–730.

Chih-Lin I and K. Sabnani. 1995b. Variable spreading gain CDMA with adaptive control for integrated traffic in wireless networks," IEEE 45th Vehicular Technology Conference, VTC95, Chicago, IL, July, 794–798.

Chih-Lin I and R. D. Gitlin. 1995. "Multi-code CDMA wireless personal communication networks," IEEE International Conference on Communications, ICC95, Seattle, WA, June, 1060–1064.

Choi, S. and K. G. Shin. 1998. "Predictive and adaptive bandwidth reservation for handoffs in QoS-sensitive cellular networks," Proc. SIGCOMM98, 155–166.

Cox, D. C. and D. O. Reudink. 1972. "Dynamic channel assignment in two-dimensional large-scale mobile communication systems," *Bell System Technology Journal*, 51, 1611–1672.

Cox, D. C. and D. O. Reudink. 1973. "Increasing channel occupancy in large-scale mobile radio systems: dynamic channel assignment," *IEEE Transactions on Vehicular Technology*, VT-22, 218–222.

Cox, D. C. 1995. "Wireless personal communications: what is it?," *IEEE Personal Communications*, 2, 2, 20–35.

Dahlman, E. and K. Jamal. 1996. "Wideband services in a DS-CDMA based FPLMTS system," *IEEE Vehicular Technology Conference*, VTC96, 1656–1660.

DeVille, J. M. 1993. "A reservation-based multiple access scheme for a future universal telecommunications system," *Mobile and Personal Communications*, IEE Conference Publication No. 387, 13–15 December, 210–215.

Epstein, B. and M. Schwartz. 1998. QoS-Based Predictive Admission Control for Multi-Media Traffic, in *Broadband Wireless Communications*, ed. M. Luise and S. Pupolin, Berlin, Springer-Verlag, pp. 213–224.

Epstein, B. and M. Schwartz. 2000. "Predictive QoS-based admission control for multiclass traffic in cellular wireless networks," *IEEE Journal on Selected Areas in Communications*, 18, 3 (March), 523–534.

Egli, J. J. 1957. "Radio propagation above 40 Mc/s over irregular terrain," *Proc. IRE*, October, 1383–1391.

Everitt, D. and D. Manfield. 1989. "Performance analysis of cellular mobile communication systems with dynamic channel assignment," *IEEE Journal on Selected Areas in Communications*, 7, 8 (October), 1172–1180.

Fattah, H. and C. Leung. 2002. "An overview of scheduling algorithms in wireless multimedia networks," *IEEE Wireless Communications*, 9, 5 (October), 76–83.

Forney, G. D. 1966. *Concatenated Codes*, Cambridge, MA, MIT Press.

Foschini, G. J. 1996. "Layered space-time architecture for wireless communication in a fading environment when using multiple antennas," *Bell Laboratories Technical Journal*, 1, 2, 41–59.

Foschini, G. J. and Z. Miljanic. 1993. "A simple distributed autonomous power control algorithm and its convergence," *IEEE Transactions on Vehicular Technology*, 42, 4 (November), 641–646.

Frenkiel, R. 2002. "A brief history of mobile communications," *IEEE Vehicular Technology Society News*, May, 4–7.

Furuskar, A. *et al.* 1999a. "EDGE: enhanced data rates for GSM and TDMA/136 evolution," *IEEE Personal Communications*, 6, 3 (June), 56–66.

Furuskar, A. *et al.* 1999b. "Capacity evaluation of the EDGE concept for enhanced data rates in GSM and TDMA/136," *Proc. IEEE Vehicular Technology Conference*, VTC99, 1648–1652.

Garg, V. K. 2000. *IS-95 CDMA and cdma2000*, Upper Saddle River, NJ, Prentice-Hall PTR.

Garg, V. K. and J. E. Wilkes. 1999. *Principles and Applications of GSM*, Upper Saddle River, NJ, Prentice-Hall PTR.

Gibson, J. D. ed. 1996. *The Mobile Communication Handbook*, Boca Raton, FL, CRC Press/IEEE Press.

Gilhausen, K. H. *et al.* 1991. "On the capacity of a cellular CDMA system," *IEEE Transactions on Vehicular Technology*, 40, 2 (May), 303–312.

Goodman, D. J. 1997. *Wireless Personal Communication Systems*, Reading, MA, Addison-Wesley.

Goodman, D. J. and A. A. M. Saleh. 1987. "The near-far effect in local Aloha radio communications," *IEEE Transactions on Vehicular Technology*, VT-36, 1 (February), 19–27.

Goodman, D. J. and S. X. Wei. 1991. "Efficiency of packet reservation access," *IEEE Transactions on Vehicular Technology*, 40, 1 (February), 170–176.

Goodman, D. J. *et al.* 1989. "Packet reservation multiple access for local wireless communication," *IEEE Transactions on Communications*, 37, 8 (August), 885–890.

GPRS. 1998. Digital cellular communications system (Phase 2+): general packet radio service description, Stage 1, EN 301 113 v6.1.1 (1998–11) (GSM 02.60 version 6.1.1 Release 1997), European Telecommunications Institute (ETSI), Sophia Antipoli, Valbonne, France.

Grandhi, S. A. *et al.* 1994. "Distributed power control in cellular radio systems," *IEEE Transactions on Communications*, 42, 2/3/4 (February–April), 226–228.

Grillo, D. *et al.* 1993. "A performance analysis of PRMA considering speech, data, co-channel interference and ARQ error recovery," *Mobile and Personal Communications*, IEEE Conference Publication No. 387, December 13–15, 161–171.

GSM World, www.gsmworld.com

Gudmundson, B. *et al.* 1992. "A comparison of CDMA and TDMA systems," *Proc. IEEE Vehicular Technology Conference*, Denver, CO, May, 732–735.

Guerin, R. A. 1987. "Channel occupancy time distribution in cellular radio systems," *IEEE Transactions on Vehicular Technology*, VT-35, 3 (August), 89–99.

Hall, E. K. and S. G. Wilson. 1998. "Design and analysis of turbo codes on Rayleigh fading channels," *IEEE Journal on Selected Areas in Communications*, 16, 2 (February), 160–174.

Heegard, C. *et al.* 2001. "High-performance wireless internet," *IEEE Communications Magazine*, 39, 11 (November), 64–73.

Heller, J. A. and I. M. Jacobs. 1971. "Viterbi decoding for space and satellite communications," *IEEE Transactions on Communication Technology*, COM-19, 5, part II (October), 835–848.

Hong, D. and S. S. Rappaport. 1999. "Traffic model and performance analysis for cellular mobile radio telephone systems with prioritized and non-prioritized handoff procedures," *IEEE Transactions on Vehicular Technology*, VT-35, 3 (August), 77–92. See also CEAS Technical Report No. 773, 1 June 1999, College of Engineering and Applied Sciences, State University of New York, Stony Brook, NY 11794.

Ho, T.-S. and K.-C. Chen. 1996. "Performance analysis of 802.11 CSMA/CA medium access control protocol," IEEE PRMC, Taipai, Taiwan, October, 407–411.

IEEE. 1999. "The evolution of TDMA to 3G," *IEEE Personal Communications*, 6, 3 (June).

IEEE 802.11. IEEE Standard for Information technology-Telecommunications and information exchange between systems-Local and metropolitan area networks-Specific requirements-Part 11: Wireless LAN Medium Access Control (MAC) and Physical Layer (PHY) Specifications, ANSI/IEEE Std 802.11, 1999 Edition, IEEE Standards Board, Piscataway, NJ.

IEEE 802.11a. Supplement to IEEE Standard for Information technology-Telecommunications and information exchange between systems-Local and metropolitan area networks-Specific requirements-Part 11: Wireless LAN Medium Access Control (MAC) and Physical Layer (PHY) specifications: High-Speed Physical Layer in the 5 GHz Band, IEEE Std 802.11a-1999, IEEE, New York, NY, 16 September 1999.

IEEE 802.11b. Supplement to IEEE Standard for Information technology-Telecommunications and information exchange between systems-Local and metropolitan area networks-Specific requirements-Part 11: Wireless LAN Medium Access Control (MAC) and Physical Layer (PHY) specifications: Higher-Speed Physical Layer Extension in the 2.4 GHz Band, IEEE Std 802.11b-1999, IEEE, January 2000.

IEEE 802.11g. IEEE Standard for Information technology-Telecommunications and information exchange between systems-Local and metropolitan area networks-Specific requirements-Part 11: Wireless LAN Medium Access Control (MAC) and Physical Layer (PHY) specifications; Amendment 4: Further Data Rate Extension in the 2.4 GHz Band, IEEE Std 802.11g-2003, IEEE, New York, NY, 25 June 2003.

IEEE 802.15. IEEE Standard for Information technology-Telecommunications and Information exchange between systems-Local and metropolitan area networks-Specific requirements, Part 15.1: Wireless Medium Access Control (MAC) and Physical Layer (PHY) Specifications for Wireless Personal Area Networks (WPANs), IEEE Std 802.15.1, IEEE-SA Standards Board, 15 April, 2000.

Jabbari, B. 1996. "Teletraffic aspects of evolving and next-generation wireless communication networks," *IEEE Personal Communications*, 3, 6 (December), 4–9.

Jain, R. and Y.-B. Lin. 1995. "An auxiliary location strategy employing forwarding pointers to reduce network impacts of PCS," *Wireless Networks*, 1, 2 (July), 197–210.

Jakes, W. C. ed. 1974. *Microwave Mobile Communications*, AT&T, 1995 edition, New York, IEEE Press.

Jayant, N. 1992. "Signal compression: technology targets and research directions," *IEEE Journal on Selected Areas in Communications*, 10, 5 (June), 796–818.

Kalden, R. *et al.* 2000. "Wireless internet access based on GPRS," *IEEE Personal Communications*, 7, 2 (April), 8–18.

Kalia, M. *et al.* 2000. "Data scheduling and SAR for Bluetooth MAC," IEEE Vehicular Technology Conference, Spring, Tokyo, 716–720.

Karol, M. J. *et al.* 1995a. "An efficient demand-assignment multiple-access protocol for wireless packet (ATM) networks," *Wireless Networks*, 1, 5 (October), 267–279.

Karol, M. J. *et al.* 1995b. "Distributed-queueing request update multiple access (DQRUMA) for wireless packet (ATM) networks," IEEE International Conference on Communications, ICC'95, Seattle, WA, June, 1224–1231.

Katzela, I. and M. Naghshineh. 1996. "Channel assignment schemes: a comprehensive survey," *IEEE Personal Communications*, 3, 3 (June), 10–31.

Koodli, R. and M. Punskari. 2001. "Supporting packet data QoS in next-generation cellular networks," *IEEE Communications Magazine*, 39, 2 (February), 180–188.

Kurose, J. F. and K. W. Ross. 2002. *Computer Networking*, Boston, MA, Addison-Wesley.

Kwok, M.-S. and H.-S. Wang. 1995. "Adjacent-cell interference analysis of reverse link in CDMA cellular radio system," IEEE PIMRC95, Toronto, Canada, September, 446–450.

Lee, W. C. Y. 1993. *Mobile Communications Design Fundamentals*, New York, Wiley-Interscience.

Lee, Y.-Z. *et al.* 2002. "An efficient and fair polling scheme for Bluetooth," *Proc. IEEE Milcom*, 1062–1068.

Levine, D. *et al.* 1997. "A resource estimation and admission control algorithm for wireless multimedia networks using the shadow cluster concept," *IEEE/ACM Transactions on Networking*, 5, 1 (February), 1–12.

Lin, Y.-B. 1997. "Reducing location update costs in a PCS network," *IEEE/ACM Transactions on Networking*, 5, 1 (February), 25–33.

Linnartz, J. P. 1993. *Narrowband Land-Mobile Radio Networks*, Boston, MA, Artech House.

Liu, Z. *et al.* 1996. "A demand-assignment access control for multi-code DS-CDMA wireless packet (ATM) networks," *Proc. IEEE Infocom'96*, San Francisco, CA, March, 713–721.

Lozano, A. and D. C. Cox. 1999. "Integrated dynamic channel assignment and power control in TDMA wireless communication systems," *IEEE Journal on Selected Areas in Communications*, 17, 11 (November), 2031–2040.

Lucky, R. W. 1965. "Automatic equalization for digital communication," *Bell System Technical Journal*, 44 (April), 558–570.

Mark, J. W. and W. Zhuang. 2003. *Wireless Communications and Networking*, Upper Saddle River, NJ, Pearson Education Inc.

Matsumoto, T. *et al*. 2003. "Overview and recent challenges of MIMO systems," *IEEE Vehicular Technology News*, 50, 2 (May), 4–9.

Meyerhoff, H. J. 1974. "Method for computing the optimum power balance in multibeam satellites," *Comsat Technology Review*, 4, 1 (Spring).

Michelson, A. M. and A. H. Levesque. 1985. *Error-Control Techniques for Digital Communication*, New York, John Wiley & Sons.

Miller, B. A. and C. Bisdikian. 2002. *Bluetooth Revealed*, 2nd edn, Upper Saddle River, NJ, Prentice-Hall PTR.

Milstein, L. 2000. "Wideband code division multiple access," *IEEE Journal on Selected Areas in Communications*, 18, 8 (August), 1344–1354.

Mitra, D. 1994. "An asynchronous distributed algorithm for power control in cellular systems," in *Wireless and Mobile Communications*, ed. J. M. Holtzman, Boston, MA, Kluwer/Academic.

Mitrou, N. M. *et al*. 1990. "A reservation multiple access protocol for microcellular mobile communication systems," *IEEE Transactions on Vehicular Technology*, 39, 4 (November), 340–351.

Moberg, J. *et al*. 2000. "Throughput of the WCDMA random access channel," *Proc. IST Mobile Communication Summit*, Galway, Ireland, October.

Modaressi, A. R. and R. S. Skoog. 1990. "Signaling system number 7: a tutorial," *IEEE Communications Magazine*, 28, 7 (July), 19–35.

Mohan, S. and R. Jain. 1994. "Two user location strategies for personal communication services," *IEEE Personal Communications*, 1, 1 (1st Qtr.), 42–50.

Naghshineh, M. and M. Schwartz. 1996. "Distributed call admission control in mobile/wireless networks," *IEEE Journal on Selected Areas in Communications*, 14, 4 (May), 711–717.

Nanda, S. *et al*. 1991. "Performance of PRMA: a packet voice protocol for cellular systems," *IEEE Transactions on Vehicular Technology*, 40, 3 (August), 584–598.

Narasimhan, P. and R. P. Yates. 1996. "A new protocol for the integration of voice and data over PRMA," *IEEE Journal on Selected Areas in Communications*, 14, 4 (May), 621–631.

Noble, B. and J. H. Daniel. 1988. *Applied Linear Algebra*, 2nd edn, Englewood Cliffs, NJ, Prentice-Hall.

Nyquist, H. 1928. "Certain topics in telegraph transmission theory," *Transactions of the AIEE*, 47 (April), 617–644.

Olafsson, H. *et al*. 1999. "Performance evaluation of different random access power ramping proposals for the WCDMA system," IEEE Personal Indoor Mobile Radio Conference.

O'Neill, E. F. ed. 1985. *A History of Engineering and Science in the Bell System: Transmission Technology (1925–1975)*, AT&T Bell Laboratories.

Oppenheim, A. V. and A. S. Willsky. 1997. *Signals and Systems*, 2nd edn, Upper Saddle River, NJ, Prentice-Hall.

Paetsch, M. 1993. *Mobile Communications in the US and Europe: Regulation, Technology, and Markets*, Boston, MA, Artech House.

Papoulis, A. 1991. *Probability, Random Variables, and Stochastic Processes*, 3rd edn, New York, McGraw-Hill.

Parekh, A. K. and R. G. Gallager. 1993. "A generalized processor-sharing approach to flow control in integrated services networks: the single-node case," *IEEE/ACM Transactions on Networking*, 1, 3 (June), 344–357.

Pasupathy, S. 1979. "Minimum-shift keying: a spectrally efficient modulation," *IEEE Communications Magazine*, 17, 4 (July), 14–22.

Pearson, R. 2001. "Complementary code keying made simple," Application Note AN9850.2, Intersil Americas Inc., November.

Pecen, M. and A. Howell. 2001. "Simultaneous voice and data operation for GPRS/EDGE: Class A dual transfer mode," *IEEE Personal Communications*, 8, 2 (April), 14–29.

Price, R. and P. E. Green, Jr. 1958. "A communication technique for multipath channels," *Proc. IRE*, 46, 555–570.

Priscoll, F. Della *et al.* 1997. "Application of dynamic channel allocation strategies to the GSM cellular network," *IEEE Journal on Selected Areas in Communications*, 15, 8 (October), 1558–1567.

Proakis, J. G. 1995. *Digital Communications*, 3rd edn, New York, McGraw-Hill.

Rabiner, L. and B.-H. Hwang. 1993. *Fundamentals of Speech Recognition*, Englewood Cliffs, NJ, Prentice-Hall.

Ramakrishna, S. and J. M. Holtzman. 1998. "A comparison between single code and multiple code transmission schemes in a CDMA system," VTC98, May, 791–795.

Rappaport, T. S. 1996. *Wireless Communications, Principles and Practice*, Upper Saddle River, NJ, Prentice-Hall.

Rappaport, T. S. 2002. *Wireless Communications, Principles and Practice*, 2nd edn, Upper Saddle River, NJ, Prentice-Hall.

Raymond, P.-A. 1991. "Performance analysis of cellular networks," *IEEE Transactions on Communications*, 37, 12 (December), 1787–1793.

Rental, C. H. *et al.* 2002. "Comparative forward link traffic channel performance evaluation of HDR and 1EXTREME systems," *Proc. 55th IEEE Vehicular Technology Conference*, VTC 2002, Spring, 160–164.

Rose, C. and R. Yates. 1995. "Minimizing the average cost of paging under delay constraints," *Wireless Networks*, 1, 2 (July), 211–219.

Sarikaya, B. 2000. "Packet mode in wireless networks: overview of transition to third generation," *IEEE Communications Magazine*, 38, 9 (September), 164–172.

Schroeder, M. R. and B. S. Atal. 1985. "Code-excited linear prediction (CELP): High quality speech at very low bit rates," *Proc. IEEE Int. Conf. Acoust. Speech, Signal Process*, March, 937–950.

Schwartz, M. 1987. *Telecommunication Networks: Protocols, Modeling, and Analysis*, Reading, MA, Addison-Wesley.

Schwartz, M. 1990. *Information Transmission, Modulation, and Noise*, 4th edn, New York, McGraw-Hill.

Schwartz, M. 1996. *Broadband Integrated Networks*, Englewood Cliffs, NJ, Prentice-Hall.

Schwartz, M. and L. Shaw. 1975. *Signal Processing: Discrete Spectral Analysis, Detection, and Estimation*, New York, McGraw-Hill.

Schwartz, M., W. R. Bennett, and S. Stein. 1966. *Communication Systems and Techniques*, New York, McGraw-Hill; reprinted, IEEE Press, 1996.

Sköld, J. *et al.* 1995. "Performance and characteristics of GSM-based PCS," *Proc. 45th IEEE Vehicular Technology Conference*, 743–748.

Solana, A. H. *et al.* 2002. "Performance analysis of packet scheduling strategies for multimedia traffic in WCDMA," *Proc. 55th IEEE Vehicular Technology Conference*, VTC 2002, Spring, 155–159.

Steele, R. ed. 1992. *Mobile Radio Communications*, London, Pentech Press; New York, IEEE Press.

Stüber, G. L. 1996. *Principles of Mobile Communications*, Boston, MA, Kluwer Academic.

Stüber, G. L. 2001. *Principles of Mobile Communications*, 2nd edn, Boston, MA, Kluwer Academic.

Sutivong, A. and J. Peha. 1997. "Call admission control algorithms: proposal and comparison," *Proc. IEEE Globecom*.

TIA. 1992. EIA/TIA Interim standard, cellular system dual-mode mobile station–base station compatibility standard, IS-54-B, Telecommunications Industry Association.

Tiedmann, E. G., Jr. 2001. "cdma20001X: new capabilities for CDMA networks," *IEEE Vehicular Technology Society News*, 48, 4 (November), 4–12.

Varsou, A. C. and H. V. Poor. 2002. "Scheduling algorithms for downlink rate allocation in heterogeneous CDMA networks," *Journal of Communications and Networks*, 4, 3 (September), 199–208.

Verdu, S. 2002a. "Spectral efficiency in the wideband regime," *IEEE Transactions on Information Theory*, 48, 6 (June), 1319–1343.

Verdu, S. 2002b. "Recent results on the capacity of wideband channels in the low-power regime," *IEEE Wireless Communications*, 9, 4 (August), 40–45.

Viterbi, A. J. 1967. "Error bounds for convolutional codes and an asymptotically optimum decoding algorithm," *IEEE Transactions on Information Theory*, IT-13 (April), 260–269.

Viterbi, A. J. *et al.* 1994. "Soft handoff extends CDMA cell coverage and increases reverse link coverage," *IEEE Journal on Selected Areas in Communications*, 12, 8 (October), 1281–1288.

Viterbi, A. J. 1995. *CDMA, Principles of Spread Spectrum Communication*, Reading, MA, Addison-Wesley.

Vucetic, B. and J. Yuan. 2000. *Turbo Codes: Principles and Applications*, Boston, MA, Kluwer Academic.

Wang, J. Z. 1993. "A fully distributed location registration strategy for universal personal communications systems," *IEEE Journal on Selected Areas in Communications*, 11, 6 (August), 850–860.

WCDM. 2000. *WCDMA for UMTS*, ed. Harri Holma and Antti Toskala, John Wiley & Sons.

Willenegger, S. 2000. "cdma2000 physical layer: an overview," *Journal of Communications and Networks*, 2, 1 (March), 5–17.

Wozencroft, J. M. and I. M. Jacobs. 1965. *Principles of Communication Engineering*, New York, Wiley.

Xu, L. *et al.* 2002. "Dynamic bandwidth allocation with fair scheduling for WCDMA systems," *IEEE Wireless Communications*, 9, 2 (April), 26–32.

Yacoub, M. D. 1993. *Foundations of Mobile Radio Engineering*, Boca Raton, FL, CRC Press.

Yeung, K. L. and T.-S. P. Yum. 1995. "Cell group decoupling analysis of a dynamic allocation strategy in linear microcell radio systems," *IEEE Transactions on Communications*, 43, 2/3/4 (February–April), 1289–1292.

Yu, O. and V. Leung. 1997. "Adaptive resource allocation for prioritized call admission in ATM-based wireless PCN," *IEEE Journal on Selected Areas in Communications*, 15, 9 (September), 1208–1225.

Zander, J. 1992. "Distributed cochannel interference control in cellular radio systems," *IEEE Transactions on Vehicular Technology*, 41, 3 (August), 304–311.

Zhang, M. and T.-S. P. Yum. 1989. "Comparisons of channel assignment strategies in cellular telephone systems," *IEEE Transactions on Vehicular Technology*, 38, 4 (November), 211–218.

Zhang, H. 1995. "Service disciplines for guaranteed performance service in packet-switching networks," *Proc. IEEE*, 53 (October), 1374–1396.

Zorzi, M. and R. R. Rao. 1994. "Capture and retransmission control in mobile radio," *IEEE Journal on Selected Areas in Communications*, 12, 8 (October), 1289–1298.

Zürbes, S. *et al.* 2000. "Radio network performance of Bluetooth," IEEE International Conference on Communications, ICC, 1563–1567.

Index